Organic and Molecular Electronics

Organic and Molecular Electronics

From Principles to Practice

Second Edition

Michael C. Petty

Department of Engineering

and

Centre for Molecular and Nanoscale Electronics
Durham University, UK

Registered Offices
John Wiley & Sons, Inc., 111 River Street, Hoboken, NJ 07030, USA
John Wiley & Sons Ltd, The Atrium, Southern Gate, Chichester, West Sussex, PO19 8SQ, UK

Editorial Office
The Atrium, Southern Gate, Chichester, West Sussex, PO19 8SQ, UK

For details of our global editorial offices, customer services, and more information about Wiley products visit us at www.wiley.com.

Wiley also publishes its books in a variety of electronic formats and by print-on-demand. Some content that appears in standard print versions of this book may not be available in other formats.

Library of Congress Cataloging-in-Publication Data

Names: Petty, Michael C.
Title: Organic and molecular electronics : from principles to practice /
    Michael C Petty, Department of Engineering and Centre for Molecular and
    Nanoscale Electronics, Durham University, UK.
Description: Second edition. | Hoboken, NJ, USA : Wiley, [2019] | Includes
    bibliographical references and index. |
Identifiers: LCCN 2018025508 (print) | LCCN 2018028736 (ebook) | ISBN
    9781118879276 (Adobe PDF) | ISBN 9781118879252 (ePub) | ISBN 9781118879283 (pbk.)
Subjects: LCSH: Organic electronics. | Molecular electronics.
Classification: LCC TK7874.8 (ebook) | LCC TK7874.8 .P47 2019 (print) |
    DDC 621.381–dc23
LC record available at https://lccn.loc.gov/2018025508

Cover Design: Wiley
Cover Image: © iStock.com/agsandrew

Set in 10/12pt Warnock by SPi Global, Pondicherry, India

Printed in Singapore by C.O.S. Printers Pte Ltd

10 9 8 7 6 5 4 3 2 1

For

Anne

Contents

Preface

Organic and molecular electronics are exciting and challenging spheres of activity. Both disciplines exploit carbon-based materials. Organic electronics (or plastic electronics) enables the fabrication of a wide range of electronic devices based around organic materials and low-cost technologies. Molecular electronics explores further into the future; taken to its limit, the field offers enormous computational power. Since the first edition of this book was published in 2007, organic electronics has progressed in leaps and bounds. Many new companies have been established to exploit the new possibilities. Televisions based on organic electroluminescent display technology are in the marketplace (although currently too expensive for a university professor!). Field effect transistors using organic semiconductors can now be fabricated with charge carrier mobilities that are competitive with amorphous and polycrystalline silicon. In terms of materials, the past 10 years has seen the emergence of graphene as a fascinating new compound, with the award of the 2010 Nobel Prize for Physics.

The second edition of this book reflects progress in both organic and molecular electronics over the last 10 years. I have kept to the original format, with an introduction into the physics and chemistry of organic materials, followed by a discussion of the means to process the materials into a form (in most cases, a thin film), where they can be exploited in electronic and optoelectronic devices. I have written the book from the viewpoint of a final-year science or engineering undergraduate. Although the second edition has a reduced introduction to organic chemistry, I hope it is accessible to readers from a wide range of backgrounds.

Examples of 'applications' have been taken from the portfolio of research that was available as the second edition of this book was being written (2017). These form a vision of organic and molecular electronics and indicate important directions in research. One major change is the inclusion of a few problems at the end of each chapter, with solutions available on the web page for the book. To add some fun, the answer to the final problem in each chapter also contains a clue. Solving these will point towards a final answer. You can submit your solutions on the web. Those submitting the correct answer will have their names entered into prize draw; details are provided on the book companion web site.

Michael C. Petty
Durham University, UK

Acknowledgements

I am indebted to my friend and colleague, Gareth Roberts, who sadly died as the first edition of this book was nearing completion. Some 40 years ago, Gareth steered me in the direction of organic materials research. Without his influence and enthusiasm, this book would never have been possible. I must also acknowledge Safa Kasap, who talked me into the original project and was a significant influence in getting the second edition started. I much appreciate all the encouragement, help and support from the staff at Wiley to see this project through to completion: particularly from Jenny Cossham, Emma Strickland and Elsie Merlin.

I am grateful to my academic colleagues at Durham, who have provided invaluable input to both editions of this book: Jas Pal Badyal, David Bloor, Martin Bryce, Chris Groves, Stewart Clark, Karl Coleman, Graham Cross, Ken Durose, Jim Feast, Gordon Love, Andy Monkman, Chris Pearson, Richard Thompson, David Wood, Jack Yarwood and Dagou Zeze. Some of the results presented are from students and post docs in my own research group: Jin Ahn, Paul Barker, John Batey, Duncan Cadd, Riccardo Casalini, Phil Christie, Mike Cousins, John Cresswell, Ajaib Dhindsa, Yesul Jeong, Carole Jones, Dan Kolb, Igor Lednev, Mohammed Mabrook, Marco Palumbo, Shashi Paul and Youngjun Yun.

Further afield, many colleagues have kindly provided me with their original data: Geoff Ashwell, Phaedon Avouris, Jeremy Burroughes, Wolfgang Knoll, Graham Leggett, Yuri Lvov, David Morris, Kosmas Prassides, Tim Richardson, Campbell Scott, Fraser Stoddart, Martin Taylor, Richard Tredgold, Dimitris Tsoukalas and Harold Zandvliet. I must also acknowledge the many other workers worldwide, whose data are reproduced in this book.

Last and by no means least, the biggest thank you to my wife Anne for her patience, never ending support and reminding me of the important things in life.

Symbols and Abbreviations

A	Hamaker constant
A	acceptor
AC	alternating current
AES	Auger electron spectroscopy
AFLC	antiferroelectric liquid crystal
AFM	atomic force microscopy
ALD	atomic layer deposition
AmI	ambient intelligence
amu	atomic mass unit
ANN	artificial neural network
ATP	adenosine triphosphate
ATR	attenuated total reflection
au	arbitrary units
B, $\mathbf{B}$	magnetic field (T)
BEDT-TTF or ET	bis(ethylenedithio)tetrathiafulvalene
BHJ	bulk heterojunction
BCS	Bardeen, Cooper and Schrieffer
c	molecular concentration ($\mathrm{mol\ m^{-3}}$)
C	capacitance (F)
CAD	computer-aided drawing
CB	conduction band
CCD	charge-coupled device
cd	candela
CFM	chemical force microscope
CGS	centimetre gram second
CIE	Commission International de l'Eclairage
CMOS	complementary metal-oxide-semiconductor
CRI	colour-rendering index
CRN	continuous random network
CT	charge transfer
CVD	chemical vapour deposition
d_{hkl}	interplanar spacing (m)
D	donor
D	diffusion coefficient ($\mathrm{m^2\ s^{-1}}$)
D, $\mathbf{D}$	electric displacement ($\mathrm{C\ m^{-2}}$)
DC	direct current
DMeFc	decamethyl-ferrocene
DMSO	dimethyl sulfoxide

DNA	dioxyribonucleic acid
DOBAMBC	*p*-decyloxybenzylidene-*p'*-amino-2-methylbutylcinnamate
DPN	dip-pen nanolithography
DRAM	Dynamic Random Access Memory
DVD	Digital Video (or Versatile) Disc
E	energy (J)
E, **E**	electric field (V m^{-1})
E_c	conduction band edge (J or eV)
E_F	Fermi energy (J or eV)
EIL	electron injection layer
EL	electroluminescent (or electroluminescence)
EM	electromagnetic
EML	emissive layer
ENFET	enzyme field effect transistor
ENIAC	Electronic Numerical Integrator and Computer
ESR	electron spin resonance
ETL	electron transport layer
E_v	valence band edge (J or eV)
EXAFS	extended X-ray absorption fine structure
FET	Field Effect Transistor
FLC	ferroelectric liquid crystal
FLOPS	FLoating-point Operations Per Second
FMM	force modulation microscope
FPGA	field-programmable gate array
G	conductance (S or Ω^{-1})
GMR	giant magnetoresistance
GPS	Global Positioning System
GSM	Global System for Mobile communications
H, **H**	magnetizing field (A m^{-1})
H_c	coercive field
HCl	hydrochloric acid
HDTTF	hexadecanolyl tetrathiafulvalene
HIL	hole injection layer
HOMO	highest occupied molecular orbital
HTL	hole transport layer
I	electric current (A)
IC	integrated circuit
IDT	indacenodithiophene
IID	isoindigo
IMFET	immuno field effect transistor
IoT	Internet of Things
IR	infrared
ISFET	ion-sensitive field effect transistor
ITO	indium–tin-oxide
ITRS	International Technology Roadmap for Semiconductors
J	joule
j	$\sqrt{-1}$
J	electric current density (A m^{-2})

k	wavevector (m^{-1})
k_B	Boltzmann's constant
LASER	light amplification by stimulated emission of radiation
LB	Langmuir–Blodgett
LbL	Layer-by-Layer
LCD	liquid crystal display
LEC	light-emitting electrochemical cell
LED	light-emitting diode
LFM	lateral force microscope
lm	lumen
LSI	large-scale integrated circuit
LUMO	lowest unoccupied molecular orbital
M	molecular weight
M, $\mathbf{M}$	magnetization $(A\ m^{-1})$
MBE	molecular beam epitaxy
MEH-PPV	poly[2-methoxy-5-(2-ethylhexyloxy)-1,4-phenylenevinylene]
MEMS	Micro-Electro-Mechanical System
MFM	magnetic force microscope
MIM	metal-insulator-metal
MIS	metal–insulator–semiconductor
MISFET	metal–insulator–semiconductor field effect transistor
MLD	molecular layer deposition
$\mathbf{m}_m$	magnetic dipole moment $(A\ m^{-2})$
MOCVD	metalorganic chemical vapour deposition
MOS	metal–oxide–semiconductor
MOSFET	metal–oxide–semiconductor field effect transistor
MPP	maximum power point (for a PV device)
MPU	microprocessor unit
M_r	remanent magnetization $(A\ m^{-1})$
MRAM	magnetic random access memory
mRNA	messenger RNA
MTDATA	4,4′,4″-tris[phenyl(m-tolyl)amino]triphenylamine
MWNT	multiwall (carbon) nanotube
n	refractive index
n	number per unit volume or concentration (m^{-3})
n'	real part of refractive index
n''	imaginary part of refractive index
NAD	nicotinamide adenine dinucleatide
NADP	nocotinamide adenine dinucleotide phosphate
NDR	negative differential resistance
NEMS	Nano-Micro-Electro-Mechanical System
NEXAFS	near edge X-ray absorption fine structure
NiCd	nickel–cadmium
NIL	nanoimprint lithography
NiMH	nickel metal hydride
NLO	nonlinear optics
NMR	nuclear magnetic resonance
NSOM	near-field scanning optical microscopy

NPB or NPD	N,N'-**di**(1-naphthyl)-N,N'-diphenyl-(1,1'-biphenyl)-4,4'-diamine
OFET	organic field effect transistor
OLED	organic light-emitting device
OPV	organic photovoltaic
p, **p**	electric dipole moment or transition dipole moment (C m)
P, **P**	polarization (C m^{-2})
PANi	polyaniline
PC	personal computer
PC	polycarbonate
PCA	principal component analysis
PCBM	phenyl-C_{61} butyric acid methyl ester
PDA	personal digital assistant
PDMS	poly(dimethylsiloxane)
PE	polyethylene/polyelectrolyte
PECVD	plasma-enhanced CVD
PEDOT/PEDT	poly(3,4-ethylenedioxythiophene)
PEDOT-PSS	poly(3,4-ethylenedioxythiophene)-poly(styrene sulfonic acid)
PEEK	poly(etheretherketone)
PEI/PMAE	polyethyleneimine/poly(ethylene-co-maleic acid)
PMMA	poly(methyl methacrylate)
poly-DADMAC	poly(diallyldimethylammonium chloride)
PP	polypropylene
ppb	parts per billion
ppm	parts per million
PPP	poly(p-phenylene)
PPV	poly(p-phenylenevinylene)
PQ	plastoquinione
PS	polystyrene
PSS	poly(styrene sulfonic acid)
PTFE	polytetrafluoroethylene
PV	photovoltaic
PVA	poly(vinyl alcohol)
PVC	poly(vinyl chloride)
PVDF	poly(vinylidene difluoride)
PZT	lead zirconate titanate
q	charge (C)
R	resistance (Ω)
RAIRS	reflection absorption infrared spectroscopy
RAM	random access memory
Re	Reynolds number
RF	radiofrequency
RFID	radiofrequency identification
RHEED	reflection high energy electron diffraction
RLC	resistor-inductor-capacitor
RMS	root-mean-squared
ROM	read only memory
RRAM or ReRAM	resistive random access memory
rRNA	ribosomal RNA

S/N	signal-to-noise ratio (dB)
SAW	surface acoustic wave
SCALPEL	SCattering with Angular Limitation Projection Electron-beam Lithography
SCM	scanning capacitance microscopy
SEM	scanning electron microscopy
SET	single electron transistor
SI	Système International d'Unités
SIMS	secondary-ion mass spectrometry
SiP	system-in-package
SNOM	scanning near-field optical microscopy
SoC	system-on-chip
SP	surface plasmon
SPM	scanning probe microscope
SPP	surface plasmon polariton
SPR	surface plasmon resonance
SRAM	static random access memory
STEM	scanning transmission electron microscope
STM	scanning tunnelling microscope
STP	standard temperature and pressure
SWNT	single wall (carbon) nanotube
t	time (s)
T	temperature (K or °C)
TADF	thermally activated delayed fluorescence
TAZ	3-(biphenyl-4-yl)-5-(4-*tert*-butylphenyl)-4-phenyl-4H-1,2,4-triazole
T_C	Curie temperature
TCNE	tetracyanoethylene
TCNQ	tetracyanoquinodimethane
TE	transverse electric
TED	transmission electron diffraction
TEM	transmission electron microscopy
TFE	tetrafluoroethylene
T_g	glass transition temperature (K or °C)
TGS	triglycine sulphate
THF	tetrahydrofuran
TIPS	triisopropylsilylethynyl
T_K	Kraft temperature
T_m	melting point (K or °C)
TM	transverse magnetic
TMTSF	tetramethyltetraselenafulvalene
TrFE	trifluoroethylene
tRNA	transfer RNA
TTF	tetrathiafulvalene
UV	ultraviolet
v	phase velocity of light in material (m s^{-1})
V	voltage (V)
V	potential energy (J or eV)
VB	valence band
VCR	video cassette recorder

v_d	drift velocity (m s^{-1}) (charge carrier velocity resulting from an applied electric field)
VDF	vinylidene difluoride
v_t	thermal velocity (m s^{-1}) (charge carrier velocity resulting from temperature)
XANES	X-ray absorption near-edge structure
XGA	eXtended Graphics Array
XPS	X-ray photoelectron spectroscopy
XUV	extreme ultraviolet
α	absorption coefficient (m^{-1})
β	phase change (°)
γ	surface tension (N m^{-1})
ε_r, $\hat{\varepsilon}_r$	relative permittivity
η	viscosity (P)
η	efficiency
θ	angle (°)
θ	fraction of surface sites occupied (Langmuir isotherm)
θ_B	Brewster angle (°)
θ_c	critical angle (°)
λ	wavelength (m)
μ	charge carrier mobility (m^2 V^{-1} s^{-1})
μ_r $\hat{\mu}_r$	relative permeability
ν	frequency (Hz)
$\Delta\nu$	frequency shift (Hz)
Π	surface pressure (N m^{-1})
ρ	density (kg m^{-3})
ρ	resistivity (Ω m)
σ	electrical conductivity (S m^{-1}) or (Ω^{-1} m^{-1})
τ	lifetime (s)
Φ	work function (J or eV)
χ_e, $\hat{\chi}_e$	electric susceptibility
χ_m, $\hat{\chi}_m$	magnetic susceptibility
X	electron affinity (J or eV)
ω	angular frequency or velocity (rad s^{-1})

About the Companion Website

This book is accompanied by a companion website:

www.wiley.com/go/petty/molecular-electronics2

The website includes:

- PPTs of all figures from the book
- Solutions manual

Scan this QR code to visit the companion website.

1

Scope of Organic and Molecular Electronics

What's in a name?

1.1   Introduction

The title of this introductory chapter (and the title of the book) suggests two distinct topics. However, the subjects are intimately related. *Organic electronics* has its origins in materials science and concerns the development of electronic and opto-electronic devices that exploit the unique macroscopic properties of organic materials. The most successful commercial product to date is the liquid crystal display (LCD). However, following many years of research, organic light-emitting devices based on dyes and polymers, organic solar cells, organic electronics circuitry, and biochemical sensors are beginning to make their technological marks. The Nobel Prize in Chemistry for 2000 was awarded to three scientists working in the area of organic electronics: Alan Heeger, Alan MacDiarmid, and Hideki Shirakawa, who have made significant contributions to the development of electrically conductive polymers. Much of the current industrially-oriented organic electronics work is being pursued under names such as *plastic electronics* or *printable electronics*, referring to the materials being exploited and the processing technology, respectively.

More challenging is *molecular electronics*. Here, the focus is on the behaviour of individual organic molecules or groups of molecules, and the precise three-dimensional positional control of individual atoms and molecules. Topics as diverse as molecular switching, DNA-electronics, and molecular manufacturing have all been described in the literature. Much of the research activity is directed towards computational architectures that may, one day, rival silicon

Organic and Molecular Electronics: From Principles to Practice, Second Edition. Michael C. Petty.
© 2019 John Wiley & Sons Ltd. Published 2019 by John Wiley & Sons Ltd.
Companion website: www.wiley.com/go/petty/molecular-electronics2

microelectronics. However, even the most optimistic researchers recognize that this is going to be some time away!

Molecular electronics also falls under the umbrella of nanotechnology. In particular, it exemplifies the 'bottom-up' theme of nanotechnology, which refers to making nanoscale structures by building organic and inorganic architectures atom-by-atom, or molecule-by-molecule. The physicist Richard Feynman was one of the first to predict a future for molecular-scale electronics. In a lecture in December 1959, at the annual meeting of the American Physical Society, entitled 'There's Plenty of Room at the Bottom', he described how the laws of physics do not limit our ability to manipulate single atoms and molecules. Instead, it was our lack of the appropriate methods for doing so. Feynman correctly predicted that the time would come in which atomically precise manipulation of matter would be possible. Several advances have now been made to suggest that the prophecy was correct. In this respect, a key invention has been scanning probe microscopy.

1.2 Organic Materials for Electronics

Liquid crystals represent a remarkable molecular electronics success story. However, the transformation of these organic compounds into the established display technology of today took many decades. In the latter half of the nineteenth century, researchers discovered several materials whose optical properties behaved in a strange way near their melting points. In 1922, George Friedel presented a liquid crystal classification scheme, but it took until the 1960s for the potential of liquid crystals in display devices to be recognized. From this point, research into liquid crystals and their applications burgeoned. It is encouraging for workers in organic electronics that the relatively unstable (both thermally and chemically) liquid crystal compounds came to form the foundation of such a substantial worldwide industry.

The interest in organic electronics derives from the intriguing electrical and opto-electrical behaviour of organic materials. Two distinct groups of compounds have been studied – low molecular weight crystalline compounds (molecular crystals) and polymers. In the former category, the photoconductivity of anthracene was discovered in 1906 [1]. However, systematic study of the electrical behaviour of organic molecular solids did not begin until the 1950s [2–4]. The phthalocyanine compounds were one of the first classes of organic molecular crystals to be investigated [2, 3]. These large flat ring-shaped structures are relatively stable organic materials and demonstrate that the words 'organic' and 'thermally unstable' need not always go hand-in-hand. A further interesting group of organic conductive compounds are charge-transfer (CT) complexes. In 1954, researchers reported CT complexes with low resistivities in combinations of perylene with iodine or bromine (perylene itself is an insulator) [5]. In 1962, the well-known electron acceptor molecule tetracyanoquinodimethane (TCNQ) was reported by workers at Dupont. Tetrathiafulvalene (TTF) was synthesized in 1970 and found to be a strong electron donor. In 1973, it was discovered that a combination of these components form a strong CT complex, referred to as TTF-TCNQ. The solid exhibits almost metallic electrical conductivity.

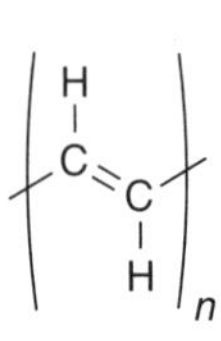

The first synthetic polymers were produced in the late nineteenth century. These were eventually developed into useful products in the 1940s and 1950s (exploiting their toughness, strength to weight ratio, low cost, and ease of fabrication). At this time, polymeric materials were all good insulators, and the idea that a plastic material might conduct electricity was not generally considered. Polyacetylene is the simplest conductive polymer. Its chemical structure, Figure 1.1, consists of a hydrocarbon chain

Figure 1.1 Chemical structure of polyacetylene.

with the carbon atoms connected together by a system of alternating single and double bonds. This particular chemical bonding arrangement confers polyacetylene with its conductive properties.

Reports on acetylene polymers date back to the nineteenth century. However, the polymer was first prepared as a linear, high molecular weight polymer of high crystallinity and regular structure in 1958, by Giulio Natta (Nobel Prize for Chemistry in 1963) and co-workers [6]. The material was discovered to be a semiconductor, with an electrical conductivity between 7×10^{-11} and $7 \times 10^{-3} \, S \, m^{-1}$, depending on how the polymer was produced. Some years layer, in 1967, a student of Hideki Shirakawa, working at the Tokyo Institute of Technology produced, by accident, a polyacetylene film with a conductivity value similar to the best Natta material. Furthermore, the conductivity could be increased by a factor of about one billion by exposure to halogens. In the form in which the polymer was discovered, the material was practically useless; it was insoluble in any solvent and so could not be processed into any kind of useable structure, such as a thin film or a wire. The problem was solved by Jim Feast, a chemist at Durham University. This involved making a soluble precursor. Rather than directly synthesizing polyacetylene, a polymer with easily removable groups is first produced. These groups have the effect of making the polymer soluble, so the material can be formed into a film, a fibre, or a wire. Only then, after processing, are the groups removed to leave the conductive polymer in the desired form.

In the 1980s, conductive polymers based on polyheterocyclic compounds (e.g. polyaniline, polypyrrole, polythiophene) were produced that were soluble in organic solvents and, consequently, could be processed into the form needed for many applications. Figure 1.2 depicts the very wide range of conductivity values that can now be found in conductive polymers.

Polymeric semiconductors can have significant advantages over their inorganic counterparts. For example, thin layers of polymers can easily be made by low-cost methods such as spin-coating or roll-to-roll processing. High temperature deposition from vapour reactants is generally needed for inorganic semiconductors. Since polymers are lightweight and can be manufactured into many different shapes, obvious uses are as components in portable batteries or as electrostatic or electromagnetic shielding, conducting adhesives, printed circuit boards and replacements for conventional electrolytes in electrolytic capacitors. Other electroactive properties of these organic compounds can also be exploited in electronic devices. Examples include components in photocopying machines, organic light emitting displays, chemical sensors, and plastic transistors. The electrical properties of the last devices cannot be directly compared to those based on single crystal silicon and gallium arsenide. The mobilities (carrier velocity per unit electric field) of the charge carriers in organic field

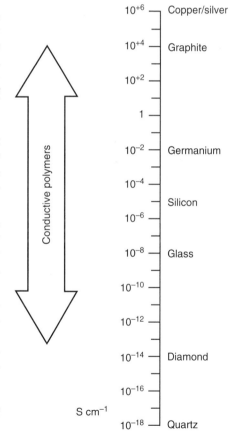

Figure 1.2 Range of conductivities for conductive polymers compared to various inorganic materials.

effect transistors are low and similar to those found in amorphous silicon. Nevertheless, the simple fabrication techniques for polymers have attracted several companies to work on polymer transistor applications, such as data storage and thin film device arrays to address LCDs. The long-term stability of organic materials has often been considered as a limitation to their exploitation in electronic devices. To some extent, this difficulty has now been resolved and some materials are now displaying adequate storage and operating lifetimes, e.g. estimated operating lifetimes of over a million hours for some organic light emitting devices. This is certainly sufficient for these structures to be exploited in current commercial products.

Certain organic compounds are superconducting, but at very low temperatures. These materials are usually based on donor-acceptor CT complexes. A challenge has been to try and increase the transition temperature. Although there have been some theoretical predictions on the type of organic molecules that would be superconductive at room temperature, this 'Holy Grail' remains elusive. Non-conductive polymers can also play an important role in molecular electronics. Pyroelectric, piezoelectric, and ferromagnetic materials may find use in infra-red detection, intruder alarms, and non-linear optics (e.g. second-harmonic generation). Similarly, some effort is being focused on organic magnetic materials. Due to the high density of iron, such molecular devices are generally inferior to conventional magnets on a 'magnet per unit weight or unit volume' basis. However, for more specialist applications, such as data storage, molecular systems may become attractive.

The development of effective devices for the identification and quantification of chemical and biochemical substances for process control and environmental monitoring is a growing need. Many sensors do not possess the specifications to conform to existing or forthcoming legislation; some systems are too bulky and/or expensive for use in the field. Inorganic materials such as the oxides of tin and zinc have traditionally been favoured as sensing elements. However, one disadvantage of sensors based on metallic oxides is that they usually have to be operated at elevated temperatures, limiting some applications. As an alternative, there has been considerable interest in trying to exploit the properties of organic materials. Many such substances, in particular phthalocyanine derivatives, are known to exhibit a high sensitivity to gases. A significant advantage of organic compounds is that their sensitivity and selectivity can be tailored to a particular application by modifications to their chemical structure. Moreover, thin film technologies, such as self-assembly or layer-by-layer electrostatic deposition, enable ultra-thin layers of organic materials to be engineered at the molecular level.

This exploitation of materials at the nano-scale, rather than the micro-scale, offers several advantages. First, an enormous increase in surface area may be achieved; the physical and chemical properties become controlled by the surface. A material can be made light-absorbing by coating its constituent particles with a dye. The particle size reduction also induces both mechanical advantages and quantum effects. For instance, the hindered propagation of lattice defects leads to strong and hard metals, and enhanced diffusional creep leads to 'super-plastic' ceramics during processing at elevated temperatures. Quantum confinement allows for control over material 'constants', as demonstrated by the blue shift of the optical spectrum of nanoparticles. Colours can be controlled and manipulated. For example, titanium dioxide in nanomaterial form absorbs much more UV light than its bulk counterpart.

1.3 Molecular Electronics

The name 'molecular electronics' dates back to the 1950s in the USA, although its meaning was somewhat different to that accepted today. Although all branches of the US Military were interested in miniaturization, the form that interest took varied among the Services. The Air Force supported an idea known as molecular electronics. The concept was radically different to any other attempt at miniaturization – this was to build a circuit in the solid without reproducing individual component function. The whole was to do more than the sum of the parts. Molecular electronics demanded a technological leap beyond even integrated circuits, which were really just discrete components formed in the same piece of silicon (with the two important problems of isolation and interconnection overcome). The idea was ahead of its time and certainly in advance of the technology available, and no significant progress was made. A detailed history of this era has been documented by Choi and Mody [7].

Serious research on molecular electronics started in the USA in the 1970s with pioneering work by Ari Aviram at IBM, who proposed a structure for a molecular rectifier [8]. Impetus was provided by the enthusiasm of Forrest Carter working in the US Naval Research Laboratory [9] and the field was epitomized by the elegant experiments on monolayer films undertaken by Hans Kuhn in Göttingen [10].

However, emerging technologies must compete with existing ways to doing things. The microelectronics industry, based largely on the inorganic semiconductor silicon, has been developing since the 1950s and is continuing to make dramatic progress. To put molecular electronics into context, a review of these developments is provided in the next section.

1.3.1 Evolution of Microelectronics

The workhorse of today's electronic computer is the metal-oxide-semiconductor field effect transistor, or MOSFET; the basic structure of a silicon MOSFET is shown in Figure 1.3. This is a three terminal device, comprising the source, drain, and gate. A conductive channel in the silicon semiconductor is formed beneath the insulating gate oxide and between the source and drain contacts. The length of the conductive channel $= L$, while the thickness of the insulating gate oxide is given by d. Application of an electric field to the gate influences the conductivity of the channel; this is called the field effect. Because there are charge carriers with positive as well as negative charges, there are two kinds of metal-oxide-semiconductor (MOS) transistor – n-channel and p-channel. The technology is therefore referred to as complementary MOS or complementary metal-oxide-semiconductor (CMOS). Table 1.1 shows some of the significant dates in the evolution of electronics during the twentieth century [11–13]. Although a patent for a MOSFET was filed in 1930, most of the developments in electronics around this time exploited the valve (or vacuum tube) as a signal processing device. By the late 1940s, after 40 years of development, valve technology was mature.

The first large-scale digital electronic computer that could be reprogrammed, the Electronic Numerical Integrator and Computer (ENIAC), was built in 1946, in the Moore School of Electronic Engineering, University of Pennsylvania (although earlier computers had been built with some of these properties). ENIAC could add 5000 numbers in 1 second. It could calculate the trajectory of an artillery shell in only 30 seconds (in contrast, a human would need about 40 hours).

Figure 1.3 Schematic diagram of a metal-oxide-semiconductor field effect transistor, or MOSFET. The length of the conductive channel $= L$, while the thickness of the insulating gate oxide is given by d.

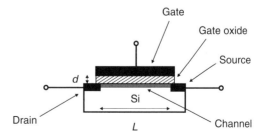

Table 1.1 Dates of key inventions in microelectronics [11–13].

Date	Milestone
1930	Metal-oxide-semiconductor field effect transistor (MOSFET) concept patent (Lilienfield, University of Liepzig, Germany).
1946	Stored-program computer (Electronic Numerical Integrator and Computer (ENIAC) – University of Pennsylvania).
1948	Bipolar transistor (Bardeen, Brattain, Shockley, Ball Laboratories, USA).
1952	IC concept (Dummer, Royal Radar Establishment, UK)
1959	Planar process (Hoerni, Fairchild, USA).
1959	IC (Kilby, Texas Instruments; Noyce, Fairchild Semiconductor, and others).
1960	MOSFET (Kahng and Atalla, Bell, USA).
1962	MOS IC (Hofstein and Heiman, RCA, USA).
1968	CMOS (Westinghouse, GT&E, RCA, Sylvania, USA).
1969	Internet (ARPAnet, USA).
1971	Microprocessor (Hoff, Intel, USA).
1972	1024 bit dynamic random access memory (DRAM) (Intel, USA).
1980	256 k DRAM (NEC-Toshiba, NTT-Musashino, Japan).
1981	MS-DOS (Gates, Microsoft, USA).

ENIAC contained about 17 000 valves, weighed 27 000 kg, occupied about 500 m^3 and consumed 174 kW. Its program was wired into the processor and had to be mechanically altered.

The junction transistor (a different device architecture to the field effect transistor described above) was invented in the Bell laboratories, USA in 1947 by John Bardeen, Walter Brattain, and William Shockley. The three inventors shared the 1956 Nobel Prize for Physics. Even though transistors as discrete devices had significant advantages over vacuum tubes, and progress on transistors was steady during the 1950s, the directors of many large electronics companies believed that vacuum tubes held an unassailable competitive position. Arguably, the most significant breakthrough for technology came late in the 1950s, when the integrated circuit, IC, was introduced. There is no consensus on who 'invented' the IC. A patent for an IC was filed in February 1959 by Jack Kilby of Texas Instruments. Another key contributor to integrated circuits was Robert Noyce of Fairchild Semiconductor. Noyce died in 1990 and Kilby was awarded the Nobel Prize for Physics in 2000. The microprocessor became a reality in the mid-1970s with the introduction of the large-scale integrated circuit (LSI) and later the very large-scale integrated circuit (VLSI). These structures consisted of many thousands of inter-connected transistors etched into a single silicon substrate.

1.3.2 Moore's Laws

A common barrier to sustaining a high rate of growth in any industry is the existence of a fundamental limitation to the basic technologies on which the industry depends. A good example of this is the commercial aerospace industry, where fundamental limitations in the strength of materials and the cost of energy have resulted in a current generation of aircraft no faster or more comfortable than those of 50 years ago. Improvements have instead been evolutionary in nature and focused largely around operating cost reduction. By contrast, no fundamental limitation in the technologies enabling the electronics industry has yet arisen as a significant barrier to its continued growth.

The first microprocessor chip (the 4004), manufactured in 1971 by Intel using 10 μm process technology on a 2-inch diameter silicon wafer, had an initial clock speed of 108 kHz and contained 2300 transistors. By 2010, an Intel processor using a 32 nm technology held 1.16 billion transistors and could operate at 3.8 GHz.

Gordon Moore of Intel was the first to quantify the steady improvement in gate density when he noticed that the number of transistors that could be built on a chip increased exponentially with time. His observation, made in 1965, was that the number of transistors per unit area on integrated circuits, or functionality per chip, had doubled every year since the integrated circuit was invented. Moore predicted that this trend would continue for the foreseeable future. This is Moore's Law, or Moore's First Law, and is illustrated in Figure 1.4, which shows data for Intel microprocessor chips [13]. Each experimental point represents a change in the device feature size used in the processing. Although not strictly a 'law', this rule has been a consistent trend and key indication of successful leading-edge semiconductor products. In recent years, the pace slowed down a little: the scaling of the microprocessor unit, MPU, is based more on a 2.5 year cycle (i.e. the functionality per chip doubles every 2.5 years), while the scaling of a dynamic random access memory, DRAM, is a 3-year cycle. Other scaling laws have been proposed, such as *Dennard scaling*, named after Robert Dennard, who worked for the IBM company. Originally formulated in 1974 for MOSFETs, it states, roughly, that as transistors get smaller their power density stays constant, so that the power use stays in proportion with area: both voltage and current scale (downward) with length. Since around 2005–2007, Dennard scaling appears to have broken down. As of 2016, transistor counts in integrated circuits are still growing, but the resulting improvements in performance are more gradual than the speed-ups resulting from significant frequency increases.

Intel started production of three-dimensional transistors, known as FinFETs and based on a 22 nm process, in late 2011. The distinguishing characteristic of the FinFET, illustrated in

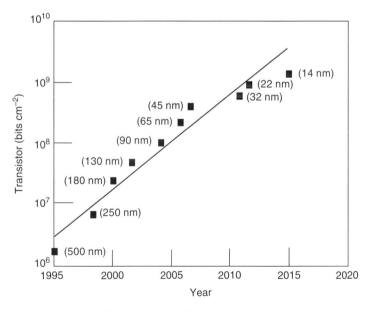

Figure 1.4 Density of transistors in Intel processors. The processing feature size is indicated adjacent to the data points. *Source*: Taken from Intel [13].

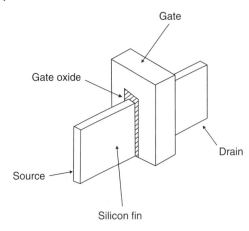

Gate

Gate oxide

Source

Drain

Silicon fin

Figure 1.5 FinFET. The gate wraps around the silicon 'fin', which forms the source and drain.

Figure 1.5, is that the gate wraps around a silicon 'fin', which forms the source and the drain. The thickness of the fin (measured in the direction from source to drain) determines the effective channel length of the device. Tri-gate FinFETs produced in 2014 were based on 14 nm processing. Just as Moore's Law reached its 50th anniversary (in 2015), IBM announced that it had produced a working 7 nm chip. The semiconductor industries have produced an International Technology Roadmap for Semiconductors (ITRS) on the future of CMOS technology [14]. The prediction is for a 3.5 nm width FinFET fin in 2020 decreasing to 2.1 nm in 2025. The ITRS figures are regularly updated.

There are many recognized factors that could bring the Moore's Law scaling to an end. Some of the technical issues are explored in the next section. However, one significant factor is the economics of chip production. The cost of building fabrication facilities to manufacture chips has been increasing exponentially, by a factor of two for every chip generation. This is some-times known as Moore's Second Law. This significant increase in cost is due to the extremely sophisticated tools that will be needed to form the increasingly small features of the devices.

1.3.3 Beyond Moore

As the dimensions of MOSFET devices decrease, the effect of the operating characteristics can be calculated [15]. The scaling can be done either to keep the electric field constant or to main-tain the same operating voltage. For a scaling factor K ($K > 1$), Table 1.2 shows the effect on the operating parameters of the MOSFET for a constant field and device resistance. The benefits of shrinking the device in terms of packing density, speed, and power dissipation are evident. The scaling of device depletion widths is achieved indirectly by scaling up the doping concentrations (which keeps the resistance unchanged). For ideal scaling, power supply voltages should also be reduced to keep the internal electric fields reasonably constant from one device generation to the next. In practice, however, power supply voltages are not scaled with the device dimensions, partly because of other system-related constraints. The Semiconductor Industries Roadmap predicts that transistor operating voltages will only reduce from 0.75 in 2020 to 0.68 V in 2025.

One parameter that does not move in a favourable direction as the device dimensions are reduced is the current density, which is increased with the scale factor. Metal conductors have an upper current density limit imposed by the process of electromigration. This is a diffusive process in which the atoms of a solid move under the influence of electrical forces. This effect limits the maximum current that can be carried by a conductor without its rapid destruction.

Table 1.2 Scaling rules for metal-oxide-semiconductor field effect transistors (MOSFETs) according to a scale factor K. The scaling keeps the electric field and device resistance constant [15].

Parameter	Scaling Factor
Device dimensions (channel length, gate width, oxide thickness)	$1/K$
Current, voltage	$1/K$
Current density	K
Impurity concentration	K
Gate capacitance	$1/K$
Time constant	$1/K$
Switching energy	$1/K^3$
Power per device	$1/K^2$

For example, the current density for aluminium conductors in integrated circuits must be kept below $10^{10}\,\mathrm{A\,m^{-2}}$.

Limits that are closely related to the basic physical laws are called fundamental limits. A basic concept of quantum mechanics is that a physical measurement performed in a time Δt (and computing may be considered as a 'measurement') must involve an energy ΔE:

$$\Delta E \geq \frac{h}{\Delta t} \tag{1.1}$$

where h is Planck's constant ($= 6.63 \times 10^{-34}\,\mathrm{J\,s}$). The energy is dissipated as heat. The power P (= energy per unit time) dissipated during the measurement, or switching, process is

$$P = \frac{\Delta E}{\Delta t} \geq \frac{h}{(\Delta t)^2} \tag{1.2}$$

This can be considered as the lower band of power dissipation per unit operation. Eq. (1.1) predicts a minimum energy dissipated in a nanosecond switching device (i.e. operating at $1\,\mathrm{GHz}$) is of the order $10^{-25}\,\mathrm{J}$, which is currently many orders of magnitude below the actual switching energy in an MOS device.

The switching energy must generally be greater than the thermal energy, otherwise the device will switch on and off randomly. This requires

$$\Delta E > k_{\mathrm{B}} T \tag{1.3}$$

where k_{B} is Boltzmann's constant ($1.38 \times 10^{-23}\,\mathrm{J\,K^{-1}}$) and T is the absolute temperature. At room temperature, $k_{\mathrm{B}} T = 4 \times 10^{-21}\,\mathrm{J}$, once again many orders of magnitude below the actual switching energy in MOSFETs.

Equation (1.3) is not regarded as a fundamental limit to computing operations, as many methods are available to reduce noise in electronic systems [16]. An assertion, put forward by John von Neumann, in a 1949 lecture, was that a computer must dissipate an energy per reversible operation (or per bit in a binary computer) given by

$$\Delta E = k_{\mathrm{B}} T \log_e 2 \tag{1.4}$$

which is about $3 \times 10^{-21}\,\mathrm{J}$ at room temperature. This can be considered as a fundamental limit.

A further issue is the time taken for a charge carrier to acquire energy from the applied electric field. The maximum rate of energy transfer (power) to an electron with an electrical charge e ($= 1.6 \times 10^{-19}$ C) moving at a speed v in an electric field E is evE (i.e. = (force × distance)/time). The electron energy must be greater than the thermal energy $k_B T$. The time to achieve this is therefore given by

$$\text{time} = \frac{\text{energy}}{\text{power}} = \frac{k_B T}{evE} \tag{1.5}$$

The maximum electron velocity in silicon is about 10^5 m s^{-1} for an applied field of 5×10^7 V m^{-1}. This gives a limiting response time of approximately 5×10^{-15} s (5 femtoseconds) at room temperature. Devices will not respond so quickly as the electrons will need to acquire energies $>> k_B T$ (10 or $100 k_B T$). However, this principle imposes a restriction on the operating frequency of silicon MOSFETs.

There are also a number of technological issues that will need to be overcome for the predictions for the CMOS-based roadmap to be realized. Not least, are the materials limitations of the silicon/silicon dioxide system. For example, charge leakage becomes a problem when the insulating silicon dioxide layers are thinned to a few nm. Quantum mechanical tunnelling through the oxide can become a serious problem if the tunnel currents become comparable to the other circuit currents. Alternative insulating materials, such as hafnium oxide, are currently under investigation by the large electronics companies.

Heat dissipation is a further important practical factor. A microprocessor chip, with around 10^8 transistors operating at the nanosecond rate can emit up to 100 watts of heat. Although the simple scaling shown in Table 1.2 predicts that the power per device decreases with the square of the scaling factor, the power density (power per unit area) will remain constant. This power density p is given by

$$p = \Delta E v n X \tag{1.6}$$

where n is the device density, v is the frequency of operation, and X is the probability that the device switches in a clock cycle, typically $X \sim 0.1$. The maximum tolerable power density (without sophisticated heat sinking) is about 100 W cm^{-2}. This means that the switching energy, operating frequency, and device density are limited by

$$\Delta E v n \approx 1000 \text{ Wcm}^{-2} \tag{1.7}$$

At high frequency, a high device density is desirable, and therefore a low energy per bit is necessary. These three limits are depicted in the switching energy – time diagram in Figure 1.6.

With the eventual demise of the top-down lithographic technologies which are currently employed, bottom-up nanomaterials fabrication for nanoelectronics has the potential to move electronic materials, computers, and devices into a new era of sustainable growth.

The promise of molecular electronics is compelling. The device density offered by silicon technology and outlined above is remarkable and offers the means to store significant amounts of data, Table 1.3 [17]. However, molecular electronics has the potential for further increases in device density. For example, using 1–3 nm organic molecules as the processing elements, 10^{13}–10^{14} 'devices' could be fitted into 1 cm^2. Discrete organic processing devices (diodes and transistors) and simple circuits based on these already exist. What is now needed is to make them work faster (the maximum operational speed of such plastic circuits is around a few MHz) and more reliably.

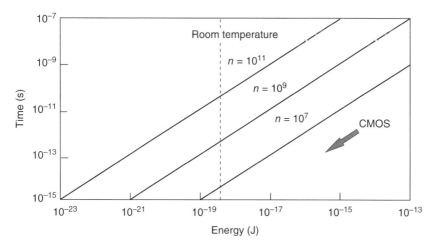

Figure 1.6 Switching time versus switching energy for metal-oxide-semiconductor field effect transistors (MOSFETs) based on Eq. (1.6). The three different curves correspond to different device densities n (in units of cm^{-2}). The dashed vertical line corresponds to $100\,k_B T\,(=4\times10^{-19}\,\mathrm{J})$, the optimum value for the energy to write a bit of information at room temperature.

Table 1.3 Information content of various sources [17]. (1 byte = 8 bits).

Application	Typical information content (bytes)
Colour photograph	10^5
Average book	10^6
Desktop computer	10^8
Genetic code	10^{10}
Human brain	10^{13}
Library of Congress	10^{15}

Molecular-scale technology will, of course, also need to address all the problems of the silicon microelectronics industry – plus many more! New approaches are needed. It will not be sufficient to reproduce the functionality of discrete silicon devices in molecules and then attempt to connect all the devices together in the same way that a silicon integrated circuit is built.

As computational devices shrink in size, the tolerance of their architectures to defects in the individual processing elements becomes more important. Even if the rate of defects in a chemically fabricated molecular circuit were only one per billion components, which exceeds the best practices in chip fabs, it would still result in a million defects in a system containing 10^{15} components. A significant defect-tolerant computer was the experimental machine known as Teremac, built by the company Hewlett Packard. Although Teremac was constructed by means of conventional technology, many of its problems resemble the challenges that face scientists who are exploring molecular electronics.

Teremac was built from a large number of components that had significant defect probabilities. To keep the construction costs reasonable, the machine was built using elements that were defective and inexpensive. Furthermore, the techniques used to connect all the components together were error-prone. Teremac is a reconfigurable multi-architecture computer with 10^6 gates that operate at 1 MHz or a total of 1 trillion bit operations per second (hence 'tera'). It is based on field-programmable gate arrays (FPGAs). These are essentially lookup tables connected by a huge number of wires and switches that are arranged to form crossbars, which permit the connection of any input to any output. In principle, FPGAs substitute memory for logic whenever possible. Perhaps the most significant fact about Teremac is that it was comatose at birth. Three-quarters of the FPGAs contained defects that would be fatal to an isolated chip. Teremac contained 220 000 wiring and gate defects – a total of 3% of all its resources. For the first 24 hours of its existence, Teremac was connected to a workstation to undergo a series of tests in order to find out where the defective resources were. These locations were then written to a configuration table as being 'in-use', to ensure that the defective components would not be assessed by a running program.

1.4 The Biological World

Scientists and technologists working in the field of molecular electronics frequently draw attention to the analogy between the devices they are working on and those found in nature. There are very many examples. For instance, the brain is nature's computer. Although the brain works in a different way to silicon microprocessors (exploiting ions rather than electrons, and utilizing parallel processing, instead of the serial approach in silicon systems), it is nonetheless constructed from a large number of individual processing elements. Nature's 'gates' are the neurons. There are about 10^{11} neurons in the human brain, and each is connected to 10^3–10^4 others. This gives a crude 'bit-count' of 10^{11}–10^{15}. An equivalent artificial 'brain' might therefore be built from $10^5 \times 8$ Gbit chips.

Living systems are able to exist because of the vast amount of highly ordered molecular machinery from which they are built. The information required to build a living cell or organism is stored in the DNA and is then transferred to the proteins by the processes called transcription and translation. These are all executed by various bio-molecular components, mostly protein and nucleic acids. Such molecular-scale machinery is highly sophisticated and has evolved over millions of years. For example, bacteria may be considered as 'micro-robots', moving throughout the human body and taking part in highly complex biochemical processes.

Electronics at the molecular scale is concerned with atomic precision and molecular manipulation and, in principle, has much to learn from nature. One rapidly developing field is that of biomimetics. However, it might be argued that attempts to reproduce nature's complex molecular architectures in the laboratory or factory using mechanical and chemical processes are too ambitious and doomed to failure. (After all, man's most successful flying machines are constructed from metal and not feathers!). A better approach for molecular electronics is to learn from nature rather than to copy it. For example, fundamental studies into the operation of nature's molecular motors may, one day, lead to a new technology for energy production.

The other field in which molecular-scale manipulation of matter is receiving considerable attention is medicine. Since all living organisms are composed of molecules, molecular biology has become the primary focus of biotechnology. Countless diseases have been cured by the ability to synthesize small molecules – drugs – that interact with the protein molecules which make up the molecular machinery that keeps us alive. The understanding of how proteins interact with DNA, phospholipids, and other biological molecules is fundamental to progress.

1.5   Future Opportunities

The growth of the electronics industry has been driven by the development of disruptive technologies in parallel with evolutionary improvements in enabling device technologies. The development of personal computers, PCs, was a disruptive applications technology. Its progress, however, was greatly accelerated by rapid evolutionary improvements in technology for manufacturing semiconductor devices (increasing significantly the number of transistors on a chip). Other 'killer' applications have included mobile phones, personal electronics (e.g. games), digital imaging and video, along with plasma, and LCD flat panel displays. Digital Video (or Versatile) Discs, DVDs, have enjoyed the most rapid and unexpected acceptance of any new home entertainment technology. The introduction of advanced optical discs with the required storage capacity, along with increasingly widespread adoption of high-definition television will help drive a high rate of growth in this sector of consumer electronics for the immediate future. Indeed, management of the current menagerie of discrete personal electronic devices may well be the most important driver for early implementation of some of the concepts embodied in 'ambient intelligence'.

The concept of ambient intelligence, AmI, was first outlined by Marc Weiser [18]. The idea was that small computer chips would be embedded in everyday objects all around us ('ubiquitous computing') and would respond to our presence, needs, and wishes by way of wireless connectivity. It was imagined that these devices would undertake important tasks without active manipulation and so unobtrusively that we would notice only their effects. Weiser referred to this as 'calm technology', since it would allow us to focus on our work and on our social and recreational activities without requiring that we 'interact' directly with these devices as we currently do today with PCs, TVs, VCRs (video cassette recorders), and other home and workplace electronic items and appliances. AmI has the potential to fulfil an important societal need – the requirement to simplify human interactions with the plethora of electronic devices that surround us today and to take advantage of the unfulfilled capabilities of electronic devices to make work and home life easier and more productive. If this technology is widely accepted and adapted, it will certainly establish rapid and sustainable growth in the electronics industry.

Since the original concept paper by Weiser, the vision of calm technology has expanded to include context awareness (i.e. analysing context, adapting to people who live in it, learning from their behaviour and eventually recognizing and demonstrating emotion). Such AmI systems would be aware of your physical and emotional state and react accordingly; for example, suggesting the need for, or requesting, medical attention, or simply adjusting ambient light, sound and colour to suit your mood.

A related area that is attracting considerable interest is the Internet of Things (IoT). Kevin Ashton, cofounder and executive director of the Auto-ID Centre at MIT, first mentioned the IoT in a presentation he made in 1999 [19]. The IoT is the network of physical objects or 'things' embedded with electronics, software, sensors, and connectivity to enable objects to exchange data with the manufacturer, operator and/or other connected devices. The IoT allows objects to be sensed and controlled remotely across existing network infrastructure, creating opportunities for more direct integration between the physical world and computer-based systems, and resulting in improved efficiency, accuracy, and economic benefit. Each thing is uniquely identifiable through its embedded computing system but is able to interact with other things the existing internet infrastructure. AmI and autonomous control are not part of the original concept of the IoT, as these do not necessarily require internet structures. However, there is a shift in research to integrate the concepts of these ideas.

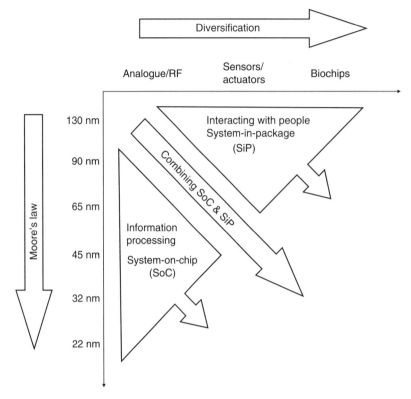

Figure 1.7 Moore's Law and diversification. *Source*: Taken from *the International Technology Roadmap for Semiconductors* [14].

The *International Technology Roadmap for Semiconductors* offers a similar viewpoint on the way ahead [14]. Figure 1.7 shows how non-CMOS devices can be integrated with the Roadmap. The essential functions of a silicon integrated circuit – or system-on-chip (SoC) – are those of data storage and digital signal processing. However, many functional requirements, such as wireless communication, sensing and actuating, and biological functions, do not scale with Moore's Law. In these cases, non-CMOS solutions can be used. In the future, the integration of CMOS and non-CMOS technologies within a single package – or system-in-package (SiP) – will become increasingly important.

A good example of the diversity of electronic systems is the 'lab-on-a-chip' technology, under development by many commercial organizations [20]. There is considerable interest in reducing the size of sensing systems and integrating them with some functionality. This has been made possible by the emergence of novel sensors. A completed system would, in essence, reduce the functionality of a complete chemistry lab to a single chip. For example, Sandia's handheld μChemlab system is about the size of a palmtop computer, but is capable of using both gas and liquid chromatographic techniques to separate the various constituents of complex chemical mixtures [21]. Samples are drawn into the device by a micropump and then passed through a series of channels where the mixture is separated into its basic components. A microcomputer measures and identifies the components and displays results on the device screen. Advanced microsensor technology is used to perform chemical analysis after separation of the mixture. The system incorporates miniaturized surface acoustic wave detectors, laser-induced fluorescence analysis, and electrochemical detection components.

Future lab-on-a-chip technologies may allow the manipulation and measurement of individual molecules. This would provide many new opportunities for studying biochemical reactions and life processes [22].

1.6 Conclusions

Organic compounds possess a wide range of fascinating physical and chemical properties that make them attractive candidates for exploitation in electronic and opto-electronic devices. It is not, however, anticipated that these materials will displace silicon in the foreseeable future as the dominant material for fast signal processing. It is much more likely that organic materials will find use in other niche areas of electronics, where silicon and other inorganic semiconductors cannot compete. Examples already exist, such as liquid crystal and organic light emitting displays. Organic photovoltaic solar cell architectures, biosensors, and plastic circuitry for identification tags and smart cards are also likely to make a major impact in the market place over the next 10 years.

Over the first decades of the twenty-first century, classical CMOS technology will come up against a number of technological barriers. The bottom-up approach to molecular electronics provides an alternative and attractive way forward and, as such, it is currently an area of exciting interdisciplinary activity. However, the challenges in fabricating molecular switches and connecting them together are formidable. Living systems use a different approach. These assemble themselves naturally from molecules and are extremely energetically efficient when compared with man-made computational devices. More radical approaches to materials fabrication and device design, exploiting self-organization, may be needed to realize fully the potential offered by molecular-scale electronics.

Problems

1.1 Compile a table contrasting the room temperature electrical conductivity and density of the following metals: silver, copper, gold, aluminium. Which element has the highest electrical conductivity and which has the highest conductivity:weight ratio? Assuming that conductive polymers possess densities of around $1\,\mathrm{g\,cm^{-3}}$, what value of electrical conductivity do they need to exhibit a conductivity:weight ratio higher than the inorganic metals?

1.2 The switching energy, E_{sw}, of a MOSFET can be approximated by the formula for the electrical energy of a parallel plate capacitor $E_{sw} = \frac{1}{2}CV^2$, where C is the capacitance and V is the applied voltage. Show that, as the linear dimensions of the MOSFET are reduced by a factor K ($K > 1$), the switching energy is *reduced* by a factor K^3 but the current density *increases* by K. State clearly any assumptions that you make. Briefly discuss the implications of these results.

1.3 Assuming that a single text character can be represented by 1 byte (8-bits) of digital information, estimate the number of bytes needed to store the complete works of Shakespeare. Compare this to the raw information content of a high-resolution $10 \times 20\,\mathrm{cm}$ colour photograph.

1.4 In computing, the term FLOPS refers to the number of FLoating-point Operations Per Second (or bit operations per second) and is a measure of computer performance.

The computing power of the human brain is based on the 'firing' of neurons. If there about 10^{11} neurons in the brain and a neuron can fire about once every 5 ms (about 200 times per second) and each neuron is connected to 10^3–10^4 other neurons, estimate the computation power of the brain in FLOPS. Compare your estimate to the power of modern supercomputers.

1.5 The energy needed to switch an individual MOSFET operating at 100 MHz is 10^{-17} J. If the maximum tolerable power density for the integrated circuit is $100\,\mathrm{W\,cm^{-2}}$, estimate the maximum device density per cm^2.

References

1 Pocchetino, A. (1906). Sul comportamento foto-elettrico dell'antracene. *Atti. Acad. Lincei. Rendiconti* 15: 355–368.

2 Eley, D.D. (1948). Phthalocyanines as semiconductors. *Nature* 162: 819.

3 Varanyan, A.T. (1948). Poluprovodnikovye svoistva organicheskikh krasitelei. 1. Ftalotsianiny. *Zhurnal Fizicheskoi Khimii* 22: 769–789.

4 Akamatsu, H. and Inokuchi, H. (1950). On the electrical conductivity of violanthrone, iso-violanthrone, and pyranthrone. *J. Chem. Phys.* 18: 810–811.

5 Akamatsu, H., Inokuchi, H., and Matsunaga, Y. (1954). Electrical conductivity of the perylene–bromine complex. *Nature* 173: 168.

6 Natta, G., Mazzanti, G., and Corradini, P. (1958). Polimerizzazione stereospecifica dell'acetilene. *Atti. Acad. Naz. Lincei, Cl. Sci. Fis. Mat. Tend.* 25: 3–12.

7 Choi, H. and Mody, C.C.M. (2009). The long history of molecular electronics: microelectronics origins of nanotechnology. *Soc. Stud. Sci.* 39: 11–50.

8 Aviram, A. and Ratner, M.A. (1974). Molecular rectifiers. *Chem. Phys. Lett.* 29: 277–283.

9 Carter, F.L. (1983). Molecular-level fabrication techniques and molecular electronic devices. *J. Vac. Sci. Technol. B* 1: 959–968.

10 Kuhn, H. (1989). Organized monolayers – building blocks in constructing supramolecular devices. In: *Molecular Electronics: Biosensors and Computers Symposium* (ed. F.T. Hong), 3–24. New York: Plenum Press.

11 Dummer, G.W.A. (1997). *Electronic Inventions and Discoveries*, 4e. Bristol: Institute of Physics.

12 Braun, E. and MacDonald, S. (1977). *Revolution in Miniature*. Cambridge: Cambridge University Press.

13 Intel: Available from URL: http://www.intel.com. Accessed 1 November 2017.

14 *International Technology Roadmap for Semiconductors*. Semiconductor Industries Association. Available from URL: http://www.itrs2.net. Accessed 1 November 2017.

15 Streetman, B.G. and Banerjee, S. (2014). *Solid State Electronic Devices*, 7e. London: Pearson.

16 Landauer, R. (1990). Computation: a fundamental physical view. In: *Maxwell's Demon: Entropy, Information, Computing* (ed. H.S. Leff and A.F. Rex), p260–p267. Adam Hilger: Bristol.

17 Tour, J.M. (2003). *Molecular Electronics*. New Jersey: World Scientific.

18 Weiser, M. (1991). The computer for the 21st Century. *Sci. Am.* 265: 66–75.

19 Wood A. The Guardian. http://www.theguardian.com/media-network/2015/mar/31/the-internet-of-things-is-revolutionising-our-lives-but-standards-are-a-must. Accessed 1 November 2017.

20 Janasek, D., Franzke, J., and Manz, A. (2006). Scaling and the design of miniaturized chemical-analysis systems. *Nature* 442: 374–380.

21 Research and technology development performed by Sandia's California laboratory for the Department of Energy. Available from URL: http://www.sandia.gov/biosystems/docs/RapiDx.pdf. Accessed 1 November 2017.

22 Craighead, C. (2006). Future lab-on-a-chip technologies for interrogating individual molecules. *Nature* 442: 387–393.

Further Reading

Amos, M. (2006). *Genesis Machines*. London: Altantic books.

Aswai, D.K. and Yakhmi, J.V. (eds.) (2010). *Molecular and Organic Electronics Devices*. Hauppauge: Nova Science Publishers.

Bar-Cohen, Y. (ed.) (2011). *Biomimetics: Nature-Based Innovation*. Boca Raton: Taylor and Francis.

Canatore, E. (ed.) (2013). *Applications of Organic and Printed Electronics*. New York: Springer.

Cicoira, F. and Santato, C. (eds.) (2013). *Organic Electronics – Emerging Concepts and Technologies*. Weinheim: Wiley.

Cuevas, J.C. and Scheer, E. (2017). *Molecular Electronics: An Introduction to Theory and Experiment*, 2e. Singapore: World Scientific.

Cuniberti, G., Fagas, G., and Richer, K. (eds.) (2005). *Introducing Molecular Electronics*. Berlin: Springer.

Drexler, K.E. (1990). *Engines of Creation: The Coming Era of Nanotechnology*. New York: Anchor Books.

Feast, W.J., Tsibouklis, J., Pouwer, K.L., and Groenendaal, M.E.W. (1996). Synthesis, processing and material properties of conjugated polymers. *Polymer* 37: 5017–5047.

Ferraro, J.R. and Williams, J.M. (1987). *Introduction to Synthetic Electrical Conductors*. Orlando: Academic Press.

Feynman RP. *There's Plenty of Room at the Bottom*. Lecture of the American Physical Society at California Institute of Technology, December 1959. Available from URL: http://www.zyvex.com/nanotech/feynman.html. Accessed 1 November 2017.

Flood, A.H., Stoddart, J.F., Steuerman, D.W., and Heath, J.R. (2004). Whence molecular electronics. *Science* 306: 2055–2056.

Gershenfeld, N. (2000). *The Physics of Information Technology*. Cambridge: Cambridge University Press.

Geoghegan, M. and Hadziioannou, G. (2013). *Polymer Electronics*. Oxford: Oxford Master Series in Physics.

Ghosh, A. (2016). *Nanoelectronics: A Molecular View*. Singapore: World Scientific.

Jones, R.A.L. (2004). *Soft Machines: Nanotechnology and Life*. Oxford: Oxford University Press.

Kagan, C.R., Fernandez, L.E., Gogotsi, Y. et al. (2016). Nano day: celebrating the next decade of nanoscience and nanotechnology. *ACS Nano* 10: 9093–9103.

Klauk, H. (ed.) (2012). *Organic Electronics II: More Materials and Applications*. Weinheim: Wiley-VCH.

Leff, H.S. and Rex, A.F. (eds.) (1990). *Maxwell's Demon: Entropy, Information, Computing*. Bristol: Adam Hilger.

Mansoori, G.A. (2005). *Principles of Nanotechnology*. New Jersey: World Scientific.

Mody, C.C.A. (2017). *The Long Arm of Moore's Law*. Cambridge, MA: MIT Press.

Petty, M.C., Bryce, M.R., and Bloor, D. (eds.) (1995). *An Introduction to Molecular Electronics*. London: Edward Arnold.

Rao, G.R. (ed.) (2017). Stretchable and Ultraflexible organic electronics. *MRS Bull.* 42 (2).

Rogers, B., Adams, J., and Pennathur, S. (eds.) (2015). *Nanotechnology: Understanding Small Systems,* 3e. Boca Raton: CRC Press.

So, F. (ed.) (2010). *Organic Electronics: Materials, Processing, Devices and Applications.* Boca Raton: CRC Press.

Sun, S.-S. and Dalton, L.R. (eds.) (2008). *Introduction to Organic Electronic and Optoelectronic Materials and Devices.* Boca Raton: CRC Press.

Wallace, G.G., Spinks, G.M., Kane-Maguire, L.A.P., and Teasdale, P.R. (2002). *Conductive Electroactive Polymers: Intelligent Materials Systems.* Boca Raton: CRC Press.

Whitesides, G.M. (2001). The once and future nanomachines. *Sci. Am.* 285: 78–83.

Whitesides, G.M. and Love, J.C. (2001). The art of building small. *Sci. Am.* 285: 38–47.

Wright, J.D. (1995). *Molecular Crystals.* Cambridge: Cambridge University Press.

2

Materials' Foundations

Organic and Molecular Electronics: From Principles to Practice, Second Edition. Michael C. Petty.
© 2019 John Wiley & Sons Ltd. Published 2019 by John Wiley & Sons Ltd.
Companion website: www.wiley.com/go/petty/molecular-electronics2

In nature's infinite book of secrecy

2.1 Introduction

All physical properties of materials ultimately depend on how the constituent atoms and molecules are held together – the chemical bonds. Therefore, this chapter begins by looking at the various types of bond that can form between atoms. These ideas are then developed with a particular emphasis on the forces that hold organic materials together. The morphological properties of crystalline and nanocrystalline materials are also described and the common defects that pervade materials (all too often ignored) are introduced.

2.2 Electronic Structure

2.2.1 Atomic Structure

Each atom in a material consists of a very small nucleus composed of *protons* and *neutrons*, which is encircled by mobile *electrons*. Both electrons and protons are electrically charged. The magnitude of the charge, *e*, is 1.60×10^{-19} C, negative in sign for the electrons and positive for the protons. Neutrons are electrically neutral. Although there is *Coulombic repulsion* between the protons, all the protons and neutrons are held together in the nucleus by the *strong force*, which is a powerful, fundamental natural force between particles. The force has a very short range of influence, typically less than 10^{-15} m. Masses for the subatomic particles are infinitesimally small: protons and neutrons have approximately the same mass, 1.67×10^{-27} kg, significantly larger than that of an electron, 9.11×10^{-31} kg.

Each chemical element is characterized by the number of protons in the nucleus or the *atomic number, Z*. For an electrically neutral or complete atom, the atomic number also equals the number of electrons. This atomic number ranges from 1 for hydrogen to 92 for uranium, the highest of the naturally occurring elements.

The atomic mass of a specific atom can be expressed as the sum of the masses of protons and neutrons within the nucleus. Although the number of protons is the same for all atoms of a given element, the number of neutrons may be variable. Thus atoms of some elements have two or more different atomic masses, called *isotopes*. The *atomic weight* of an element corresponds to the weighted average of the atomic masses of the atom's naturally occurring isotopes. The *atomic mass unit*, amu, may be used for computations of atomic weight. A scale has been established whereby 1 amu is defined as 1/12 of the atomic mass of the most common isotope of carbon, carbon-12 (^{12}C). Within this scheme, the masses of protons and neutrons are slightly greater than unity. The *atomic weight* of an element or the *molecular weight* of a compound (also called *molar weight* or *molar mass*) may be specified on the basis of amu per atom (molecule) or mass per *mole* of material. In 1 mole of substance there are 6.02×10^{23} (*Avogadro's number*) of atoms or molecules (or 6.02×10^{26} kmol^{-1}).

2.2.2 Electrons in Atoms

During the latter part of the nineteenth century, it was realized that many phenomena involving electrons in solids could not be explained in terms of classical mechanics. What followed was the establishment of a set of principles and laws that govern systems of atomic and subatomic entities which came to be known as *quantum mechanics*. One early outcome of quantum mechanics was the simplified *Bohr atomic model*, in which the electrons are assumed to revolve

Figure 2.1 Schematic representation of the Bohr atom.

Electron

Nucleus

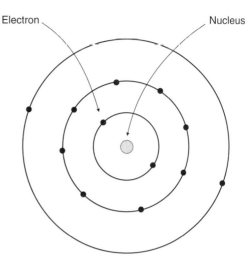

around the atomic nucleus in discrete orbitals. An important assumption in the Bohr model is that only certain orbits with fixed radii are stable around the nucleus. For example, the closest orbit of the electron in the hydrogen atom can only have a radius of 0.053 nm. A Bohr atom is depicted in Figure 2.1.

The Bohr model was eventually found to have some significant limitations. Since the electron is constantly moving around an orbit with a given radius, over a long period of time (perhaps 1 ps on the atomic time scale) the electron would appear as a spherical negative-charge cloud around the nucleus and not as a single dot representing a finite particle. The electron can therefore be viewed as a charge contained within a spherical shell at a given radius. This *wave-mechanical model* of an atom describes the position of an electron in terms of a probability distribution or electron cloud.

The electronic orbital is therefore a region of space in an atom or molecule where an electron with a given energy may be found. Due to the requirement of stable orbits, the electrons occupy well-defined spherical regions, distributed in various *shells* and *subshells*.

In this model of the atom, the shells and subshells are spatial regions around the nucleus where the electrons are most likely to be found. Using wave mechanics, every electron in an atom is characterized by four parameters called *quantum numbers*. The size, shape, and spatial orientation of an electron's probability density are specified by these numbers.

The first three quantum numbers are the principal, n, orbital (angular momentum), l, and magnetic, m_l, numbers (Table 2.1). The principal quantum number is related to the distance of the electron from the nucleus and gives the total energy of an electron in an atom, the energy increasing with n. This number may take on integral values beginning with unity and specifies the main shell where the electron is located. The shells are denoted by the capital letters K, L, M, N, O, and so on, which correspond, to $n = 1, 2, 3, 4, 5, ...,$ as indicated in Table 2.2. The orbital quantum number specifies the magnitude of the orbital momentum of the electron; the l values 0, 1, 2, 3 indicate the electron subshell and are labelled lowercase s, p, d, f states. This second quantum number is related to the shape of the electron subshell. Furthermore, the number of these subshells is restricted by the magnitude of n; the allowable subshells for several n values are shown in Table 2.2. The number of energy states for each subshell is determined by the third quantum number m_l. For an s subshell, there is a single energy state, whereas for p, d, and f subshells, three, five, and seven states exist, respectively.

Table 2.1 Summary of quantum numbers of electrons in atoms.

Name	Symbol	Permitted values	Property
principal	n	positive integers (1,2,3...)	orbital energy (size)
angular momentum	l	integers from 0 to $(n-1)$	orbital shape (the l values 0,1,2, and 3 correspond to s, p, d, and f orbitals, respectively)
magnetic	m_l	integers from $-l$ to 0 to $+l$	orbital orientation
spin	m_s	$+1/2$ or $-1/2$	direction of electron spin

Table 2.2 Maximum number of available electrons in the shells and subshells of an atom.

Principal quantum number n	Shell designation	Subshells	Number of states	Number of electrons per subshell	per shell
1	K	s	1	2	2
2	L	s	1	2	8
		p	3	6	
3	M	s	1	2	18
		p	3	6	
		d	5	10	
4	N	s	1	2	32
		p	3	6	
		d	5	10	
		f	7	14	

In the absence of an external magnetic field, the states within each subshell are identical. However, when a magnetic field is applied, these subshell states split, each state assuming a slightly different energy. For this reason, m_l is known as the magnetic quantum number (in fact, m_l determines the orientation of the orbital magnetic moment relative to the magnetic field). Electrons also have a fourth quantum number m_s associated with the characteristics of spin, which may have a value of either $+\frac{1}{2}$ or $-\frac{1}{2}$ (Table 2.1).

A complete energy level diagram for the various shells and subshells using the wave-mechanical model is shown in Figure 2.2. Several features of the diagram are worth noting. First, the smaller the principal quantum number, the lower the energy level; for example, the energy of the 1s state is less than that of the 2s state, which in turn is less than the 3s state. Second, within each shell, the energy of a subshell level increases with the value of the l quantum number. For example, the energy of a 3d state is greater than a 3p state, which is larger than that of a 3s state. Finally, there may be overlap in energy of a state in one shell with states in an adjacent shell, which is especially true of d and f states; for example, the energy of a 3d state is greater than that of a 4s state.

Figure 2.2 Relative electron energies for shells and subshells.

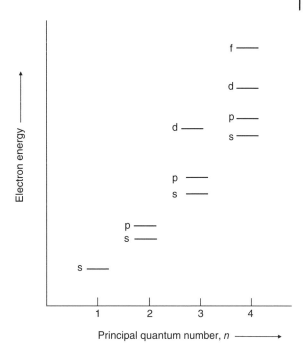

Table 2.3 Relationship between type of orbital and the number of nodes in the corresponding electron wavefunction.

Orbital	Nodes
1s	no nodes
2s	one spherical node
$2p_x$	one node, the *y-z* plane
$2p_y$	one node, the *x-z* plane
$2p_z$	one node, the *x-y* plane

In quantum mechanics, an atomic orbital is defined as a one-electron *wavefunction*, ψ (*x*, *y*, *z*, *t*), which is function of both position (*x*, *y*, and *z* using Cartesian co-ordinates) and time (*t*). Wavefunctions are discussed further in Chapter 3, Section 3.3.1. For each point in space, there is associated a number whose square is proportional to the probability of finding an electron at that point, i.e. $|\psi(x, y, z, t)|^2$ is the probability of finding the electron per unit volume at *x*, *y*, *z*, and at time *t*.

Only $|\psi|^2$ has a physical meaning, as the function ψ can be a complex mathematical quantity, with real and imaginary parts. The probability distribution has all the properties associated with waves. It has a numerical magnitude (its amplitude), which can be either positive or negative (corresponding to a wave crest or a wave trough) and nodes. A node is a region where a crest and a trough meet. In three-dimensional waves characteristic of electron motion, the nodes are two-dimensional surfaces at which ψ = 0. Consequently, atomic orbitals may be characterized by their corresponding nodes, related to the quantum numbers. For example, the nodes associated with the 1s, 2s, and 2p orbitals (there are three of these) are given in Table 2.3.

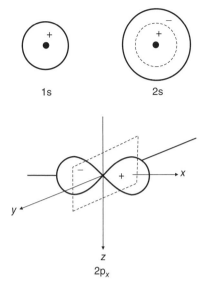

Figure 2.3 Schematic representations of s and p orbitals.

The 1s orbital has no nodes. The wavefunction is a spherically symmetric function whose numerical value decreases exponentially from the nucleus.

The orbitals in Table 2.3 are depicted schematically in Figure 2.3. The plus and minus signs in the figure have no relationship to electric charge. They are simply the arithmetic signs associated with the wavefunction, e.g. a positive sign denotes a wave crest and a negative sign a wave trough. The signs associated with the wavefunctions determine how two or more wavefunctions will combine when they interact (Section 2.4).The dashed lines in Figure 2.3 represent nodal surfaces. These nodes are a sphere for the 2s orbital and a plane for the 2p orbital.

In summary, the various n quantum numbers define electron shells, within which the electrons are most likely to be located. The shells may be further divided into subshells, according to their l values. The first shell contains just 1s orbital, denoted the 1s. The second shell has one spherical orbital, the 2s, and 3p orbitals, the 2p orbitals, which are arranged at right angles to each other. The third shell has 1s orbital, the 3s, 3p orbitals, the 3p orbitals, and a group of five d orbitals, the 3d orbitals. The pattern is that each successive shell comprises all the orbital types of the previous shell (but of greater size), plus a new group of orbitals that the previous shell does not possess. The energy of an electron depends both on the shell and, in general, on the nature of the orbital.

2.2.3 Filling of Orbitals

To determine the way in which the electron states are filled with electrons, the *Pauli Exclusion Principle* is used. This states that no two electrons in an atom can have identical quantum numbers, i.e. each electron in an atom has a unique set of quantum numbers. If this principle did not hold, all of the electrons in an atom would be located in the lowest energy state (the K shell). The Pauli Exclusion Principle holds for not just electrons but for any *fermions* (half-integer spin particles like electrons, protons, neutrons, muons, and many more). It does not apply to particles of integer spin (*bosons*). In filling up the various orbitals, *Hund's Rule* is also important. This states that when several orbitals of equal energy are available, electrons enter singly with parallel spins. In other words, one can add one electron to each orbital of equal level (such as the five d orbitals) and these will all have the same spin. Electrons are only paired when there are no available orbitals of the same energy. When all the electrons occupy the lowest possible energies according to the foregoing restrictions, an atom is said to be in its *ground state*. However, electron transitions to higher energy states are possible. The electron configuration or structure of an atom represents the manner in which the electron states are occupied. In the conventional notation, the number of electrons in each subshell is indicated by a superscript after the shell-subshell designation. For example, the electron configurations for hydrogen, carbon, and silicon are $1s^1$; $1s^2 2s^2 2p^2$; and $1s^2 2s^2 2p^6 3s^2 3p^2$.

The outermost filled shell is occupied by the *valence electrons*. These participate in the bonding between atoms and are responsible for many of the physical and chemical properties of the resulting solids. In addition, some atoms have what are termed 'stable electron configurations', i.e. the states within the outermost or valence electron shells are completely filled. Such elements are neon, argon, krypton, and helium. These are the inert gases and are very unreactive chemically.

2.2.4 The Periodic Table

All the elements have been classified according to their electron configurations in the *periodic table*, as shown in Figure 2.4. The similarity of elements in each group (vertical column) of the periodic table is the result of similarity in the outer electron or valence electron configuration. All the alkali metals – Group 1A – (lithium, sodium, potassium, rubidium, caesium, and francium) have the 'inert-gas-plus-one-configurations'. The alkaline-earth metals make up Group 2A; these elements (beryllium, magnesium, calcium, strontium, barium, and radium) have 'inert-gas-plus-two' configurations, while the halogens that make up Group 7B (fluorine, chlorine, bromine, iodine, and astatine) possess 'inert-gas-minus-one' structures. The horizontal rows in the periodic table are the periods. The elements in the three long periods Groups 3A to 2B are termed the transition metals, which have partially filled d electron levels and, in some cases, one or two electrons in the next higher energy shell. Groups 3B, 4B, and 5B (boron, silicon, germanium, arsenic, etc.) are metalloids, displaying characteristics that are intermediate between the metals and non-metals by virtue of their outer electron structures. Most of the elements in the periodic table really come under the metal classification. These are sometimes called *electropositive* elements, indicating that they are capable of giving up their few outermost electrons to become positively charged ions. However, the elements situated on the right-hand side of the periodic table are *electronegative*, i.e. they readily accept electrons to form negatively charged ions, or sometimes they share electrons with other atoms. Figure 2.5 shows how the *electronegativity* of the elements varies across the periodic table. As a general rule, electronegativity increases in moving from left to right and from bottom to top. Atoms are more likely to accept electrons if their outer shells are almost full, and if they are less 'shielded' from (i.e. closer to) the nucleus. On the Pauling scale of electronegativity (Linus Pauling, 1932), the most electronegative chemical element (fluorine – see Figure 2.4) is given an electronegativity value of 3.98 (textbooks often state this value to be 4.0); the least electronegative element (francium) has a value of 0.7, and the remaining elements have values in between. On this scale, hydrogen is arbitrarily assigned a value of 2.1 or 2.2. The *valence*, *valency*, or *valency number*, is a measure of the number of chemical bonds formed by the atoms of a given element.

A slight complication occurs with the M and N shells because the 3d and the 4s subshells have similar energies, shown in Figure 2.2 (a result of electrostatic screening of the outermost electron by the inner ones). Argon ($Z = 18$) has all the 1s, 2s, 3s, and 3p subshells filled, but in potassium ($Z = 19$) the additional electron goes into a 4s energy level rather than a 3d level because the 4s level has a slightly lower energy.

Something similar happens for elements with atomic number $Z = 57$ to $Z = 71$, which have two electrons in the 6s subshell but only partially filled 4f and 5d subshells. These are the rare-earth elements, the lanthanides (L); they all have similar properties. Yet another such series, called the actinide series (A), starts with $Z = 89$.

Figure 2.4 The periodic table of the elements. The symbol for each element is shown together with its atomic number.

Legend:
- Metals
- Metalloids
- Non-metals
- Transition metals

Period	1A	2A	3A	4A	5A	6A	7A	8	1B	2B	3B	4B	5B	6B	7B	0
1	1 H															2 He
2	3 Li	4 Be									5 B	6 C	7 N	8 O	9 F	10 Ne
3	11 Na	12 Mg									13 Al	14 Si	15 P	16 S	17 Cl	18 Ar
4	19 K	20 Ca	21 Sc	22 Ti	23 V	24 Cr	25 Mn	26 Fe	27 Co	28 Ni	29 Cu	30 Zn	31 Ga	32 Ge	33 As	34 Se	35 Br	36 Kr
5	37 Rb	38 Sr	39 Y	40 Zr	41 Nb	42 Mo	43 Tc	44 Ru	45 Rh	46 Pd	47 Ag	48 Cd	49 In	50 Sn	51 Sb	52 Te	53 I	54 Xe
6	55 Cs	56 Ba	L	72 Hf	73 Ta	74 W	75 Re	76 Os	77 Ir	78 Pt	79 Au	80 Hg	81 Tl	82 Pb	83 Bi	84 Po	85 At	86 Rn
7	87 Fr	88 Ra	A															

L: 57 La | 58 Ce | 59 Pr | 60 Nd | 61 Pm | 62 Sm | 63 Eu | 64 Gd | 65 Tb | 66 Dy | 67 Ho | 68 Er | 69 Tm | 70 Yb | 71 Lu

A: 89 Ac | 90 Th | 91 Pa | 92 U | 93 Np | 94 Pu | 95 Am | 96 Cm | 97 Bk | 98 Cf | 99 Es | 100 Fm | 101 Md | 102 No | 103 Lr

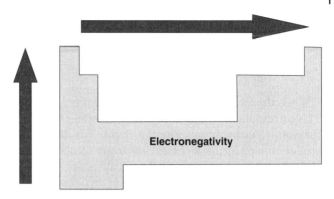

Figure 2.5 Variation of electronegativity for elements across the periodic table.

2.3 Chemical Bonding

In this section, the main types of chemical bonding that are responsible for the cohesive forces between atoms and molecules are described. This list is not exhaustive. The particular types of bond that form in organic compounds are described in more detail in Section 2.4.

2.3.1 Bonding Principles

All of the mechanisms that cause bonding between atoms derive from electrical attraction and repulsion. The general principle is illustrated in Figure 2.6, which shows what happens to the energy of the system as two isolated atoms approach each other. As the atoms become closer they exert attractive and repulsive forces on each other as a result of mutual electrostatic interactions. Initially, the attractive force dominates over the repulsive force but as the atoms become close the electron shells of the individual atoms overlap and the

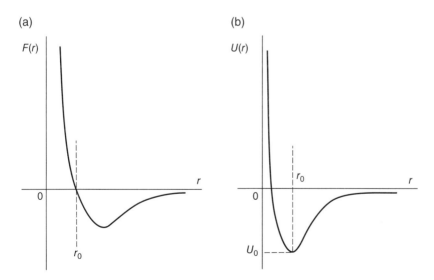

Figure 2.6 (a) Interaction force $F(r)$ between two atoms as a function of their interatomic separation r. (b) Potential energy $U(r)$ as a function of interatomic separation r. The equilibrium separation is r_0.

repulsive force dominates. Figure 2.6 shows that the force is zero at an interatomic separation of $r = r_0$. The potential energy U is related to the force F by

$$F = \frac{dU}{dr} \tag{2.1}$$

When the force between the atoms is zero ($r = r_0$), the potential energy will be a minimum. This corresponding minimum energy U_0 defines the bond energy of the molecule, representing the energy needed to separate the atoms. The differing strengths and differing types of bond are determined by the particular electronic structures of the atoms involved. The weak *van der Waals* bond provides a universal weak attraction between closely spaced atoms. Generally, the effect is dominated by *ionic*, *covalent*, or *metallic* bonding.

The existence of a stable bonding arrangement (whether between a pair of otherwise isolated atoms or throughout a large, three-dimensional crystalline array) implies that the spatial configuration of positive ion cores and outer electrons has less total energy than any other configuration (including infinite separation of the repulsive atoms). The energy deficit of the configuration compared with isolated atoms is called the *cohesive energy* and ranges from 0.1 eV per atom for solids that are held together by van der Waals forces to 7 eV per atom in some covalent and ionic compounds, and in some metals. (An electron volt, or eV, is a unit of energy equal to the energy acquired by an electron falling through a potential difference of one volt, approximately 1.60×10^{-19} J.)

In covalent bonding, the angular placement of bonds is very important, while in some other types of bonding it is important to have the largest possible *coordination number* (number of nearest neighbours). For some solids, two or more quite different structures would result in nearly the same energy. A change in temperature or pressure can then provoke a change, a *phase change*, from one form of the solid to another.

2.3.2 Ionic Bond

The ionic bond is perhaps the simplest chemical bond to understand. This results from the mutual attraction of positive and negative charges. The atoms of certain elements lose or gain electrons very easily because, in doing so, they acquire a completely filled outer electron shell which provides the atom with great stability. For example, sodium (Na), which has only one electron in its outer occupied shell (3s shell), will readily lose this electron to become a singly charged positive ion (Na^+). Similarly, calcium (Ca) will readily lose the two electrons in its outer occupied 4s shell to become the doubly charged calcium ion (Ca^{2+}).

Atoms like those of chlorine, which have an almost completely filled outer shell, readily accept additional electrons to complete their outer shell and become negative ions. The formation of a molecule of sodium chloride can be depicted, as shown in Figure 2.7. In fact, the negative charge associated with the chlorine ion possesses an attraction for all positive charges in its vicinity; this will also be the case for the positive sodium ion. Consequently, Na^+ ions will surround themselves with Cl^- ions and vice versa. The result, shown in Figure 2.8, is a regular array of sodium and chlorine ions in a crystalline lattice (two interpenetrating cubic networks of sodium and chorine ions are formed, Section 2.5.4).

As shown in Figure 2.5, the elements are arranged in the periodic table in order of their ability to loose their electrons or *ionization energy*. Generally ionic bonds are formed between elements with a large difference in their ionization energies, e.g. between elements of Groups IA (metal) band 7B (non-metal) and between elements of Groups 2A (metal) and 6B (non-metal), for example LiF and MgO. These ionic crystals show many similar physical properties. They are strong brittle materials with high melting points compared to metals.

Figure 2.7 Formation of an ionic bond between Na and Cl atoms in NaCl. The equilibrium separation is r_0.

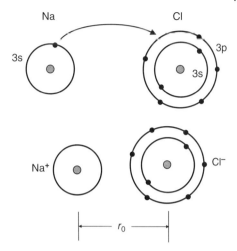

Figure 2.8 Solid NaCl crystal with Na^+ and Cl^- ions arranged close to one another.

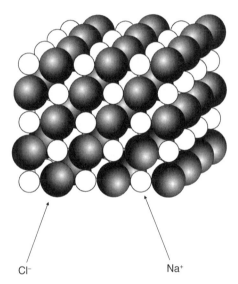

Most become soluble in polar liquids, such as water. Since all the electrons in ionic crystals are within rigidly positioned ions, there are no free electrons to move in response to an applied electric field and to contribute to the electrical conductivity (in contrast to metals). Therefore, ionic solids are typically electrical insulators. Moreover, compared to metals, ionic solids have poor thermal conductivity, since the constituent ions cannot readily pass vibrational kinetic energy to their neighbours.

2.3.3 Covalent Bond

Another way in which the outer electron shell of atoms can be effectively filled to achieve a stable configuration is by sharing of electrons; this results in the covalent bond. A simple example is that of the hydrogen molecule, shown in Figure 2.9. When the 1s subshells overlap, the electrons are shared by both atoms and each atom now has a complete subshell. As shown in the figure, the two electrons must now orbit both atoms; they therefore cross the overlap region

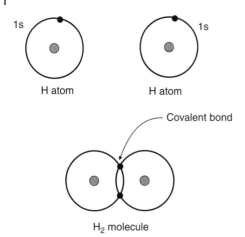

1s   1s

H atom   H atom

Covalent bond

H_2 molecule

Figure 2.9 Formation of a covalent bond between two hydrogen atoms.

more frequently, indeed twice as often. Thus, electron sharing results in a greater concentration of negative charge in the region between the two nuclei, which keeps the nuclei bonded to each other.

In the ground state, the two electrons will both occupy the same state and can be described by the same mathematical form of the wavefunction. This, of course, is allowable in terms of the Pauli Exclusion Principle. The combination of two waves having the same sign is reinforcing. This is true for all types of wave motion, from sound waves to the waves in an ocean. The covalent bond represents a combination of two 1s atomic orbitals to give a new wavefunction. The increased magnitude of the wavefunction between the atoms corresponds to higher electron density in this region. Electrons are attracted electrostatically to both nuclei and the increased electron density between the nuclei counterbalances the internuclear repulsion. In contrast, when two waves of opposite sign interact, they interfere, or cancel each other. At the point of interference, the wavefunction has the value zero, i.e. a node has been created.

A mathematical method known as the *Linear Combination of Atomic Orbitals* may be used to find the form of the electron wavefunctions in the hydrogen molecule. If the 1s atomic orbitals of two nuclei A and B are denoted ψ_A and ψ_B then, as the nuclei are brought together to form the molecule, two linear combinations of ψ_A and ψ_B are found to give satisfactory molecular orbitals. These are

$$\psi_S = N_S\left(\psi_A + \psi_B\right)$$

$$(2.2)$$

$$\psi_{AS} = N_{AS}\left(\psi_A - \psi_B\right)$$

where N_S and N_{AS} are normalizing constants. The two wavefunctions, ψ_A and ψ_B, are respectively, symmetrical and antisymmetrical about the mid-point of a line between the two nuclei A and B. As noted earlier, the square of the wavefunctions gives the probability of finding an electron or the electron density. This is shown in Figure 2.10. It is evident that the electron density corresponding to the antisymmetric wavefunction is low in the region between the nuclei. The energy of this orbital increases continuously with decreasing separation of the nuclei and the absence of a minimum leads to a repulsive energy and what is called an *antibonding* orbital. The electron in the ground state of the positively charged *hydrogen molecule ion* is therefore located in the *bonding* ψ_S orbital. These orbitals are cylindrically symmetric with respect to the axis of the molecule and are referred to a *sigma bond*, or *σ–bond*.

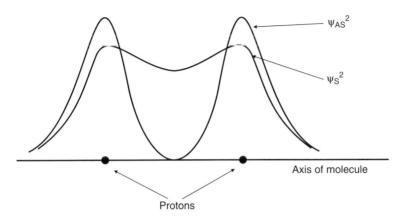

Figure 2.10 A comparison of the electron densities of the ψ_S and ψ_{AS} wavefunctions in the hydrogen molecule. The densities are symmetrical about the axis of the molecule.

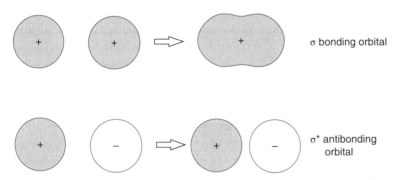

Figure 2.11 The formation of the σ bonding orbital and the σ^* antibonding orbital for the hydrogen molecule. The + and the – refer to the signs of each orbital.

A further illustration of the bonding and antibonding situations in the hydrogen molecule is given in Figure 2.11. In the ground state of the hydrogen molecule, the two electrons will both occupy the symmetrical wavefunction. This will have a lower energy than the asymmetrical orbital and in accordance with the Pauli Exclusion Principle, the electrons will have opposite spins. Each wavefunction corresponds to two quantum states so that the total number of states for the two atoms has remained unaltered.

The bonding combination corresponds to a decrease in energy (greater stability); the antibonding combination corresponds to an increase in energy (lower stability). Two atomic orbitals give rise to two molecular orbitals. The two paired electrons of opposite spin available for the bond can be put into the bonding molecular orbital. The energy relationships are summarized in Figure 2.12. Note how the energies of the two starting orbitals separate or spread apart when they interact. The amount of the separation depends on the degree to which the orbitals overlap. A slight overlap gives two molecular orbitals that differ little in energy; a large overlap results in strong separation. For axially symmetric orbitals, such as p orbitals, the greatest overlap occurs when the bonds are allowed to interact along the nuclear axis.

The atoms taking part in covalent bonding need not necessarily be alike. For example, hydrogen and chlorine may be combined via covalent bonding to form hydrogen chloride.

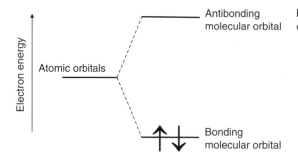

Figure 2.12 Energy relationships of atomic orbitals before (left) and after bonding (right).

Most bonding *within* organic compounds is covalent bonding; this is discussed in some detail in Section 2.4.

Because of the strong Coulombic attraction between the shared electrons and the positive nuclei, the covalent bond energy is usually the highest for all bond types. This leads to high melting temperatures and to very hard solids; for instance diamond, a covalently bonded solid is one of the hardest known materials. Covalently bonded solids are also insoluble in nearly all solvents. The directional nature and strength of the covalent bond also makes these materials nonductile (or nonmalleable). Under a strong force they exhibit little fracture. Furthermore, since all the valence electrons are locked in the bonds between the atoms, they are not free to drift in the crystal when an electric field is applied. Consequently, the electrical conductivity of such materials is very poor.

It is possible to have interatomic bonds of intermediate character between ionic and covalent. Complete ionic bonding requires the presence of an extremely electropositive component (which can be ionized easily to form a cation) and of an extremely electronegative component (for which the electron affinity to form an anion is as large as possible). These requirements are well satisfied in the alkali halides, in which there is a strong encouragement for electron transfer. In compounds with less extreme electropositive and electronegative character, however, there is less than 100% charge transfer from cation to anion. For example, the noble metals have larger ionization energies than alkali metals, and silver halides are less ionic in nature than the corresponding alkali halides.

There is a continuous progression from purely ionic character to purely covalent character, as the electronegativity difference in the resulting compounds becomes smaller. When there is a partial tendency towards electron sharing, the optimum bonding can be considered as arising from a resonance between ionic and covalent charge configurations. The resulting time-averaged wavefunction for a bonding electron is then

$$\psi = \psi_{cov} + \lambda \psi_{ion} \tag{2.3}$$

where ψ_{cov} and ψ_{ion} are normalized wavefunctions for completely covalent and ionic forms, and λ is a parameter which determines the degree of ionicity:

$$\% \text{ ionicity} = \left[\frac{100\lambda^2}{1+\lambda^2} \right] \tag{2.4}$$

It is possible to estimate the percentage ionic character of a bond between elements A and B (A being the most electronegative) from the following expression:

$$\% \text{ ionicity} = 100 \left\{ 1 - \exp[-(0.25)(X_A - X_B)^2] \right\} \tag{2.5}$$

where X_A and X_B are the electronegativities of the respective elements (Section 2.2.4).

2.3.4 Metallic Bonding

In contrast to ionic and covalent bonds, the properties of the metallic bond cannot be inferred from the nature of bonding in isolated atoms or molecules and a model for bonding in metals is not easy to construct. In some respects, it may be regarded as intermediate in character between that of ionic and covalent bonds.

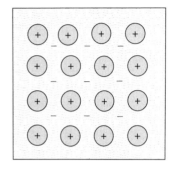

The atoms of metallic elements readily lose their valence electrons (thereby acquiring the stable closed shell electron configuration of the inert gases) to form positive ions. These valence electrons become delocalized and may be regarded as being shared by all of the ions in the crystal. The free valence electrons thereby form an electron 'gas', which may be regarded as permeating the entire crystal, as shown in Figure 2.13. Metallic bonding is essentially an electrostatic attraction between the array of positive ion cores and the electron gas; it is non-directional. Consequently, the metal ions try to get as close as possible, which leads to close-packed crystal structures. The free electrons of this model explain the high electrical and thermal properties of metals, together with their high ductility.

Figure 2.13 Schematic diagram of metallic bonding showing the positive ion cores of the metal atoms surrounded by negative electrons.

2.3.5 Van der Waals Bonding

Since the atoms of inert elements have full shells and cannot accept any electrons or share any electrons, it might be thought that no bonding is possible between them. However, a solid form of argon exists at low temperatures. The forces that hold the argon atoms together are due to van der Waals bonding. This is electrostatic in nature and results from an attraction between the electron distribution of one atom and the nucleus of another.

When averaged over time, the centre of mass of electrons in the closed shell of an element such as argon coincides with the location of the positive nucleus. However, at any instant, the centre of mass is displaced from the nucleus as a result of the various motions of the individual electrons around the nucleus. This creates an instantaneous separation of the centres of positive and negative charge in the atom – an *electric dipole*. The macroscopic manifestation of an atomic dipole on the surface of a material is called the *polarization*. This is defined as the dipole moment (Chapter 4, Section 4.3) per unit volume (or charge per unit surface area).

When two such atoms approach each other, the rapidly fluctuating electronic charge distribution on one atom can induce a dipole on another atom, as shown in Figure 2.14. This type of attraction between two atoms is due to induced synchronization of the electron motions around the nucleus. This is referred to as dipole–dipole attraction. The magnitude of the attractive force drops very sharply with atomic separation, scaling as $(distance)^{-7}$. The energy of van der Waals bonds is at least an order of magnitude lower than those of ionic, covalent and metallic bonding, which explains why the inert gases only solidify at very low temperatures.

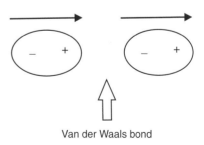

Van der Waals bond

Figure 2.14 Van der Waals bonding between two dipoles.

(a)

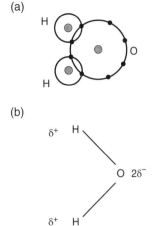

(b)

(c)

Figure 2.15 Hydrogen bonding in water. (a) Relative positions of hydrogen and oxygen atoms in an H$_2$O molecule. (b) Charges on hydrogen and oxygen atoms. (c) Resulting electric dipole in a molecule of water.

2.3.6 Hydrogen Bonding

It is also possible to have an attraction and bonding between molecules that already possess a permanent electric dipole, such as H$_2$O. This is depicted in Figure 2.15. Each hydrogen atom in water shares its electrons with the oxygen atom, which has four unshared electrons. As a result, the negative charge distribution is shifted more towards the oxygen atom and the positive hydrogen proton is relatively exposed. This means that the H$_2$O molecule has a permanent charge separation, resulting in an electric dipole, shown in Figure 2.15c.

When two water molecules are close together, the unshared electrons on the oxygen will be attracted to the exposed proton of the hydrogen in the neighbouring molecule, leading to what is known as *hydrogen bonding*. The true attraction is actually between two permanent dipoles. Hydrogen bonding is very important in organic materials. For example, it is responsible for holding the carbon chains in polymers together. Although the carbon–carbon bond in a chain is due to covalent bonding, the interaction between the chains arises from hydrogen bonds. For example, in poly(vinyl chloride) (PVC), the Cl atom has five unshared electrons and therefore has a slightly more negative charge distribution than the hydrogen atom, in which the proton is relatively exposed. Consequently, a hydrogen bond develops between the Cl and the H of the neighbouring chain. These bonds are weak and can easily be stretched or broken. Polymers, therefore, have lower elastic moduli and melting temperatures than metals or ceramics.

Table 2.4 contrasts the energies involved in the five mains types of chemical bonding discussed in this section.

Table 2.4 Comparison between different types of chemical bonding.

Bond type	Examples	Bond energy (eV per atom)	Melting temperature (°C)
Ionic	NaCl	3.3	801
	MgO	5.2	2852
Covalent	Si	4.0	1410
	C (diamond)	7.4	3550
Metallic	Cu	3.1	1083
	Al	3.4	660
Van der Walls	Ar	0.08	−189
	Cl$_2$	0.32	−101
Hydrogen	H$_2$O	0.52	0
	NH$_3$	0.36	−78

2.4 Bonding in Organic Compounds

2.4.1 Hybridized Orbitals

Carbon has an atomic number of six and is found in Group IVB of the periodic table. Its electron configuration is $1s^2, 2s^2, 2p^2$, i.e. the inner s shell is filled and the four electrons available for bonding are distributed two in s orbitals and two in p orbitals. As the s orbital is spherically symmetrical it can form a bond in any direction. In contrast, the p orbitals are directed along mutually orthogonal axes and will tend to form bonds in these directions (Figure 2.3).

When two or more of the valence electrons of carbon are involved in bonding with other atoms, the bonding can be explained by the construction of *hybrid orbitals* by mathematically combining the wavefunctions of the 2s and 2p orbitals. In the simplest case, the carbon 2s orbital hybridizes with a single p orbital. Two *sp hybrids* result by taking the sum and difference of the two orbitals, shown in Figure 2.16, and 2p orbitals remain. The sp orbitals are constructed from equal amounts of s and p orbitals; they are linear and 180° apart.

Other combinations of orbitals lead to different hybrids. For example, in the case of three groups bonded to a central carbon atom, three equivalent *sp^2 hybrids* may be constructed from the 2s orbital and two p orbitals (e.g. a p_x and a p_y). Each orbital is 33.3% s and 66.7% p. The three hybrids, shown in Figure 2.17, lie in the *x-y* plane (the same plane defined by the two p orbitals) directed 120° from each other, and the remaining *p* orbital is perpendicular to the sp^2 plane. Four *sp^3 hybrids* may be derived from an s orbital and 3p orbitals. These are directed to the corners of a tetrahedron with an angle between the bonds of 109.5°; each orbital is 25% s and 75% p. Figure 2.18 shows the tetrahedral structure of methane, CH_4. Each bond is derived by the interaction of a C(sp^3) hybrid orbital with a hydrogen 1s and may be described as a C (sp^3) – H(1s) bond.

Long-chain organic compounds, such as many of those referred to this book (e.g. in Chapter 7, Section 7.3), involve C—H bonds that are all approximately C(sp^3)—H(s) with a bond length of

Figure 2.16 Mathematical combination of s and p orbitals to yield two sp hybridized orbitals.

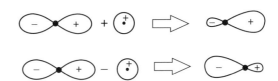

Figure 2.17 Three sp^2 hybridized orbitals.

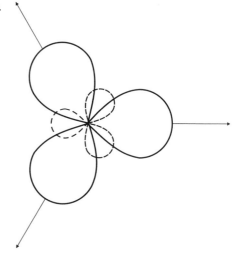

(a)

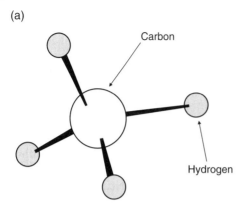

Figure 2.18 (a) Tetrahedral arrangement of carbon and hydrogen atoms in a methane CH_4 molecule. (b) One of the four $C(sp^3) - H(1s)$ bonds in methane.

(b)

C ⟨shape⟩ H

$C(sp^3) - H(1s)$

0.110 nm and a strength of 4.26 eV (98×10^3 kcal kmol^{-1}). All the C—C bonds are approximately $C(sp^3)$—$C(sp^3)$ with a length of 0.154 nm. The tetrahedral geometry of the sp^3 orbitals leads to an alkyl chain with a zig-zag conformation.

It is important to note that a chemical bond can be formed from a mixture of the above hybrid orbitals. In the case of a molecule of ammonia, NH_3, the H—N—H bond angle of 107.1° does not correspond to any simple hybrid. Thus, sp, sp^2, and sp^3 hybrids must be considered as limiting cases only. In addition to its three N—H bonds, ammonia also has a *nonbonding* pair (or lone pair) of electrons on the nitrogen. These electrons are in an orbital which is slightly more s in character than in a simple sp^3 orbital. Consequently, the three hybrid orbitals that overlap with the three hydrogen atoms contain slightly less s character than in a sp^3 orbital (in fact 23%s and 77%p).

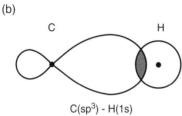

$CH_3CH_2CH_2CH_2CH_3$

Pentane

CH_3
|
$CH_3CH_2CHCH_3$

Isopentane

2.4.2 Isomers

Isomeric compounds are those that have identical chemical formulas (e.g. C_4H_{10}), but differ in the nature or sequence of bonding of their atoms or in the arrangement of atoms in space. Isomers may be subdivided into *constitutional* (or structural) isomers and *stereoisomers*. The former group differ in the order in which the atoms are connected, so they contain different functional groups and/or bonding patterns. For example, ethyl methyl ether and isopropyl alcohol can both be described by the chemical formula C_3H_8O. There are three structural isomers of the alkane pentane, all with the chemical formula C_5H_{12}; these are depicted in Figure 2.19. The isomers – pentane, isopentane, and

CH_3
|
CH_3 — C — CH_3
|
CH_3

Neopentane

Figure 2.19 Isomers of pentane. The compounds have the same chemical formula C_5H_{12} but different bonding arrangements between the atoms.

neopentane – differ in the spatial arrangement of the atoms. As the number of carbons in these alkanes increases so does the number of possible isomers, for example there are nine possible heptanes (C_7H_{16}) and 75 decanes ($C_{10}H_{22}$). In general, structural isomers have different physical and chemical properties. Branched-chain hydrocarbons are more compact than straight-chain isomers and tend to possess lower boiling points and higher densities.

Stereoisomers are compounds that possess the same sequence of covalent bonds but differ in the arrangements of their atoms in space. For this group, a distinction may be made between isomers that rapidly interconvert at ordinary temperatures, known as *conformational isomers* or *conformations*, and those that interconvert only at elevated temperatures or not at all, called *configurational isomers*.

Conformational Isomers

Figure 2.20 shows the most stable structure of pentane. In this representation, a dashed bond projects away from the reader, a heavy wedge bond projects towards the reader, and a normal bond lies in the plane of the page. Rotation about one or more of the single bonds in this molecule leads to different conformations. These isomers cannot usually be isolated as they interconvert too rapidly. The conformation in Figure 2.20 is called *staggered* and has all the carbon–hydrogen bonds as far away from each other as possible. The least stable conformation is called *eclipsed* and has the carbon–hydrogen bonds as close as possible. Figure 2.21 gives a way of representing both the eclipsed and staggered conformations of ethane (C_2H_6). These diagrams are called *Newman projections*. The C—C bond is viewed end-on; the nearest carbon is represented by a point and the three groups attached to the carbon radiate as three lines from this point. The farthest carbon is represented by a circle with its bonds radiating from the edge of a circle.

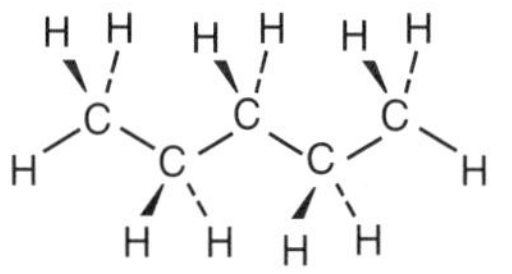

Figure 2.20 Covalent bonding in pentane. In the representation shown, a dashed bond points away from the viewer, a heavy wedge bond projects towards the viewer, and a normal bond lies in the plane of the page.

For longer chain alkanes, the conformational situation becomes complex. There are different staggered and eclipsed conformations, not all of equal energy. The lowest energy conformation is the one in which the two large methyl groups are as far apart as possible – 180°; this is the *anti conformation*. A higher energy *gauche conformation* exists when the methyl groups are 60° apart. Newman projections are shown for both these arrangements in Figure 2.22. Note

Figure 2.21 Eclipsed and staggered conformations of ethane. The Newman projections are shown on the right.

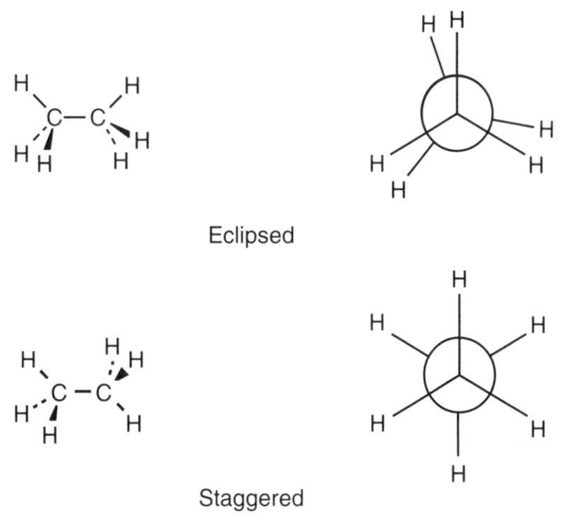

Eclipsed

Staggered

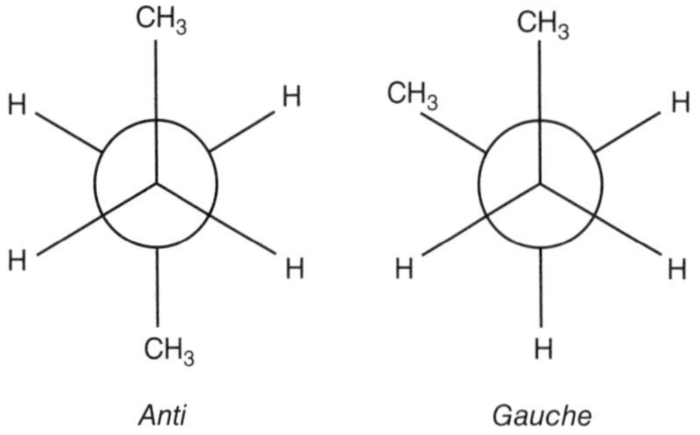

Figure 2.22 *Anti* and *gauche* conformations of alkanes.

Anti

Gauche

that there are no eclipsing interactions for the *gauche* conformation; the increased energy simply arises because the large methyl groups are close together.

Configurational isomers

Interconversion between configurational isomers requires the breaking of bonds rather than rotation about bonds and requires high temperatures or special catalysts. Further examples of configurational isomers – *trans* and *cis isomers* – are described in the following section dealing with double bonds.

Some configurational isomers differ in their three-dimensional relationship of the substituents about one or more atoms; these are called optical isomers. *Enantiomers* are optical isomers that are non-superimposable mirror images. These exhibit the property of *chirality*. Human hands are the most universally recognized example of chirality: the left hand is a non-superposable mirror image of the right hand. Irrespective of how the two hands are oriented, it is impossible for all the major features of both hands to coincide. Figure 2.23 shows the example of the two amino acids (Chapter 12, Section 12.2.1) glycine and alanine. Alanine, which has four different chemical groups bonded to the central carbon atom, is a chiral molecule – it cannot be superimposed on its mirror image. There are two enantiomers of alanine. Molecules that contain one such asymmetric carbon atom are always chiral. In contrast, glycine with only three different groups attached to the carbon atom is achiral. Glycine is the only amino acid that is not chiral.

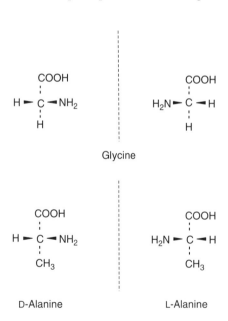

Glycine

D-Alanine   L-Alanine

Figure 2.23 Stereoisomerism. Top – the glycine molecule is superimposable on its mirror image; glycine is achiral. Bottom – the molecule of alanine cannot be superimposed on its mirror image; alanine is chiral. The two stereoisomers (enantiomers) of alanine are D-alanine and L-alanine.

The easiest way to understand the above is to think of carbon as a central atom in a pyramid with other atoms at each apex. When the four atoms are all different (W, X, Y, and Z, say) it is impossible for the original structure to be superimposed on its mirror image, as illustrated in Figure 2.24a. However, when two of the groups attached to the central carbon are the same, Group W in Figure 2.24b, then the mirror image can be superimposed on the original structure by selecting the other W group to be the top of the pyramid.

(a)

(b)

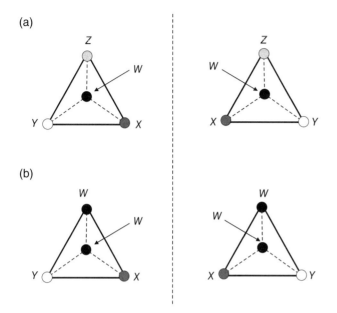

Figure 2.24 Schematic representation of an asymmetric carbon atom. In the figures, the carbon atom is not shown but assumed to be at the centre of the pyramids. The vertical dashed line represents a mirror plane. In (a), four different groups are attached to the carbon. The carbon is asymmetric, as this particular molecular arrangement cannot be superimposed on its mirror image (shown on the right). Different stereoisomers are therefore possible. In (b), the carbon with two identical groups is not asymmetric, as either clockwise sequence of the groups in the pyramid's base can be obtained by choosing which *W* group to be at the top. No isomerism is possible.

Most of the physical properties (boiling points, densities, refractive indices, etc.) of the two of stereoisomers of alanine are identical. However, they differ in one important respect, the way in which they interact with polarized light (Chapter 4, Section 4.2). One enantiomer will rotate the plane of polarized light to the left (anti-clockwise) and is called the *levorotatory* isomer, while the other will rotate the plane to the same extent to the right, and is called the *dextrorotarory*. By convention, the two stereoisomers of alanine, shown in Figure 2.23, are designated L and D (small capital letters). These L and D symbols refer to the absolute configuration of the stereoisomer rather than to the direction of rotation of the plane-polarized light, i.e. some L isomers can be levorotatory and some dextrorotatory.

Good examples of the marked differences in the chemical behaviour of chiral molecules can be found in the biological world. The D-isomer of glucose occurs widely in nature, but the L-isomer (L-glucose) does not (although it can be synthesized in the laboratory). L-Glucose is indistinguishable in taste from D-glucose, but cannot be used by living organisms as a source of energy. Many medicines are produced as a mixture of enantiomers, only one of which is pharmacologically active, as it can prove costly to separate the isomers. One of the enantiomers of salbutamol, used in the treatment of asthma, is 68 times more effective than the other. One of the most dramatic effects of chirality was seen with the disastrous introduction of thalidomide in the 1950s and 1960s. This drug was intended as a sleeping aid for pregnant women, and to combat morning sickness and other symptoms. It was later found to be responsible for causing birth defects; this was related to one of the activities of one of the enantiomers of thalidomide.

2.4.3 Double and Triple Bonds

Despite the propensity of carbon atoms to form four covalent bonds, in the ethylene molecule, C_2H_4, each of the two carbons is attached to just three atoms. The apparent deficiency in bonds to carbon is avoided by the bonding of two carbon atoms by two bonds, or what is more properly called a *double bond*. The bonding for hydrocarbons with carbon–carbon double bonds (*alkenes*) and *triple bonds* (*alkynes*) involve sp^2 and sp hybrids, respectively.

In ethylene ($CH_2{=}CH_2$), two sp^2 hybrids on each carbon bond with the hydrogens. A third sp^2 hybrid on each carbon forms a $C(sp^2)$—$C(sp^2)$ single bond, leaving a p orbital 'left over' on each carbon. This orbital lies perpendicular to the plane of the six atoms. The 2p orbitals are parallel to each other and have regions of overlap above and below the molecular plane. This type of bond, in which there are two sideways bonding regions above and below a nodal plane, is called a *pi-* or *π-bond*. In contrast, the bond formed by the head-on overlap of the two carbon sp^2 orbitals is the σ-bond, already described in Section 2.3.3. The π-bonds are the result of overlap of dumbell-like p orbitals lying adjacent to each other, as shown in Figure 2.25. One consequence of this structure is that the orbitals of the π-bond act like struts to stop the atoms at each end from rotating; the molecule stays more or less flat.

The C=C double bond distance in ethylene is 0.133 nm, less than the value of 0.154 nm given above for a C—C single bond, while the C—H bond is 0.108 nm long. Strengths of the C=C and C—H bonds in ethylene are 6.61 eV (152×10^3 kcal kmol^{-1}) and 4.48 eV (103×10^3 kcal kmol^{-1}), respectively.

As one might expect, double bonds are stronger than single ones: to break the carbon–carbon link in ethylene requires more energy than is needed to do the same in ethane, C_2H_6, in which two CH_3 groups are linked by a single bond between carbons. The double bond is, however, considerably less than twice as strong; breaking open the π component of the double bond is easier than breaking a single bond. For this reason, ethylene reacts with other compounds more readily than does ethane. Carbon compounds that contain π-bonds are said to be *unsaturated*, meaning that the carbon atoms, while having formed the requisite number of bonds, are not fully saturated in terms of their number of potential neighbours. *Saturated* carbon molecules, on the other hand, contain only single bonds.

The noncylindrically symmetric electron density about the C=C bond axis in ethylene, shown in Figure 2.26, and related compounds leads to a barrier to rotation about this axis. Therefore, two isomers exist. These particular isomers are not easily interconverted and are configurational isomers, as discussed in Section 2.4.2. These are known by the prefixes *cis-* (from the Latin for 'on this side' – both bonds are on the same side of the alkene plane) and *trans-* ('across' – the bonds are on opposite sides of the alkene). Figure 2.27 illustrates the *cis* and *trans* forms for a long-chain compound containing a C=C double bond.

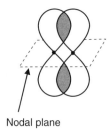

Figure 2.25 σ- and π-bonds.

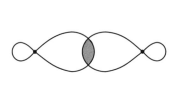

σ bond

Nodal plane

π bond

Figure 2.26 σ- and π-bonding in ethylene, CH_2=CH_2.

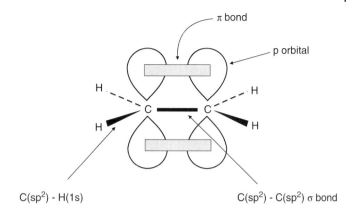

C(sp^2) - H(1s)   C(sp^2) - C(sp^2) σ bond

Figure 2.27 The *cis*- and *trans*-isomers for a long chain compound containing a C=C double bond.

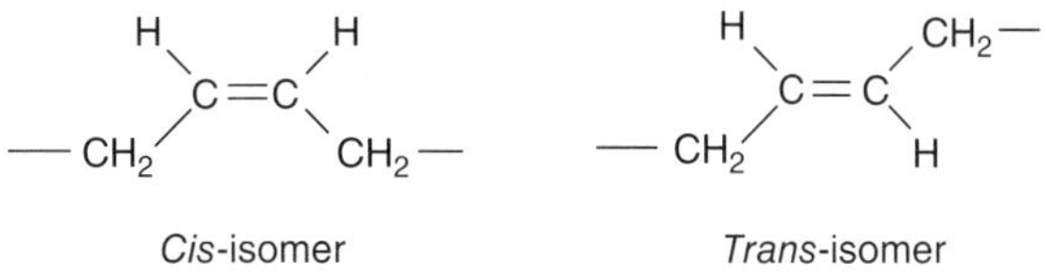

Cis-isomer   *Trans*-isomer

Figure 2.28 Bonding in acetylene CH≡CH. Top: σ-bonding. Bottom: π-bonding. The electrons in the two orthogonal p orbitals form a cylindrically symmetric torus of doughnut-like electron distribution.

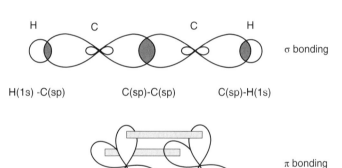

H(1s) -C(sp)   C(sp)-C(sp)   C(sp)-H(1s)

σ bonding

π bonding

In acetylene (HC≡CH), which contains a C≡C triple bond, the C—H bonds are shorter than in acetylene (0.106 nm) and are C(sp)—H(s) in character. The bond strength is 5.43 eV (125 kcal kmol^{-1}). The molecule has a linear structure with a C≡C bond of strength 8.7 eV (200 × 10^3 kcal kmol^{-1}) and length 0.120 nm, the shortest carbon–carbon bond distance known. A σ-bond is formed from the head-on overlap of the carbon sp hybrid orbitals. The two orthogonal p orbitals give rise to cylindrically symmetrical π-bonds, as shown in Figure 2.28.

Table 2.5 compares the carbon–carbon and carbon-hydrogen bond characteristics for ethane, ethylene, and acetylene. In unsaturated alkyl chains, both —C= and —C≡ bonds are associated with small, but finite, dipole moments (Section 2.3.5). This is simply because of the different amounts of s character of the carbon orbitals making up the bond. Symmetrically substituted alkenes and alkynes, of course, possess no overall moment.

Benzene C_6H_6 is a particularly interesting molecule, with an important relevance to organic and molecular electronics. X-ray crystallography reveals that this compound has a regular

Table 2.5 Comparison of carbon–carbon and carbon–hydrogen bonds in ethane, ethylene, and acetylene.

Molecule	Bond	Bond strength (eV)	Bond length (nm)
Ethane CH_3CH_3	$C(sp^3)—C(sp^3)$ $C(sp^3)—H(1s)$	3.83 4.26	0.154 0.110
Ethylene $H_2C=CH_2$	$C(sp^2)=C(sp^2)$ $C(sp^2)—H(1s)$	6.61 4.48	0.133 0.108
Acetylene $HC\equiv CH$	$C(sp)=C(sp)$ $C(sp)\equiv H(1s)$	8.70 5.43	0.120 0.106

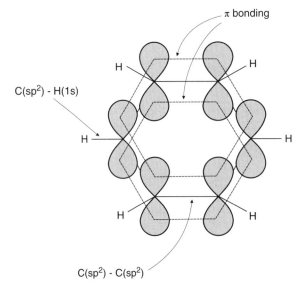

Figure 2.29 Orbital structure of benzene.

π bonding

$C(sp^2)$ - $H(1s)$

$C(sp^2)$ - $C(sp^2)$

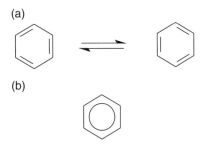

(a)

(b)

Figure 2.30 (a) Two equivalent resonance structures for benzene. (b) Method of indicating partial double bond character of the carbon–carbon bonds in benzene.

hexagonal structure and that the carbon–carbon bond distance is 0.140 nm, intermediate between those for a single bond (0.154 nm) and a double bond (0.133 nm) (Table 2.5). The carbon atoms form a hexagon with one hydrogen atom attached to each carbon, as depicted in Figure 2.29. The bonding between the carbon atoms takes the form of $C(sp^2)=C(sp^2)$ σ-bonds and a π-bond formed from the p orbitals extending around the hexagon. Benzene can be considered as two equivalent structures, shown in Figure 2.30a, in resonance. The resulting *resonance hybrid* is conveniently shown with a circle (or dashed hexagonal line) inside a hexagon to indicate the partial double bond character of the bonds.

2.5   Crystalline and Noncrystalline Materials

2.5.1   States of Matter

The three most common states, or phases, of matter, gases, liquids, and solids are very familiar. Phases that are not so well known are plasmas and liquid crystals (described in Chapter 8). All these states are generally distinguished by the degree of translational and orientational order of the constituent molecules. On this basis, some phases may be further subdivided. For example, solids, consisting of a rigid arrangement of molecules, can be crystalline or amorphous. In an amorphous solid (a good example is a glass), the molecules are fixed in place, but with no pattern in their arrangement. As shown in Figure 2.31, the crystalline solid state is characterized by long-range translational order of the constituent molecules (the molecules are constrained to occupy specific positions in space) and long-range orientational order (the molecules orient themselves with respect to each another). The molecules are, of course, under a constant state of thermal agitation, with a mean translational kinetic energy of $3\,k_{\mathrm{B}}T/2$ ($\frac{1}{2}\,k_{\mathrm{B}}T$ for each component of their velocity, where k_{B} = Boltzmann's constant – Chapter 3, Section 3.2.1). However, this energy is considerably less than that associated with the chemical bonds in the material and the motion does not disrupt the highly ordered molecular arrangement.

In the gaseous state, the intermolecular forces are not strong enough to hold the molecules together. These are therefore free to diffuse about randomly, spreading evenly throughout any container they occupy, no matter how large this is. The number of molecules and the size of the container determine the average interatomic distance. A gas is easily compressed, as it takes comparatively little force to move the molecules closer together.

On the microscopic level, the liquid state is generally thought of as a phase that is somewhere between that of a solid and that of a gas. The molecules in a liquid neither occupy a specific average position nor remain oriented in a particular way. They are free to move around and, as in the gaseous state, this motion is random. The physical properties of both liquids and gases are *isotropic*; i.e. they do not depend upon direction (directional dependent properties are called *anisotropic*). Therefore, there is similarity between liquids and gases and, under certain conditions, it is impossible to distinguish between these two states. When placed in a container, the liquid will fill it to the level of a free surface. Liquids flow and change their shape in response to weak outside forces. The forces holding a liquid together are much less than those in a solid. Liquids are highly incompressible, a characteristic that is exploited in hydraulic systems.

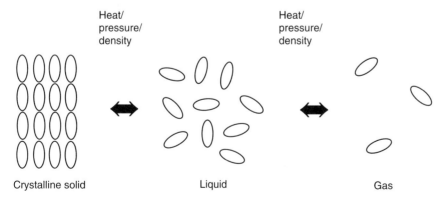

Figure 2.31 The three most common bulk phases of matter.

Almost all elements and chemical compounds possess a solid, a liquid, and a gaseous phase. High temperatures and low pressures favour the gaseous phase. A transition from one phase to another can be provoked by a change in temperature, pressure, density, or volume.

2.5.2 Phase Changes and Thermodynamic Equilibrium

When an arbitrary *thermodynamic system* (e.g. a fixed amount of matter) is left to itself, its properties (pressure, volume, temperature, etc.) will generally change. However, after a sufficiently long period, the properties will be invariant with time. This final state will depend on the nature of the system. For example, a simple mechanical arrangement, such as a ball rolling down a hill, consistently comes to rest in the state of lowest energy (with the ball at the bottom of the hill). Any thermodynamic system, left for long enough, will reach a state of *thermodynamic equilibrium*. Here, thermodynamic equilibrium is achieved when the *free energy* is minimized. For systems at constant volume, this is given by the *Helmholtz function F*:

$$F = U - TS \tag{2.6}$$

and for constant pressure changes, the free energy is given by the *Gibbs* function *G*:

$$G = H - TS = U + PV - TS \tag{2.7}$$

In both the above equations, U is the *internal energy*, the sum of all the potential and kinetic energies of the system, H is the *enthalpy* or *latent heat* of transformation (e.g. heat associated with a process such as solidification or vaporization), and S is the *entropy*. Classically, entropy is the heat into, or out of, the system divided by temperature:

$$S = \frac{H}{T} \tag{2.8}$$

Entropy is also a measure of the disorder of a thermodynamic system. The specific molar entropy change ($\Delta S/n$, where n is the number of moles) for vaporization is normally much greater than that for solidification, confirming the increased disorder produced in the former phase transformation.

Equations (2.6) and (2.7) reveal that free energy can be minimized either by reducing the internal energy or by increasing the entropy term. At low temperature, the internal energy of the molecules makes the greater contribution to the free energy, and so the solid phase is the favoured state. At higher temperature, the entropy of the system becomes the predominant influence. As a result, the fluid phases are stable at elevated temperatures, although they constitute higher internal energy configurations than the solid phase.

Many states of matter that are commonly encountered are not in a true state of thermodynamic equilibrium. Equilibrium may be approached very slowly (e.g. an amorphous glass will eventually crystallize over a period of many hundreds of years) or the system may be in a *metastable* state (e.g. a supercooled liquid or vapour). However, as such systems will have directly measurable properties (*P*, *H*, *S*, etc.) that are stable during an experiment, it is appropriate to assume these properties can be related in the same way as for a true equilibrium state.

Thermodynamically, phase changes of matter can be divided into *first-order* and *higher-order transitions*. In the former case, the specific molar Gibbs function g (= G/n) is continuous but there is a discontinuity in the first derivative of g across the transition. An example of such a change is that of melting or evaporation: in these cases the discontinuity in $\frac{\partial g}{\partial T}$ is simply equal to the entropy of transformation ($\Delta H/T$). For second-order thermodynamic transitions, both the specific Gibbs function and its first derivative are continuous, but the second derivative changes discontinuously. In such transitions, the enthalpy of transformation is zero and the molar specific volume does not change. The transition of a liquid to a vapour at the critical point, ferromagnetic to paramagnetic transitions (Chapter 5, Section 5.7) and the change of a superconductor from the superconducting state to the normal state in zero magnetic field, are examples of second-order phase transitions.

2.5.3 The Crystal Lattice

An ideal crystal contains atoms arranged in a repetitive three-dimensional pattern. If each repeat unit of this pattern, which may be an atom or group of atoms, is taken as a point then a three-dimensional *point lattice* is created. A *space lattice*, such as that shown in Figure 2.32, is obtained when lines are drawn connecting the points of the point lattice. The space lattice is composed of box-like units, the dimensions of which are fixed by the distances between the points in the three non-coplanar directions x, y, and z. These are known as unit cells and the crystal has a periodicity (based on the contents of these cells) represented by the translation of the original unit of pattern along the three directions x, y, and z. These directions are called the *crystallographic axes*. Any directions may, in principle, be chosen as the crystallographic axes, i.e. not necessarily orthogonal axes. It is most convenient to select a set of axes which bears a close resemblance to the symmetry of the crystal. In Figure 2.32, the angle between the y and z axes is designated α, between the z and x axes, β, and between the x and y axes, γ. The measured edge lengths of the unit cell along the x, y, and z axes are commonly given the symbols a, b, and c.

2.5.4 Crystal Systems

As noted above, the unit cells of which a space lattice is composed do not necessarily have their three axes at right angles. The lengths of the sides can also vary from the case where

Figure 2.32 A unit cell with x, y, and z coordinate axes, showing cell parameters a, b, and c and interaxial angles α, β, and γ.

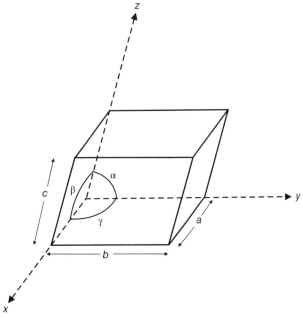

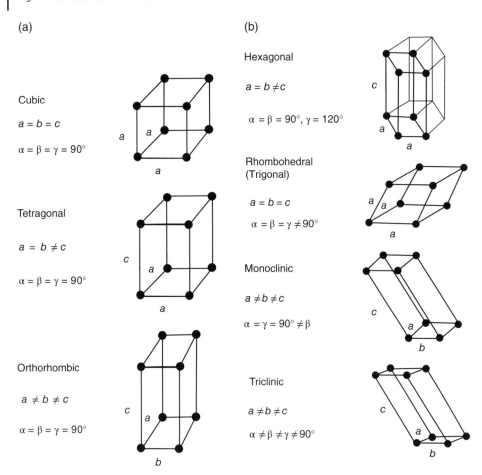

(a)

Cubic

$a = b = c$

$\alpha = \beta = \gamma = 90°$

Tetragonal

$a = b \neq c$

$\alpha = \beta = \gamma = 90°$

Orthorhombic

$a \neq b \neq c$

$\alpha = \beta = \gamma = 90°$

(b)

Hexagonal

$a = b \neq c$

$\alpha = \beta = 90°, \gamma = 120°$

Rhombohedral
(Trigonal)

$a = b = c$

$\alpha = \beta = \gamma \neq 90°$

Monoclinic

$a \neq b \neq c$

$\alpha = \gamma = 90° \neq \beta$

Triclinic

$a \neq b \neq c$

$\alpha \neq \beta \neq \gamma \neq 90°$

Figure 2.33 Lattice parameter relationships for the seven crystal systems. (a) Cubic, tetragonal and orthorhombic. (b) Hexagonal, rhombohedral (trigonal), monoclinic and triclinic.

they are all equal to the case where no two of them are the same. Crystals can belong to seven possible *crystal systems*, characterized by the geometry of the unit cell. These are listed in Figure 2.33: *cubic, tetragonal, orthorhombic, hexagonal, rhombohedral, monoclinic,* and *triclinic*. In the simplest lattices based on these crystal systems (primitive unit cells) the lattice points are positioned at the corners of the cell. However, the monoclinic, orthorhombic, tetragonal, and cubic systems can also have cells that possess additional lattice points. These can occur at the centre of faces or in the middle of the body diagonal, leading to *based-centred* (C), *face-centred* (F), and *body-centred* (I) unit cells. There are seven of these lattices and with the seven primitive (P) lattices they constitute 14 distinct *Bravais lattices*. Whereas a primitive cell possesses one lattice point per unit cell (each of the eight corners being shared by eight cells), body-centred and base-centred cells possess two points and face-centred cells four points.

The crystal system described above is the point group of the lattice (the set of rotation and reflection symmetries which leave a lattice point fixed) not including the positions of the atoms in the unit cell. In contrast, the *crystal structure* is one of the lattices with a unit cell that contains atoms at specific coordinates, at every lattice point. Because it includes the unit cell, the symmetry of the crystal can be more complicated than the symmetry of the lattice. The *crystallographic point group* or *crystal class* is the set of non-translational symmetries that leave a point in the crystal fixed. There are 32 possible crystal classes (i.e. 32 possible combinations of symmetry elements in three dimensions). Finally, the *space group* of the crystal structure is composed of the translational symmetries in addition to the symmetries of the point group. There are 230 distinct space groups.

2.5.5 Miller Indices

Many sets of planes can be drawn through the lattice points of a crystal structure and diffraction of X-rays (Chapter 6, Section 6.3) by the crystal can be treated as reflections of the X-ray beam by these planes. It is desirable, therefore, to be able to describe each set of planes uniquely. This is accomplished using *Miller indices*, which were originally derived to describe crystal faces, but can be applied equally well to any plane or set of planes in a crystal. Miller indices are allocated to a plane in a crystal by first specifying its intercepts on the three crystal axes, in terms of the lattice constants. The reciprocals of these numbers are then taken and reduced to three integers, usually the smallest three integers, having the same ratio. The result is then enclosed in parenthesis: (*hkl*). For example, Figure 2.34 shows a crystal plane intercepting the *x*, *y*, *z* axes at 3*a*, 2*b*, 2*c*. The reciprocals of these numbers are $\frac{1}{3}, \frac{1}{2}, \frac{1}{2}$. The smallest three integers having the same ratio are 2, 3, 3. Therefore, the Miller indices of the plane are (233). If one or more of the intercepts are at infinity (i.e. for planes parallel to a crystallographic axis) then the corresponding index is zero. Figure 2.35 shows some important planes in a cubic crystal. Other conventions that are used include the following:

$\left(hk\bar{l} \right)$: for a plane that intercepts the *x*-axis on the negative side of the origin.

{*hkl*}: for planes of equivalent symmetry, e.g. {100} for (100), (010), (100), (010), and (001) in a cubic crystal.

[*hkl*]: for a direction in a crystal, e.g. [100] is the *x*-axis in a cubic crystal.

<*hkl*>: for a full set of equivalent directions.

Figure 2.34 Representation of a (233) crystallographic plane.

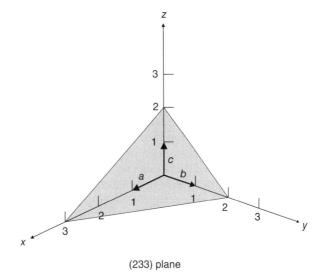

(233) plane

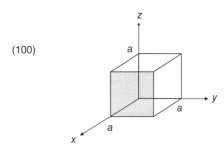

(100)

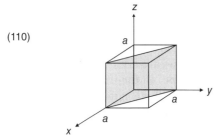

(110)

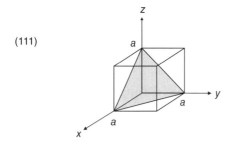

(111)

Figure 2.35 Miller indices of some important planes in a cubic crystal (lattice constant = a).

2.5.6 Distance between Crystal Planes

In crystallography, it is often necessary to calculate the perpendicular distances between successive planes in a series of planes (hkl). This is called the d_{hkl} spacing. For a cubic crystal system, this is easily obtained from simple geometry:

$$d_{hkl} = \frac{a_0}{\sqrt{\left(h^2 + k^2 + l^2\right)}} \tag{2.9}$$

The equivalent relationship for the orthorhombic systems is.

$$d_{hkl} = \frac{1}{\sqrt{\left(h/a_0\right)^2 + \left(k/b_0\right)^2 + \left(l/c_0\right)^2}} \tag{2.10}$$

2.5.7 Defects

No crystal is perfect. Thermodynamic equilibrium requires that certain defects are always present in crystals. The surface of a crystal also sets an upper limit on its perfection. Macroscopic samples of most solids comprise many crystallites, randomly oriented, with grain boundaries separating one crystallite from the next. Such grain boundaries can act as locations for many mobile forms of impurity. Within each crystalline region, there will certainly be finite concentrations of point and line defects.

Point defects.

The point defects in a solid can be grouped into two principal categories. These originate either from foreign atoms (impurities) or from native atoms. Certain atomic sites may not be occupied. This results in point defects known as *vacancies*, illustrated for a cubic lattice in Figure 2.36. Figure 2.36a shows a perfect lattice. Removal of an atom from one of the lattice sites produces a vacancy (Figure 2.36b). If the native atom is relocated into a non-regular atomic site, this produces an *interstitial* defect (Figure 2.36c). These two types of defect formed by native atoms are also referred to as *Schottky* (vacancy) and *Frenkel* (interstitial) defects. In both cases, the defects produce a deformation of the regular lattice.

All crystalline solids contain vacancies and it is not possible to create such a material that is completely free of these defects. The necessity of the existence of vacancies is explained using the principles of thermodynamics. In essence, the presence of vacancies increases the entropy (i.e. randomness) of the crystal. Impurity atoms in otherwise pure crystals can be also considered as point defects, and they play an important role in the electronic and mechanical properties of materials. Such atoms may either occupy a normal atomic site in the host lattice, a substitutional impurity, or a non-regular atomic site, an interstitial impurity. These are shown in Figure 2.37.

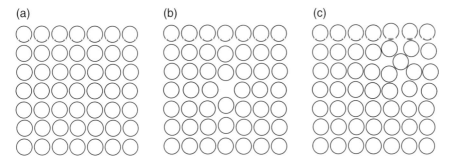

Figure 2.36 (a) Perfect cubic lattice. (b) Vacancy or Schottky defect. (c) Interstitial or Frenkel defect. The defects cause a deformation of the regular lattice.

Figure 2.37 Substitutional and interstitial impurity atoms (shown in dark) in a cubic lattice.

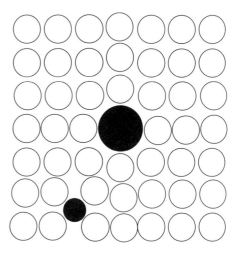

All point defects produce a local distortion in the otherwise perfect lattice, the degree of distortion depending on the crystal structure, parent atom size, impurity atom size, and crystal bonding. These local distortions act as additional scattering centres to the flow of charge carriers through the crystal, thereby increasing its electrical resistance. In the case of ionic crystals, the removal of an ion produces a local charge as well as distortion of the lattice. To conserve overall charge neutrality, the vacancies occur either in pairs of oppositely charged ions or in association with interstitials of the same sign (e.g. in Na^+Cl^-, an Na^+ ion may move into an interstitial site).

Line Defects

Line defects in a crystalline solid are called *dislocations*. There are two basic types of dislocation, *edge* and *screw* dislocations, depicted in Figure 2.38. The edge dislocation, Figure 2.38a, can be visualized as an extra plane of atoms inserted part way into the crystal lattice. The edge of this extra plane is the actual dislocation. There is severe distortion in the region around the dislocation and the lattice planes are bent. The presence of the edge dislocation greatly facilitates slip in the crystal when a shear force is applied, the slip occurring normal to the line marking the end of the extra plane of atoms. In a screw dislocation, Figure 2.38b, part of the lattice is displaced with respect to the other point, and the displacement is parallel

(a)   (b)

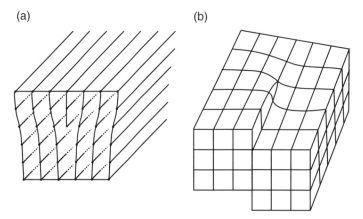

Figure 2.38 Two idealized forms of dislocation. (a) Edge dislocation. (b) Screw dislocation. In general, a dislocation follows a curved path, which has varying components of these two extremes.

to the direction of the dislocation. A general dislocation in a real crystal is likely to have both edge and screw character.

The dislocation density is an indication of crystal perfection. For single crystals of highly perfect solids such as silicon, it is possible to create crystals in which the dislocation density is less than $100 \, \mathrm{cm}^{-2}$. A more typical value for other crystals is around $10^4 \, \mathrm{cm}^{-2}$, while dislocation densities for metallic crystals are usually at least $10^7 \, \mathrm{cm}^{-2}$. Unlike vacancies, thermal equilibrium does not require the presence of dislocations in a single crystal. The application of mechanical stress encourages dislocations to sweep through a crystal, generating point defects until the movement is pinned either by impurities or by the intersection with the path of other dislocations (a process known as work hardening).

Plane Defects

The predominant plane defect is the *grain boundary*, illustrated in Figure 2.39. Most crystalline solids do not consist of a large single crystal (an exception is perhaps the semiconductor silicon, the crystal growth of which has been perfected over the latter half of the twentieth century), but of many randomly oriented crystallites. The junction of these crystallites or grains results in grain boundaries, which represent mismatches in the rows and planes in the adjoining

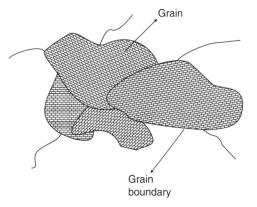

Grain

Grain
boundary

Figure 2.39 Individual grains in a polycrystalline material. The grains are separated by grain boundaries.

crystallites. Each grain itself is an individual crystal and probably contains the point and line defects described above. Various degrees of crystallographic misalignment between adjacent grains are possible. When this orientation mismatch is slight, on the order of a few degrees, then the term low-angle grain boundary is used. These boundaries can be described in terms of dislocation arrays.

The atoms are bonded less regularly along a grain boundary and there is an interfacial or grain boundary energy (similar to surface energy) as a consequence of this boundary energy. Grain boundaries are more chemically reactive than the grains themselves and impurity atoms often preferentially segregate in these regions.

A further plane defect is the *stacking fault*, which occurs when mistakes are made in the sequence of stacking the crystal planes. The plane separating two incorrectly juxtaposed layers is the stacking fault. These defects occur most readily in crystals in which the layer sequence is ABCABC ..., for example the face-centred cubic structure.

Surfaces

Crystal surfaces represent a special type of plane defect. Much of the understanding of solids is based on the fact that these are perfectly periodic in three dimensions; for example, the electronic and vibrational behaviour can be described in great detail using methods that rely on this periodicity. The introduction of a surface breaks the periodicity in one direction and can lead to structural and also to subsequent electronic changes. As technology moves further in nanoscale dimensions, surfaces are becoming increasingly important in the determination of the behaviour of related devices. In the limiting cases, devices built using nanotechnology will be dominated by surface effects.

When the crystal lattice is abruptly terminated by a surface, the atoms cannot fulfil their bonding requirements, as illustrated for a silicon surface in Figure 2.40. Each silicon atom in the bulk of the crystal has four covalent bonds, each bond with two electrons. The Si atoms at the surface are left with *dangling bonds*, that is, bonds that are half full, having only one electron. These dangling bonds look for atoms to which they can bond.

Atoms from the environment can therefore bond with the Si atoms on the surface. For example, a hydrogen atom can be captured by a dangling bond and hence become absorbed, or *chemisorbed*. A foreign atom or molecule is absorbed if it chemically bonds with the atoms on the surface. The H atom in Figure 2.40 forms a covalent bond with a silicon atom and hence becomes absorbed. However, an H_2 molecule cannot form a covalent bond, but due to hydrogen bonding, it can form a secondary bond with a surface Si atom and become *adsorbed*. Similarly, a water molecule in the air can readily become adsorbed at the surface of the crystal.

The difference between chemisorption and physisorption (adsorption) therefore lies in the form of the electronic bond between the adsorbate and the substrate. If an adsorbed molecule

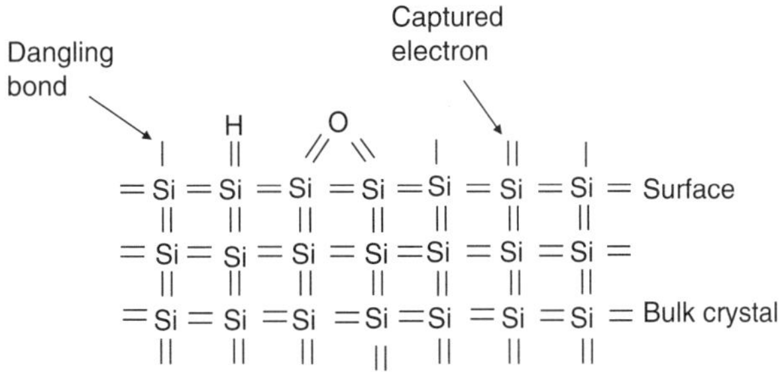

Figure 2.40 Schematic representation of possible defects occurring at the surface of silicon.

or atom suffers significant electron modifications relative to its state in the gas phase to form a chemical bond with the surface (covalent or ionic), it is said to be chemisorbed. If, on the other hand, it is held on the surface only be van der Waals forces, relying on the polarizability of the otherwise undisturbed molecule, it is said to be physisorbed. Clearly, physisorption produces weak bonds while chemisorption often produces strong bonds. It is usual to regard the upper limit of the bond strength in physisorption as around 0.6 eV per atom or molecule, or 60 kJ mole^{-1} (1 eV molecule^{-1} = 96.5 kJ mole^{-1}). Thermal considerations lead to the conclusion that such weekly bonded species would be desorbed from a surface at a temperature much in excess of 200 K. Adsorbates stable on a surface above this sort of temperature are therefore likely to be chemisorbed [1]. Also, if there are any free electrons, these can be captured by dangling bonds. The dangling bonds represent surface traps for carriers and can affect the performance of electronic devices (Chapter 9, Section 9.3).

When left unprocessed, the surface of a crystal will have absorbed and adsorbed atoms and molecules from the environment. This is one of the reasons that the surface of a silicon wafer in microelectronics technology is first etched and then oxidized to form SiO_2, a passivating layer on the crystal surface. Many substances have a natural oxide layer on their surfaces.

2.5.8 Amorphous Solids

When the periodic and repeated arrangement of atoms in a solid is perfect and extends throughout the entirety of the specimen without interruption, the result is a single crystal. Noncrystalline or *amorphous* materials also occur in nature. Such materials are often formed by rapidly cooling a liquid so that the random arrangement of the atoms or molecules in the liquid phase is frozen-in. These materials are called *glasses*. The amorphous structure may have short-range order, as the bonding requirement of the individual molecules must be satisfied. The structure is a *continuous random network*, CRN, of atoms. As a consequence of the lack of long-range order, amorphous materials do not possess such crystalline imperfections as grain boundaries and dislocations, a distinct advantage in some applications. Some materials can exist in both the crystalline and amorphous forms. Figure 2.41 contrasts these two forms of silicon dioxide, SiO_2. The amorphous phase is vitreous silica and is formed by rapidly cooling molten SiO_2.

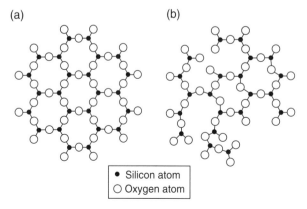

(a)   (b)

• Silicon atom
○ Oxygen atom

Figure 2.41 Silicon dioxide. (a) Crystalline form of SiO_2. (b) Amorphous SiO_2.

2.6 Polymers

Polymeric materials can have massive molecular structures in comparison to the carbon-based compounds described earlier in this chapter. Most polymers are in the form of long and flexible chains composed of structural entities called *mer* units (derived from the Greek word meros, meaning part), which are successively repeated along the chain. The term *monomer* refers to the basic molecule from which the polymer is formed. Table 2.6 shows the

Table 2.6 Some common polymers and their repeating (mer) units.

Polymer	Repeating (mer) structure
Polyethylene (PE)	
Poly(vinyl chloride) (PVC)	
Poly(vinyl alcohol) (PVA)	
Polytetrafluoroethylene (PTFE)	
Polypropylene (PP)	
Poly(methyl methacrylate) (PMMA)	
Polystyrene (PS)	
Polycarbonate (PC)	

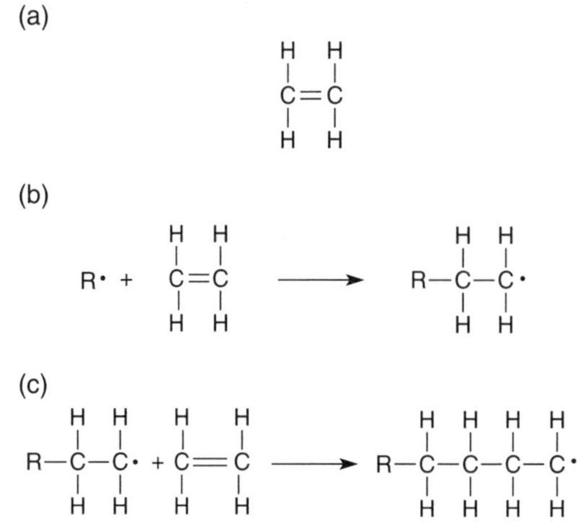

(a)

Figure 2.42 Polymerization of ethylene (C_2H_4) by the sequential addition of polyethylene monomer units. (a) Ethylene molecule. (b) Reaction between an initiator or catalyst species ($R^\bullet$) and the ethylene mer unit. (c) Further mer units are added to increase the length of the polymer chain.

(b)

(c)

list of some common monomer units and the corresponding polymers. When all the repeating units along a chain are of the same type, the resulting polymer is called a *homopolymer*. There is no restriction in polymer synthesis that prevents the formation of compounds other than homopolymers. Chains composed of two or more different mer units are termed *copolymers*.

The process of polymerization may proceed in a number of ways. For example, if ethylene gas is subjected catalytically to appropriate conditions, it will transform to polyethylene. The process, illustrated in Figure 2.42, begins when an active mer is formed by the reaction between an initiator or catalyst species ($R^\bullet$) and the ethylene mer unit. The polymer chain then forms by the sequential addition of polyethylene monomer units to this active initiator-mer centre. The active site, or unpaired electron (denoted by $^\bullet$) is transferred to each successive end monomer and it is linked to the chain. The final result is the polyethylene molecule. This is an example of *addition polymerization* (or chain reaction polymerization). The growth of the chain can be extremely rapid; the time required to grow a molecule consisting of, say, 1000 mer units is on the order of 10^{-2}–10^{-3} s. A further important method of polymerization is *condensation polymerization*. Here, the polymer is formed by stepwise intermolecular chemical reactions that normally involve more than one monomer species. There is usually a small molecular weight by-product such as water, which is eliminated.

2.6.1 Molecular Weight

Large molecular weights are to be found in polymers with very long chains. During the polymerization process, in which these large macromolecules are synthesized from smaller molecules, not all the polymer chains will grow to the same length. This results in a distribution of chain lengths or molecular weights. There are several ways in defining the average molecular weight. The number-average molecular weight $\bar{M}_n$ is expressed as

$$\bar{M}_n = \sum x_i M_i \tag{2.11}$$

where M_i represents the mean (middle) molecular weight of size range i and x_i is the fraction of the total number of chains within the corresponding size range. A weight-average molecular weight $\bar{M}_w$, based on the weight fraction of molecules within the various size ranges may also be defined.

An alternative way of expressing average chain size of a polymer is as the *degree of polymerization n*, which represents the average number of mer units in a chain. Both number-average and weight-average degrees of polymerization are possible. In the former case

$$n_n = \frac{\bar{M}_n}{\bar{m}} \tag{2.12}$$

where $\bar{m}$ is the mer molecular weight. Equation (2.12) can also be used for copolymers, but the value of $\bar{m}$ will be calculated as a summation over the different mer units.

The *polydispersity index*, or *dispersity*, is used as a measure of the broadness of a molecular weight distribution of a polymer, and is defined by

$$\text{dispersity} = \frac{\bar{M}_w}{\bar{M}_n} \tag{2.13}$$

The larger the dispersity, the broader the molecular weight. In the case of a polymer where all the chain lengths are equal (*monodisperse*), the dispersity is unity.

2.6.2 Polymer Structure

Polymers can be subdivided into three, or possibly four, structural groups. Examples of linear, branched and cross-linked polymers are shown in Figure 2.43. The molecules in linear polymers consist of long chains of monomers joined by bonds that are rigid to a certain degree – the monomers cannot rotate freely with respect to each other. Typical examples are polyethylene, poly(vinyl alcohol) (PVA), and PVC, both shown in Table 2.6.

Branched polymers have side chains that are attached to the chain molecule itself. Branching can be caused by impurities or by the presence of monomers that have several reactive groups. Chain polymers composed of monomers with side groups that are part of the monomers, such as polystyrene (PS) or polypropylene (PP) (Table 2.6), are not considered branched polymers. When branching is prevalent, it can have a marked effect on the polymer properties. For example, the polymerization of ethylene under high-pressure conditions gives a product that has many side chains. This material is softer and much less crystalline (Section 2.6.3) than the non-branched form, which is synthesized from a catalytic low-pressure process. The two forms may be distinguished by a difference in density; the more crystalline material being the more dense.

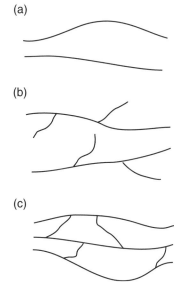

Figure 2.43 Schematic representations of polymer chains: (a) linear; (b) branched; and (c) cross-linked.

In cross-linked polymers, two or more chains are joined together by side chains. With a small degree of cross-linking, a loose network is obtained that is essentially two-dimensional.

High degrees of cross-linking result in a tight three-dimensional structure. Cross-linking is usually caused by chemical reactions. An example of a two-dimensional cross-linked structure is vulcanized rubber, in which cross-links are formed by sulfur atoms. Thermosetting plastics are examples of highly cross-linked polymers; their structure is so rigid that when heated they decompose or burn rather than melt.

2.6.3 Polymer Crystallinity

The crystalline state may exist in polymeric materials. If the chains of polymeric molecules were all of the same length and completely stretched out, or if the chains were folded back and forth in a symmetrical manner, as depicted in Figure 2.44, then these molecules would form completely crystalline structures. However, most polymeric materials comprise of regions that are both crystalline and amorphous. The density of a crystalline polymer will be greater than an amorphous one of the same material and molecular weight, since the chains are more closely packed together for the crystalline structure. The degree of crystallinity may therefore be determined from accurate density measurements.

The molecular chemistry as well as chain configuration also influences the ability of a polymer to crystallize. Crystallization is not favoured in polymers that are composed of chemically complex mer structures. On the other hand, crystallization is not easily prevented in chemically simple polymers such as polyethylene, even for very rapid cooling rates. When a polymer is stretched or drawn, the internal shearing action tends to align the long molecules preferentially in the stretch direction. Not surprisingly, such one-dimensional ordering tends to induce crystallization.

An interesting example of a polymer exhibiting various forms of crystallinity is the material poly(vinylidene difluoride) (PVDF). This polymer, shown in Figure 2.45a, consists of the mer unit CH_2CF_2. PVDF has at least four different crystalline phases, two of which, the α-phase and the β-phase, are shown in Figures 2.45b and c respectively. These are both based on an orthorhombic unit cell (Figure 2.33). The figures show the projection of the carbon (small open circles) and fluorine (large filled circles) onto the *ab* planes of the unit cells. The α-phase is the most common structure and the other forms can be obtained from this parent phase by applications of mechanical stress, heat, and electric field. The polymer chain in the α-phase results in the dipole moments associated with the carbon-fluorine bonds arranged in opposite directions so that there is no net polarization within the crystal.

When the α-phase of PVDF is mechanically deformed by stretching or rolling at temperatures below 100 °C, the β-phase of PVDF is formed. The unit cell of this structure has a net

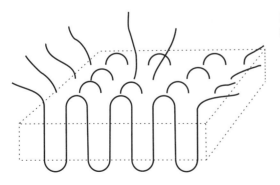

Figure 2.44 Chain-folded lamella example of polymer crystallinity.

Figure 2.45 Structural forms of poly(vinylidene difluoride) (PVDF). (a) Section of $(CH_2CF_2)_n$ chain. (b) α-phase. (c) β-phase. The projections of C atoms (small open circles) and F atoms (large full circles) onto the *a-b* planes are shown. The H atoms have been omitted.

(a)

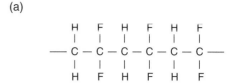

(b)

(c)

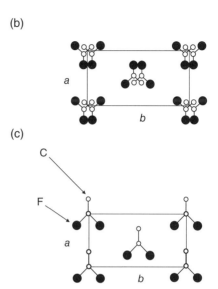

dipole moment (normal to the chain direction). However, because of the random orientation of the crystallites, there is no net polarization in the material. The application of a strong electric field, a process called *poling*, is needed to confer the PVDF with an overall dipole moment. The β-phase of PVDF can be exploited in piezoelectric and pyroelectric applications and is discussed further in Chapter 5, Section 5.6.

Because of their chain-like nature and semi-crystalline structure, the concept of defects in polymeric materials is somewhat different to those described in Section 2.5.7 for crystalline materials. However, important defects do occur in polymers, particularly in their crystalline regions. As will be discussed in Chapter 3, structural defects along the chains of conductive polymers can be very important in organic electronics applications.

When a polymer is heated towards its melting point, several distinct features may be observed. First, and in contrast to most other materials, the melting takes place over a range of temperatures. At low temperature, the amorphous regions of a polymer are in a glassy state in which the molecules are effectively frozen in place (apart from thermal vibrations). As the material is heated, it will eventually reach its *glass transition temperature*, T_g. At this point, large segments of the molecular chains can start to move. A rubbery-like phase is formed. It is important to note that the glass transition temperature is not the same as the melting point of the polymer T_m. The value of T_g for a particular polymer is related to its amorphous regions and will depend on its molecular weight, on its thermal history and on the rate of heating or cooling. The approximate glass transition temperature is 150 °C for polycarbonate (PC), whereas its melting point is 265 °C. The glass transition, indicated in Figure 2.46, is a second-order thermodynamic transition (Section 2.5.2), in contrast to melting. In a second-order transition, there is no associated heat of transformation, as in the case of melting or vaporization.

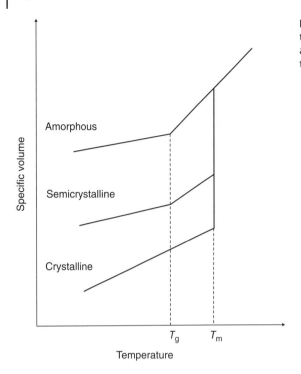

Figure 2.46 Specific volume versus temperature for amorphous, semicrystalline, and crystalline materials. T_g = glass transition temperature. T_m = melting temperature.

The volume of the polymer will change to accommodate the increased motion of the chains, but this does not happen in a discontinuous manner.

2.7 Soft Matter: Emulsions, Foams, and Gels

It is likely that some of the applications of organic electronics will make use of 'soft' materials such as rubbers and plastics. Other states of matter that might be exploited are *emulsions*, *foams*, and *gels*. The liquid crystalline state is discussed separately in Chapter 8. Many of the interesting states of soft matter are based on *colloids*. Nature, of course, makes extensive use of materials in these soft forms.

A colloid is a suspension in which the dispersed phase is so small (1–1000 nm) that gravitational forces are negligible and interactions are dominated by short-range forces, such as van der Waals attraction (Section 2.3.5) and surface charges. The molecules in suspension are in continual collision with solvent molecules and the walls of the container, and display a random motion through the liquid known as *Brownian motion* (discovered by the botanist Robert Brown in 1827). A *sol* is a colloidal suspension of solid particles in a liquid, whereas an *aerosol* describes a colloidal suspension of particles in a gas (the suspension may be called a *fog* if the particles are liquid and a *smoke* if they are solid). An emulsion is a suspension of liquid droplets in another liquid; the two common types of emulsions are oil-in-water and water-in-oil (the term 'oil' in this context is a general word denoting the water-insoluble fluid). If two pure, immiscible liquids, such as benzene and water, are vigorously shaken together, they will form a dispersion. However, on stopping the agitation, phase separation will quickly occur. A surfactant component is always needed to obtain a reasonably stable emulsion. In contrast, microemulsions are generally thermodynamically stable. The dispersed microdroplets are kinetically stabilized because they are embedded inside surfactant micelles (Chapter 12, Section 12.2.5).

A foam can be considered as a type of emulsion in which the inner phase is a gas. As with emulsions, a surfactant compound is needed to provide stability. A gel is a porous three-dimensionally interconnected solid network that expands in a stable fashion throughout a liquid medium. The sol-gel technology that is used to form nanostructured materials is described in Chapter 7, Section 7.2.6).

2.8 Diffusion

The atoms and molecules that constitute solid materials do not always stay in place. Many reactions that are important in the processing of materials rely on the transfer of mass within the solid state or from a liquid, gas, or another solid phase. The physical process governing all these phenomena is that of *diffusion*. Whenever there is a concentration gradient of particles, there is a net diffusional motion of particles in the direction of decreasing concentration. The origin of diffusion lies in the random motion of particles. This process is a type of transport phenomenon and can also apply to the spontaneous spreading of other physical properties, such as heat, momentum, or light. The different forms of diffusion can be modelled quantitatively using the diffusion equation. In all cases, the net flux of the transported quantity (atoms, energy, or electrons) is equal to a physical property (diffusivity, thermal conductivity, or electrical conductivity) multiplied by a gradient (concentration, thermal, or electric field gradient). For transport in a single (x) direction, the flux of particles, J (i.e. the number of particles diffusing through and perpendicular to a unit cross-sectional are per unit time) is given by.

$$J = -D\frac{dC}{dx} \tag{2.14}$$

where C is the particle concentration. The constant of proportionality D is called the diffusion coefficient, which has units of $m^2\,s^{-1}$. The negative sign in Eq. (2.14) indicates that the direction of diffusion is down the concentration gradient. This expression is called *Fick's first law*. Most practical situations are nonsteady state ones, where the diffusion flux and the concentration gradient at a particular point in material vary with time. Under these conditions, Eq. (2.14) is replaced a partial differential expression, *Fick's second law*.

At the atomic level, diffusion is a migration of atoms from one lattice site to the next. For atoms to move, two conditions must be met: (i) there must be an adjacent empty site, and (ii) the atom must have sufficient energy to break its existing bonds with neighbouring atoms and to produce some subsequent lattice distortion. As this energy is vibrational in nature, the temperature will have a significant effect on the diffusion coefficients and diffusion rates.

Diffusion may also occur along defects such as dislocation, grain boundaries, and surfaces. For materials that are electrically charged (e.g. ions), diffusion will be accompanied by an electric current (unless an electric field is applied, the currents due to the movement of positively and negatively charged ions will be opposite and will cancel to maintain charge neutrality). A link between the diffusion coefficient and diffusive conductivity is provided by the *Einstein relation* in Eq. (3.69), Chapter 3.

The diffusion of ions and small molecules with the chains of polymer materials can be significant, leading to swelling and/or chemical reactions. The diffusion rates are greater through the amorphous regions of these materials. Polymer membranes may be used as selective filters in order to separate chemical species. However, for applications in which a polymer material is used as a support (substrate) for organic electronics devices such as displays and transistors (i.e. the realm of plastic electronics), it is important that the diffusion rates for species such as oxygen and water are minimized.

Problems

2.1 If rigid spheres represent atoms, calculate the fraction of the volume of the unit cell actually occupied by atoms for the simple cubic, body-centred cubic and face-centred cubic structures.

2.2 Assuming that 1000 atoms can be used to construct a single molecular 'device', estimate the maximum number of devices that can be packed into a volume of $1\,cm^3$.

2.3 The number-average molecular weight of a PP polymer is $1\,000\,000\,g\,mol^{-1}$. Calculate the number-average degree of polymerization.

2.4 Draw the planes whose Miller indices are (421), (112), $(1\bar{1}0)$ and $(\bar{1}21)$. In each case, state their intercepts with the x, y, and z axes.

2.5 Which elements do the following electronic configurations correspond to?
(a) $2s^1$
(b) $2s^22p^1$
(c) $3s^23p^6$
(d) $4s^2$
(e) $2s^22p^3$

Reference

1 Woodruff, D.P. and Delchar, T.A. (1994). *Modern Techniques of Surface Science*, 2e. Cambridge: Cambridge University Press.

Further Reading

Blythe, T. and Bloor, D. (2005). *Electrical Properties of Polymers*. Cambridge: Cambridge University Press.

Callister, W.D. (2005). *Materials Science and Engineering: An Integrated Approach*, 2e. New York: Wiley.

Hamley, I.W. (2000). *Introduction to Soft Matter*. Chichester: Wiley.

Jones, R.A.L. (2004). *Soft Machines*. Oxford: Oxford University Press.

Kasap, S.O. (2008). *Principles of Electrical Engineering Materials and Devices*, 3e. Boston: McGraw-Hill.

Pauling, L. (1960). *The Nature of the Chemical Bond*, 3e. New York: Cornell University Press.

Petty, M.C., Bryce, M.R., and Bloor, D. (eds.) (1995). *An Introduction to Molecular Electronics*. London: Edward Arnold.

Schwoerer, M. and Wolf, H.C. (2006). *Organic Molecular Solids*. Berlin: Wiley-VCH.

Sears, F.W. and Salinger, G.L. (1975). *Thermodynamics, Kinetic Theory and Statistical Thermodynamics*. Reading: Addison-Wesley.

Streitwieser, A. and Heathcock, C.H. (1976). *Introduction to Organic Chemistry*. New York: Macmillan.

3

Electrical Conductivity

Organic and Molecular Electronics: From Principles to Practice, Second Edition. Michael C. Petty.
© 2019 John Wiley & Sons Ltd. Published 2019 by John Wiley & Sons Ltd.
Companion website: www.wiley.com/go/petty/molecular-electronics2

Two truths are told

3.1 Introduction

Metallic and semiconductive properties are not restricted to inorganic materials. For example, Figure 1.2, in Chapter 1, shows that the room temperature conductivity values for organic polymers can extend over the entire spectrum of electrical conductivity, from insulating behaviour, as shown in diamond, to highly conducting, as for copper. The explanation can be found in the nature of the chemical bonds that hold solids together. In this chapter, the physical principles underlying some of the important electrical conductivity processes in crystalline inorganic solids are first developed. These ideas are then extended to encompass organic solids, which are generally less ordered.

3.2 Classical Theory

The idea of electrons in a metal being free to move about as if they constituted an 'electron gas' was first put forward by Drude in 1900, soon after the discovery of the electron. The theory was later extended by Lorentz (1905), who applied classical *kinetic theory of gases* to the free electron gas in a metal. According to this hypothesis, these electrons would possess a range of speeds (or energies) distributed about a mean velocity, $\bar{v}$, as depicted in Figure 3.1. The resulting curve is known as the *Maxwell-Boltzmann distribution* and has the mathematical form

$$n(v) = Av^2 T^{-3/2} \exp\left(-Bv^2/T\right) dv \tag{3.1}$$

where $n(v)$ is the number of molecules having a velocity in the range v to $v + dv$, T is the absolute temperature, and A and B are constants. If the temperature is increased, the maximum and mean velocities shift to higher values and the Maxwell-Boltzmann curve flattens out. If the temperature is reduced to zero, $\bar{v}$ becomes zero.

In the Drude theory, the behaviour of free electrons in a metal is, in many respects, analogous to that of gaseous molecules. There is one significant difference between the two cases. In the free electron theory, collisions between electrons are neglected and the electrons only undergo collisions with the metal ions. Drude also assumed that the distance travelled by an electron between collisions, the mean free path (see the following section), is governed by the lattice spacing in the crystal and is independent of the electron's speed.

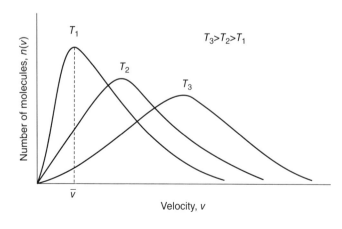

Figure 3.1 Maxwell-Boltzmann distribution of the electron velocities in a metal at different temperatures. Mean velocity $= \bar{v}$.

3.2.1 Electrical Conductivity

The *electric field strength*, **E**, is the force on a unit positive electric charge placed in that field. It is related to the gradient of the *electric potential*. For an electric field applied in one dimension, say the *x*-direction:

$$E_x = \frac{dV}{dx} \tag{3.2}$$

where *V* is the potential in Volts. The units of **E** are therefore $V\,m^{-1}$.

The *electrical conductivity* (measured in units of Siemens per metre, $S\,m^{-1}$, or reciprocal Ohms per metre, $\Omega^{-1}\,m^{-1}$) of a solid is the rate at which charge is transported across a unit area of the solid as a result of a unit applied electric field. If the current per unit area, or *current density*, is J_x (units $A\,m^{-2}$), the conductivity, σ, is given by

$$\sigma = \frac{J_x}{E_x} \tag{3.3}$$

The electric field and current density have a direction associated with them; these are therefore both vector quantities. In this book, when the direction is important, such quantities are denoted using a bold and normal typeface. Hence **J** is the current density vector, while *J* simply represents the magnitude of the current density.

The basic Drude model is that of a random motion of the electrons before any electric field is applied, i.e. as many electrons are moving in any one direction as in the opposite direction, so that there is no net flow in any direction and the current is zero. An estimate of the average *thermal velocity* of an electron, $\bar{v}_t$, can be estimated from the Maxwell-Boltzmann distribution law:

$$\frac{1}{2}m\bar{v}_t^2 = \frac{3}{2}k_B T \tag{3.4}$$

where *m* is the electron mass and k_B is a constant, known as Boltzmann's constant $(1.38 \times 10^{-23}\,J\,K^{-1})$.

The direction of electron motion may be altered when it collides with a metal ion. The average distance travelled by an electron before it experiences such a collision is called the *mean free path*, λ. For a metal at room temperature, the average electron thermal velocity is about $10^6\,m\,s^{-1}$ and the mean free path is approximately 10 nm. In some special cases, the electrons are not subject to scattering events. This situation is referred to as *ballistic* transport and occurs, for example, when the mean free path is long or when electrons are confined in ultra-small regions in semiconductor structures. Ballistic transport is determined by the electronic structure of semiconductors, and allows ultra-fast devices to be fabricated.

When an electric field is applied across the sample, the electrons will be accelerated during their free periods between collisions. This acceleration is generally taken to be in a direction opposite to that of the field, since the charge on the electron is negative. Simultaneously, collisions of the electrons with charged ions and lattice vibrations, or *phonons*, will tend to restore the condition in which all the electron velocities are random. The situation is illustrated in Figure 3.2. The randomization process is called *scattering*. In equilibrium, the situation is equivalent to the valence electrons in the sample all possessing a common *drift velocity* due to the applied electric field. This electron drift constitutes the electric current. It is important to realize that this drift velocity is superimposed on the random electron velocities due to thermal motion and its value will be many orders of magnitude less than that of the thermal electron velocities.

(a)

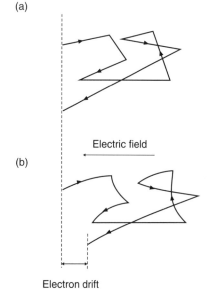

Electric field

(b)

Electron drift

Figure 3.2 Model for electron drift. (a) In the absence of an applied electric field, a conduction electron moves about randomly in a metal being frequently and randomly scattered by thermal vibrations of the atoms. There is no net electron drift in any direction. (b) With an applied field there is a net drift of the electrons along the direction of the field. After many scattering events, the electron has been displaced a small distance from its original position towards the positive electrode.

A simple relationship for the conductivity of a material may be using Newton's laws. With an electric field applied, the force, F, acting on an individual electron of charge, e (e = 1.60×10^{-19} C), will be proportional to its charge and also to the applied field, i.e. $F = eE$. Therefore, the acceleration, a, experienced by the electron (Newton's Second Law) can be expressed as

$$am = eE \tag{3.5}$$

If the average time of flight between collisions is denoted by τ, an appropriate time constant (or relaxation time), and assuming that all the extra energy gained from the applied field is lost upon collision, the average velocity in the electric field, $\bar{v}_d$, is given by $a\tau$, i.e.

$$\bar{v}_d = \frac{e}{m}E\tau \tag{3.6}$$

This represents the drift velocity of the electron. If there are a total of n electrons per unit volume taking part in the conduction, the total current density will be the product of the total charge density and the drift velocity, i.e.

$$J = ne\bar{v}_d \tag{3.7}$$

3.2.2 Ohm's Law

Substituting the value of $\bar{v}_d$ from Eq. (3.6) into Eq. (3.7) gives

$$J = \frac{ne^2\tau}{m}E \tag{3.8}$$

which can be recognized as a form of Ohm's Law – the current density is proportional to the electric field. The electrical conductivity of the material can then be calculated by noting Eq. (3.3):

$$\sigma = \frac{ne^2\tau}{m} \tag{3.9}$$

The electrical behaviour of a material may also be characterized in terms of its *resistance*, R, measured in Ohms (Ω) or its *resistivity*, ρ, measured in Ω m. The latter quantity provides a useful parameter that is independent of the dimensions of the sample and is simply the reciprocal of the conductivity, i.e.

$$\rho = \frac{1}{\sigma} = \frac{m}{ne^2\tau} \tag{3.10}$$

For a wire of cross-sectional area, A, length, l, and with a resistance, R, the resistivity is given by

$$\rho = \frac{RA}{l} \tag{3.11}$$

Noting that

$$E = \frac{V}{l}, J = \frac{I}{A}$$

it is evident that Eq. (3.8) can also be written as

$$V = IR \tag{3.12}$$

which is the most familiar form of Ohm's law. In some situations (in particular, when dealing with electrical conductivity at high frequencies, Section 3.6) a parameter called the *conductance*, G, is used; this is simply the reciprocal of the resistance (i.e. $G = 1/R$) with units Ω^{-1} or S.

For many applications in organic electronics, the conductive materials are in the form of thin films and it is the *sheet resistance* (or surface resistance) R_s that is important. This is because the sheet resistance is easier to measure than the resistivity. The sheet resistance of a layer with a thickness, t, is given by

$$R_s = \frac{\rho}{t} \tag{3.13}$$

Strictly speaking, the units of sheet resistance are Ohms. However, it is normal to quote the sheet resistance in units of Ohms per square (although this can give rise to some confusion). This nomenclature is particularly useful when the resistance of a rectangular piece of material, with length l (between the electrodes) and width w is needed. The resistance is simply the product of the sheet resistance and the number of squares, or

$$R = R_S \frac{l}{w} \tag{3.14}$$

3.2.3 Charge Carrier Mobility

Equation (3.9) can also be written as

$$\sigma = ne\mu \tag{3.15}$$

where

$$\mu = \frac{e\tau}{m} \tag{3.16}$$

The parameter μ is called the *carrier mobility* of the electrons and is equal to the average drift velocity per unit applied electric field (Eq. (3.6)), i.e.

$$\mu = \frac{\bar{v}_d}{E} \tag{3.17}$$

As noted above, the velocity of the electrons acquired as a result of their finite temperature (thermal velocity) is much larger than the carrier drift velocity.

The concept of the mobility of a charge carrier is an important one in semiconductor device physics. It is a measure of how quickly the carriers (electrons) will respond to an applied electric field and provides an indication of the upper frequency limit of the material if it is used in a device such as a transistor (Chapter 9, Section 9.4). The scattering processes mentioned above will be highly dependent upon temperature. Phonon scattering (scattering due to lattice vibrations) will increase as the temperature increases, leading to a reduction in the carrier mobility. In contrast, scattering due to ions will result in the mobility increasing with temperature as the increased thermal velocity reduces the time the carrier spends in the vicinity of the ion. The mobility of charge carriers in organic molecular and polymeric compounds is generally lower than in inorganic semiconductors such as silicon and gallium arsenide (see Table 3.1, Section 3.4.1). A high mobility in a material does not necessarily imply a high electrical conductivity. Equation (3.15) shows that σ also depends on the concentration of the charge carriers (electrons in our example).

Equation (3.17) suggests that the carrier drift velocity will continue to increase as the applied electric field increases. At very high electric fields, the carrier energy can become larger than

Table 3.1 Room temperature carrier mobilities for field effect transistors based on organic semiconductors. After Facchetti [6], Dimitrakopoulos and Mascaro [7], Pearson et al. [8], Abe et al. [9], Novoselov et al. [10], and McEuen and Park [11]. The electron mobilities in single crystal silicon and gallium arsenide are also given.

Material	Carrier mobility ($cm^2 V^{-1} s^{-1}$)
Si single crystal (electrons)	1500
GaAs single crystal (electrons)	8500
Rubrene single crystal	20
Tetracene single crystal	2.4
TCNQ single crystal	1.6
Pentacene	$10^{-3}-1$
Polyacetylene	10^{-4}
Polythiophene	10^{-5}
Phthalocyanine	$10^{-4}-10^{-2}$
Thiophene oligomers	$10^{-4}-10^{-1}$
Organometallic dmit complex	0.2
Benzothiophene derivative	7
C_{60}	0.3
Carbon nanotube	$10^{3}-2 \times 10^{4}$
Graphene	2.5×10^{5}

Figure 3.3 Carrier drift velocity versus electric field showing saturation at $v_d = v_{sat}$ at high values of E.

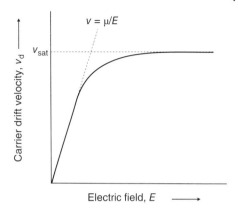

the normal thermal energy. The carriers are referred to as *hot* and this situation leads to a reduction in the carrier mobility and a saturation of the drift velocity. The effect is depicted in Figure 3.3, where the saturation velocity is shown as v_{sat}. In silicon, the maximum drift velocity for both electrons and holes is approximately $10^5 \, \text{cm} \, \text{s}^{-1}$, corresponding to an applied field of $5 \times 10^7 \, \text{V} \, \text{m}^{-1}$.

If an electron concentration gradient exists in a sample, then the carrier motion will also be affected by diffusion (Chapter 2, Section 2.8). For the one-dimensional case, the total current density is therefore given by an equation of the form

$$J = ne\mu E + eD\frac{dn}{dx} \tag{3.18}$$

where D is the electron diffusion coefficient. The first term represents the drift process while the second term reflects diffusion (Chapter 2, Section 2.8).

3.2.4 Fermi Energy

The classical theory for electrical conductivity can provide reasonable 'order of magnitude' values for the conductivity of some metals. However, the theory does fail in predicting the correct temperature dependence of conductivity.

The Drude model treats electrons as classical indistinguishable particles having velocities or energies governed by the Maxwell-Boltzmann distribution law. No restriction is placed on the number of electrons that can possess any particular value of energy. The quantum mechanical approach, first put forward by Sommerfeld in 1928, postulated that the valence electrons in a metal obey *Fermi-Dirac quantum statistics*. In accordance with the theory, the electrons obey the Pauli Exclusion Principle (Chapter 2, Section 2.2.3) and, in contrast to classical theory, they must be regarded as indistinguishable particles. The Fermi-Dirac distribution function $f(E)$ governing the occupation of the electron levels (or states) over energy may be shown to be

$$f(E) = \frac{1}{\exp\left[(E - E_F)/k_B T\right] + 1} \tag{3.19}$$

where $f(E)$ is the probability of a level at energy E being filled with an electron, and E_F, which has the dimensions of energy, is termed the *Fermi level* or *Fermi energy*.

When $E = E_F$, then from Eq. (3.19), $f(E) = 0.5$, i.e. *the Fermi level is the energy at which there is a 50 : 50 chance of finding an electron.*

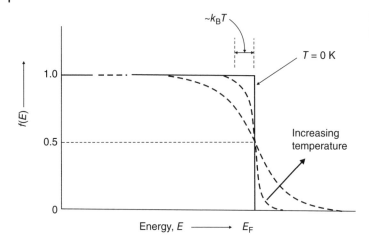

Figure 3.4 The Fermi Dirac function $f(E)$ at $T = 0$ K and at $T > 0$ K.

The Sommerfeld model predicts a Fermi energy in metals at low temperatures of several electron volts per electron. Figure 3.4 shows how the occupation of electron states $f(E)$ varies with energy. At $T = 0$ K, all energy levels below E_F are completely filled [$f(E) = 1$], while the levels are completely empty [$f(E) = 0$] above E_F. Thus, at absolute zero, the Fermi energy represents the demarcation between filled and empty states. At higher temperatures, the Fermi-Dirac distribution function becomes less step-like and some of the levels below E_F become depopulated and some above become populated. However, the effect is quite small except at high temperatures. The function $f(E)$ drops from unity to zero as the energy increases by a few $k_B T$. Since the value of $k_B T$ is only about 0.026 eV at 300 K, the transition range is very narrow.

For values of energy that are more than a few $k_B T$ from the Fermi level (i.e. $(E-E_F)/k_B T >> 1$), Eq. (3.19) reduces to

$$f(E) = \exp\left(\frac{E_F}{k_B T}\right) \exp\left(-\frac{E}{k_B T}\right) \tag{3.20}$$

which is the same form as the Maxwell-Boltzmann law given by Eq. (3.1).

The Sommerfeld model also predicts how the number of electron states will vary with energy. This is given by a function called the *density of states*, $S(E)$, where

$$S(E) = CE^{1/2} \tag{3.21}$$

and C is a constant.

The number of occupied states within an incremental energy range can be found by multiplying the density of states function by the probability of their occupancy, i.e.

$$N(E)dE = S(E)f(E)dE \tag{3.22}$$

where $N(E)dE$ is the number of occupied states in the energy range E to $E + dE$. Figure 3.5 reveals how the theoretical distribution of electrons varies with energy for the cases $T = 0$ K and $T > 0$ K.

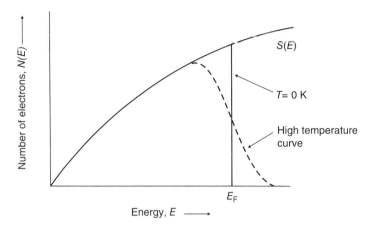

Figure 3.5 Theoretical distribution of the electrons as a function of energy at temperature = 0 K and at a higher temperature. E_F is the Fermi energy or Fermi level. $S(E)$ is the density of states function.

3.3 Energy Bands in Solids

One of the main problems with the Drude and Lorentz approaches is the postulate that the mean free path of the electrons is assumed to be of the same order as the lattice spacing. In reality, mean free paths in metals are much larger (~50 nm in copper at 300 K). The free electron theories of Drude, Lorentz, and Sommerfeld also fail to predict why some materials are metals, others are good insulators, and some are of intermediate conductivity (semiconductors). The *band theory of solids*, described in the following sections, addresses these issues.

3.3.1 Quantum Mechanical Foundations

Before band theory can be developed, some insight into some of the more important physical foundations is needed.

Electromagnetic Waves

Besides what is known as light, *electromagnetic radiation* includes radiation of longer (infrared, microwave, terahertz) and shorter (ultraviolet, X-ray) wavelengths. As the name implies, electromagnetic (EM) radiation contains both electric E and magnetic field B components. The relationship between these fields is best illustrated by considering *plane-polarized* radiation, in which the electric field is confined to a single plane. Figure 3.6 depicts such radiation of wavelength λ, frequency ν and travelling with *phase velocity* v (in a vacuum, the phase velocity $= c$, the velocity of light $= \lambda\nu = 3.00 \times 10^8 \, \mathrm{m\,s^{-1}}$) along the x-axis. The electric component of the radiation is in the form of an oscillating electric field, while the magnetic component is an oscillating magnetic field. The vectors representing these fields are orthogonal to each other and, in free space, are also orthogonal to the direction of propagation of the wave. The amplitude of each field component can be described mathematically by a simple sinusoid. For example, the variation of the electric field E may be written as

$$E = E_0 \cos(\omega t - kx) \tag{3.23}$$

where E_0 represents the maximum amplitude of the wave. Equation (3.23) indicates that the amplitude of the electric field depends both on time t and position x, as shown in Figure 3.7.

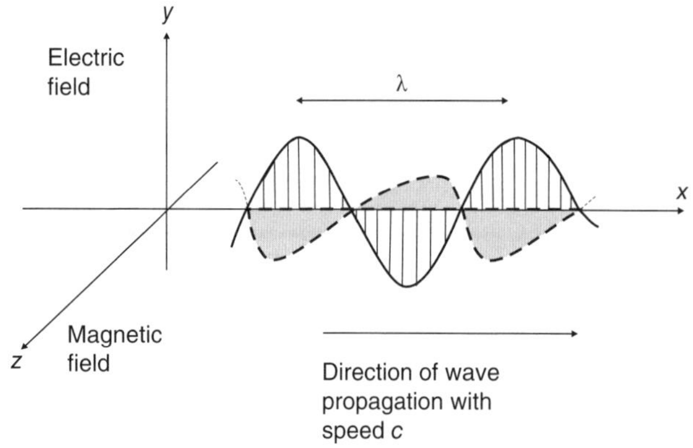

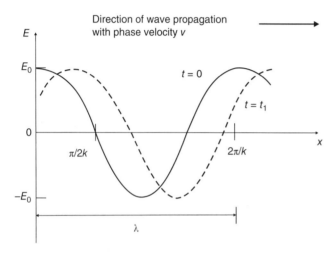

y

Electric field

λ

x

Magnetic field

z

Direction of wave propagation with speed c

Figure 3.6 Schematic representation of an electromagnetic (EM) wave travelling in free space with velocity c, the speed of light. Wavelength $= \lambda$. The wave consists of oscillating electric and magnetic field components at right angles to each other and orthogonal to the direction of propagation x.

E

Direction of wave propagation with phase velocity v

E_0

$t = 0$

$t = t_1$

0

$\pi/2k$

$2\pi/k$

x

$-E_0$

λ

Figure 3.7 Variation of the electric field E of an electromagnetic (EM) wave, given by Eq. (3.23), with position x for time $t = 0$ and a later time $t = t_1$.

The constant ω is called the *angular frequency* of the wave ($= 2\pi\nu$, where ν is in Hz), while k is known as the *wavevector* or *propagation constant*. This is related to the wavelength λ of the wave by

$$k = \frac{2\pi}{\lambda} \tag{3.24}$$

Figure 3.7 compares the form of Eq. (3.23) at time t and at a later time $t = t_1$, shown by the broken line. This can be verified by noting that the first zero of Eq. (3.23) is given by

$$(\omega t - kx) = -\pi/2$$

or

$$x = (\pi/2k) + (\omega t/k) \tag{3.25}$$

Hence, the wave representing the electric field is moving in the positive x-direction with a constant phase velocity v given by

$$v = \frac{\omega}{k} \tag{3.26}$$

Photons as Particles

The wave nature of light can account for many of its characteristics, such as interference and diffraction. Other features can be explained only if light can also be treated as particles. The most important phenomenon is the photoelectric effect by which electrons are emitted from a clean metal surface in vacuum when irradiated by light of an appropriate wavelength. An increase in light intensity is accompanied by an increase in electron emission but not in electron energy. The successful interpretation of the photoelectric effect, given by Einstein in 1905, proposed that light also consists of 'energy packets' or quanta. Each of these has an energy E of magnitude given by

$$E = h\nu \tag{3.27}$$

where h is a constant known as Planck's constant (6.63×10^{-34} J s). The light quanta also possess a momentum p given by de Broglie's relationship:

$$p = \frac{h}{\lambda} \tag{3.28}$$

The expressions for the energy and momentum of the photon $E = h\nu$ and $p = h/\lambda$ can also be written in terms of the angular frequency ω and wavevector k as follows:

$$E = \hbar\omega \quad p = \hbar k \tag{3.29}$$

where $\hbar = h/2\pi$.

The wave nature of particles such as photons cannot be represented adequately by an equation of the form of Eq. (3.23). Instead, the particle is represented by a group or packet of a large number of waves, spread over a distance Δx; each of the waves has a slightly different frequency and velocity from the other members of the group. The waves within the group interfere with one another to give a localized entity that moves through space with a wavelength λ and a *group velocity* v_g, as shown in Figure 3.8. This group velocity represents the speed at which the wave packet, or particle, propagates and is given by

$$v_g = \frac{d\omega}{dk} \tag{3.30}$$

Group velocity is therefore not generally the same as the phase velocity v of a wave given by Eq. (3.26). The group velocity is the speed at which the energy of a wave packet propagates. This velocity cannot exceed that of light in free space. In contrast, the phase velocity is the speed at which the crests (or troughs) of a wave moves through space. In certain circumstances it is possible for $v > c$. Only in the case of a linear relationship between the angular frequency ω and the wavevector k (as, for example, the case for EM waves propagating in vacuum) will $v = v_g$.

Electron Wavefunction

Following de Broglie's suggestion that electrons possessed a wave nature, Schrödinger (1926) argued that it should be possible to represent these electron waves mathematically. Such a wavefunction will depend on position and time and can be represented in one dimension by $\psi(x,t)$. The significance of $\psi(x,t)$ has been introduced in Chapter 2, Section 2.2.2. To recall, $\psi(x,t)$ will have the following properties:

$|\psi(x,t)|^2$ is the probability of finding the electron per unit length at x at time t.
and
$|\psi(x,t)|^2 \, dx$ is the probability of finding the electron between x and $x + dx$ at time t.

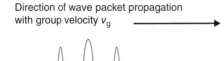

Direction of wave packet propagation with group velocity v_g

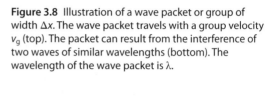

Figure 3.8 Illustration of a wave packet or group of width Δx. The wave packet travels with a group velocity v_g (top). The packet can result from the interference of two waves of similar wavelengths (bottom). The wavelength of the wave packet is λ.

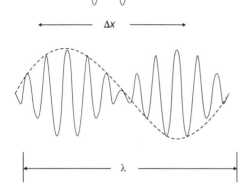

Since the electron carries a charge, the function $\left| \psi(x,t) \right|^2 dx$ also represents the charge distribution associated with the electron. In the case of three dimensions, $\left| \psi(x,y,z,t) \right|^2 dx\, dy\, dz$ is the probability of finding the electron in a small elemental volume $dx\, dy\, dz$ at x, y, z at time t.

Schrödinger Wave Equation

The Schrödinger wave equation is an important relationship involving the electron wavefunction. In essence, it is a statement of the *conservation of energy*, i.e. that the total of the kinetic energy, E, and potential energy, V, is a constant. It is useful to draw on an analogy with the classical case. Here, the appropriate equation can be written as

$$\frac{1}{2m} p^2 + V = E \tag{3.31}$$

i.e. kinetic energy + potential energy = total energy

In the quantum mechanical formulation for one dimension and for a potential energy that is only dependent on space, i.e. $V = V(x)$, the relevant equation becomes

$$\frac{d^2\psi}{dx^2} + \left(\frac{2m}{\hbar^2} \right) \psi \left[E - V(x) \right] = 0 \tag{3.32}$$

Essentially, the classical quantities of Eq. (3.31) have been replaced by quantum mechanical *operators*, which operate upon the wavefunction (representing a particular state of the system) to extract observable value quantities such as position, momentum, and energy. For example, the momentum p in Eq. (3.31) is replaced by the momentum operator $\dfrac{\hbar}{j} \dfrac{\partial}{\partial x}$, where $j = \sqrt{-1}$.

Equation (3.32) is called the *time-independent Schrödinger wave equation*. Many of the problems in solid-state physics are concerned with solving this equation for various forms of the potential energy. Once the wavefunction has been determined, then the probability distribution and energy of the electron can be determined from ψ^2. The solutions for ψ are known as *eigenfunctions* (characteristic functions) and the corresponding energies are called *eigenenergies*.

The solutions to the Schrödinger wave equation for the case of (a) free electrons and (b) electrons confined to a one-dimensional 'box' are illustrated below.

(a) Free electrons

In this simple case, the value of the potential energy is zero and Eq. (3.32) becomes

$$\frac{d^2\psi}{dx^2} + k^2\psi = 0 \tag{3.33}$$

where $k^2 = (2m/\hbar^2)E$. Solutions to this differential equation yield

$$\psi(x) = A\exp(jkx)\,\psi(x) = B(-jkx) \tag{3.34}$$

where A, B are constants. Each of the above equations represents a travelling wave. The first solution represents a wave travelling in the $+x$ direction, while the second is a wave travelling in the $-x$ direction. The total energy E of the electron is simply its kinetic energy. Therefore

$$E = \frac{\hbar^2 k^2}{2m} \tag{3.35}$$

As noted in Eq. (3.29), the wavevector k is proportional of momentum p. Thus, the energy versus momentum relationship for a free electron is essentially a parabolic one, as shown in Figure 3.9.

(b) Electrons confined to a one-dimensional box

A further important solution to the Schrödinger wave equation considers electrons confined by an infinite potential energy barrier V_∞ to a one-dimensional 'box' of length L, as shown in Figure 3.10. The solution yields eigenenergies E_n:

$$E_n = \frac{\hbar^2 \pi^2}{2mL^2} n^2 \tag{3.36}$$

where n is a quantum number (Chapter 2, Section 2.2.2). For each value of n, there will be an electron wavefunction (the eigenfunction) and a corresponding eigenenergy. This result is summarized in Figure 3.11. The wavefunction and energy of the electron both depend on the quantum number n. The picture is that of a large number of possible standing waves, analogous to standing waves on a vibrating string, each particular wavefunction having its own specific energy. The energy of the electron increases as n^2, so the minimum energy of the electron corresponds to $n = 1$. This is called the *ground state*. Note that the energy of the electron in this potential well cannot be zero, even though the potential energy is zero. The electron always has kinetic energy, even when it is in its ground state.

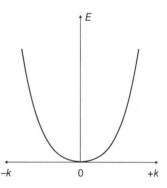

Figure 3.9 Electron energy E versus wavevector k for a free electron.

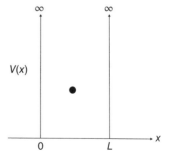

Figure 3.10 Electron confined by infinite potential energy barriers to a one-dimensional 'box' of dimension L.

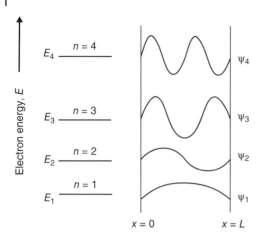

Figure 3.11 Solutions to electron in a one-dimensional 'box' problem for confinement by infinite potential energy barriers. The allowed energy levels for the electron are shown on the left and the corresponding wavefunctions on the right.

The result above demonstrates clearly that the energetic spacing of the eigenenergies depends on the dimensions of the box; the smaller the box the larger the spacing. If the length of the box is of atomic dimensions, say 0.1 nm, the energy states are widely spaced (in energy) and a large amount of energy (hundreds of electronvolts) will be required to promote an electron to a higher energy state. Electrons in a crystal or polymer (dimensions millimetres to centimetres) will have much more closely space eigenenergies.

Heisenberg's Uncertainty Principle

The wavefunction of a free electron corresponds to a travelling wave with a single wavelength λ. The travelling wave extends over all space, for instance along all the x-direction, with the same amplitude, so the probability distribution function is uniform throughout the whole of space. The 'uncertainty' Δx in the position of the electron is therefore infinite. However, the uncertainty in the momentum Δp_x of the electron is zero, because λ is well-defined, which means that p_x is known exactly from the de Broglie relationship (Eq. (3.28)).

For an electron trapped in a one-dimensional potential energy well, as described in the previous section, the wavefunction extends from $x = 0$ to $x = L$, so the uncertainty in the position of the electron is L. The electron is within the well, but cannot be located precisely. The momentum is either $p_x = \hbar k$ in the $+x$ direction or $-\hbar k$ in the $-x$ direction. The uncertainty Δp_x in the momentum is therefore $2\hbar k$, i.e. $\Delta p_x = 2\hbar k$. These results are summarized in the *Heisenberg uncertainty principle*, which states that the product of the position and momentum uncertainties must be equal to or greater than Planck's constant h:

$$\Delta x \, \Delta p_x \geq h \tag{3.37}$$

Therefore, the position and momentum of a particle along a given coordinate cannot be known simultaneously. There is a corresponding relationship between the uncertainty ΔE in the energy E (or angular frequency ω) of the particle and the time duration Δt during which it possesses the energy (or during which the energy is measured):

$$\Delta E \, \Delta t \geq h \tag{3.38}$$

The uncertainty relationships of Eqs. (3.37) and (3.38) are sometimes written in terms of $\hbar$ (i.e. $h/2\pi$). There is also a numerical factor which comes about when a Gaussian spread for all possible position and momentum values is considered. Details can be found in advanced quantum mechanics books.

Quantum Mechanical Tunnelling

An interesting result is found if the problem depicted in Figure 3.11, i.e. an electron confined by infinite potential energy barriers, is replaced by one where the potential energy is finite. Solutions for wavefunctions with $n = 1$ and $n = 2$ are shown superimposed on the one-dimensional potential well in Figure 3.12. In both cases, ψ penetrates the barrier, leading to a finite probability of finding the electron on the outside the box. This mechanism, by which the electron moves through the barrier, is known as *quantum mechanical tunnelling*. The probability of finding the electron outside the box diminishes very rapidly with distance so that tunnelling is only important over small, nanometre, dimensions. A number of electron devices can exploit the tunnelling phenomenon (see, for example, Chapter 11, Section 11.5).

3.3.2 Kronig-Penney Model

A useful model for the propagation of electron waves in a crystal was first formulated by Kronig and Penney in 1930. The periodic potential associated with the positive ion cores is represented by a series of rectangular barriers of width b, height V_0 separated by distance a, as shown in Figure 3.13. The mathematical form of the repeating energy barrier is

$$V(x) = V_0 \quad -b < x < 0$$
$$V(x) = 0 \quad 0 < x < a$$

The Kronig-Penney model solves the Schrödinger wave equation in the regions of the barriers and also between the barriers. A solution is then found that is appropriate for both of these

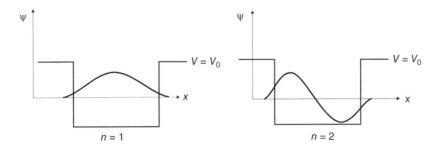

Figure 3.12 Wavefunctions for $n = 1$ and $n = 2$ are superimposed on a one-dimensional well of finite height. Penetration of the wavefunctions outside the well represent quantum mechanical tunnelling.

Figure 3.13 The Kronig-Penney model. Potential distribution in a one-dimensional lattice. The energy barriers have a height V_0, width b and spacing a.

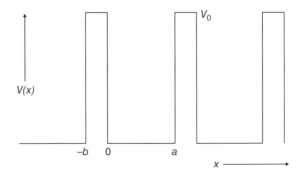

regions which will satisfy the boundary conditions at the barriers, i.e. at $x = 0$ and at $x = a$ in Figure 3.13. The detailed mathematics can be found in a number of texts (e.g. [1]). The result is the following equation, which expresses conditions for solutions to the Schrödinger wave equation to exist:

$$\cos ka = \cos \alpha a + \gamma a \left(\frac{\sin \alpha a}{\alpha a} \right) \tag{3.39}$$

where k is the wavevector, γ is a variable related to the 'strength' of the barrier and given by

$$\gamma = \frac{m V_0 b}{\hbar^2} \tag{3.40}$$

and α is a further variable related to energy:

$$\alpha = \frac{(2mE)^{1/2}}{\hbar} \tag{3.41}$$

Equation (3.39) therefore gives the functions of α that permit solutions of the electron wave to exist. The significance of this relationship can be best understood by reference to Figure 3.14. Here, the right-hand side of Eq. (3.39) is plotted as a function of αa for a particular value of γ ($= \pi/a$).

Since values of $\cos(ka)$ – the left-hand side of Eq. (3.39) – can only lie between +1 and −1, only certain ranges of αa will satisfy Eq. (3.39). These ranges are indicated in Figure 3.14. Moreover, since αa is proportional to energy, the Kronig-Penney model predicts that the motion of electrons in a periodic lattice is characterized by bands of allowed energies separated by forbidden regions (i.e. energy ranges for which no solutions to the Schrödinger wave equation exist).

The width of the energy bands is governed by the parameter γ. When γ is very small, i.e. for a very 'weak' barrier, Eq. (3.39) gives $\alpha = k$, so that

$$E = \frac{\hbar^2 k^2}{2m} \tag{3.42}$$

which is the energy versus wavevector relationship derived from the free electron theory outlined in the previous section (Eq. (3.35)).

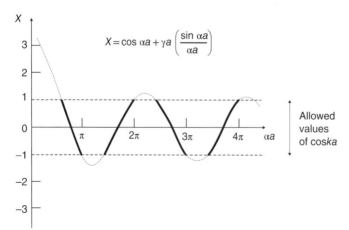

Figure 3.14 Solutions to the Kronig-Penney model. Allowed values of the parameter αa are restricted to those which yield values of the function X lying between +1 and −1. In the figure, the parameter $\gamma a = \pi$.

In contrast, when γ is large, solutions to the Schrödinger wave equation become restricted to those for which sin(αa) = 0, i.e.

$$\alpha a = \pm n\pi \tag{3.43}$$

so that in accordance with Eq. (3.41)

$$E = \frac{\hbar^2 \pi^2}{2ma^2} n^2 \tag{3.44}$$

Therefore, for a tightly bound electron, the Kronig-Penney model predicts a series of discrete levels, as determined for an electron in a one-dimensional box (Eq. (3.36)) with L replaced by a. Under such conditions, the electron may be considered to be confined to a single cell of the potential barrier model depicted in Figure 3.13.

For intermediate values of the barrier strength parameter γ, the electrons having energies within the appropriate bands can propagate freely throughout the crystal. This occurs even though the electron energy is less than the barrier height (via quantum mechanical tunnelling).

The variation of electron energy E with wavevector k is shown in Figure 3.15. This diagram again shows the formation of allowed bands of energies separated by forbidden energy ranges or *energy gaps* or *band gaps*. The figure should be contrasted to the continuous parabolic E versus k curve for a free electron, in Figure 3.9. The discontinuities in Figure 3.15 occur when

$$k = \pm \frac{n\pi}{a} \tag{3.45}$$

where n = 1, 2, 3, ... These k values define the boundaries of what are termed *Brillouin zones*. For example, the first zone contains k values lying between $-\pi/a$ and $+\pi/a$, while the second zone contains values between $-2\pi/a$ and $+2\pi/a$. At the boundary between the zones, $k = \pm n\pi/a$ and since $k = 2\pi/\lambda$ (Eq. (3.24))

$$n\lambda = 2a \tag{3.46}$$

which, as will be discussed in Chapter 6, Section 6.3, represents Bragg reflection for normal incidence of the electron waves on a set of planes with separation a. Thus, electrons with energies at the boundaries of the Brillouin zones undergo multiple reflections and the

Figure 3.15 The variation of the electron energy E with the wavevector k for the Kronig-Penney model showing the formation of bands of allowed energies separated by forbidden energy regions.

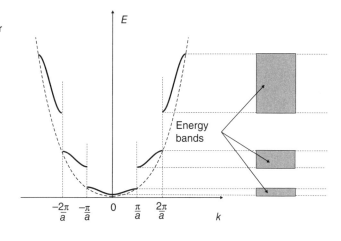

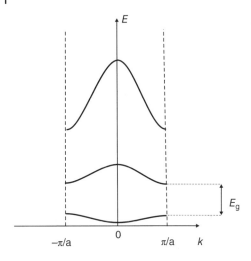

Figure 3.16 Reduced zone representation of the E versus k curves. E_g = band gap.

travelling wave is replaced by equal components of waves travelling in the positive and negative directions. The wavefunctions representing solutions to the Schrödinger wave equation become standing waves.

Within an energy band, the electron energy is a periodic function of k. This means that if k is replaced by $k + 2 \pi n/a$ where n is an integer, the left-hand side of Eq. (3.39) remains the same. For convenience, E-k curves are often depicted in what is known as the *reduced wavevector representation*, in which all the energy bands are shown in the first Brillouin zone, depicted in Figure 3.16. In three-dimensional solids, the E-k curves depend on the direction of the electron wavevector with respect to the crystallographic axis, and the shape of the Brillouin zones will reflect the symmetry of the crystal structure.

In Chapter 2, Section 2.2.3, it was shown how the electron energy levels in two isolated hydrogen atoms split (the bonding and antibonding molecular orbitals) when the atoms joined in a covalent bond to form a hydrogen molecule. An appropriate engineering analogy is that given by Kasap [2] for the resonant frequency in an RLC (resistor-inductor-capacitor) circuit. In isolation, the circuit will possess a characteristic resonant frequency ω_0. However, when two identical circuits are brought together, the circuits become coupled via mutual inductance. The effect is the formation of two resonant frequencies ω_1 and ω_2, below and above ω_0.

When many isolated atoms are brought together to form a complete crystal, multiple splitting of the electron energy levels associated with the isolated atoms occurs. Figure 3.17 illustrates the process of formation of a crystal of silicon, which shows the electron energy levels as a function of interatomic separation. The equilibrium separation of silicon atoms in the crystal is shown as r_0. As the isolated atoms become closer together, their electron orbitals overlap and discrete energy levels associated with the isolated atoms split to form bands of energies. Isolated silicon atoms possess the electronic structure $1s^2 2s^2 2p^6 3s^2 3p^2$ in their ground state (Chapter 2, Section 2.2.3). Each atom has available two 1s states, two 2s states, six 2p states, two 3s states, six 3p states, and higher states. If we consider N atoms, there will be $2N$, $2N$, $6N$, $2N$, and $6N$ states of types 1s, 2s, 2p, 3s, and 3p, respectively. As the interatomic spacing decreases, these discrete energy levels split into bands beginning with the outer ($n = 3$) shell. As the 3s and 3p bands grow, they merge into a single band composed of a mixture of energy levels. This new band of 3s and 3p levels contains $8N$ available electron states. The chemical bonding in a silicon crystal consists of four sp^3 hybridized covalent bonds associated with each atom (Chapter 2, Section 2.4.1). As the distance between the atoms approaches the interatomic

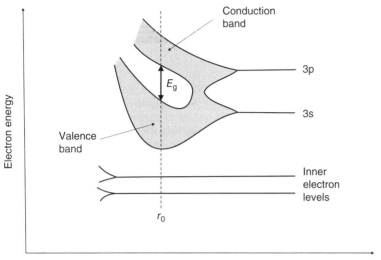

Figure 3.17 Energy levels in silicon as a function of inter-atomic spacing. The inner electron levels are completely filled with electrons. At the actual atomic spacing of the crystal r_0, the electrons in the 3s subshell and the electrons in the 3p subshell undergo sp^3 hybridization and all are accommodated in the lower valence band, while the upper conduction band is empty. The energy separation is the band gap E_g.

spacing of silicon, the band splits into two bands separated by an energy gap E_g. The upper band, called the conduction band, contains $4N$ states, as does the lower energy valence band. In general, the total number of levels in a band must be the total number of levels in an individual atom multiplied by the total number of atoms in the solid. Thus, an s band will have $2N$ levels, a p band $6N$ levels, a d band $10N$ levels, and so on (Chapter 2, Table 2.2).

The $4N$ electrons in the original isolated silicon atoms ($2N$ in 3s states and $2N$ in 3p states) must occupy states in the valence or conduction bands. At 0 K, the electrons will occupy the lowest energy states available to them. For the silicon crystal, there are exactly $4N$ states in the valence band. Thus at 0 K, every state in the valence band will be filled, while the conduction band will be completely empty.

An estimate of the average energy level separation within a band can be obtained from the fact that there are approximately 10^{26} atoms in a macroscopic (kg) quantity of a crystal. Energy bands are typically 1 eV wide, giving an energy separation of 10^{-26} eV. To a first approximation, therefore, the variation in electron energy across a band can be treated as being continuous.

3.3.3 Conductors, Semiconductors, and Insulators

Although the motion of an electron through a crystal lattice as described by the Kronig-Penney model is an oversimplification, it does explain qualitatively the electrical behaviour of most materials. Every solid has its own characteristic energy band structure. After the allowed and forbidden energy regions have been determined, the occupation of the available energy levels is decided by the Fermi-Dirac distribution function (Eq. (3.19)). At absolute zero, all the states lying below the Fermi level are filled and all those above are empty.

Either the energy bands overlap or they do not and the Fermi level may lie within one of the bands or in the forbidden energy region. This leads to essentially three cases, which are illustrated in Figure 3.18.

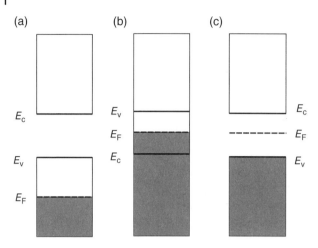

(a) (b) (c)

Figure 3.18 Possible energy band structures of crystalline solids. (a) Energy gap between the conduction and valence bands. The valence band is partly filled. (b) Overlapping valence and conduction bands. (c) Energy gap between the valence and conduction bands. The valence band is completely filled. Figures (a) and (b) represent the band structures of metals, while (c) is the band structure for an insulator. E_v = top of valence. E_c = bottom of conduction band. E_F = Fermi energy.

For conductivity to be possible, there must be available in the same band, energy states that are occupied and others that are empty. This is clearly the case in Figures 3.18a and b; these represent the energy band structures for metals. Metals generally have bands that are only half-filled. For example, in a piece of sodium comprising N atoms, there will be $2N$ available states in the uppermost occupied energy band. As sodium has one valence electron (Group IA in the Periodic Table, Chapter 2, Section 2.2.4), there are only N electrons to be accommodated in the band, which is therefore half full.

In Figure 3.18c, the filled valence band (the highest energy of which is represented by E_v in the diagram) is separated from the next highest band. Conductivity is not possible and the situation corresponds to that of an insulator. If a valence band electron were able to move as a consequence of an applied field, it would take energy from the field and change its energy state, i.e. move to a higher energy in the valence band. This is not possible, since no higher energy levels are available for the electron to move into without violating the Pauli Exclusion Principle.

It is possible for the forbidden energy gap in a crystal to be fairly narrow. At a sufficiently high temperature, some of the valence electrons can gain enough thermal energy to transfer to states lying in the lowermost empty band, the conduction band. When an external electric field is applied, these electrons can gain additional energy from the field, and conduction is possible. The currents produced in such crystals are necessarily small, so they are distinguished from metallic conductors and true insulators by being called semiconductors. The energy gaps in semiconductors are mostly in the range 0.1–3 eV at room temperature (for Si, E_g = 1.1 eV; for GaAs, E_g = 1.4 eV).

Note that the conductivity of a semiconductor will increase as the temperature increases, since more valence electrons can transfer to empty states lying in the conduction band. This is exactly opposite to the temperature dependence of conductivity in metals, in which the resistivity increases with increasing temperature. The conductivity of a semiconductor will also be affected by radiation, i.e. it will be photoconductive. Incident radiation with energy equal to or greater than the band gap will be absorbed, promoting electrons from the valence band to the conduction band and thereby increasing the conductivity.

3.3.4 Electrons and Holes

The thermal or optical excitation of an electron from the valence band to the conduction band, as illustrated in Figure 3.19, leaves behind a valence band state with a missing electron.

Figure 3.19 Model for intrinsic conduction. Thermal excitation of electrons across the band gap E_g produces an increase in electron population of the conduction band and an increase in the hole population of the valence band.

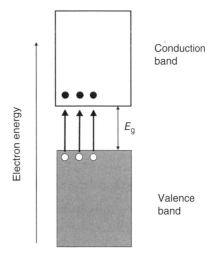

This unoccupied electron state has an apparent positive charge, because the crystal region was neutral prior to the removal of an electron. The valence band state with the missing electron is called a *hole* and can 'move' in the direction of the field by exchanging places with a neighbouring valence electron. Hence, it contributes to the conduction in the semiconductor.

From the previous discussion, it is evident that the electrons in a crystal are not completely free, but instead interact with the periodic potential of the lattice. As a result, their motion cannot be expected to be the same as for electrons in free space. In applying the usual equations of electrodynamics to charge carriers in a solid, it is reasonable to expect that the electron mass will not be the same as that in free space. The relationship between electron energy, E, and wavevector, k, for a free electron has already been given by Eq. (3.35) and may be modified by replacing the free electron mass m with an effective mass m^*, i.e.

$$E = \frac{\hbar^2 k^2}{2m^*} \tag{3.47}$$

The effective mass will be related to the second derivative of the *E-k* curve, since

$$\frac{d^2 E}{dk^2} = \frac{\hbar^2}{m^*} \tag{3.48}$$

or

$$m^* = \frac{\hbar^2}{d^2 E / dk^2} \tag{3.49}$$

Therefore, the electron's mass is determined by the curvature of the energy bands. This idea can now be applied to the energy bands shown in Figures 3.16 and 3.17. Near the bottom of a band, $d^2 E / dk^2$ is a positive quantity, and so is the effective mass. However, near the top of a band, the curvature is such that the effective mass of the electrons is now negative.

A negative effective mass implies that if electrons near the top of the valence band have their energies increased by the application of an electric field, then this results in a reversal of their momenta (i.e. the momentum transfer from the crystal lattice to the electron is opposite and larger than the momentum transfer between the applied field and the electron).

For nearly filled bands, such as the valence band, it is much easier to deal with the behaviour of vacant sites rather than that of electrons. To achieve this, the electron effective mass, which is negative near the top of a band, is replaced by a hole effective mass, which is positive. If there is an electron missing from a given state, then the state is deemed to be occupied by a positive charge. Holes lower in the valence band possess a greater energy and it follows that representation of hole energy in a band must be opposite to that of electron energy. Valence band electrons with negative charge and negative mass in an electric field move in the same direction as holes with a positive charge and positive mass. We can therefore account fully for charge transport in the valence band by considering hole motion.

3.3.5 Intrinsic and Extrinsic Conduction

As noted above, the conductivity of a semiconductor can be affected by thermal or optical processes. Heating a pure semiconductor such as Si from 0 K will produce equal numbers of electrons in the conduction band and holes in the valence band. This process is equivalent to breaking some of the covalent bonds in the silicon crystal. The vacant states or holes that are produced in the valence band correspond to incomplete bonds. On the application of an electric field, both the electrons and holes contribute to the resulting electric current. The conductivity is referred to as *intrinsic*. An intrinsic semiconductor is therefore a material in which the electronic properties (such as the DC conductivity) are not determined by (electrically active) impurities. The necessary level of purity depends upon the band gap and on the temperature. The semiconductor germanium, which has a band gap of 0.67 eV at room temperature, is intrinsic provided that the impurities are present at less than about 1 part in 10^{10}. On the other hand, at the same temperature, intrinsic silicon with the larger band gap of 1.1 eV would require a purity level of better than 1 part in 10^{13}.

At room temperature, the value of the thermal energy $k_B T$ is approximately 0.025 eV and the electrons in the crystal receive energies distributed about this mean value. Since the band gaps of semiconductors lie in the range 0.1–3 eV, the proportion of electrons excited into the conduction band will be very small. This will be particularly so for semiconductors having the larger energy gaps.

The intrinsic carrier concentration n_i will depend on both the band gap and temperature. It is given by

$$n_i = A T^{3/2} \exp\left(\frac{-E_g}{2 k_B T}\right) \tag{3.50}$$

where A is a constant. This variation of n_i with temperature will be dominated by the exponential term in the above equation. The carrier concentration can be substituted into Eq. (3.15) to provide an expression for the temperature dependence of the conductivity. As both electrons and holes are present in the semiconductor, the more general form of this relationship is

$$\sigma = n e \mu_e + p e \mu_h \tag{3.51}$$

where n, p and μ_e, μ_h are the carrier concentrations and mobilities of the electrons and holes, respectively. Of course, for intrinsic material $n = p = n_i$ and so the intrinsic conductivity σ_i can be written as

$$\sigma_i = n_i \left(\mu_e + \mu_h\right) \tag{3.52}$$

If the temperature dependences of the carrier mobilities are slowly varying functions of temperature, then the conductivity will exhibit an almost exponential increase with temperature.

n-Type Doping

Whereas the conductivity of an intrinsic semiconductor is a function only of temperature, that of an *extrinsic* semiconductor is determined by the impurity content. The process of adding small amounts of impurity to control the electrical conductivity of semiconductors is called *doping*. In an extrinsic semiconductor, a specific impurity element is added in controlled amounts; its concentration determines the conductivity of the sample and the type of carriers present. When these are electrons, i.e. negative carriers, the sample is termed *n-type*; when these are holes, i.e. positive carriers, the sample is *p-type*.

A semiconductor such as silicon has tetrahedral covalent bonding in which each atom shares its four valence electrons with each of its four nearest neighbours. If a small amount of a Group V element, such as phosphorus, is incorporated into a silicon crystal, the phosphorus impurity atom is substituted for a silicon atom on the lattice so that it is tetrahedrally bound to its nearest neighbours. In the free atom state, phosphorus has two 3s electrons and three 3p electrons, so that in the substitutional alloy the atom has one surplus electron over and above that necessary to form the covalent bonds. At very low temperatures, close to 0 K, this electron is loosely bound to its parent atom due to the Coulombic attraction of the nucleus. However, if the temperature of the semiconductor is increased, this electron is released from the attraction of the parent impurity atom by thermal energy and is then free to move throughout the crystal. Each impurity atom gives rise to one electron in the conduction band and such atoms are called *donors*. With a suitable number of donor atoms in a sample, the electron concentration may be many orders of magnitude greater than that of an intrinsic sample.

The band representation of an n-type semiconductor is shown in Figure 3.20. At room temperature, each donor atom is ionized and provides an electron in the conduction band. In addition to phosphorus, the other Group V elements, arsenic, antimony, and bismuth, also behave as donor impurities in silicon. The ionization energies, ΔE_{D}, of these donors are all around 0.05 eV.

p-Type Doping

The isolated atoms of the Group III elements – boron, aluminium, gallium, and indium – have outermost electronic shells of two electrons in s states and one electron in a p state. Consequently, when one of these elements is added to silicon, there is a deficiency of one electron for tetrahedral bonding of the impurity atom into the crystal lattice. At very low temperatures, this missing bond is localized in the Coulombic field of the impurity atom. At ordinary temperatures, however, the impurity atom, which is called an *acceptor*, receives an electron from

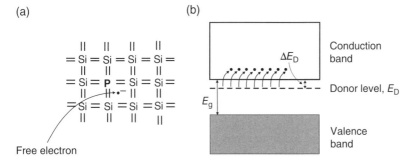

Figure 3.20 Donor impurity, such as phosphorus, in silicon. (a) A phosphorus atom substitutes for a silicon atom in the lattice, leaving a free electron. (b) Energy band diagram showing the formation of a donor level with activation energy ΔE_{D} associated with the phosphorus impurity, close to the conduction band of the silicon.

(a)   (b)

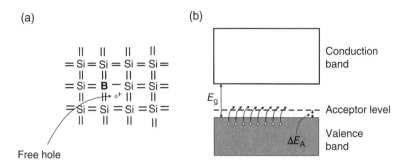

Figure 3.21 Acceptor impurity, such as boron, in silicon. (a) A boron atom substitutes for a silicon atom in the lattice, leaving a free hole. (b) Energy band diagram showing the formation of an acceptor level with an activation energy ΔE_A associated with the boron impurity, close to the valence band of the silicon.

another atom in the crystal and the missing bonding electron, or hole, becomes mobile. This band representation is shown in Figure 3.21. The acceptor atom becomes ionized, with the simultaneous creation of a mobile hole in the valence band. The ionization energies of the acceptor atoms ΔE_A are 0.05–0.16 eV in silicon.

The temperature dependence of the carrier concretion of an extrinsic semiconductor n_e depends on the relationship between the magnitude of the ionization energies for donors (or acceptors) and the absolute temperature. For example, for an n-type material at sufficiently high temperature, all the donor atoms will be ionized and

$$n_e = N_D \tag{3.53}$$

where N_D is the concentration of donor atoms. At lower temperatures, n_e is given by

$$n_e = BT^{3/4} \exp\left(\frac{-\Delta E_D}{2k_B T}\right) \tag{3.54}$$

where B is a constant. In the case for p-type material, a similar equation holds, but with the ionization for donor atoms ΔE_D replaced by ΔE_A, the acceptor ionization energy. Figure 3.22 reveals how the carrier concentration in an n-type semiconductor is expected to vary with temperature. In the figure, log (n) is plotted as a function of 1/Temperature. At low temperatures (and above 0 K) there is sufficient thermal energy to ionize the donor atoms and the carrier concentration versus temperature relationship is that of Eq. (3.54). This is the ionization region. At intermediate temperatures, all the donor atoms are ionized and the carrier concentration is constant (Eq. (3.53)). A further temperature increase will take the semiconductor into the intrinsic conduction region (Eq. (3.50)), where there is sufficient thermal energy available to promote electrons from the valence band to the conduction band.

A semiconductor may also contain simultaneously both donor and acceptor type impurity atoms. When the donor and acceptor concentrations are approximately equal, the material is referred to as being *compensated*. In other cases, the conductivity is governed by the net excess impurity of one type. For silicon, when an impurity element from a group other than III or V is present, it may give rise to one or more separate levels lying within the forbidden energy gap and such levels may lie deep within the gap.

The above principles may be applied to the doping of other inorganic semiconductors, including compound semiconductors such as GaAs. In such compounds, a departure from a strict *stoichiometric* composition (i.e. Ga: As = 1 : 1) leads to either an n-type or p-type sample,

Figure 3.22 The variation of the carrier concentration, *n*, in an n-type semiconductor with temperature, *T*. Concentration of donor atoms = N_D.

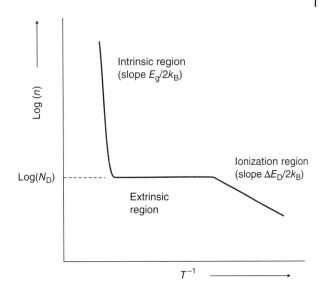

depending on which element is present in the greater concentration. The III-V semiconductor gallium arsenide can also be doped n-type by adding tellurium from Group VI as a donor or n-type by adding zinc from group II as an acceptor.

Traps and Recombination Centres
Many impurities give rise to one or more localized energy levels within the band gap of the semiconductor. These levels can exchange charge with the conduction and valence bands. For example, such states can temporarily remove electrons or holes from the conduction or valence bands, respectively, acting as carrier *traps*. Alternatively, the localized level can attract first an electron (hole) and subsequently a hole (electron). This is the process of *recombination* (also called *Shockley-Read-Hall or SRH recombination*) and the localized level is referred to as a *recombination centre*. Trapping involves the temporary removal of a carrier (which is subsequently excited back into the valence or conduction band) while, in the case of recombination, the carriers are permanently removed from the bands. Traps can be viewed as defects in the regular crystalline structure of the semiconductor; these can be the main limiting factor on the performance of many semiconductor devices.

Fermi Level Position
There is an interdependence of the position of the Fermi level and the carrier concentration for both intrinsic and extrinsic semiconductors. At 0 K, the Fermi level represents the energy below which all electron states are occupied and above which all electron states are filled (Section 3.2.4). In a metal, E_F coincides with the top occupied level in a partially filled band. When the temperature rises, some electrons will move from energy band levels below E_F to levels above E_F, thereby conforming to the distribution shown in Figure 3.5. The Fermi energy no longer separates filled and unfilled states, but is still a useful reference level. For intrinsic semiconductors (and insulators) at 0 K, the Fermi level can still be positioned somewhere between the valence band (all states filled) and the conduction band (all states empty). It can be shown that the Fermi level is located nearly midway between the valence and conduction bands. As the temperature rises, E_F remains positioned approximately at the mid-point in the energy gap, at energy E_{Fi}.

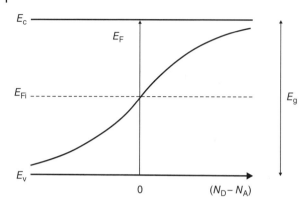

Figure 3.23 The variation in the position of the Fermi level E_F with net donor or acceptor concentration. For $N_D = N_A$ the position of the Fermi level lies close to the middle of the band gap, at energy E_{Fi}.

In the case of extrinsic material, the Fermi level will move towards the conduction band for n-type material and towards the valence band for p-type, as indicated in Figure 3.23. For samples containing both donor and acceptor impurities, the position of the Fermi level will be determined by the net donor or acceptor concentration $(N_D - N_A)$. As an extrinsic material is heated and it moves into the region of intrinsic conductivity, E_F will again move towards the middle of the band gap, to E_{Fi}.

3.3.6 Quantum Wells

A single crystal of an inorganic semiconductor contains a continuum of electron energy levels in the valence and conduction bands and discrete energy levels in the band gap arising from doping. Discrete energy levels for electrons and holes may also be obtained by the process of quantum-mechanical confinement. This process is particularly important in nanotechnology. For example, Figure 3.24 shows the spatial variation in the conduction and valence bands for a

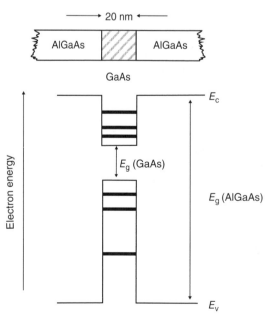

Figure 3.24 Example of a quantum well. Discontinuities in the conduction and valence band edges in the case of a thin layer of GaAs sandwiched between layers of wider band gap AlGaAs. Quantum states are formed in the valence and conduction bands of the GaAs.

multilayer structure in which a thin ($\approx 5\,\text{nm}$) layer of GaAs is sandwiched between two layers of AlGaAs, which has a wider band gap than the GaAs. The consequence of confining electrons and holes in such a thin layer is that these particles behave according to the electron in a box model (Section 3.3.1). Instead of having the continuum of states normally available in the conduction band, the conduction band electrons in the narrow-gap material are confined to discrete quantum states, as described by Eq. (3.36). Similarly, the states in the valence band available for holes are restricted to discrete levels in the quantum well.

3.3.7 Disordered Semiconductors

The energy band model for semiconductors is a consequence of the ordered arrangement of the atoms on a lattice, i.e. a crystalline structure. Interesting questions then arise: what happens if the regular periodicity of the lattice is disrupted? Will the material still possess a recognizable band structure? As noted in Chapter 2, Section 2.5.8, the atoms in an amorphous solid (Figure 2.40 b) are often arranged in a continuous random network. Amorphous pure silicon contains numerous dangling bonds similar to those found at the surface of single crystal silicon (Chapter 2, Section 2.5.7). However, so long as the short-range order present in the crystalline phase is essentially unchanged (i.e. similar bond lengths, bond angles, and local co-ordination), the main features of the density of states function is preserved. The overall result is that energy bands in amorphous materials are generally less defined than in their crystalline counterparts. *Tails* in the density of states function can extend into the band gap, as depicted in Figure 3.25, and localized states can be found within the band gap [3]. For certain materials, the tails may extend so far into the energy gap that they partially overlap. The concept of *mobility edges* within the band tails is introduced for amorphous materials. These are associated with the critical energies separating localized from extended states (in which the conductivity mechanisms are different). The separation between the energies of the mobility edges is called the *mobility gap*, and this parameter is often used instead of the energy gap.

Figure 3.25 Electronic energy versus density of states for an amorphous semiconductor showing overlapping conduction and valence band states. The shaded area represents localized states. A mobility gap separates the extended conduction and valence band states from the localized states.

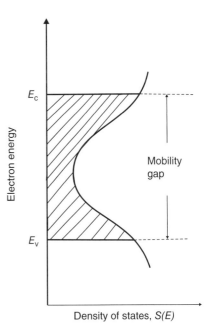

3.3.8 Conductivity in Low-Dimensional Solids

As the size associated with materials and devices moves into the nanoscale, dimensional effects become increasingly important. There are essentially two influences that need to be considered, determined by external (e.g. geometrical shaping) and internal (increase in anisotropy – Chapter 2, Section 2.5.1) factors. An example of the first would be to take a metallic wire and to draw it out until it becomes sufficiently thin to be considered one-dimensional. How thin does the wire have to be? Important parameters are the mean free path of an electron (Section 3.2.1) and the *Fermi wavelength*. The latter is the wavelength of the electrons at the Fermi energy. If the electrons in a thin metal wire can be considered by the electron in a box model (Section 3.3.1), and if the size of the box is just the Fermi wavelength, only the first eigenstate is occupied. If the energy difference to the next level is much larger than the thermal energy ($>> k_B T$), there are only completely occupied and completely empty levels and the system is an insulator.

A thin wire represents a small box for electronic motion perpendicular to the axis, but it is a very large box for motion along the wire. Hence, in two-dimensions (radially) it is an insulator, and in one-dimension (axially) it is a metal. If there are only a few electrons, the Fermi energy is small and the Fermi wavelength is large. This is the case for semiconductors at low doping concentrations. Wires of such semiconductors are already one-dimensional if their diameter is around 10 nm.

If the dimensions of a conductor are large, then its resistance will scale with its length and cross-sectional area, according to Eq. (3.11). For very small dimensions, this should lead to a zero resistance (infinite conductance). However, experiment reveals that the resistance approaches a limiting value when the length of the conductor becomes much shorter than the mean free path. The 'excess' resistance can be interpreted in terms of localization of the electron states. The effect occurs for resistance values that are greater than about $\hbar/e^2$. This combination of fundamental constants is the *von Klitzing constant* and has a value of 4.11 kΩ.

One-dimensional (and even zero-dimensional) systems may also be achieved by selecting the anisotropy of the material. For example, certain materials (conductive polymer chains and carbon nanotubes) may be regarded as natural (quasi) one-dimensional. Examples of two-dimensional solids include graphene and the phthalocyanine compounds, which are described in Chapter 5. Semiconductor *quantum dots* may be considered as zero-dimensional objects.

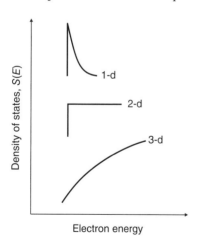

Figure 3.26 Density of states functions at the band edge in one-, two-, and three-dimensional electronic systems.

These structures are small discs of material, which are small compared to the Fermi wavelength, so that the electrons are restricted in all three dimensions. The quantum dots of the organic world are the fullerenes, regular clusters of carbon atoms (Chapter 5). A C_{60} cluster (60 carbon atoms) has a diameter of about 1 nm.

Further peculiarities of one-dimensional systems are band-edge singularities in the density of states function. The density of states function $S(E)$ (Eq. (3.21)) depends on the crystal structure and, near the band edge, it reflects the dimensionality of the system. The parabolic nature of $S(E)$ in the case of a three-dimensional crystal has been shown in Figure 3.5. Figure 3.26 contrasts the $S(E)$ function for one-, two-, and three-dimensional systems. For one-dimension, a singularity to infinity is predicted. If the Fermi level is within such a region, high-temperature superconductivity is favoured, as described in Section 3.4.4.

3.4 Organic Compounds

3.4.1 Band Structure

A periodic variation in the potential experienced by an electron moving in a solid results in discontinuities in the energy versus momentum relationship (*E-k* curve), and to a series of energy bands separated by energy gaps. Although these ideas have been developed in order to explain the electrical behaviour of inorganic materials, such as silicon or gallium arsenide, there is no reason that the physical basis is not equally applicable to other substances in the solid state. Organic solids possess varying degrees of crystallinity. Many, such as molecular crystals and charge-transfer complexes, can be grown in the form of single crystals, while an elongated polymer chain is perhaps the best natural manifestation of the Kronig-Penney model. The following sections will examine the electronic structure of a number of important groups of organic solids.

Molecular Crystals

Molecular crystals can possess a high degree of both short- and long-range order. However, they differ from crystals of metals and inorganic materials because they are made up of discrete molecules. The compound pentacene, shown in Figure 3.27, comprising a linear arrangement of five interconnected benzene rings, is a good example of a molecular crystal. The molecule itself may be regarded as a miniature lattice, with a precisely spaced series of atoms. The atoms are in close proximity, giving rise to a good overlap of their atomic orbitals. The intra-molecular interactions between the atoms lead to a splitting of the carbon $2p_z$ orbitals (Chapter 2, Section 2.4.3, Figure 2.29) and to a localization of the π electrons over the molecule. As a result, bonding orbitals take up these electrons, while the antibonding orbitals remain empty. On a downward positive electron energy scale, there is a *highest occupied molecular orbital*, known by the acronym *HOMO* and a *lowest unoccupied molecular orbital*, *LUMO*, with an energy separation of about 2 eV. In the simplest approximation of an energy-level description, internal electron–electron interactions are neglected, with the consequence that the LUMO describes both the lowest energetic position of an electron excited out of one of the occupied lower levels of a neutral molecule, and also the lowest possible energetic position of an additional electron brought in from outside the molecule.

The relatively strong intra-molecular forces in molecular crystals such as pentacene must be contrasted to the weaker van der Waals' inter-molecular forces that hold the molecules together in the solid crystalline state. As a consequence, the qualitative description provided by the molecular orbital model is largely unaffected. In the organic crystalline solid, there is a moderate splitting of the molecular energy levels by these inter-molecular interactions into narrow 'bands'. However, there is an important shift in these levels as neutral molecules become embedded in the solid-state environment. The resulting energy bands are very different from those found in inorganic semiconductors such as silicon. Their small width and high associated effective mass lead to very small mobilities for the charge carriers. The electronic energy levels in an isolated pentacene molecule (gas phase) and in the solid crystalline phase are contrasted in Figure 3.28 [4]. The energy required to remove an electron from the HOMO level in the crystal is *I*, the ionization energy (Chapter 2, Section 2.3.2), while the empty LUMO level can be similarly characterized by a binding energy or

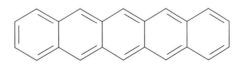

Figure 3.27 Chemical structure of pentacene.

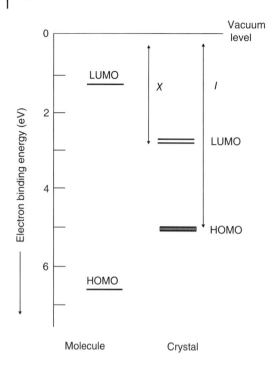

Figure 3.28 Electronic energy levels associated with pentacene. The levels on the left are for free molecules in the gaseous state, while those on the right are for the crystal. I = ionization energy. χ = electron affinity [4].

electron affinity, X, which, like I, is measured with respect to the vacuum level. Clearly, the HOMO-LUMO separation or energy gap ΔE is given by

$$\Delta E = I - X \tag{3.55}$$

An approximate, but very common, way to estimate HOMO and LUMO energies is by using electrochemical measurements (cyclic voltammetry experiments), usually in combination with optical spectroscopy. Such measurements provide good information on energy variations rather than on absolute energy values. The value of ΔE varies with the molecular weight of the organic compound. For example, for the acene series from benzene to pentacene, this energy separation decreases with the number of benzene rings, and hence the size of the π electron system.

It is important to note that the Fermi level of an organic semiconductor is often not located halfway between the HOMO and LUMO levels (NB, as noted in Section 3.3.5, the position of E_F in an intrinsic semiconductor is about the mid-point of the band gap). This is the result of inadvertent doping.

Polymers

A linear arrangement of atoms, as might be found in a polymer chain, provides an excellent experimental basis for the application of the Kronig-Penney model. An important feature of the band model is that the electrons are delocalized or spread over the lattice. The strength of the interaction between the overlapping orbitals determines the extent of delocalization that is possible for a given system. For many polymeric organic materials, the molecular orbitals responsible for bonding the carbon atoms of the chain together are the sp^3 hybridized σ bonds, which do not give rise to extensive overlapping. The resulting band gap is large, as the electrons involved in the bonding are strongly localized on the carbon atoms and cannot contribute to the conduction process. This is why a simple saturated polymer, such as polyethylene, $-(CH_2)_n-$, is an electrical insulator.

A significant increase in the degree of electron delocalization may be found in unsaturated polymers, i.e. those containing double and triple carbon–carbon bonds. If each carbon atom along the chain has only one other atom, e.g. hydrogen, attached to it, the spare electron in a p_z-orbital of the carbon atom overlaps with those of carbon atoms on either side, forming delocalized molecular orbitals of π-symmetry.

Suppose that a polymer is composed of a linear chain of atoms, with N atoms, each separated by a distance d, then the total length of the chain is $(N\text{-}1)d$, which for a large number of atoms approximates to Nd. The eigenenergies, given by Eq. (3.36), are then

$$E_n = \frac{\hbar^2\pi^2 n^2}{2m(Nd)^2} \tag{3.56}$$

Assuming that the π electrons from the N p orbitals are available, with two electrons per molecular orbital (according to Pauli) the HOMO will be that given by $n = N/2$, and the corresponding energy will be

$$E_{HOMO} = \left(\frac{N}{2}\right)^2 \left(\frac{\hbar^2\pi^2}{2m(Nd)^2}\right) \tag{3.57}$$

The LUMO has the energy

$$E_{LUMO} = \left(\frac{N}{2}+1\right)^2 \left(\frac{\hbar^2\pi^2}{2m(Nd)^2}\right) \tag{3.58}$$

The energy required to excite an electron from the HOMO to the LUMO level is the band gap of the polymer, E_g, i.e.

$$E_g = E_{LUMO} - E_{HOMO} = (N+1)\left(\frac{\hbar^2\pi^2}{2m(Nd)^2}\right) \approx \left(\frac{\hbar^2\pi^2}{2md^2}\right)\left(\frac{1}{N}\right) \tag{3.59}$$

The band gap is therefore predicted to decrease with increasing length of the polymer chain, and will practically vanish for macroscopic dimensions. For example, if $d = 0.3\,nm$ and $N = 100$, then $E_g = 42\,meV$.

From the above, it might be expected that a linear polymer backbone consisting of many strongly interacting coplanar p_z orbitals, each of which contributes one electron to the resultant continuous π-electron system, would behave as a one-dimensional metal with a half-filled conduction band. In chemical terms, this is a *conjugated chain* and may be represented by a system of alternating single and double bonds. It turns out that, for one-dimensional systems, such a chain can more efficiently lower its energy by introducing bond alternation (alternating short and long bonds). This limits the extent of electronic delocalization that can take place along the backbone. The effect is to open an energy gap in the electronic structure of the polymer. All conjugated polymers are large band gap semiconductors, with band gaps of more than about 1.5 eV, rather than metals. The effect, known as *Peierls distortion*, is described in the following section.

Peierls Distortion

In 1955, Peierls showed that a monatomic metallic chain is unstable and will undergo a metal-to-insulator transition at low temperature. In Figure 3.29a, such a chain, for example sodium atoms, is shown. (This is simply a thought experiment, because sodium atoms will not arrange in chains – they tend to form clusters.) An S-shaped energy versus wavevector relation, parabolic on both sides, will result. The density of states has square root singularities at the top and at the bottom of the band (which are of no significance in this particular context). The band is half-filled because each sodium atom contributes one delocalized electron to the solid and two electrons – spin-up and spin-down – can be accommodated in each state. The Fermi energy E_F is at the band centre (the corresponding Fermi wavevector, k_F, is shown in Figure 3.29a).

Now suppose that the arrangement of the atoms is changed and every second atom is displaced by a small amount, δ, so that the atoms are no longer equidistant; short spacings, $a - \delta$, and long spacings, $a + \delta$, alternate (Figure 3.29b). Again, the arrangement is periodic, but now with a repeat distance $2a$ instead of a. In the E-k diagram of Figure 3.29a, the boundary of the first Brillouin zone occurs at $k = \pi/a$; in the case of Figure 3.29b it is at half that distance, just at the Fermi wavevector k_F. The system has thereby been transformed from a metal with no gap at the Fermi level into a semiconductor with a gap at $\pi/2a$. All states below the gap are filled at absolute zero, all above are empty.

The important question is what is the most stable (lowest energy) state for the one-dimensional array of atoms? It turns out that the answer is the distorted case. This is because the states in the gap have been accommodated above and below the gap. States cannot disappear and there are as many states as atoms in the lattice (because the states are formed from atomic orbitals). Consequently, when summing the electronic energies from zero to E_F, it is evident that the creation of the gap had reduced the electronic energy. However, to achieve this, work has to be done (against the interatomic forces) to displace the atoms. A full analysis

(a)

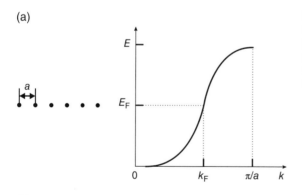

Figure 3.29 Peierls distortion in an isolated chain of equidistant monatomic sodium atoms. Energy versus wavevector diagrams (a) without distortion and (b) with distortion. The distorted lattice has a periodicity of one-half of that of the undistorted lattice.

(b)

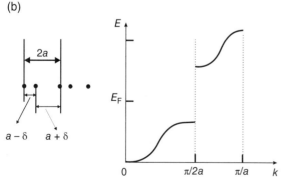

reveals that the electronic energy is approximately linear in δ, whereas the elastic energy depends quadratically on δ. For sufficiently small displacements, the gain in electronic energy predominates over the elastic term. Consequently, under Peierls assumptions, there is always a gap at absolute zero. Electron-lattice coupling will drive the one-dimensional metal into an insulator.

The link between the preceding discussion for a linear array of sodium atoms and a chain of carbon atoms is perhaps unclear. The polymer polyethylene consists of long chains of carbon atoms with two hydrogen atoms per carbon atom. As noted above, the sp^3 hybridized σ bonds do not give rise to extensive overlapping and the band gap is large. Removing one hydrogen atom from each carbon (i.e. to provide an unsaturated polymer chain) would leave unbonded electrons everywhere. The result is a chain of CH˙ *radicals*. The superscript dot denotes a chemical group carrying an odd number of electrons. It is these radicals that have some similarity with an alkali atom – both have an extra electron. Hence, the electrons in polyacetylene are not completely delocalized along the chain. There is an alternation of short and long bonds between the carbon atoms (where short bonds are drawn as double bonds and long bonds are single bonds), which leads to a semiconductive rather than a metallic band structure.

In summary, a completely delocalized electron system in one dimension is expected to lead to the metallic state. However, as depicted in Figure 3.30a, the Peierls transition leads to a bond-alternation, a doubling of the unit cell, and a semiconductive state.

Figure 3.30b shows the electronic energy band structure for *trans*-polyacetylene; the band structure for the *cis* isomer (Chapter 2, Section 2.4.3) is somewhat different. The valence and conduction bands of semiconductive polymers are often referred to as the π and π* bands, respectively. In theory, the π-electron band structure extends over a band width W, given by

$$W = 2zt \tag{3.60}$$

where z is the number of nearest neighbours and t is known as the *transfer integral*. This is a measure of the π wavefunction overlap between neighbouring carbon sites along the backbone,

Figure 3.30 (a) The electrons associated with the CH˙ radicals in *trans*-polyacetylene will delocalize over the chain. The Peierls distortion leads to single/double bond alternation. (b) Resulting energy bands.

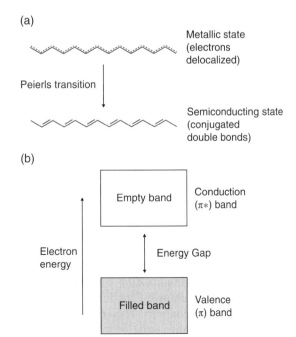

(a)

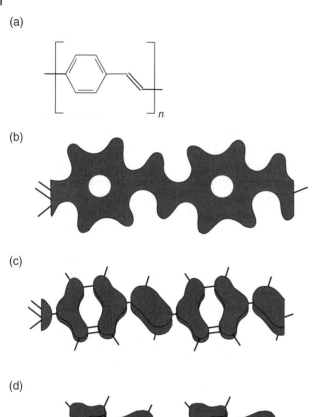

(b)

(c)

(d)

Figure 3.31 (a) Repeat unit of poly (*p*-phenylene vinylene), PPV. (b) Total charge density. (c) highest occupied molecular orbital (HOMO) charge density. (d) Lowest unoccupied molecular orbital (LUMO) charge density. *Source*: Reprinted from Zheng et al. [5]. Copyright (2004), with permission from IOP Publishing Limited.

i.e. the degree of delocalization. For linear polymers such as polyacetylene, $z = 2$ and $t \approx 2.5$ eV, giving $W \approx 10$ eV.

The distribution of electronic charges in an isolated chain of a conductive polymer can be illustrated with reference to some band structure calculations for poly(*p*-phenylenevinylene), PPV [5]. The repeat unit is shown in Figure 3.31a (Chapter 5, Section 5.3). There are eight carbon atoms in a PPV repeat unit but only one electron per atom takes part in the conjugated π-bonding. Since each band can hold exactly two electrons, the four π valence bands with the lowest energy are filled and the four π^* conduction bands at higher energies are empty. Three of the four π bands belong to the benzene ring, and the corresponding wavefunctions are localized along the chain. The other π band derives mainly from the vinylene states, and the corresponding wavefunctions are localized on the vinyl C=C double bond.

The total electronic charge density for an isolated PPV chain is shown in Figure 3.31b. As expected, the total charge density in the plane of the backbone is equally distributed in the benzene ring and the vinyl linkage. The pattern is consistent with the delocalization of electrons suggested in the bonding diagram (Figure 3.31a). The wavefunctions of the uppermost valence band and the lowest conduction band are expected to be pure π and π^*, i.e. bonding and antibonding combinations, respectively, of carbon p orbitals directed perpendicular to the plane of the backbone (assumed to contain the phenyl ring).

The charge densities for the lowest conduction band and highest valence bands are shown in Figures 3.31c and d, respectively. (These represent where an electron would be if it were in the first excited state.) In the case of the highest valence band, the charge density is spread over the molecule, therefore delocalized over the entire chain. The largest density is in the double bond of the vinylene group. The charge density of the lowest conduction band (Figure 3.31d) is highest on those bonds where it is least concentrated in the highest valence band (Figure 3.31c). However, the charge is delocalized along the PPV chain. For both the valence and conduction bands, there is a smaller charge density on the hydrogen atoms.

The energy gap of *trans*-polyacetylene is about 1.4 eV, which is comparable to the values 1.1 eV for single crystal silicon and 1.4 eV for GaAs. However, the electrical conductivity in inorganic crystals and organic polymers is very different. This is because the conductivity also depends on the mobility of the charge carriers (Eq. (3.52)). The band model of semiconductors predicts that the greater the degree of electron delocalization, the larger the width of the bands (in energy terms) and the higher the mobility of the carriers within the band. For inorganic semiconductors such as silicon or gallium arsenide, the three-dimensional crystallographic structure provides for extensive carrier delocalization throughout the solid, resulting in relatively high carrier mobilities.

Electrical conduction in polymers not only requires carrier transport along the polymer chains but some kind or transfer, or 'hopping', between these chains, which tend to lie tangled up like a plate of spaghetti. The charge carrier mobilities in organic polymers are therefore quite low, making it difficult to produce very high-speed electronic computational devices that are competitive to those based on silicon and gallium arsenide. However, some improvement in the carrier mobility can be achieved by both increasing the degree of order of the polymer chains and by improving the purity of the material. Table 3.1 contrasts the room temperature carrier mobility values (in the commonly used units of $cm^2 V^{-1}s^{-1}$) for Si, GaAs, pentacene, polyacetylene and a number of other organic compounds (which will be described further in Chapter 5) [6–11]. Most conductive organic compounds are predominantly p-type conductors and the mobility values will refer to holes. Although the mobility values for the organic materials listed in Table 3.1 are quite low (as expected, the mobilities are somewhat higher for materials in the form of single crystals), other features make them attractive for certain types of electronic device, as discussed in later chapters. Notable exceptions are the very high mobility values for carbon nanotubes and graphene. However, both materials are challenging to process into electron device architectures. Moreover, graphene is a zero band gap material and it is difficult to exploit its high mobility in current logic circuits [10]. Such aspects are discussed further in Chapter 11, Section 11.10.

Charge-Transfer Complexes

A further important class of electrically conductive organic materials are *charge-transfer* compounds. These materials are formed by the combination of two (or more) types of neutral molecules, one of which is an electron donor, D, i.e. has a low ionization energy and can be easily oxidized (electron removal) and the other is an electron acceptor, A, i.e. has a high electron affinity and can easily be reduced (electron addition). The transfer of an electron from the donor molecule to the acceptor molecule can be represented as

$$D + A \rightarrow \left[D^{+\bullet} \right]\left[A^{-\bullet} \right] \tag{3.61}$$

The transfer process leaves behind an organic cation $[D^{+\bullet}]$ that has a 'free' electron, i.e. an electron that is not strongly involved in the chemical bonding. At the same time, the acceptor molecule gains an electron to become the anion radical $[A^{-\bullet}]$. Under certain electronic and

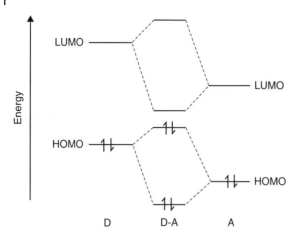

Figure 3.32 Molecular orbitals formed in a charge-transfer complex. The highest occupied molecular orbital (HOMO) and lowest unoccupied molecular orbital (LUMO) levels in the isolated donor D and acceptor A molecules interact to produce new energy levels in the D-A complex.

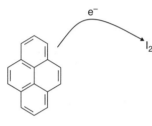

Figure 3.33 Charge-transfer reaction between pyrene and iodine to give a high-conductivity complex.

structural criteria, these electrons can become conduction electrons, just like those of traditional metals. In terms of the energy band diagram, Figure 3.32 shows the energy level arrangement in the separated donor and acceptor molecules compared to the D-A complex. The LUMO and HOMO levels of both the D and A molecules split. As a consequence, there is a smaller energy gap for excitation of an electron from the HOMO to the LUMO level in the complex. A good example of a charge-transfer process is the reaction between pyrene (conductivity $10^{12}\,S\,m^{-1}$) and iodine (conductivity $10^{-7}\,S\,m^{-1}$) to give a complex with a conductivity of about $1\,S\,m^{-1}$, as shown in Figure 3.33. In many cases, partial transfer of charge occurs between the donor and acceptor molecules; say six electrons in every 10 donor atoms are transferred. This leads to *mixed valence states* in the complex. Charge-transfer complexes generally pack closely in their crystalline phase through the formation of rigid multi-sandwich stacks, resulting in a rather brittle solid.

The stacking can be of two types: mixed stacks in which the donors and acceptors stack alternately ...*ADADADAD*...or segregated stacks in which the donors and acceptors form separate donor stacks (...*DDDDDDDD*...) and acceptor stacks (...*AAAAAAAA*...). These are illustrated in Figure 3.34. Molecular compounds with mixed stacks are not highly conductive because of electron delocalization on the acceptor species. However, for segregated stacks, the π overlap and charge-transfer interaction between adjacent molecules in the stacking directions are strong, causing the unpaired electrons to delocalize partially along these one-dimensional molecular stacks, resulting in a high conductivity in this direction. This overlap is different from the *p*-orbital overlap forming the π bands in conjugated polymers. Overlapping in conjugated polymers occurs sideways, in the direction of the polymer axis, and leads to very wide bands, ≈ 10 eV. Overlapping in charge-transfer salts is top to bottom, along the stacking axis and leads to rather narrow bands, with widths of the order of 1 eV. Conjugated polymers are *intra-molecular* one-dimensional conductors; charge-transfer salts are *inter-molecular* conductors (there is also an inter-molecular contribution in polymers, for example inter-chain overlapping with bandwidth < 1 eV, and an inter-stack overlap in charge-transfer salts, with a bandwidth << 1 eV).

Well-known donor and acceptor molecules are tetrathiafulvalene (TTF) and tetracyanoquinodimethane (TCNQ) (Figure 3.35a) (Chapter 5, Section 5.4). The latter compound is a very

Figure 3.34 Possible stacking sequences for donor D and acceptor A molecules. (a) In mixed stacks, the electrons become localized on the acceptor molecules and are unable to move through the stack. (b) For segregated stacks, the electrons are delocalized throughout both stacks resulting in high conductivity.

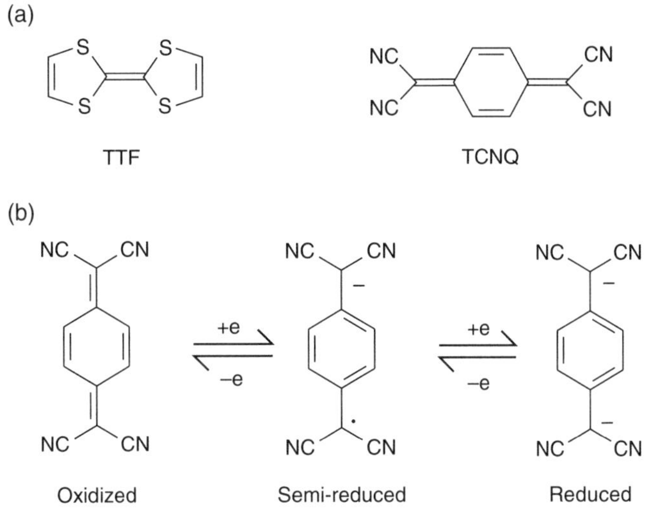

(a)

Electrons localized at acceptor molecules

(b)

Electrons delocalized throughout both stacks

Figure 3.35 (a) Charge-transfer compounds tetrathiafulvalene, TTF, and tetracyanoquinodimethane, TCNQ. (b) Oxidized, semi-reduced, and reduced forms of TCNQ.

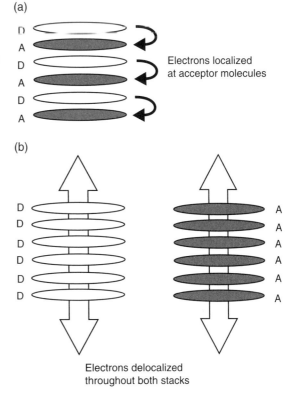

(a)

TTF   TCNQ

(b)

Oxidized   Semi-reduced   Reduced

strong acceptor forming first the radical anion and then the dianion (Figure 3.35b). The stability of the semi-reduced radical ion with respect to the neutral molecule mainly arises from the change from the relatively unstable quinoid structure to the aromatic one (Chapter 5, Section 5.2.6), allowing extensive delocalization of the π electrons over the carbon skeleton. As a consequence, TCNQ not only forms typical charge-transfer complexes but is also able to form true radical-ion salts, incurring complete one-electron transfer. Thus, on addition of lithium iodide to a solution of TCNQ, the simple lithium TCNQ salt is formed:

$$TCNQ + LiI \Leftrightarrow Li^+ TCNQ^{-\bullet} + \frac{1}{2}I_2 \tag{3.62}$$

Following removal of the free iodine precipitate, the TCNQ salt may be crystallized. The crystals show an electronic conductivity of about $10^{-5}\,S\,cm^{-1}$. A 1 : 1 TCNQ:TTF salt exhibits a high room temperature conductivity $(5 \times 10^2\,S\,cm^{-1})$ and metallic behaviour is observed as the temperature is reduced to 54 K.

3.4.2 Doping

The principle of doping in organic semiconductors is similar to that their inorganic counterparts. Impurities are added, which either transfer an electron to the electron conducting (LUMO or π^*) states (n-type doping) or remove an electron from the hole conducting (HOMO or π) states to generate a free hole (p-type doping). However, the term 'doping' can be a misnomer, as it tends to imply the use of minute quantities, parts per million or less, of impurities introduced into a crystal lattice, as is the case for inorganic semiconductors. In the case of conductive polymers, typically 1–50% by weight of chemically oxidizing (electron withdrawing) or reducing (electron donating) agents are used to alter physically the number of π-electrons on the polymer backbone, leaving oppositely charged counter ions alongside the polymer chain. These processes are *redox* (reduction-oxidation) chemistry. For example, the halogen doping process that transforms polyacetylene to a good conductor is oxidation (or p-doping):

$$[CH]_n + \frac{3ny}{2}I_2 \rightarrow \left[(CH)^{y+} \left(I_3^-\right)_y \right]_n \tag{3.63}$$

Reductive doping (n-doping) is also possible, e.g. using an alkali metal:

$$[CH]_n + nyNa \rightarrow \left[(CH)^{y-}\, yNa^+ \right]_n \tag{3.64}$$

In both cases, the doped polymer is a salt. The counter ions, I_3^- or Na^+, are fixed in position while the charges on the polymer backbone are mobile and contribute to the conductivity. The doping effect can be achieved because a π electron can be removed (or added) without destroying the σ backbone of the polymer. In this way, the charged polymer remains intact. The resulting increase in conductivity can be many orders of magnitude.

A completely different type of doping is possible with the polymer polyaniline (PANi). This material, the chemical structure of which is shown in Figure 3.36a, can be considered as being derived from a polymer which consists of alternating reduced (Figure 3.36b) and oxidized (Figure 3.36c) repeat units [12, 13]. The average oxidation state can be varied continuously from the completely reduced polymer to the fully oxidized material. Complete protonation of the imine nitrogen atoms in the emeraldine base form of polyaniline (for which $x = 0.5$ in Figure 3.36) by aqueous HCl, for example, leads to a structural change with one unpaired spin per repeat unit, but with no change in the number of electrons. The result is a half-filled band and a conductive state, where there is a positive charge in each repeat unit (from protonation) and an associated negative charge (e.g. Cl$^-$). The protonation is accompanied by an increase in conductivity of about 11 orders of magnitude.

Doping of molecular crystals can be achieved in a similar way to that described above. For example, high conductivities can be achieved when organic dyes with a weak donor character

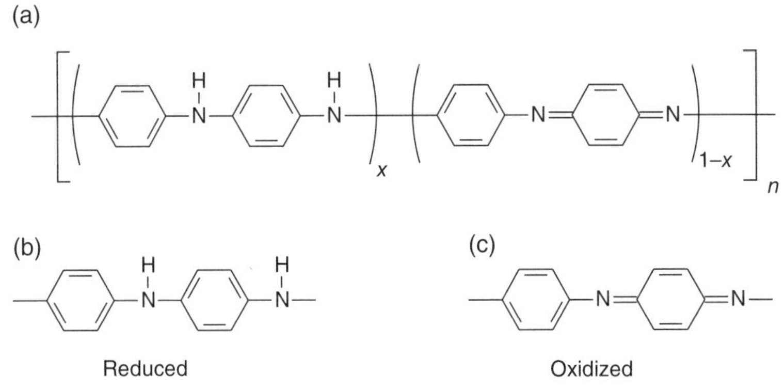

(a)

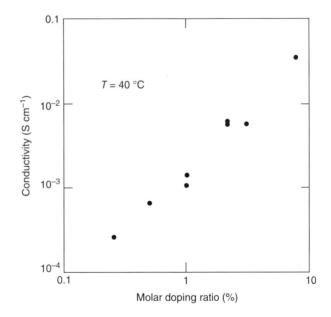

(b)

Reduced

(c)

Oxidized

Figure 3.36 (a) Generic form of polyaniline. The polymer consists of alternating reduced (b) and oxidized (c) units. The average oxidation state $(1-x)$ can be varied continuously from 0 (completely reduced) to 1 (fully oxidized).

Figure 3.37 Conductivity of p-doped zinc phthalocyanine as a function of the doping concentration with the molecular dopant F_4-TCNQ. *Source*: Reprinted from Pfeiffer et al. [14]. Copyright (2003), with permission from Elsevier.

$T = 40\,°C$

(such as phthalocyanine, Chapter 5, Section 5.4) are exposed to strongly oxidizing gases such as iodine or bromine. Alternatively, relatively large aromatic molecules, which are strong π-electron donors or π-electron acceptors, can be used. Figure 3.37 shows the conductivity of zinc phthalocyanine doped with the strong acceptor F_4-TCNQ (p-type doping) as a function of the molecular doping ratio [14]. The conductivity can be controlled over two orders of magnitude by changing the amount of dopant; furthermore, the conductivity is many orders of magnitude higher than the background conductivity of undoped zinc phthalocyanine $(\sim 10^{-10}\,S\,cm^{-1})$.

Finally, it should be noted that the doping of organic semiconductive materials can be achieved in two further ways: photo-doping and charge-injection. In the former case, the material is locally oxidized and reduced by photo-absorption. In the second case, electrons and holes can be injected from suitable metal contacts directly into the π^*- and π-bands, respectively.

3.4.3 Solitons, Polarons, and Bipolarons

In Chapter 2, Section 2.5.7, some of the defects that can be associated with crystalline materials were introduced. There are also a number of specific defects associated with polymers that can influence their electrical and optical properties. An overview of these is given in this section.

Solitons

First, consider the bonding structure in the *trans* form of polyacetylene. The double and single bonds can be interchanged without affecting the overall electron energy. This was also seen to be the case for benzene (Chapter 2, Section 2.4.3, Figure 2.29). Therefore, there are two *degenerate* lowest energy states A and B, shown in Figure 3.38, possessing distinct bonding structures. A simple defect can occur when these two degenerate forms of the polymer are joined, as shown in Figure 3.39. In this instance, a bond mismatch occurs, i.e. a carbon site with one too few π electrons, so that it cannot form a double bond. This results in a dangling bond. This concept has already been introduced when dealing with defects occurring at the surfaces of crystalline semiconductors (Chapter 2, Section 2.5.7) and in amorphous silicon (Section 3.3.7). The defect results in one unpaired π-electron, but as the entire system is electrically neutral, it has the same number of protons as electrons.

The topological defects described above are called *solitons*, due to their non-dispersive nature (an everyday manifestation of a soliton, or solitary wave, is the bow wave of a boat). The soliton will be mobile due to the translational symmetry of the chain. Although the soliton shown in Figure 3.39 is shown to occur as an abrupt change from the A to the B form of polyacetylene, the evidence is that the defect extends over about seven carbon atoms [14]. Solitons can become positively or negatively charged. If the localized state contains one electron, the soliton is neutral and can be associated with an energy level half way between the valence and conduction bands, as depicted in Figure 3.40a. The unpaired electron will have a spin of ½ and can be detected by *electron spin resonance*, ESR. When the electron in the localized state is removed, for example by p-doping, the soliton is positively charged with spin = 0, Figure 3.40b, and is no longer detectable by ESR. Similarly, if n-doping occurs, a negative soliton is obtained with spin = 0, (Figure 3.40c). The ESR studies reveal that neutral solitons are highly mobile, whereas positive and negative solitons are believed to be localized over a number of carbon atoms.

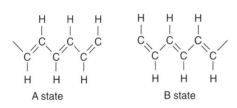

A state   B state

Figure 3.38 Two degenerate forms (A and B states) of *trans*-polyacetylene.

Soliton

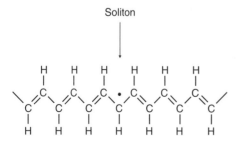

Figure 3.39 Soliton formation in a *trans*-polyacetylene chain.

Polarons and Bipolarons

As noted above, polyacetylene is a conjugated polymer with degenerate ground states. However, most conjugated polymers have *non-degenerate* ground states. In such materials, solitons are not stable and double-conjugational defects are found instead. Such a species is termed a *polaron* if it is singly charged or a *bipolaron* if it is doubly charged. In essence, a polaron can be thought of as the bound state of a charged soliton and a neutral soliton, whose mid-gap energy states hybridize to form bonding and antibonding levels (e.g. a 'combination' of the levels shown in Figures 3.40a and c). Consequently, a polaron is characterized by two states in the gap. Polarons are also encountered in inorganic semiconductor physics: an electron moves through a lattice and affects the constituent ions. However, the resulting distortion of the lattice in inorganic semiconductors is small compared to the polaron defect in conjugated polymers

The polaron and bipolaron are illustrated schematically for poly(p-phenylene) (PPP) in Figure 3.41. The positive (negative) polaron is a radical cation (anion), an entity consisting of a single electronic charge associated with a local geometrical relaxation of the bond lengths. Similarly, a bipolaron is a bound state of two charged solitons of like charge (or two polarons whose neutral solitons annihilate each other) with two corresponding mid-gap levels. The formation of these defect states leads to new localized energy levels in the band gap; consequently, characteristic optical absorption signatures will be observed (Chapter 4, Section 4.4). For further information, the reader is referred to the excellent book by Roth, which contains a more detailed description of the 'menagerie' of conjugational defects found in conductive polymers [15].

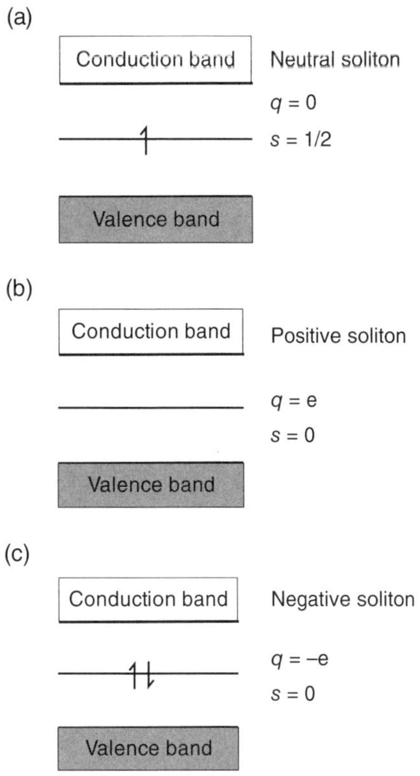

Soliton and polaron states play an important role in the conductivity of conjugated polymers. For example, it is generally believed that electrical conduction in doped polyacetylene proceeds by a hopping mechanism that may involve the capture of a mobile electron from a neutral soliton in an adjacent chain. But due to disorder, solitons and polarons lose many of their characteristic features in real polymer samples.

Figure 3.40 Energy levels associated with (a) a neutral soliton, (b) a positively charged soliton, and (c) a negatively charged soliton. The charge q and spin s of the defects are indicated.

3.4.4 Superconductivity

Only a decade after the discovery of the electron (by Thompson in 1887) and the introduction of Drude's model of the electron gas (1900), Onnes discovered that the resistivity of mercury suddenly vanishes when the sample is cooled slightly below the boiling point of liquid helium. A large number of *superconductors* have now been identified, including elements, alloys and intermetallic compounds, organic charge-transfer salts, fullerenes, and oxides. The highest superconducting transition temperatures have been achieved with the oxides (around 135 K at atmospheric pressure). The Bardeen, Cooper, and Schrieffer (BCS) theory for superconductivity, formulated in 1957, assumes an attractive interaction between the electrons. This is provided by the exchange of *phonons* (lattice vibrations), which bind two electrons together to form *Cooper pairs*. These pairs of electrons are correlated over large distances (100 nm to 1 μm) compared to the average distance between electrons. Consequently, the Cooper pairs interpenetrate highly. The two electrons involved in a Cooper pair have opposite spin and the quasi-particle representing the pair has no net spin. Hence, the Cooper pairs do not obey

(a)

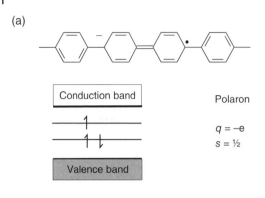

Conduction band

Polaron

$q = -e$
$s = ½$

Valence band

Figure 3.41 Schematic diagram of (a) a polaron and (b) a bipolaron in poly(*p*-phenylene) (PPP). The charge *q* and spin *s* of the defects are indicated.

(b)

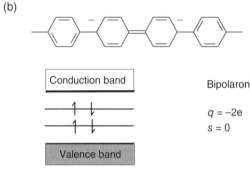

Conduction band

Bipolaron

$q = -2e$
$s = 0$

Valence band

Fermi-Dirac statistics (Section 3.2.4). They can therefore all 'condense' to the lowest energy state and possess one single wavefunction that describes the entire assembly of Cooper pairs. Since the electrons operate as a pair, an individual electron cannot gain or loose small amounts of energy from an applied field or through collisions. Cooper pairs are able to move through the lattice without any energy exchange in collisions, which of course is the origin of electrical resistance.

A superconductor below its critical temperature expels all the magnetic field from the bulk of the sample, as if it were a perfectly diamagnetic substance (Chapter 5, Section 5.7.1). This phenomenon is called the *Meissner* effect and is often used as a test for superconductivity.

Organic superconductors can be subdivided into one-, two-, and three-dimensional solids and encompass the charge-transfer salts, discussed briefly above in Section 3.4.1, and the fullerenes. Perhaps the 'Holy Grail' in superconducting materials research is to identify a compound that exhibits the superconducting state at room temperature. In 1964, Little envisaged a superconducting polymer with a conjugated backbone and dye groups attached regularly to the chain, as shown in Figure 3.42 [16]. The dye groups are characteristically polarizable. Little suggested that formation of Cooper pairs amongst the π-electrons of the chain might occur through such polarization of the side groups. Taking into account that the electronic mass is much smaller than any atomic mass, the coupling might be expected to be very large indeed, in comparison with that in conventional semiconductors. As a consequence, Little suggested a critical temperature for the transition from the superconducting state to the normal state of about 2000 K. This dramatic prediction – a room temperature organic superconductor – met with considerable criticism and has not yet been realized in practice. Organic superconductors do exist, but with transition temperatures considerably below room temperature. The highest transition temperature for a charge-transfer salt is about 13 K and about 38 K for fullerenes [17] (Chapter 5, Section 5.5.2).

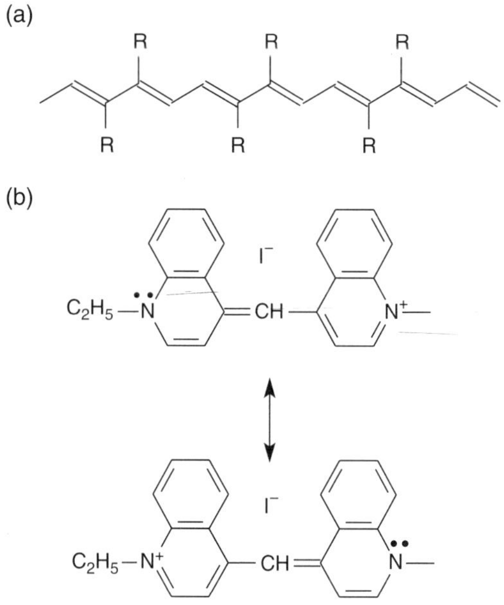

Figure 3.42 (a) Proposed structure of a room temperature polymeric superconductor by Little [16]. Side groups, designated R are attached to a polyacetylene backbone. (b) Suggestion for R substituent. These side groups are resonating hybrids of the two extreme structures depicted.

3.5 Low-Frequency Conductivity

In the following sections, the theoretical ideas that have been presented to this point are extended to include the effects that are observed in the laboratory when a voltage is applied to macroscopic samples and a resulting electrical current is produced. At a fundamental level, the presence of a DC voltage will lead to a drift in the free charges within the specimen. For those materials with no or very few such charges, the main effect of the applied field is to polarize the solid. The result is a separation of the centres of positive and negative charge and electric dipoles (Chapter 2, Section 2.3.5) are produced. For such cases, the application of a step DC voltage across the sample can produce an initial *displacement current* that can dominate over the small ionic and electronic contributions for a long time. In extreme cases, a steady current reading will never be obtained and a direct measurement of conductivity will not be possible. It can therefore be more appropriate to use AC methods to study the electrical processes in the sample; this is discussed in Section 3.6 on AC conductivity.

Some of the important low frequency electrical conductivity processes observed in organic solids are outlined in the following sections. These processes can occur simultaneously in the samples and each may dominate at different values of the applied electric field and/or over different temperature ranges.

3.5.1 Electronic Versus Ionic Conductivity

Although, most of the electrical processes described focus on electrons as the transporters of electric charges, nature favours ionic conduction. Ions possess shape as well as charge (and thereby carry more 'information') and can be exploited very effectively in natural processes involved in the transfer and storage of information.

The preparation methods of many organic solids involve the use of aqueous solvents and other substances containing ions. Ionic conduction can also be a significant factor in polymeric materials, particularly those that contain counterions to balance the charges on the polymer backbone. An ionic contribution to the measured conductivity is therefore to be expected. It is crucial to be able to separate these ionic processes from electronic conduction.

The most definitive evidence for ionic conduction is the detection of electrolysis products formed on discharge of the ions as they arrive at the electrodes. However, the very low level of conductivity in most organic materials usually precludes such detection. Even at a conductivity of $10^{-9}\,\Omega^{-1}\,m^{-1}$ (and many organic compounds possess lower ionic contributions to their conductivity), $100\,V$ applied across a sample $100\,mm^2$ in area and $1\,mm$ thick will only produce about $10^{-11}\,m^3$ of gas at standard temperature and pressure (STP) per hour [18].

Ions travelling through a sample under the influence of an applied electric field will accumulate at defects (e.g. grain boundaries) or at one of the solid electrodes. The resulting polarization will reduce the ionic current to zero over a period. Ionic conductivity is expected therefore to give rise to time-dependent currents. Furthermore, ions do not respond as readily as electrons to high-frequency fields and the ionic contribution to conductivity should be reduced at high frequencies. For organic materials based on charge-transfer complexes, the presence of a charge-transfer band in the infra-red part of the EM spectrum can be used to identify electronic conductivity.

A strong correlation between the measured conductivity of a sample and its permittivity indicates the presence of ionic conductivity [18]. This can be explained by the reduction of the Coulomb forces between ions in a high permittivity medium. For example, the absorption of water, which has a relatively high permittivity, generally enhances the conductivity of a polymer significantly and polymer-water systems frequently conform to the equation:

$$\log \sigma = -\frac{A}{\varepsilon_s} + B \tag{3.65}$$

where A, B are constants and ε_s is the low frequency (or static) value of the real part of the relative permittivity (Section 3.6.1).

Further evidence of ionic conduction may be obtained from studies of the dependence of current on applied voltage. A simple model, based on the work of Mott and Gurney, gives [18]

$$J \propto \sinh \frac{eaE}{2k_BT} \tag{3.66}$$

where a is the distance between potential wells associated with the ionic movement.

3.5.2 Quantum Mechanical Tunnelling

If the energy of an electron is less than the interfacial potential barrier at a metal-insulator interface upon which it is incident, classical physics predicts reflection of the electron at the interface. The electron cannot penetrate the barrier and its passage from one electrode to the other is inhibited. Quantum mechanics contradicts this view. The wave nature of the electron allows penetration of the forbidden region of the barrier. This is the origin of quantum mechanical tunnelling described earlier in Section 3.3.1. The wavefunction associated with the electron decays rapidly with depth of penetration from the electrode-insulator interface and, for barriers of macroscopic thickness, is essentially zero at the opposite interface, as shown in Figure 3.43a for a simple metal-insulator-metal (MIM) structure. This indicates a zero probability of finding the electron here. However, if the barrier is very thin (< 5 nm) the wavefunction has a non-zero value at the opposite interface. For this case, there is a finite probability that the electron can pass from one electrode to the other by penetrating the barrier (Figure 3.43b).

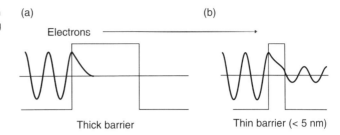

Figure 3.43 Schematic representation of the quantum mechanical tunnelling of an electron wave: (a) thick barrier; (b) thin barrier.

The current versus voltage relationships for tunnelling are quite complex and depend on the magnitude of the applied voltage and whether the tunnel barrier is symmetric or asymmetric (i.e. whether the two electrodes are similar or different metals) [19]. For very low applied voltages (much less than the energy barrier height divided by the electronic charge), the tunnelling probability varies exponentially with the barrier thickness, and the tunnelling conductivity, σ_t, may be given by an equation of the form

$$\sigma_t = A\exp(-Bd) \tag{3.67}$$

where d is the thickness of the tunnel barrier and A, B are constants. However, at higher voltages, the conductivity data for insulating films have been shown to deviate from linearity. For example, one theoretically predicted current versus voltage dependence for a symmetrical rectangular barrier is [20]

$$I = 2I_0\left(\frac{\pi CkT}{\sin(\pi CkT)}\right)\exp(-BV^2)\sinh\left(\frac{CV}{2}\right) \tag{3.68}$$

where I_0 is a constant, and B and C are coefficients related to the tunnelling barrier height.

The ability to form thin organic films with precisely defined thickness, such as Langmuir–Blodgett films (Chapter 7, Section 7.3.1), should offer an excellent basis for studying quantum mechanical tunnelling. Even so, experimental data must be treated with caution because of the presence of oxide and other layers on the metallic electrodes that will have comparable thicknesses with the organic film that is supposedly being tested [21]. However, with careful attention to experimental detail, it is possible to observe some of the theoretical predictions for tunnelling. Figure 3.44 shows the results of experiments using monolayer films of different chain length fatty acids sandwiched between metallic electrodes [22]. The dependence of the tunnelling conductivity (measured at an applied voltage of 10 mV) on film thickness is clearly of the form of Eq. (3.67). For larger applied voltages, the current versus voltage behaviour for fatty acid monolayer films is also similar to that predicted by theory [23].

3.5.3 Variable Range Hopping

Polymers contain both crystalline and amorphous regions (Chapter 2, Section 2.6). Disorder in a doped conductive polymer can also arise from the random distribution of doping ions, or the clustering of doping ions. Consequently, the electronic band structure may be more like that of an amorphous semiconductor rather than its crystalline counterpart. States in the band tails are localized and there exists a mobility edge (Section 3.3.7), which separates the region of the extended states in the interior of the band from the region of the localized states. Whether the doped polymer behaves as a metal or an insulator is determined by the relative position of the Fermi level with respect to the mobility edge.

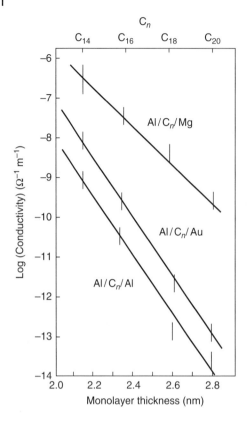

Figure 3.44 Logarithm of conductivity versus monolayer thickness for different metal-monolayer-metal structures. C_n is the number of carbon atoms in the monolayer compound. *Source*: Reprinted with permission from Polymeropoulos [22]. Copyright 1977, American Institute of Physics.

At high temperature, the conductivity will be determined by electrons (or holes) thermally activated to extended states above (below) the mobility edge (E_c for electrons and E_v for holes, Figure 3.25). In this case, the conductivity can be expressed in the form of Eq. (3.15), i.e. as for the crystalline case. However, for disordered materials, the carrier mobility μ cannot be calculated assuming the electrons are free (with their paths interrupted only by occasional scattering events), because the mean free path is of the order of an interatomic distance. It is generally assumed that the transport is diffusive (Chapter 2, Section 2.8) and the mobility is given by the *Einstein relation*:

$$\mu = \frac{eD}{k_B T} \tag{3.69}$$

where D, the diffusion coefficient, is $\nu_{el} a^2/6$. Here, ν_{el} is an electronic frequency which, from quantum theory, is roughly $\hbar/ma^2$, where a is taken to be an interatomic spacing or as the coherence length of an electron wave. Therefore,

$$\mu \approx \frac{e\hbar}{6mk_B T} \approx 6\,\mathrm{cm}^2\,\mathrm{V}^{-1}\,\mathrm{s}^{-1} \tag{3.70}$$

Carrier mobilities of around $1\,\mathrm{cm}^2\,\mathrm{V}^{-1}\,\mathrm{s}^{-1}$ are measured in organic molecular materials in polycrystalline form; the values for conductive polymers are usually much less (Table 3.1).

If the localized energy levels in a disordered material are unoccupied, these provide a possible alternative conduction path for carriers, in parallel with the band (extended state) conduction discussed above. Clearly, the electron motion in localized states is much smaller than in extended states, but at low temperatures the number of electrons activated to the conduction band can be sufficiently small to allow localized state conduction to become the predominant conduction mechanism.

Electrons can move between localized states in three different ways. First, they can be thermally activated over the potential barrier separating the two states. However, the barrier height is usually of the same order as the energy separating the localized states from the extended states. If the temperature is sufficiently high to facilitate jumping, extended state conduction will be the more favoured process. The emission at lower temperatures can be aided by the electric field and will be discussed in Section 3.5.5. Second, electrons may tunnel through the potential barrier and, third, they can move by a combination of activation and tunnelling. Since in the disordered solid, the localized states will not be degenerate, some measure of activation will generally be required to allow the full state and the empty state to resonate and hence allow tunnelling. This thermally assisted tunnelling is called *hopping* and is an important feature of conduction in non-crystalline materials.

There have been many theoretical treatments of the tunnelling process [3, 24]. It is generally assumed that the states are distributed randomly in energy and space and that the Fermi level lies within these localized states. The *variable range hopping* model of Mott [3] gives the following equation for the temperature dependence of conduction:

$$\sigma = \sigma_0 \exp\left[-\left(\frac{T_0}{T}\right)^{1/4}\right] \tag{3.71}$$

where σ_0 and T_0 are constants. The term variable range hopping comes from the fact that the average hopping distance varies with temperature and is not necessarily to the closest site. If the dimensionality of the transport is reduced, then the factor ¼ in Eq. (3.71) is replaced by ⅓ in the two-dimensional case and ½ for one-dimensional hopping.

3.5.4 Fluctuation-induced Tunnelling

There is a variety of disordered materials, such as some conductor-insulator composites, disordered semiconductors, and doped organic semiconductors, in which most of the conduction electrons are delocalized and free to move over distances that are very large compared to atomic dimensions. For these random systems, the electrical conduction is dominated by electron transfer between large conductive segments rather than by hopping between localized states. An appropriate model to describe the electrical conductivity of these materials is *fluctuation-inducted tunnelling* [25]. For this, thermally activated voltage fluctuations across insulating gaps play an important role in determining the temperature and electric field dependence of the conductivity. The relationship between conductivity and temperature approximates to

$$\sigma = \sigma_0 \exp\left(-\frac{T_1}{T + T_0}\right) \tag{3.72}$$

where σ_0, T_0, and T_1 are constants. Since T_0 is an additive constant to T, it may be considered as the temperature above which the fluctuation effects become significant. Figure 3.45 shows experimental data for the normalized resistivity of doped polyacetylene thin films [25]. The solid line is the full fluctuation-induced tunnelling model (rather than the approximate form given by Eq. (3.72)), with $T_1 = 150\,\text{K}$ and $T_0 = 13.6\,\text{K}$. The relatively good fit between experiment and theory suggests that the nonmetallic nature of the resistivity, which increases with temperature, is consistent with the intrinsic metallic nature of the doped polymer chains.

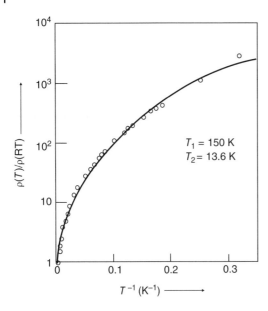

Figure 3.45 Normalized resistivity of doped polyacetylene plotted as a function of $1/T$. $\rho(RT)$ denotes the value of resistivity at room temperature. The solid line is calculated from the model for fluctuation-induced conduction [25]. *Source*: Reprinted with permission from Sheng [25]. Copyright (1980). American Physical Society.

3.5.5 Space-Charge Injection

If the electrical contacts to an insulating or semiconducting sample are Ohmic, these allow electron transfer between the electrodes and the sample and the resulting current is proportional to the applied voltage (Eq. 3.8). Under certain conditions, however, the contacts can become 'super-Ohmic' and the current is only limited by the *space-charge* between the electrodes. This conductivity regime is called *space-charge-limited* [26]. In the simplest case, the current density J varies with applied voltage V and sample thickness d as follows:

$$J \propto \frac{V^2}{d^3} \tag{3.73}$$

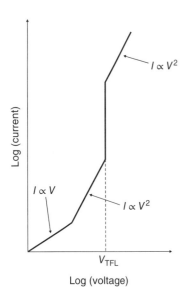

Figure 3.46 Space-charge-limited current versus voltage characteristics for a semiconductor (or insulator) containing traps. V_{TFL} is the trap-filled-limit.

Figure 3.46 shows the expected current versus voltage behaviour under these conditions. The lowest voltage region of the curve corresponds to the situation in which the injection of excess carriers in negligible. At these voltages, the volume conductivity dominates (i.e. Ohm's law). Only when the injected carrier density exceeds the volume generated carrier density, will space-charge effects be seen and the quadratic current versus voltage dependence become evident.

If the insulator or semiconductor contains traps, then the injected charge will fill these. When sufficient charge has been injected, the traps will become saturated. The voltage at which this occurs corresponds to *the trap-filled limit* and is shown as V_{TFL} in Figure 3.46. Beyond V_{TFL}, the material behaves as if it was trap-free and the quadratic current versus voltage relationship given in Eq. (3.73) is again observed.

Figure 3.47 Current versus voltage characteristics of a Au-copper phthalocyanine-Au sandwich at different temperatures. The linear dependence of current on voltage at low biases is the result of Ohmic conductivity. The super-Ohmic regions at higher biases indicate space-charge limited conductivity. The value of *m* from Eq. (3.74) is indicated for each curve. *Source*: Reprinted from Gould [27]. Copyright (1996), with permission from Elsevier.

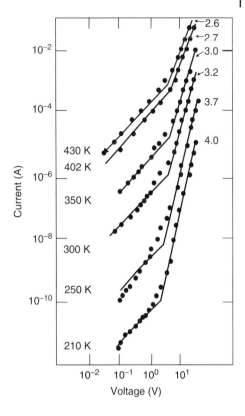

Equation (3.73) is a special case of the general scaling law for bulk for bulk space-charge currents in a homogeneous medium [24]:

$$J \propto d \left(\frac{V}{d^2} \right)^m \tag{3.74}$$

where *m* is a constant, which need not necessarily be an integer. For example, in the trap free insulator case above, $m = 2$. In the case of 'double' injection, i.e. electrons injected at one electrode and holes at the other, $m = 3$, and for recombinative space-charge injection, $m = \frac{1}{2}$. Figure 3.47 shows current versus voltage data, in the form of a $\log(I)$ versus $\log(V)$ plot, a sample of a thin film copper phthalocyanine, a molecular crystalline compound (Chapter 5, Section 5.4), sandwiched between gold electrodes [27]. The transitions from Ohmic conductivity, where $m = 1$ to regions for which $m > 1$ are clearly evident. The values of *m* in the high voltage regions are shown on the figure. The dependence of *m* on the temperature is an indication that the organic solid contains an exponential distribution of trapping centres within the energy band gap.

3.5.6 Schottky and Poole-Frenkel Effects

When an insulator (or semiconductor) is placed in contact with a metal, a redistribution of charge will occur in the interface region (to minimize the free energy of the system). This will produce a distortion or bending of the energy bands. Abrupt changes in the potential energy do not occur, as these would imply infinite electric fields. The potential step changes smoothly

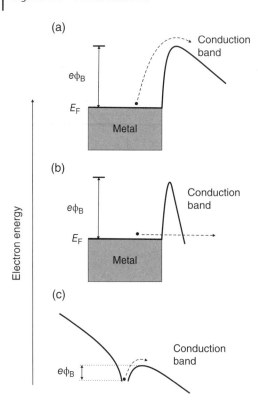

Figure 3.48 (a) Schottky emission of an electron from the Fermi level E_F in a metal into the conduction band of a semiconductor (or insulator); (b) Fowler-Nordheim tunnelling; (c) Poole-Frenkel effect–the applied electric field lowers the barrier surrounding a trapped carrier and helps it to move into the conduction band.

because of the *image force* effect. This arises because of the metal's surface becoming positively charged by an escaping electron. The resulting potential barrier at the metal-insulator interface is called a *Schottky barrier* (Figure 3.48a). The electrons in the metal are filled to the Fermi level E_F and they must gain enough energy to overcome the barrier, ϕ_B, to get into the conduction band of the semiconductor. This process is called *Schottky emission* and has a current density versus electric field dependence [19]

$$J \propto \exp\left(\frac{\beta E^{0.5}}{k_B T}\right) \tag{3.75}$$

where β is a constant.

If the applied electric field is large enough, the barrier to electrons at the interface becomes very thin and electrons can tunnel directly from the metal to the conduction band of the insulator. This conductivity mechanism is termed *field emission* and is illustrated in Figure 3.48b. The resulting current density versus voltage behaviour is [19]

$$J \propto E^2 \exp\left(-\frac{\gamma}{E}\right) \tag{3.76}$$

where γ is another constant. The process is also referred to as *Fowler-Nordheim tunnelling*. This process can also be thermally assisted, the electrons being excited to an energy state where they are then able to tunnel through the barrier.

A similar process to Schottky emission can take place at impurity centres in the bulk of the material. This is the *Poole-Frenkel* effect and is illustrated in Figure 3.48c. The application of a high electric field will result in the lowering of the potential barrier around the centre, allowing a carrier to escape into the conduction band of the insulator. The current versus voltage behaviour for Poole-Frenkel conduction has the form [19]

$$J \propto E \exp\left(\frac{\delta E^{0.5}}{k_B T}\right) \tag{3.77}$$

where δ is a constant. This equation is similar to Eq. (3.75) for Schottky emission (although the β and δ coefficients differ) and experimentally it can be difficult to distinguish between the two processes. However, Poole-Frenkel conduction is essentially a bulk effect and should show little dependence upon electrodes or upon the polarity of the field. Poole-Frenkel and Schottky emission processes are usually observed at high applied electric fields ($> 10^7 \, \text{V m}^{-1}$).

3.6 Conductivity at High Frequencies

Direct current and high frequency phenomena are essentially separate and, for the most part, independent quantities. However, at low frequencies, both can contribute to a measured current. The frequency-dependent conductivity $\sigma(\omega)$ of a sample may be expressed as

$$\sigma(\omega) = \sigma_{DC} + \sigma_{AC} \tag{3.78}$$

where σ_{DC} is the DC conductivity discussed in the previous sections and σ_{AC} is the AC component.

The AC conductivity of a sample is generally measured by applying a voltage at an appropriate frequency and then measuring both the in-phase and 90° out-of-phase components of the current. The equipment required depends on the frequency of measurement: simple bridge circuits can be used over the frequency range $1-10^7$ Hz; for lower frequencies, step-response methods are used (e.g. a voltage step is applied and the subsequent current monitored over time).

3.6.1 Complex Permittivity

The AC conductivity is related to the *permittivity* of the material. The *relative permittivity* ε_r of a material is a complex quantity, i.e.

$$\varepsilon_r = \varepsilon_r' - j\varepsilon_r'' \tag{3.79}$$

where ε_r' and ε_r'' are the real and imaginary parts, respectively, both frequency-dependent. Often, the real part of the relative permittivity, ε_r', is called the *dielectric constant*. Application of an electric field to a material will create electric dipoles (Chapter 2, Section 2.3.5). Some materials already contain dipoles by virtue of their chemical bonding; these are *polar* compounds. In this case, the application of the electric field tries to align the dipoles in the field direction. This is opposed by thermal motion. The real part of the permittivity is a measure of the ability of an electric field to polarize the medium, while the imaginary part represents losses in the material as an AC field attempts to orient the electric dipoles within the material.

The polarization mechanism described above, i.e. the reorientation of dipoles in an electric field, is called *orientational* (or *dipolar*) *polarization*. In general, a dielectric medium will also exhibit other polarization mechanisms on application of a field. These are *electronic polarization*, the displacement of the electron clouds around the nucleus (described more fully in Chapter 4,

Section 4.3), *ionic polarization*, due to the displacement of positive and negative ions, and *interfacial polarization*, resulting from the accumulation of charges at interfaces within the material. Each process will produce a change in ε_r' and a peak in ε_r'' over a particular frequency range. For example, in ionic solids this, corresponds to the infrared part of the EM spectrum, where the energy stored in the induced dipoles is the same as the lattice vibrational frequency. At higher frequencies, the positive and negative ions in the solid will not be able to respond sufficiently fast to the applied AC electric field and the ionic contribution to the material's permittivity is lost.

Figure 3.49 shows the frequency dependence of the real and imaginary parts of the permittivity. Although the figure shows distinct peaks in ε_r'' and transitions in ε_r', in real materials, these features can be broader and overlap with one another. For single crystals, the polarization features also depend on the orientation of the electric field with respect to the crystallographic axes. At low frequencies, the interfacial (or space-charge) features are very broad, because there can be a number of different conduction mechanisms contributing to the charge accumulation. At high frequencies – usually microwave and beyond – the polarization processes that take place are undamped and are invariably observed as resonances.

The real part of permittivity is the value that is used when calculating the capacitance of a parallel plate capacitor, e.g. for a parallel plate capacitor

$$C = \frac{\varepsilon_r \varepsilon_0 A}{d} \tag{3.80}$$

where ε_0 is a constant called the *permittivity of free space* ($= 8.85 \times 10^{-12}\,\mathrm{F\,m^{-1}}$) and A, d are the plate area and dielectric thickness, respectively, of the capacitor. Application of an AC voltage of RMS amplitude V to this capacitor will result in an AC current, I, given by

$$I = YV = \frac{V}{Z} = j\omega CV \tag{3.81}$$

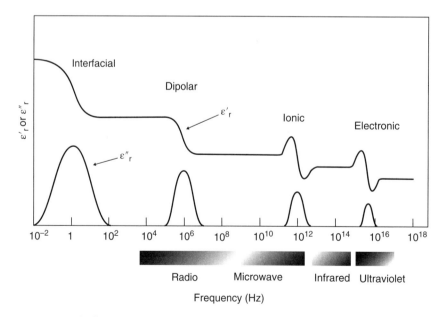

Figure 3.49 The frequency dependence of the real, ε_r', and imaginary, ε_r'', parts of the permittivity in the presence of interfacial, orientational, ionic, and electronic polarization mechanisms.

where Y is called the electrical *admittance* and Z is the electrical *impedance*. The presence of j in the above equation indicates a 90° phase shift between the applied voltage and the current. Substituting Eq. (3.80) into Eq. (3.81) and replacing with the complex quantity of Eq. (3.79) then gives

$$I = V\left(j\omega C + G\right) \tag{3.82}$$

where

$$C = \frac{A\varepsilon_0\varepsilon_r'}{d} \text{ and } G = \frac{\omega A\varepsilon_0\varepsilon_r''}{d} \tag{3.83}$$

Therefore, the admittance of the dielectric medium is a parallel combination of an ideal, lossless capacitor C, with a relative permittivity ε_r' and a conductance G (or resistance $R = 1/G$) proportional to ε_r''. This is illustrated in Figure 3.50

From the above, it is expected that the measured AC conductivity, σ_{AC}, will increase with frequency:

$$\sigma_{AC} = \omega\varepsilon_0\varepsilon_r'' \tag{3.84}$$

If the sample is lossless over the frequency range, then ε_r'' will be constant and the AC conductivity will increase linearly with frequency. However, many inorganic and organic solids exhibit a simple power-law relationship of the form

$$\sigma_{AC} \propto \omega^n \tag{3.85}$$

where n is less than unity (usually in the range $0.7 < n < 1$). This is often referred to as the *universal law* for the response of dielectrics [28]. Many theories have been developed to relate this observed AC electrical behaviour to the microscopic properties of the solid. Such models are often based on the hopping of charge carriers.

Figure 3.51 shows the AC conductivity data for a thin film of an anthracene derivative [29]. The low frequency behaviour $\sigma \propto \omega^{0.85}$ agrees with that expected from the discussion above. However, at frequencies above 10^4 Hz, the power dependence changes to approximately 1.8. The explanation is that at these high frequencies contact resistances associated with the

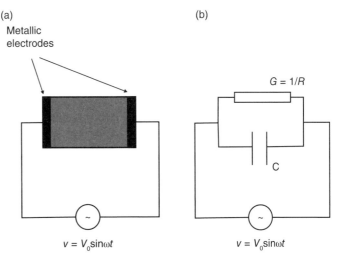

Figure 3.50 (a) A dielectric stimulated by an AC voltage. (b) Equivalent electrical circuit.

(a) Metallic electrodes

(b)

$G = 1/R$

C

$v = V_0\sin\omega t$

$v = V_0\sin\omega t$

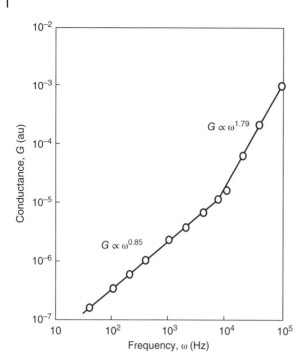

Figure 3.51 AC conductivity data showing the variation in conductance G with frequency ω for a thin film of an anthracene derivative. *Source*: Reprinted from Roberts et al. [29]. Copyright (1980), with permission from Elsevier.

metallic electrodes become important. A full analysis of the equivalent circuit of the sample, including these additional resistances, reveals the $\sigma \propto \omega^{2n}$ relationship. In the case $n = 1$, a square law dependence of conductance on frequency is seen at high frequencies.

3.6.2 Impedance Spectroscopy

The measurement of the electrical impedance of a sample as a function of frequency is often referred to as *impedance spectroscopy* (or *admittance spectroscopy* in the case of a measurement of Y with frequency). To investigate the equivalent electrical circuit of the material under investigation, the measured data are plotted in the form of the real part of the impedance Z' versus the imaginary part Z'', a so-called *Cole-Cole diagram*. Semicircles are revealed in the complex impedance plane. These result from the classical *Debye equations*, which relate the real and imaginary parts of the permittivity to frequency:

$$\varepsilon_r' = \varepsilon_\infty + \frac{\varepsilon_s - \varepsilon_\infty}{1 + \omega^2 \tau^2} \text{ and } \varepsilon_r'' = \frac{(\varepsilon_s - \varepsilon_\infty)\omega\tau}{1 + \omega^2 \tau^2} \tag{3.86}$$

where ε_s and ε_∞ represent the low and high frequency values of the real parts of the relative permittivity, respectively, and τ is a characteristic time constant of the system.

Figures 3.52a and b show two simple resistor-capacitor circuits and their resulting impedance spectra. The arrows on the figures indicate the direction of increasing frequency. The circuit with two resistor-capacitor combinations (Figure 3.52b) will possess two distinct time constants, R_1C_1 and R_2C_2, and shows distinct two semicircles in the impedance plot (Figure 3.52d), if these time constants are well-separated, i.e. $R_1C_1 >> R_2C_2$. The high and low frequency limits can be used to determine some of the circuit components.

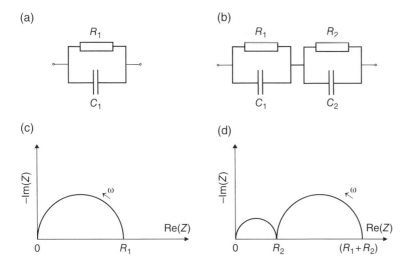

Figure 3.52 Diagrams (a) and (b) represent two simple resistor-capacitor circuits, while (c) and (d) are plots in the complex impedance plane of the imaginary part of impedance, Im(Z), versus the real part, Re(Z), for circuits (a) and (b), respectively.

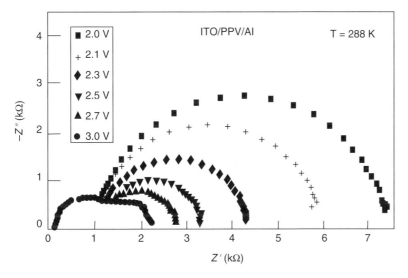

Figure 3.53 Imaginary (Z'') versus real (Z') parts of the complex impedance for an ITO-PPV-Al sandwich structure for various values of the forward bias. The frequency increases around the semicircles from right to left. The left semicircles are assigned to the polymer bulk and the right semicircles to a Schottky barrier. *Source*: Reprinted with permission from Scherbel et al. [30]. Copyright (1998), American Institute of Physics.

Experimental data are shown in Figure 3.53 for a layer of the conductive polymer poly (p-phenylenevinylene) or PPV between aluminium and indium-tin-oxide (ITO) electrodes [30]. This type of structure forms an organic light emitting device (Chapter 9, Section 9.6) and will emit visible light when a negative voltage, or forward bias, is applied to the Al contact. The impedance spectra are shown for different values of forward bias voltages. Two semicircles are evident, suggesting that the equivalent circuit of the Al-PPV-ITO device is more complex than that of a lossy parallel plate capacitor, i.e. a parallel combination of a resistor and capacitor, as depicted in Figure 3.52a. The small semicircle close to the origin in Figure 3.53 (corresponding to the high frequency region) is independent of voltage, whereas the larger right-hand semicircle (low frequency) shrinks as the applied bias is increased. The interpretation is that the device comprises of two distinct parts, a voltage-independent bulk region accounting for the left semicircle in the figure and a Schottky barrier (Section 3.5.5) at the Al-PPV interface. The resistance of the latter will depend on the applied voltage, accounting for the behaviour of the right semicircle in Figure 3.53. The equivalent circuit of the Al-PPV-ITO sandwich structure can therefore be modelled by the circuit shown in Figure 3.52b.

Conductivity measurements over a wide frequency range provide a very useful insight into the electrical networks associated with these and other devices. However, the interpretations are not all straightforward. In some cases, impedance spectroscopy reveals arcs of semicircles rather than full semicircles. This can be attributed to a distribution of relaxation times for the material under study.

Problems

3.1 Compare the value of the average thermal velocity with the drift velocity of an 'electron gas' at 300 K. The applied electric field is $10\,V\,m^{-1}$ and the relaxation time is 10^{-14}s.

3.2 A current of $1\,\mu A$ flows through a cylindrical wire conductor of 1 mm diameter and length 1 m. The insulation is $100\,\mu m$ of plastic with a resistivity of $10^{16}\,\Omega\,m$. What fraction of the current is leaked if the conductor is at 100 V potential above the outer environment?

3.3 Polymers, in common with all materials, can exhibit polarization due to displacement of their electrons. Additionally, many polymer molecules contain polar groups such as

$$\overset{+}{C} = \overline{O} \text{ or } \overset{+}{C} - \overline{\overline{Cl}}$$

Displacement of these by applied fields gives further scope for polarization.

How would these two sorts of polarization be affected by:
i) the closeness of the polymer's dielectric glass transition to its operating temperature; and
ii) the frequency of the applied field?

3.4 In a sample of a conductive polymer, the doping is such that there are 10^{18} holes m^{-3} and 5.0×10^{10} electrons m^{-3}. What fraction of the conduction current is carried by the minority species (electrons)? Mobilities μ_h and μ_e are 1.0×10^{-6} and $8.0 \times 10^{-9}\,m^2\,V^{-1}s^{-1}$, respectively.

3.5 Measurements of the conductivity of an intrinsic semiconductor reveal that the conductivity varies with temperature as $\exp(-6493/T)$. Calculate the energy gap of the semiconductor in eV. State any assumptions that you make.

References

1 Blakemore, J.S. (1969). *Solid State Physics*. Philadelphia: Saunders.
2 Kasap, S.O. (2008). *Principles of Electrical Engineering Materials and Devices*, 3e. Boston: McGraw-Hill.

3 Mott, N.F. and Davis, E.A. (1979). *Electronic Properties in Non-crystalline Materials*, 2e. Oxford: Clarendon Press.

4 Karl, N. (2001). Low molecular weight organic solids. Introduction. In: *Organic Electronic Materials* (ed. R. Farchioni and G. Grosso), 215–239. Berlin: Springer.

5 Zheng, G., Clark, S.J., Brand, S., and Abram, R.A. (2004). First-principles studies of the structural and electronic properties of poly-*para*-phenylene vinylene. *J. Phys. Condens. Matter* 16: 8609–8620.

6 Facchetti, A. (2008). Molecular semiconductors for organic field-effect transistors. In: *Introduction to Organic and Optoelectronic Materials and Devices* (ed. S.-S. Sun and L.R. Dalton), 287–318. Boca Raton: CRC Press.

7 Dimitrakopoulos, C.D. and Mascaro, D.J. (2001). Organic thin-film transistors: a review of recent advances. *IBM J. Res. Dev.* 45: 11–27.

8 Pearson, C., Moore, A.J., Gibson, J.E. et al. (1994). A field effect transistor based on Langmuir-Blodgett films of an $Ni(dmit)_2$ charge transfer complex. *Thin Solid Films* 244: 932–935.

9 Abe, M., Mori, T., Osaka, I. et al. (2015). Thermally, operationally, and environmentally stable organic thin-film transistors based on bis[1]benzothieno[2,3-*d*:2′,3′-*d*′]naphtho[2,3-*b*:6,7-*b*′] dithiophene derivatives: effective synthesis, electronic structures, and structure-property relationship. *Chem. Mater.* 27: 5049–5057.

10 Novoselov, K.S., Fal'ko, V.I., Colombo, L. et al. (2012). A roadmap for graphene. *Nature* 490: 192–200.

11 McEuen, P.L. and Park, J.Y. (2004). Electron transport in single-walled carbon nanotubes. *MRS Bull.* 29: 272–275.

12 Heeger, A.J. (2002). Semiconducting and metallic polymers: the fourth generation of polymeric materials. *Synth. Met.* 125: 23–42.

13 MacDiarmid, A.G. (2002). Synthetic metals: a novel role for organic polymers. *Synth. Met.* 125: 11–22.

14 Pfeiffer M, Leo K, Zhou X, Huang JS, Hofmann M, et al. (2003). Doped organic semiconductors: physics and application in light emitting diodes. *Org. Electron.* 4: 89–103.

15 Roth, S. (1995). *One-dimensional Metals*. Weinheim: VCH.

16 Little, W.A. (1964). Possibility of synthesizing an organic superconductor. *Phys. Rev A.* 134: 1416–1424.

17 Saito, G. and Yoshida, Y. (2011). Organic Superconductors. *The Chemical Record* 11: 124–145.

18 Blythe, T. and Bloor, D. (2005). *Electrical Properties of Polymers*. Cambridge: Cambridge University Press.

19 Simmons, J.G. (1971). *DC Conduction in Thin Films*. London: Mills & Boon.

20 Stratton, R. (1962). Volt-current characteristics for tunnelling through insulating films. *J. Phys. Chem. Solids.* 23: 1177–1190.

21 Petty, M.C. (1996). *Langmuir-Blodgett Films*. Cambridge: Cambridge University Press.

22 Polymeropoulos, E.E. (1977). Electron tunnelling through fatty-acid monolayers. *J. Appl. Phys.* 48: 2404–2407.

23 Petty, M.C. (1992). Characterization and properties. In: *Langmuir-Blodgett Films* (ed. G.G. Roberts), 133–221. New York: Plenum Press.

24 Roberts, G.G., Apsley, N., and Munn, R.W. (1980). Temperature dependent electronic conduction in polymers. *Phys. Rep.* 60: 59–150.

25 Sheng, P. (1980). Fluctuation-induced tunnelling conduction in disordered materials. *Phys. Rev. B.* 21: 2180–2195.

26 Tredgold, R.H. (1966). *Space Charge Conduction in Solids*. Amsterdam: Elsevier.

27 Gould, R.D. (1966). Structure and electrical conduction properties of phthalocyanine thin films. *Coord. Chem. Rev.* 156: 237–274.

28 Jonscher, A.K. (1983). *Dielectric Relaxation in Solids*. London: Chelsea Dielectric Press.

29 Roberts, G.G., McGinnity, T.M., Barlow, W.A., and Vincett, P.S. (1980). AC and DC conduction in lightly substituted anthracene Langmuir films. *Thin Solid Films* 68: 223–232.

30 Scherbel, J., Nguyen, P.H., Paasch, G. et al. (1998). Temperature dependent broadband impedance spectroscopy on poly-(p-phenylene-vinylene) light-emitting diodes. *J. Appl. Phys.* 83: 5045–5055.

Further Reading

Barford, W. (2005). *Electronic and Optical Properties of Conjugated Polymers*. Oxford: Oxford University Press.

Bredas, J.-L. and Marder, S.R. (eds.) (2016). *The WSPC Reference on Organic Electronics: Basic Concepts*, vol. 1. World Scientific.

Cuevas, J.C. and Scheer, E. (2017). *Molecular Electronics: An Introduction to Theory and Experiment*, 2e. Singapore: World Scientific.

Cuniberti, G., Fagas, G., and Richer, K. (eds.) (2005). *Introducing Molecular Electronics*. Berlin: Springer.

Dressel, M. and Grüner, G. (2002). *Electrodynamics of Solids: Optical Properties of Electrons in Matter*. New York: Cambridge University Press.

Farchioni, R. and Grosso, G. (eds.) (2001). *Organic Electronic Materials*. Berlin: Springer.

Ferraro, J.R. and Williams, J.M. (1987). *Introduction to Synthetic Electrical Conductors*. Orlando: Academic Press.

Harrop, P.J. (1972). *Dielectrics*. London: Butterworth.

Hummel, R.E. (1992). *Electronic Properties of Materials*, 2e. Berlin: Springer-Verlag.

Macdonald, J.R. (ed.) (1988). *Impedance Spectroscopy: Emphasizing Solid Materials and Systems*. New York: Wiley.

Nabook, A. (2005). *Organic and Inorganic Nanostructures*. Boston: Artech House.

Rudden, M.N. and Wilson, J. (1993). *Elements of Solid State Physics*. Chichester: Wiley.

Salaneck, W.R., Lundström, I., and Rånby, B. (eds.) (1993). *Conjugated Polymers and Related Materials*. Oxford: Oxford University Press.

Stallinger, P. (2009). *Electrical Characterization of Organic Electronic Materials and Devices*. Chichester: Wiley.

Stubb, H., Punkka, E., and Paloheimo, J. (1993). Electronic and optical properties of conducting polymer thin films. *Mater. Sci. Eng.* 10: 85–140.

Sun, S.-S. and Dalton, L.R. (2008). *Introduction to Organic and Optoelectronic Materials and Devices*. Boca Raton: CRC Press.

Wright, J.D. (1995). *Molecular Crystals*. Cambridge: Cambridge University Press.

4

Optical Phenomena

What light through yonder window breaks?

4.1 Introduction

The response of materials to high frequency electromagnetic radiation is generally very different to their response to lower frequency (e.g. DC) electric fields. The behaviour is conveniently described in terms of a refractive index, rather than a conductivity or permittivity. In this chapter, the properties of electromagnetic radiation, introduced in Chapter 3, Section 3.3.1, will be further discussed. The linear and nonlinear interactions of EM radiation with organic materials and, in

Organic and Molecular Electronics: From Principles to Practice, Second Edition. Michael C. Petty.
© 2019 John Wiley & Sons Ltd. Published 2019 by John Wiley & Sons Ltd.
Companion website: www.wiley.com/go/petty/molecular-electronics2

particular, the effects that occur at the various interfaces, which may be present in organic and molecular electronic devices, are then examined. Selected specialist topics – waveguiding, surface plasmon resonance (SPR), and photonic crystals – are introduced towards the end of this chapter.

4.2   Electromagnetic Radiation

In free space, the electric and magnetic components of an EM wave are at right angles to each other and also to the direction of propagation. *Plane polarized* EM radiation is radiation in which the electric (and magnetic) field is confined to a plane. The plane of polarization is conventionally taken to be the plane containing the direction of the electric field. Unpolarized radiation, or radiation of an arbitrary polarization, can always be resolved into two orthogonally polarized waves. If the two electric field components possess a constant phase difference and equal amplitudes, the resultant EM wave is said to be *circularly polarized.* If the amplitudes differ, then the wave is *elliptically polarized.* (Plane and circular polarizations are special cases of elliptical polarization.) Plane polarized radiation can be produced from unpolarized radiation using specials sheets of material – usually in the form of aligned long-chain molecules – that absorb one polarized direction almost completely.

In many experiments, plane polarized EM radiation is used to study organic solids. These are often in the form of thin films deposited onto planar substrates (Chapter 6, Section 6.6). Two important measuring arrangements may be distinguished: p-*polarized* or *transverse-magnetic* (TM) incident radiation, in which the electric field vector is in the plane of incidence of the EM wave; and s-*polarized* or *transverse-electric* (TE) incident radiation, where the electric vector is perpendicular to the plane of incidence. These are contrasted in Figure 4.1

(a)

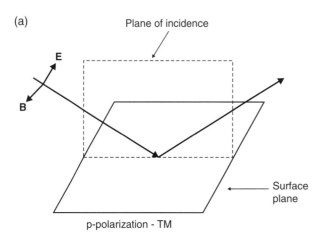

p-polarization - TM

Figure 4.1 Diagrams showing s- and p-polarized EM waves incident on a surface. (a) For p-polarized (TM–transverse magnetic) waves, the electric field vector is in the plane of incidence. (b) For s-polarized (TE–transverse electric) waves, the electric field is perpendicular to the plane of incidence of the EM wave.

(b)

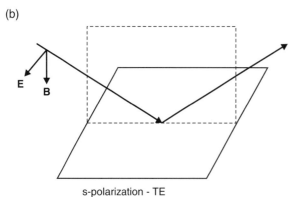

s-polarization - TE

4.3 Refractive Index

The speed of light in a vacuum is $3.00 \times 10^8 \, \text{m s}^{-1}$. However, when passing through a material, the light appears to slow down. The origin of this effect is the interaction of the EM wave with the atoms and molecules that make up the medium, and is described in terms of the *refractive index*.

A static electric field can induce an electric dipole (Chapter 2, Section 2.3.5) in a solid material; this is depicted in Figure 4.2, which illustrates the application of an electric field to a neutral atom. Before the electric field is applied, the centres of positive and negative charge coincide (Figure 4.2a). The field then produces a very small shift (much exaggerated in Figure 4.2b) in the electron cloud with respect to the nucleus. This results in the formation of an induced dipole. This is the origin of the electronic polarization, described in Chapter 3, Section 3.6.1 (although the discussion concerned AC fields). The induced dipole can be modelled as two equal and opposite charges ($\pm\Delta q$) at a distance apart of d. The *dipole moment*, **p**, can then be defined:

$$\mathbf{p} = \mathbf{d}\Delta q \tag{4.1}$$

where the use of the bold typeface indicates that dipole moment is a vector quantity, directed from negative to positive charge. The distance d (a scalar quantity) is therefore replaced by the vector **d** from negative to positive charge in Eq. (4.1). As noted in Chapter 3 (Section 3.6.1), the application of an electric field to a polar compound (i.e. one that already contains dipoles by virtue of the chemical bonding) will cause the dipoles to attempt to orient in the field direction. This *ground-state* dipole results from the electronegativities of the constituent atoms. A simple example is the carbonyl group (C=O), which has a small positive charge on the carbon atom and a small negative charge on the oxygen (the latter is more electronegative than the former).

The oscillating electric field of an EM wave produces oscillating dipoles in any matter with which it is able to interact. The electron cloud in an atom is relatively light with respect to the nucleus and can respond to EM fields up to very high frequencies (visible to ultraviolet light). The oscillating dipoles, in turn, produce their own oscillating electric and magnetic fields which radiate EM waves–the dipoles acting as 'molecular aerials'. These combine with the incident wave to produce an electromagnetic wave that travels through the material with a velocity slower than the velocity of light in vacuum. If v is the phase velocity of light in the material and c is the velocity of light in vacuum, then

Figure 4.2 Formation of a dipole moment. (a) The centre of the negative electron cloud in a neutral atom coincides with the nucleus. (b) Application of an electric field, **E**, results in a shift in the centre of positive and negative charge and creates an electric dipole, which can be represented by an equal positive and negative charge Δq at a distance d apart.

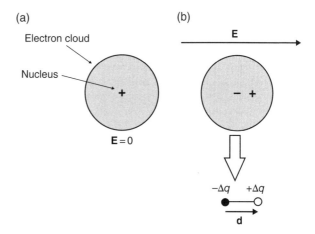

$$n = \frac{c}{v} \tag{4.2}$$

where n is the refractive index. Typical values are 1.0003 for air, 1.33 for water, and about 1.5 for glass. Refractive index may also be linked to the relative permittivity ε_r (Chapter 3, Section 3.6.1) and relative *permeability* μ_r (Chapter 5, Section 5.7.1) of the medium:

$$n = \sqrt{\mu_r \varepsilon_r} \tag{4.3}$$

For nonmagnetic materials $\mu_r = 1$ and $n = \sqrt{\varepsilon_r}$. (It will be shown in the next section that both the permittivity and permeability are tensor quantities.)

Like the permittivity, the refractive index of a material is also a complex quantity:

$$n = n' - jn'' \tag{4.4}$$

where the real part n' controls effects such as reflection, refraction (by *Snell's law*) and the velocity of propagation, and the imaginary part n'' represents loss (i.e. absorption in the material). Because light of various wavelengths produces different effects on the charges present in matter, the velocity at which light propagates through a material depends on the wavelength of the incident light. This means that the refractive index also depends on wavelength (the frequency dependence of ε_r has already been noted in Chapter 3, Section 3.6.1).

4.3.1 Permittivity Tensor

The response of a material to the electric and magnetic components of an EM wave is determined by its permittivity and permeability. For most of the organic compounds discussed in this book, the relative permeability μ_r is equal to unity (magnetic organic compounds are discussed in Chapter 5, Section 5.7), so the interaction with EM waves is due solely to the dielectric properties.

The electric field, **E**, induces an *electric displacement*, **D**, in a medium such that

$$\mathbf{D} = \hat{\varepsilon}_r \varepsilon_0 \mathbf{E} \tag{4.5}$$

The electric field and resulting displacement are both vectors quantities, each with three components along three mutually orthogonal directions. This makes the permittivity a *second rank tensor* (formally written as $\hat{\varepsilon}_r$) and its components may be represented by a 3×3 matrix array.

For any non-optically active dielectric, it is possible to choose the x, y, and z axes so that the off-diagonal elements in the 3×3 permittivity tensor array are zero. This process is called *diagonalizing* the matrix and the resulting directions are called the *principal axes*. In matrix terms, Eq. (4.5) can then be written as

$$\begin{pmatrix} D_x \\ D_y \\ D_z \end{pmatrix} = \varepsilon_0 \begin{pmatrix} \varepsilon_{11} & 0 & 0 \\ 0 & \varepsilon_{22} & 0 \\ 0 & 0 & \varepsilon_{33} \end{pmatrix} \begin{pmatrix} E_x \\ E_y \\ E_z \end{pmatrix} \tag{4.6}$$

i.e.

$$\begin{aligned} D_x &= \varepsilon_0 \varepsilon_{11} E_x \\ D_y &= \varepsilon_0 \varepsilon_{22} E_y \\ D_z &= \varepsilon_0 \varepsilon_{33} E_z \end{aligned} \tag{4.7}$$

In an isotropic (Chapter 2, Section 2.5.1) medium the induced polarization is independent of the electric field direction, so that $\varepsilon_{11} = \varepsilon_{22} = \varepsilon_{33}$. Therefore, all propagation directions experience the same refractive index.

For an anisotropic medium, the situation changes. In general, there are two possible values of the phase velocity for a given direction of propagation. These are associated with mutually orthogonal polarization of the light waves. The two polarizations define the *ordinary* and *extraordinary* rays and possess distinct refractive indices, and therefore different angles of refraction at an interface. Hence, when light of an arbitrary polarization propagates through an anisotropic medium, it can be considered to consist of two independent waves, which travel with different velocities. Media exhibiting such effects are said to be *birefringent*. Two levels of anisotropy may be distinguished.

1) uniaxial medium: $\varepsilon_{11} = \varepsilon_{22} \neq \varepsilon_{33}$
   For most ray directions and polarizations, both ordinary (refractive index = $\sqrt{\varepsilon_{11}} = \sqrt{\varepsilon_{22}}$) and extraordinary rays (refractive index = $\sqrt{\varepsilon_{33}}$) are generated. The exception is for light travelling in the z direction (the optic axis), where all polarizations are governed by the ordinary refractive index.
2) biaxial medium: $\varepsilon_{11} \neq \varepsilon_{22} \neq \varepsilon_{33}$
   Here, there are two optic axes of transmission along which the velocities of the two orthogonally polarized waves are the same. Otherwise, ordinary and extraordinary rays are produced.

Many liquid crystalline molecules (Chapter 8, Section 8.2) and organic molecules that are used for self-assembly (Chapter 7, Section 7.3) are rod-like. The films will possess a biaxial symmetry. However, $\varepsilon_{11} \approx \varepsilon_{22}$ and the situation will approximate to case 1, with two refractive indices – along the rod and perpendicular to it.

4.3.2 Linear and Nonlinear Optics

A second definition of electric displacement, **D**, to that provided by Eq. (4.5) is

$$\mathbf{D} = \varepsilon_0 \mathbf{E} + \mathbf{P} \tag{4.8}$$

where **P** is the *polarization* induced in the material by the electric field. Polarization is the induced charge per unit area or the dipole moment per unit volume; the units are $C\,m^{-2}$.

In *linear optics*, the electric field is linearly related to the polarization, i.e.

$$\mathbf{P} = \varepsilon_0 \hat{\chi}_e \mathbf{E} \tag{4.9}$$

where $\hat{\chi}_e$ ($= \hat{\varepsilon}_r - 1$) is the second rank **susceptibility** tensor. For *nonlinear optics* (NLO), the above equation is replaced by a series expansion in the electric field:

$$\frac{\mathbf{P}}{\varepsilon_0} = \hat{\chi}_e^{(1)}\mathbf{E} + \hat{\chi}_e^{(2)}\mathbf{E}\cdot\mathbf{E} + \hat{\chi}_e^{(3)}\mathbf{E}\cdot\mathbf{E}\cdot\mathbf{E} + \ldots\ldots \tag{4.10}$$

where $\hat{\chi}_e^{(2)}$ is called the second-order susceptibility tensor (SI units = $C\,V^{-2}$) and $\hat{\chi}_e^{(3)}$ the third-order susceptibility tensor. The second-order term is responsible for effects such as *second-harmonic generation* and the *linear electro-optic effect* (*Pockels' effect*), while the third-order term is responsible for *third-harmonic generation* and the *quadratic electro-optic effect* (*Kerr effect*). An important consequence of crystal symmetry is that in a material possessing a *centre of inversion* (a centrosymmetric material) the second-order nonlinear susceptibility $\hat{\chi}_e^{(2)}$ is zero.

Nonlinear properties are normally measured on macroscopic samples that consist of many individual molecules. On the microscopic level, a molecule placed in an electric field experiences a polarizing effect through a change in its dipole moment, **p**. Its relation to electric field may also be expressed by a power series:

$$\mathbf{p} = \hat{\alpha}\mathbf{E} + \hat{\beta}\mathbf{E}\cdot\mathbf{E} + \hat{\gamma}\mathbf{E}\cdot\mathbf{E}\cdot\mathbf{E} + \dots \tag{4.11}$$

where $\hat{\alpha}$ is the linear *polarizability* and $\hat{\beta}$, $\hat{\gamma}$, etc. (all tensor quantities) are the higher-order polarizabilities, or *hyperpolarizabilities*.

The SI units for the β coefficients are $[\mathrm{C\,m^3\,V^{-2}}]$. However, workers in the NLO often quote values using the CGS (centimetre gram second) system (in common with many workers on magnetic materials – Chapter 5, Section 5.7.1), which can be obtained by multiplying the SI values by 2.7×10^{20}. Similarly, the second order nonlinear susceptibility coefficients $\chi_e^{(2)}$ measured in the SI units of $[\mathrm{C\,V^{-2}}]$ can be converted to CGS units by multiplying by 2.7×10^{14}.

Centrosymmetric molecules will respond to the electric field of the optical wave to give equal and opposite polarization as the phase of the wave changes through 180° and therefore will have zero β coefficients. Large β coefficients arise through the presence of an asymmetric mobile π-electron system and low lying charge-transfer states in molecules. A substantial proportion of NLO materials is represented by organic donor and acceptor moieties linked via aromatic, alkenyl, or alkynyl spacer groups. A generalized model of a typical second-order NLO molecule is shown in Figure 4.3.

The donor unit, which should be electron-rich, normally pushes electrons towards the electron-deficient acceptor via a π-resonance effect. Thus, the donor moiety is usually loaded with heteroatoms bearing lone pairs of electrons (preferred to alkyl donating groups, which are weaker electron contributors). Typical examples of strong electron-donating groups include alkoxy (RO—), alkylthio (RS—), and amino (R_2N—) subunits. An efficient acceptor is one that is electron-deficient and can easily accommodate a negative charge. Common electron-withdrawing groups include nitro (—NO_2), cyano (—CN), and carbonyl (—C(O)—) functionalities. An example of an efficient second-order NLO material is 4-(dimethylamino)-4'-nitrostilbene, shown in its two resonance forms in Figure 4.4. The arrows indicate how electrons move along the length of the molecule.

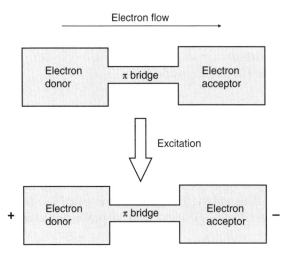

Figure 4.3 A generalized second-order NLO organic material. Electron-rich (donor) and electron-deficient (acceptor) groups are joined by a π-conjugated bridge.

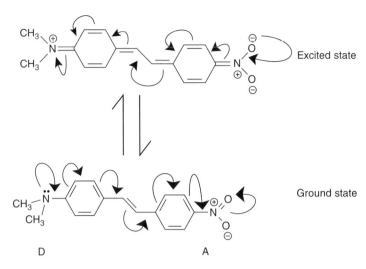

Figure 4.4 Structure of 4-(dimethylamino)-4′-nitrostilbene in its ground (bottom) and excited state (top). The arrows indicate a 'push-pull' sequence of π-electron movement.

4.4 Interaction of EM Waves with Organic Molecules

4.4.1 Absorption Processes

The intensity of light passing through an absorbing material is reduced according to *Beer's law*:

$$I = I_0 \exp(-\alpha d)) \tag{4.12}$$

where I is the measured intensity after passing through the material, I_0 is the initial intensity, α is the *absorption coefficient* (units m^{-1}) and d is the path length. This is often written in a more convenient form as

$$A = -\log_{10}\left(\frac{I_0}{I}\right) \tag{4.13}$$

where A is the *absorbance* of the sample and is given by $A = (\log_{10}e)\alpha d = 0.43\alpha d$. The absorption coefficient of a solid is related to the imaginary part of the complex refractive index by

$$\alpha = \frac{4\pi n''}{\lambda} \tag{4.14}$$

A straightforward test of the reproducibility of the transfer of a dye chromophore from the water surface to a solid substrate, say by the Langmuir–Blodgett (LB) technique (Chapter 7, Section 7.3.1), is to monitor the optical absorption (at a particular wavelength) as a function of the number of LB layers deposited. Figure 4.5 shows the result of such an experiment, using an organotransition metal complex [1]. The increase in absorbance with increasing film thickness agrees with Beer's law.

The *Beer–Lambert law* introduces the concentration of an absorbing species into the above relationship and is used in work with solutions. The absorbance of a sample A can be expressed as

$$bcd = -\log_{10}\left(\frac{I_0}{I}\right) = A \tag{4.15}$$

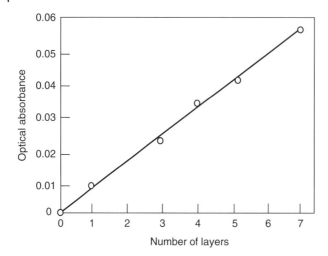

Figure 4.5 Optical absorbance at 332 nm versus number of Langmuir–Blodgett layers for an organotransition metal complex. *Source*: Reprinted from Richardson et al. [1]. Copyright (1988), with permission from Elsevier.

where b is the molar absorption coefficient (litre mol^{-1} m^{-1}) and c is the molar concentration of the solution (mol litre^{-1} or M). The value of b is usually quoted at an absorption maximum. Since the absorbance is proportional to the concentration of the solution, a plot of A versus c should yield a straight line. This relationship is observed in many materials. However, at high concentrations, changes in the position or intensity of the absorption maximum, caused by the formation of molecular aggregates such as *dimers* or *trimers* (Section 4.4.2) often results in a deviation from the Beer–Lambert law. The interaction between the dye molecules and the solvent can also give deviations from this law as a result of *solvatochromism*, which is the ability of a chemical substance to change colour due to a change in solvent polarity. Negative solvatochromism corresponds to a shorter wavelength or *hypsochromic* shift, while positive solvatochromism corresponds to a longer wavelength, or *bathochromic*, shift with increasing solvent polarity. The sign of the solvatochromism depends on the difference in dipole moment of the molecule of the dye between its ground state and excited state.

In terms of electronic energy levels, a molecule in its ground state can absorb a photon of light if the photon energy is equal to the difference between the two energy levels in the system. When this occurs, the molecule is excited into a higher energy state so that

$$E_1 - E_0 = h\nu \tag{4.16}$$

where ν is the frequency of the radiation in Hz, and E_1 and E_0 are the excited and ground states of the molecule, respectively. Thus, for molecular crystals, the minimum photon energy required for absorption will be when E_0 corresponds to the highest occupied molecular orbital (HOMO) level and E_1 to the lowest unoccupied molecular orbital (LUMO) level. In the case of a conductive polymer, absorption will occur when the incident radiation has sufficient energy to promote electrons from the valence (π) band to the conduction (π^*) band. The optical absorption spectra for *trans*-polyacetylene are shown in Figure 4.6. The data are shown for both the undoped polymer and following doping with iodine [2]. For the undoped material, there is a rapid increase in the optical absorption when the incident radiation has an energy exceeding 1.4 eV, the band gap of the polymer. On iodine doping, a mid-gap soliton state (Chapter 3, Section 3.4.3) is formed, accounting for the absorption peak at about 1 eV.

Optical absorption experiments can also reveal information about the charge-transfer bands in certain complexes. For example, Figure 4.7 shows the polarized absorption spectra of a perylene/tetracyanoethene crystal [3]. The lowest energy (lowest wavenumber) transition is

Figure 4.6 Optical absorption of *trans*-polyacetylene as a function of doping level. As the doping increases a mid-gap soliton level forms below the band edge. *Source*: Reproduced from Friend et al. [2]. Copyright (1985), with permission of The Royal Society.

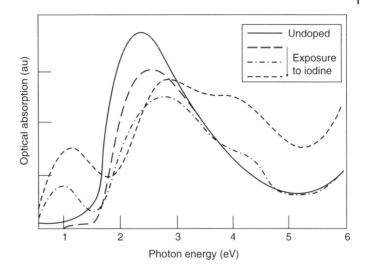

Figure 4.7 Polarized optical absorption spectra along different crystallographic axes for the charge-transfer compound perylene/tetracyanoethene. *Source*: Reprinted from Kuroda et al. [3]. Copyright (1967), with permission from Elsevier.

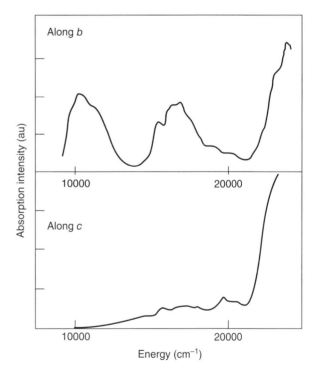

largely charge transfer in character and is strongly polarized along the *b*-axis, the stacking axis, in the crystal. The higher energy absorption consists of several transitions of similar energy, some of which have mixed charge-transfer and local-excitation-of-donor characters. Since local excitations of perylene are polarized predominantly in the molecular plane, whereas charge-transfer transitions are polarized along the stacking axis, these mixed character transitions are less strongly polarized than the lowest energy transition.

The dipole moment that is induced by the applied oscillating EM field is called the *transition dipole moment*. It can be calculated from an integral taken over the product of the wavefunctions of the initial and final states of a spectral transition. Light will not be absorbed completely unless the oscillating electric field is parallel to the transition moment. This dipole moment is fixed relative to the molecular structure. In ethylene, the $\pi \rightarrow \pi^*$ transition, which is associated with the C=C bond, has its transition moment polarized along the bond. Complex molecules may have more than one transition moment.

For dye molecules on a solid plate (substrate), the orientation (if any) can be conveniently investigated by using polarized light. For instance, if a wavelength at which a molecule absorbs is chosen, then direct comparison of the absorption intensities of the s- and p-polarized radiation (at the same angle of incidence) enables the average orientation of the transition moments to be determined.

Most organic molecules have an even number of electrons, with all electrons paired. Within each pair, the opposing electron spins cancel and the molecule has no net electronic spin. Such an electronic structure is called a *singlet* state. When a ground state singlet absorbs a photon of sufficient energy, it is converted to an excited singlet state in which the spin of the excited electron is not altered. The process is so fast that the excited state has the same geometry of bond distances and bond angles. This is the *Franck-Condon principle*. However, the most stable geometry of the excited state often differs from that of the ground state, so that the excited electronic state is formed in an *excited vibrational state*.

Shapes of absorption bands can be explained by reference to *Morse curves*. These are plots of potential energy as a function of the nuclear distance r. In Figure 4.8, Morse curves

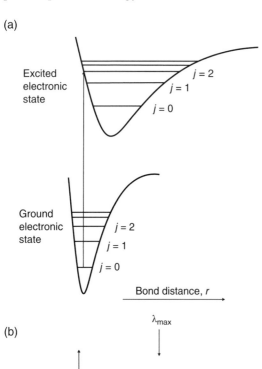

Figure 4.8 (a) Morse curves for the ground and excited states of a molecule. (b) The origin of the shape of absorption bands.

corresponding to a ground and an electronically-excited state of a particular molecule are shown. The horizontal lines represent the vibrational levels of the electronic states and each of these has a vibrational quantum number associated with it, j = 0, 1, 2, 3, ..., n; these levels may have energy spacings of $\approx$ 0.01 eV. The excited state is represented by a Morse curve that is displaced vertically (higher energy) and horizontally (due to vibration increased bond length) from that of the ground state. By the Franck-Condon principle, electronic transitions are represented on the Morse curve by vertical lines.

The probability and intensity of an electronic transition will not be a maximum between the j = 0 levels of the ground and excited states. Because of the relative displacement of the two Morse curves, the maximum absorption will occur between the j = 0 level of the ground state and a higher energy level in the excited state (j = 2 in Figure 4.8a). This corresponds to the wavelength for maximum absorption, λ_{max}. On each side of this, the absorption intensities will decrease to zero, producing the familiar bell-shaped absorption band shown in Figure 4.8b. In the case of polyatomic molecules, the Morse curves are replaced by poly-dimensional surfaces and the number of allowed transitions will become very large. Consequently, the absorption bands become smooth curves. Sometimes, however, vibrational fine structure can be seen.

4.4.2 Aggregate Formation

The absorption spectrum of an organic compound may well differ in the solid and solution states. Although the forces between molecules in crystals are weak and short range (van der Waals interactions) and the overlap between adjacent molecules in the lattice is small, there is still a substantial difference between the electronic spectra of molecular crystals and free molecules. Some of these differences are caused by interactions between the electronic states of molecules in the vicinity; others are due to crystal lattice properties. Crystal spectra have absorption bands that are broader than those in solution. This is because the molecular interactions are affected by thermal vibrations as well as the relative orientation of the molecules. Bands are often shifted and split when compared to solution.

The formation of aggregates is common in concentrated dye solutions and is often difficult to avoid when the molecules are in the solid state. The interactions between dye molecules can be explained by considering energetically delocalized states, termed *excitons*. In Figure 4.9, $E_{excited}$ and E_{ground} are the unperturbed energy levels of an isolated molecule in dilute solution. When these molecules are brought close together, as shown in Figure 4.9a, the dipoles interact and multiple excitation energy levels are observed. The splitting of energy levels, *Davydov splitting*, is determined by the difference between interacting transition dipole moments, their relative orientations, and the number of interacting molecules. For a simple dimer, the energies are given by

$$\Delta E_{dimer} = \Delta E_{monomer} + D \pm \varepsilon \tag{4.17}$$

where ΔE represents the transition energies to the excited state of the dimer or monomer, ε is the exciton interaction energy, and D is a *dispersion energy* term, given by $W-W'$ in the figure; this depends on van der Waals interactions between the molecules.

Equation (4.17) shows that for a dimer, the excited energy state has two possible levels. These can be attributed to the phase relationship between the transition dipole moments. If the dipole moments of the two molecules are parallel, as shown in Figure 4.9b, the net transition moment of the dimer is zero in the lower energy state E' and this transition is forbidden. Only transitions to the higher energy state E'' are allowed, leading to a hypsochromic shift, *H-band*, which is often observed in organized thin films. If the transition dipoles are in-line rather than parallel, Figure 4.9c, transitions to the higher energy state are forbidden and a bathochromic shift is

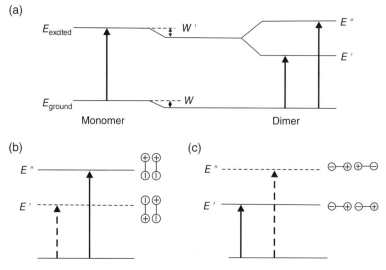

Figure 4.9 (a) Aggregation effects on the energy levels of molecules. (b) Parallel transition dipoles (hypsochromic shift). (c) In-line transition dipoles (bathochromic shift).

observed in the spectrum. Higher aggregates of this type exhibit a large bathochromic shift, with an intense, narrow absorption; these are termed *J-aggregates*. Of course, parallel and in-line are the two extreme forms of molecular aggregation and, in general, the transition dipoles are oriented at some angle to each other, resulting in more complex spectral changes.

The formation of J-aggregates is illustrated in Figure 4.10a, which shows the absorption spectrum of a monolayer of an amphiphilic cyanine dye mixed in a 1:1 M ratio with octadecane [4]. A very narrow absorption band is due to an in-phase relationship between the oscillators corresponding to the dye chromophores. The closest packing of the chromophores is achieved with the brickwork arrangement of dye molecules, in which the octadecane chains fit into the cylindrical holes left by the hydrocarbon chains of the dye (Figure 4.10b).

4.4.3 Excitons

Extending the above dimer model to an infinite three-dimensional lattice is a similar exercise to the extension of a covalent bond between two atoms to the delocalized band structure of a metal or semiconductor (Chapter 3, Section 3.3). Sharply defined energy levels associated with the two-body system become a band of energy levels in the infinite lattice, with a bandwidth dependent on overlap within the lattice. Thus Figure 4.9 can be applied to a crystal if each of the levels E' and E'' is replaced by a narrow band of levels. These bands are known as exciton bands. In solids with strong intermolecular interactions, an exciton can be delocalized over a number of molecules. Depending on the degree of delocalization, the excitons are identified as *Frenkel*, charge-transfer, or *Wannier-Mott*. These two extreme cases are depicted in Figure 4.11. The attraction between the negative and positive charges provides a stabilizing energy balance. Consequently, excitons have slightly less energy than an unbound electron and hole. The Frenkel exciton, shown in Figure 4.11a, corresponds to a correlated electron–hole pair localized on a single molecule. Its radius is therefore comparable to the size of the molecule (< 0.5 nm) or is smaller than the intermolecular distance. A Frenkel exciton can be considered as a neutral particle that can diffuse from site to site, perhaps moving hundreds of molecules away from their origins. Excitons can therefore transport energy without involving the migration of net

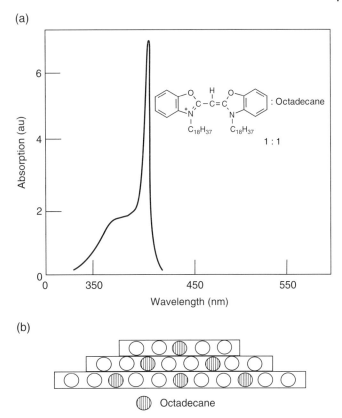

Figure 4.10 (a) Narrow absorption band resulting from J-aggregate formation in a mixed monolayer of an amphiphilic cyanine dye and octadecane. (b) The molecules are assumed to be packed in a brickwork arrangement shown (viewed from above). *Source*: Reproduced with permission from Kuhn et al. [4]. Copyright (1972), John Wiley & Sons Inc.

electric charges. In contrast, Wannier-Mott excitons (Figure 4.11b) occur in crystalline materials in which overlap between neighbouring lattice atoms reduces the Coulombic interaction between the electron and the hole of the exciton (leading generally to higher permittivity values). This results in a large radius, 4–10 nm, many times the size of the lattice constant. This type of exciton is not found in van der Waals-bonded organic solids, but is more typically found in inorganic semiconductors such as silicon or gallium arsenide. Figure 4.11 also indicates the energy levels associated with the different types of excitons. The Frenkel exciton can be considered as two distinct localized states, located above the HOMO and below the LUMO levels of the molecule, whereas the Wannier-Mott excitons are associated with the extended band structure of the crystal in which they are formed. Electronic excitation energy can be conducted within organic solids by excitons moving (hopping) from one location to another. This is an important and characteristic process.

4.4.4 Effect of Electric Fields on Absorption

Changes in the optical properties of organic molecules may be produced by applying large electric fields. *Electroabsorption* is one technique that has been used to study materials. This is a particular branch of the group of experimental methods known as *modulation spectroscopy* and involves monitoring small changes in a sample's optical transmission that result from the application of an external field. For molecular compounds, these changes are usually the result of the *Stark effect*. The change in potential energy ΔV that arises when a molecule is placed in an electric field may be written as

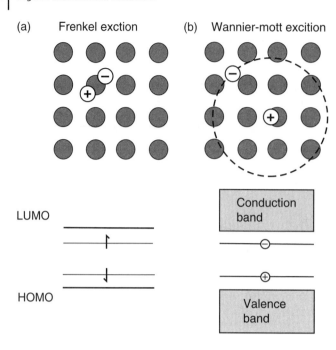

(a) Frenkel exction   (b) Wannier-mott excition

Figure 4.11 (a) Frenkel and (b) Wannier-Mott excitons with their respective electron energy diagrams.

$$\Delta V = -pE\cos\theta + \frac{\alpha}{2}E^2 \tag{4.18}$$

where the dipole moment p makes an angle θ to the applied field and α is the polarizability (Eq. (4.11)). The first term on the right-hand side of Eq. (4.18) represents the linear Stark effect and the second term the quadratic effect. In general, non-centrosymmetric crystals with polar space-groups exhibit a first-order Stark effect. This is usually recognized in electroabsorption spectra as the second derivative of the zero-field absorption curve. Crystal structures having centrosymmetric crystal structures may still exhibit a second-order Stark effect. This process is observed in electro-absorption as a first-order derivative of the zero-field absorption curve.

4.4.5 Emission Processes

Luminescence is a term that is used to describe the emission of light by a substance caused by any process other than a rise in temperature. Molecules may emit a photon of light when they decay from an electronically excited state to the ground state. The excitation leading to this emission may be caused by a photon (*photoluminescence*), an electron (*electroluminescence* or *cathodoluminescence*), or a chemical reaction (*chemiluminescence*). Other types of luminescence include *bioluminescence*, *thermoluminescence*, and *mechanoluminescence*.

The processes leading to emission can be considered in terms of the energy levels in the molecule. Figure 4.12, a *Jablonski diagram*, shows a very simplified energy diagram (e.g. omitting internal conversion) illustrating some processes that can occur. Each molecular state in the diagram corresponds to a bonding or anti-bonding molecular orbital. The orbitals associated with a carbon–carbon bond can be either σ or π type, with corresponding antibonding orbitals σ^* and π^*, respectively. There are also valence-shell electrons that do not participate in the formation of molecular bonds. These nonbonding orbitals are designated as n.

Following promotion to an electronically excited singlet excitonic state (S_1 or S_2 in the figure), the molecule usually relaxes to the lowest vibrational level of S_1, the energy being lost as

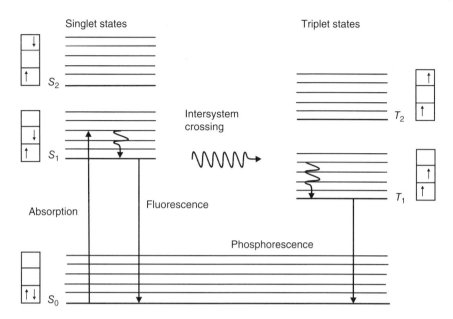

Figure 4.12 Jablonski diagram showing the radiative and nonradiative decay of an excited molecule. The orientations of the electron spins are shown in the boxes next to each state.

heat via intermolecular collisions. This process takes about 10^{-11}s, or approximately 10^2 vibrations of the molecule. The lifetime of S_1 in its lowest vibrational state is longer, 10^{-8}–10^{-7}s. This state may then decay to the ground state with the emission of a photon.

This emission is termed *fluorescence*. Alternatively, the energy may be transferred by *intersystem crossing* to a *triplet excitonic state* (the *multiplicity* of such a state, $= 2S + 1$, where S is the algebraic sum of the electron spins, is 3). In such a state, the spins of the two electrons are parallel and a transition to the ground state, with the emission of a photon, involves a change of spin. This is a *spin-forbidden* transition; quantum mechanical selection rules forbid transitions in which the electron spin changes, i.e. only singlet to singlet or triplet to triplet transitions are allowed. Triplet states are fairly long lived, with lifetimes of greater than 10^{-5}s. The triplet state is typically at a lower energy than the corresponding singlet state. The triplet state can decay to S_0, emitting a photon. This process is known as *phosphorescence* and may persist for seconds or even longer after the incident excitation has ceased. Nonradiative transitions from the excited singlet and triplet states to the ground state are also possible. In these instances the original light quantum is converted into heat.

Triplet excitons in aromatic molecules are more effective energy conductors than singlet excitons. Their lifetimes are longer and times to move between sites in the material (hopping times) are not much shorter than those in the singlet state. Triplet excitons can also combine their energy to form a singlet exciton, resulting in *delayed fluorescence* or *triplet annihilation.* The former gives rise to the emission of fluorescence light emitted with a delay (determined by the lifetime of the triplet excitons), whereas triplet annihilation gives reduced emission intensity as two triplet states have been combined into one.

The energy of an emitted photon is lower than that of the absorbed photon because the excited state has lost energy before the emission occurs. This is evident by reference to the Morse curves, as shown in Figure 4.13. The energetic red-shift is often referred to as the *Stokes' shift*. Almost all molecular organic compounds show this energy shift. For example, Figure 4.14 compares the absorption and emission spectrum in a thin film of the organic dye compound

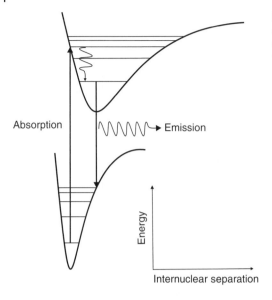

Figure 4.13 Absorption and emission processes compared using the Morse curve. The emission occurs at a lower energy (longer wavelength) than the absorption.

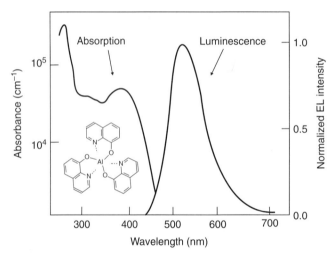

Figure 4.14 Absorption and luminescence of aluminium *tris*(8-hydroxyquinoline), with the chemical formula shown. *Source*: Reproduced from Bulović et al. [5]. Copyright (2001), with kind permission of Springer Science and Business Media.

aluminium *tris*-(8-hydroxyquinoline) (Alq$_3$) [5]. There is almost no overlap between these spectra and the Alq$_3$ is therefore transparent to its own emission radiation. The absorption and emission spectra of organic solids often exhibit a mirror-image relationship to one another.

Another important interaction of excited states is *excimer* formation. An excimer is an excited state dimer and results from the interaction between an excited singlet state and an unexcited molecule. Excimer emission is usually broad, structureless, and somewhat red-shifted from the normal molecular fluorescence. In molecular crystals, excimer emission can be observed when the crystal structure is such that adjacent molecules form parallel-plane dimers, which allow appreciable overlap of the delocalized π-electrons. An excellent example of the correlation of crystal structure to excimer emission is provided by work with perylene [6]. Perylene exists in two crystal forms, the dimeric α-form as shown in Figure 4.15a and the monomer β-form shown in Figure 4.15b. The emission spectra of perylene from solution and the two crystal forms are shown in Figure 4.15c. In solution, the fluorescence emission is highly structured and corresponds to a mirror image of the solution absorption spectrum. The β crystal

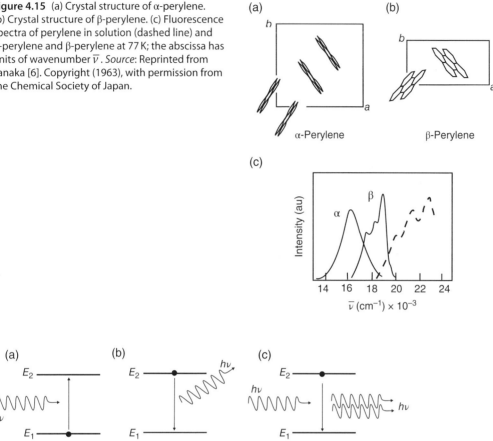

Figure 4.15 (a) Crystal structure of α-perylene. (b) Crystal structure of β-perylene. (c) Fluorescence spectra of perylene in solution (dashed line) and α-perylene and β-perylene at 77 K; the abscissa has units of wavenumber $\bar{\nu}$. *Source*: Reprinted from Tanaka [6]. Copyright (1963), with permission from the Chemical Society of Japan.

Figure 4.16 Absorption and emission processes associated with two electron energy levels E_1 and E_2. (a) Absorption of a photon with energy $h\nu$ promotes an electron from level E_1 to E_2. (b) Spontaneous emission, the photons are emitted randomly (c) Stimulated emission, the emitted photons are coherent with the stimulating photons.

form also shows a structured emission and a good mirror image relationship with the crystal absorption spectrum. The excimer emission spectrum of the α crystal form is in marked contrast; it is broad, structureless, and considerably red-shifted from the monomer emission (note the abscissa in Figure 4.15c is in terms of wavenumber $\bar{\nu}$ – Chapter 6, Section 6.6).

The light emitted from an organic solid as described above will generally be *incoherent*, which means that there is no ordered phase relationship between the EM waves emitted. The electron in the excited state E_1 spontaneously falls back into the ground state E_2, with the emission of a photon of energy $h\nu = E_2-E_1$. If there are many such excited electrons, they will emit photons at random times and so generate incoherent light. This is *spontaneous emission*, illustrated in Figure 4.16. However, it is possible for an incoming photon to trigger the emission process. This situation is called *stimulated emission* (Figure 4.16c). The emitted photon is in phase with the incoming photon, it is going in the same direction, and it has the same frequency (since it must also have the same energy E_2-E_1). The result is the emission of *coherent* radiation. Stimulated emission is the basis for the operation of the *laser* (an acronym for *l*ight *a*mplification by *s*timulated *e*mission of *r*adiation). This device acts as a photon amplifier, since one incoming photon results in two outgoing photons. An important requirement is that there are more electrons in energy state E_2 than in E_1, a situation referred to as *population inversion*. It is not possible to achieve this state with only two energy levels as, in the steady state, the incoming photon flux will cause as many upward transitions as downward stimulated transitions. To create a population inversion, an additional level(s) is needed and energy is *pumped* into the lasing medium by some process such as the passage of a current, the creation of an electrical discharge, or illumination with EM radiation.

4.4.6 Energy Transfer

The energy of excited molecules can also be transferred to other molecules. In general terms, during the energy transfer process, an excited donor molecule D* transfers its energy to an acceptor A which, in turn is promoted to an excited state:

$$D^* + A \rightarrow D + A^* \tag{4.19}$$

For example, in some doped organic films, the exciton energy can be transferred from the host to the guest molecules, quenching the luminescence of the host while increasing that of the guest. Two important and distinct energy transfer mechanisms are recognized: *Förster transfer* and *Dexter transfer*. The former is a mechanism of excitation transfer that can occur between molecular species that are separated by relatively large distances (i.e. exceeding their van der Waals radii). It is described in terms of a Coulombic interaction between the transition dipole moments. The energy transfer rate K_{ET} depends on the distance R between the donor and acceptor molecules:

$$K_{ET} = \left(\frac{1}{\tau}\right)\left(\frac{R_0}{R}\right)^6 \tag{4.20}$$

The constant R_0 is called the Förster radius and τ is the average donor excitation lifetime for recombination in the absence of energy transfer. When $R = R_0$, the probability that an exciton will recombine at the donor is equal to its transfer probability. In Förster energy transfer, the spin of both D and A is conserved. Dexter excitation transfer is the result of an electron exchange mechanism. This requires an overlap of the wavefunctions of the donor and acceptor. Dexter transport occurs over short distances, typically 0.1 nm, and is effectively restricted to neighbouring molecules. In the Dexter transfer process, only the total spin of the D*A system is conserved and triplet-triplet energy transfer, which is forbidden in the Förster process, is now allowed. Schematic representations of both Förster and Dexter processes for the case of singlet-singlet transfer are shown in Figure 4.17.

An elegant demonstration of Förster energy transfer is provided by work with multilayer films, such as a Langmuir–Blodgett film [4]. The film deposition method allows molecular assemblies to be built up, in which individual layers of molecules are positioned at a very precise distance from one another (Chapter 7, Section 7.3.1). Figure 4.18 illustrates the type of experiment that can be undertaken. A donor (sensitizer) molecule within a monolayer is represented by D and an acceptor molecule by A. The D molecules absorb in the ultraviolet part of the EM spectrum and fluoresce in the blue, while molecules of A absorb in the blue and fluoresce in the yellow. If there is sufficient distance between A and D, the fluorescence of D only appears. However, if the distance between the monolayers of D and A is reduced enough, A absorbs the fluorescence of D and yellow fluorescence is observed.

Figure 4.17 Schematic representation of Förster and Dexter energy transfer processes from an excited donor D* to acceptor A in single-singlet energy transfer.

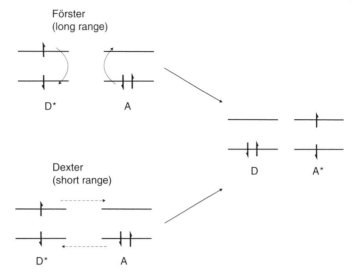

Figure 4.18 (a) Monolayers of a sensitizing donor D and an acceptor molecule A are separated by monolayers of a spacer material. In (b) the spacing between D and A is sufficiently close for energy transfer to occur between the D and A molecules.

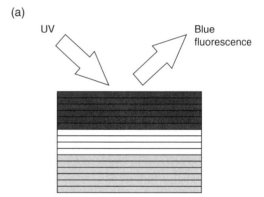

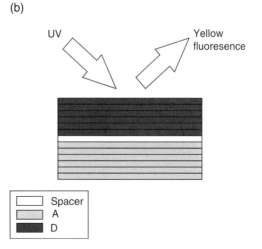

4.5 Transmission and Reflection from Interfaces

4.5.1 Laws of Reflection and Refraction

When electromagnetic waves are incident on the interface between two dielectrics, the familiar phenomena of reflection and refraction take place. Figure 4.19 shows a simple case in which a p-polarized EM wave is incident, at an angle θ_i, to the boundary between two dielectric media of refractive indices (real quantities) n_1 and n_2. Usually there will be a reflected beam at an angle θ_r and a transmitted (refracted) beam at θ_t. Application of the boundary conditions that the normal components of the magnetic fields and the tangential component of the electric fields are continuous at the interface gives the following relations:

$$\theta_i = \theta_r \tag{4.21}$$

$$\frac{\sin\theta_i}{\sin\theta_t} = \frac{n_2}{n_1} \tag{4.22}$$

Equations (4.21) and (4.22) are, of course, familiar as the *law of reflection* and *Snell's law of refraction*. (These two equations also hold for s-polarized incident radiation.)

4.5.2 Fresnel Equations

The proportions of the incident electric field amplitude that are reflected may also be evaluated. For p-polarized (TM) radiation, the ratio $\frac{E_r}{E_i}$ defines the *reflection coefficient* $r_{||}$ and $\frac{E_t}{E_i}$ is the *transmission coefficient* $t_{||}$

$$r_{||} = \frac{E_r}{E_i} = \frac{\tan(\theta_i - \theta_t)}{\tan(\theta_i + \theta_t)} \tag{4.23}$$

$$t_{||} = \frac{E_t}{E_i} = \frac{2\cos\theta_i\sin\theta_t}{\sin(\theta_i + \theta_t)\cos(\theta_i - \theta_t)} \tag{4.24}$$

Equations (4.23) and (4.24) are known as *Fresnel equations* and apply at optical frequencies to transparent nonmagnetic media where the refractive indices are real quantities (i.e. in Eq. (4.4), $n'' = 0$). Equivalent relationships for $r_\perp$ and $t_\perp$ may be obtained for s-polarized incident radiation. The r and t coefficients also give the phases of the beams. If positive, there is no change in phase; if negative, the phase changes by π.

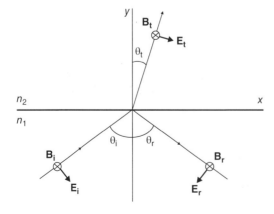

Figure 4.19 Reflection and refraction of an electromagnetic wave at a boundary between two dielectric media. The incident wave is p-polarized with its electric vector $\mathbf{E}_i$ in the plane of incidence. $\otimes$ represents a magnetic vector perpendicular to the plane of incidence.

There is a particular angle of incidence at which $r_\parallel$ becomes zero, while $r_\perp$ remains finite. This is called the *Brewster angle* θ_B and is given by

$$\tan\theta_B = \frac{n_2}{n_1} \tag{4.25}$$

For incidence from air to glass ($n_1 = 1$; $n_2 = 1.5$), $\frac{n_2}{n_1} = 1.5$ and θ_B is approximately 56°. At this angle, any incident light can only be reflected with the electric field vector perpendicular to the plane of incidence, giving a method of producing linearly polarized light from an unpolarized beam.

The amount of power (or intensity) reflected or transmitted from an interface is proportional to the square of the electric field amplitude. *Reflecting R or transmitting power coefficients* (or intensities) *T* may be defined as

$$R = r^2 \text{ and } T = t^2 \tag{4.26}$$

where *T* and *R* are also known as the *transmittance* and *reflectance*, respectively. At normal incidence (i.e. $\theta_i = 0$), and for the case $n'' = 0$, *R* and *T*, are given for both states of polarization by

$$R = \left(\frac{n_1 - n_2}{n_1 + n_2}\right)^2 \tag{4.27}$$

$$T = 1 - R = \frac{4n_1 n_2}{\left(n_1 + n_2\right)^2} \tag{4.28}$$

Equation (4.27) predicts that *R* is approximately 4% for radiation incident from air to a glass surface.

When radiation is incident from a dense to a less dense medium, Snell's law gives a *critical angle of incidence* θ_c at which there is 100% reflection (for both polarizations). The situation is referred to as *total internal reflection*.

Figure 4.20 summarizes the main features for the reflection of p- and s-polarized EM radiation at a vacuum/dielectric interface ($n_1 = 1$; $n_2 = n$).

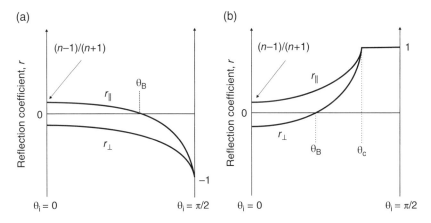

Figure 4.20 Variation of reflected wave amplitude with angle of incidence for: (a) waves incident from vacuum ($n = 1$) to a dielectric with refractive index n; and (b) waves incident from inside the dielectric to a vacuum. $r_\parallel$ – reflection coefficient for p-polarized incident radiation; $r_\perp$ – reflection coefficient for s-polarized incident radiation; θ_B – Brewster angle; θ_c – critical angle.

If the EM radiation is incident from a transparent to an absorbing medium, then the real refractive index n_2 in Eq. (4.22) is replaced by $n_2' - jn_2''$, i.e.

$$\sin\theta_t = \frac{n_1\sin\theta_i}{n_2' - jn_2''} \tag{4.29}$$

Therefore, θ_t is complex and does not represent the angle of refraction, except for the special case $\theta_i = \theta_t = 0$. Here, the Fresnel refection coefficients (identical for both components of polarization) are given by

$$r_\parallel = r_\perp = \frac{n_1 - n_2' + jn_2''}{n_1 + n_2' - jn_2''} \tag{4.30}$$

4.5.3 Ellipsometry

As shown above, the reflectivity from a surface differs in both amplitude and phase for s- and p-polarized incident radiation. *Ellipsometry* is a comparison of these reflectivities and it is the basis for a powerful method for measuring the thickness and optical constants of thin films. The presence of a surface layer alters the ratio of the electric field vectors vibrating in the plane of incidence and perpendicular to it, also their difference in phase Δ. In the former case, an angle ψ is defined as $\tan^{-1}$ (reflectivity amplitude ratio). The theory of ellipsometry correlates the parameters Δ and ψ with the optical thickness of the layer and the optical constants of the surface.

If there is no mixing between the polarizations, the reflection behaviour can be fully described by the two Fresnel coefficients above for p-polarized input to p-polarized output $r_\parallel$ and the equivalent for s-polarization:

$$\frac{r_\parallel}{r_\perp} = \tan\psi\,\exp j\Delta \tag{4.31}$$

It is usual to measure indirectly the ratio of these quantities by finding ψ and Δ. Using the Fresnel equations, it is relatively straightforward to calculate ψ and Δ. However, to evaluate the refractive index and thickness of a thin film from measured values, the situation is more complicated. Computer programs do this for commercial ellipsometers.

4.5.4 Thin Films

The above ideas can now be applied to the situation in which there is more than one interface, for example that of a thin film on a solid substrate. Figure 4.21 shows a schematic diagram for this. Radiation will first be reflected at the air/film boundary; the Fresnel equations for the appropriate polarization will govern the amounts of light transmitted and reflected. The transmitted radiation will undergo further reflection and transmission at the film/substrate interface. The overall air/film/substrate electric field amplitude reflection coefficient r is obtained by combining the Fresnel coefficients (for the appropriate polarization) after allowing for a change of phase β across the film (thickness d). The expression is given by

$$r = \frac{r_{01} + r_{12}\exp(-2j\beta)}{1 + r_{01}r_{12}\exp(-2j\beta)} \tag{4.32}$$

Figure 4.21 Multiple transmission and reflection from an air/thin film/substrate combination. T = transmitted intensity; R = reflected intensity.

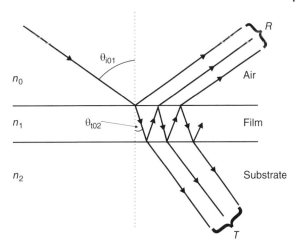

where r_{01}, r_{12} are the Fresnel coefficients for the air/film and film/substrate boundary, respectively. The phase change β across the film of thickness d is

$$\beta = 2\pi\left(\frac{d}{\lambda}\right)n_1\cos\theta_{t01} = 2\pi\left(\frac{d}{\lambda}\right)\left(n_1^{\,2} - n_0^{\,2}\sin^2\theta_{i01}\right)^{0.5} \tag{4.33}$$

The transmission coefficient t is given by

$$t = \frac{t_{01}t_{12}\exp(-j\beta)}{1 + r_{01}r_{12}\exp(-2j\beta)} \tag{4.34}$$

Equations (4.32) and (4.34) are generally valid. For non-normal incidence, each takes two possible forms, depending on the state of polarization of the incident EM radiation. If the film is absorbing, or if it is bound by absorbing media, then the values of n_0, n_1, n_2 are replaced by the corresponding complex quantities. The resulting equations for r and t become somewhat cumbersome, although readily calculable using a computer.

The oscillatory nature of Eqs. (4.32) and (4.34) is a result of the constructive and destructive interference that results from the multiple reflected and transmitted EM waves. By appropriate choice of the thin film thickness ($d = \lambda/4$) and refractive index ($n_1 = \sqrt{n_2}$), it is possible to obtain zero reflectance at normal incidence for a particular wavelength (the principle of an anti-reflection coating).

Under certain conditions the equations for the transmitted and reflected intensities (powers) from a thin absorbing film/transparent substrate combination may be considerably simplified. For normal incidence and if multiple reflections and interference can be neglected (which will be the case if the film is sufficiently absorbing):

$$T = (1 - R)^2 \exp\left(\frac{-4\pi n''}{\lambda}\right) \tag{4.35}$$

Independent measurements of T and R then enable values of n' and n'' to be determined. Interference effects are found with EM radiation of wavelengths other than visible; the interference of X-rays passing through thin films provides a precise means to measure their thickness (Chapter 6, Section 6.3.2). Figure 4.22 shows a set of interference fringes that result from the second-harmonic generation in a thin organic film of a dye, which possesses the general chemical

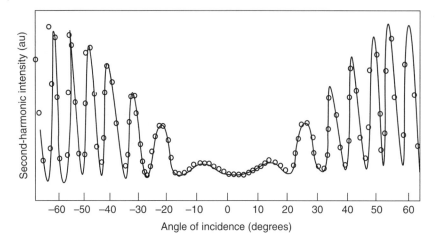

Figure 4.22 Experimental (points) and theoretical (solid curve) data for second-harmonic intensity of an oligomeric Langmuir–Blodgett film as a function of the angle of incidence of the fundamental laser beam. *Source*: Reproduced from Allen et al. [7]. Copyright (1993), with permission from the Royal Society of Chemistry.

structure shown in Figure 4.3 [7]. In the figure, the intensity of the generated second-harmonic radiation is plotted as a function of the angle of incidence of the fundamental laser beam. In this case, the fringing effect is due to interference between the harmonic signals produced from the front and back surfaces of the organic film. The amplitude of the fringes can be used to determine the electro-optic coefficient of the film.

4.5.5 Transmission through Conductive Thin Films

The optical transmission through thin, electrically conductive films is an important issue for a number of organic electronic devices. In the case of the light-emitting displays and photovoltaic structures described in Chapter 9, a compromise needs to be reached between the sheet resistance (which needs to be as low as possible) of the semi-transparent electrode used as one of the device electrodes and its optical transmission (which needs to be as high as possible). In the case of a thin metallic film in air, and assuming that the film thickness, d, is much less than the wavelength of the light, the transmission is given by [8–10]

$$T = \frac{1}{\left(1 + \dfrac{Z\sigma_{AC}d}{2}\right)^2} = \frac{1}{\left(1 + \dfrac{Z}{2R_s}\dfrac{\sigma_{AC}}{\sigma_{DC}}\right)^2} \tag{4.36}$$

where σ_{DC} and σ_{AC} are the low-frequency and high frequency conductivities, respectively, and R_s is the sheet resistance, as discussed in Chapter 3. The term Z represents the *electrical impedance of free space* $\left(= \sqrt{\dfrac{\mu_0}{\varepsilon_0}} \approx 377\,\Omega\right)$. Figure 4.23 shows data obtained for thin films of carbon nanotubes [9, 10]. The full line shows the best fit to Eq. (4.36), assuming that the ratio of the AC and DC conductivities remains constant for nanotubes of different densities in the measured optical frequency range. The best fit is obtained with $\sigma_{AC} = \sigma_{DC}$.

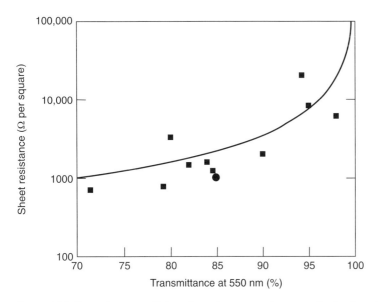

Figure 4.23 Transmittance at 550 nm for carbon nanotube network of various sheet resistances. The solid line is a fit to Eq. (4.36). The best fit is obtained with $\sigma_{AC} = \sigma_{DC}$. *Source*: Reproduced with permission from Hu, Hecht, and Grüner [9]. Copyright (2004) American Chemical Society.

4.6 Waveguiding

Electromagnetic waves can be confined (or guided) within thin films or channels of material. Such *waveguides* play an important role in optical communications. A well-known example is that of an optical fibre shown in Figure 4.24. The central region is called the *core*, whilst the surrounding regions are referred to as the *cladding*. To confine an EM wave fully within the core, it must be unable to escape, i.e. it must be totally internally reflected. The conditions for total internal reflection are that the core medium must be of the greater refractive index, i.e. $n_1 > n_2$ in Figure 4.24. However, the total internal reflection condition does not specify fully the requirements for waveguiding.

Figure 4.25 shows an example of a planar waveguide formed by a thin film of refractive index n_f on top of a substrate of refractive index n_s. The topmost layer (refractive index n_0) is often air and has a much lower refractive index than the other two. Waveguiding will occur in the film providing that its thickness d *is* given by

$$d \geq \frac{\lambda\left(m + 0.5\right)}{2\sqrt{\left(n_f^2 - n_s^2\right)}}$$

(4.37)

Figure 4.24 The basic structure of an optical fibre. A core region (refractive index n_1) is surrounded by a cladding region (refractive index n_2) where the refractive index of the core material is greater than that of the cladding.

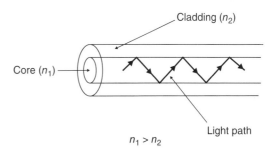

Cladding (n_2)

Core (n_1)

Light path

$n_1 > n_2$

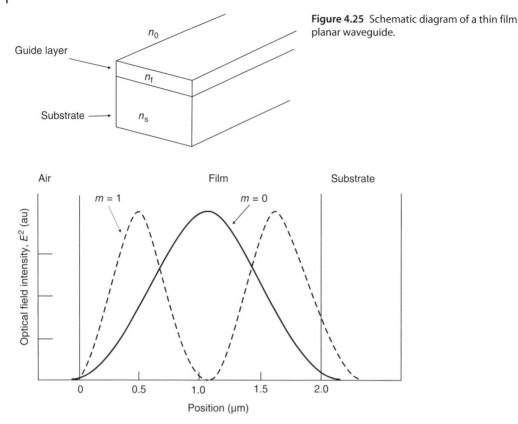

Figure 4.25 Schematic diagram of a thin film planar waveguide.

Figure 4.26 The optical electric field intensity E^2 for two modes propagating in a 2 μm thick film. The solid line is the $m = 0$ mode. The $m = 1$ mode (dashed line) has a minimum in the electric field intensity near the middle of the waveguide [8].

where λ is the wavelength in the guide and m is an integer. Each value of m is associated with a distinct wave pattern or *mode* within the waveguide. Transverse magnetic and transverse electric radiation (Section 4.2) give rise to two types of mode; these are called TM_m and TE_m. The cut-off thickness for waveguiding in the asymmetric structure shown in Figure 4.25 is usually about one-third of the wavelength of the light. Figure 4.26 shows how the optical electric field intensity, E^2, varies with position in a 2 μm thick guiding layer for the TE_0 and TE_1 modes [11]. The $m = 1$ mode has two maxima in the field intensity within the guided layer and a minimum near the middle of the guide. It is also important to note that the electric field penetrates both the air and substrate regions. These fields are *evanescent* in nature and will be discussed more fully in the following section. Different modes propagate along a waveguide with different velocities, even if they are generated by monochromatic radiation. This phenomenon is called *mode dispersion*.

4.7 Surface Plasmons

Surface plasmons are collective oscillations of the free charges at a metal boundary that propagate along the interface; these therefore represent a special type of guided wave. The intensity maximum of these waves is a maximum in the surface and decay exponentially perpendicular to the surface. Surface plasmons can be produced by electrons and by light.

4.7.1 The Evanescent Field

As has been noted in the Section 4.5.2, for light inci dent from a dense to a less dense medium at an angle of incidence greater than the critical angle θ_c there will be total internal reflection. This situation is sketched in Figure 4.27. If the reflectance R is recorded as a function of the angle of incidence, R becomes unity at the critical angle. At angles greater than θ_c, the reflected beam has the same amplitude as the incident beam, but a phase difference between 0 and π. The effect on the transmitted beam is rather strange. This takes on the form of a wave that travels along the boundary with an amplitude that decays perpendicular to the boundary. The wave carries no energy away from the boundary and is the evanescent wave already encountered in the discussions on waveguiding (Section 4.6). Figure 4.28 shows how both the ampli-tude and intensity (square of the amplitude) of a typical evanescent wave vary with distance from a glass/air interface [11]. It is evident that the penetration depth (into the air medium in this example) is similar to the wavelength of light. The electric field associated with evanescent waves is therefore a convenient way of exciting molecules close to a surface.

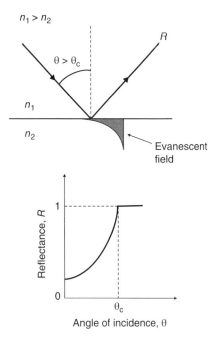

Figure 4.27 The total internal reflection from light incident from a medium 1 with refractive index n_1 to an interface with a medium 2 with refractive index n_2, where $n_1 > n_2$. For angles of incidence θ greater than the critical angle θ_c the reflectance is unity. However, an evanescent field penetrates into medium 2.

Figure 4.28 Decay of amplitude E and intensity E^2 of an evanescent wave at a glass/air interface [8].

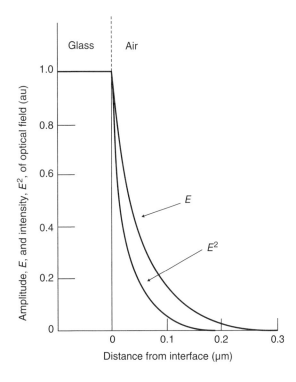

4.7.2 Surface Plasmon Resonance

Under certain conditions, electromagnetic surface waves can propagate along the interface between two media. The requirement is a combination of a lossless medium with a positive permittivity with another of negative real part and positive imaginary part. Although this is an idealized situation (as all materials exhibit loss), it is close to the practical system of a dielectric and metal at frequencies below the *plasma frequency*, ω_p. This frequency corresponds to that at which the permittivity of the metal is equal to zero; for $\omega < \omega_p$, the ε_r' is negative. The modes are called *surface plasmons* or *surface plasmon polaritons* (SPPs) – plasmon because the optical properties of the metal are consistent with its electrons behaving as a free electron plasma and polariton to suggest the coupling of photons with polar excitations in the metal.

Surface plasmons can be generated on a metal surface using an electron beam. However, photons do not possess sufficient energy to couple directly to these surface oscillations. Optical excitation of SPPs requires matching of both energy (i.e. optical frequency) and momentum. The latter quantity is related to the wavelength of the light by de Broglie's relationship in Eq. (3.28) (Chapter 3, Section 3.3.1), $p = h/\lambda$. Figure 4.29 contrasts the wavevector versus frequency relationships for light in air [ω/k = constant = phase velocity of light; Chapter 3, Section 3.3.1, Eq. (3.26)] and surface plasmons; such curves are called *dispersion curves*. These two curves do not intersect at any frequency. To be able to couple the light to the SPPs, the momentum of the photons at a particular frequency must be increased. This is accomplished by decreasing the velocity of the light, i.e. slowing the photons. As Figure 4.29 reveals, the dispersion curve for light in glass now intersects that of the surface plasmons.

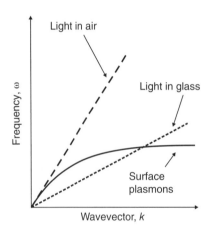

Figure 4.29 Dispersion curves for light in air and glass and for surface plasmons.

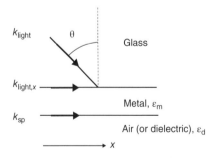

Figure 4.30 Coupling conditions between light incident from glass to a boundary with a thin metal film. k_{light} is the wavevector of the incident light; $k_{light,x}$ is the component of the wavevector in the *x*-direction; and k_{sp} is the surface plasmon wavevector.

An experimental arrangement to realize the above conditions is shown schematically in Figure 4.30. This is known as the *Kretschmann* configuration and exploits evanescent electromagnetic waves. Monochromatic light (p-polarized) travels through glass and is reflected from a metal coating. If this coating is sufficiently thin, the resulting evanescent field penetrates to the opposite side of this layer to an interface between the metal and air (or a dielectric coating). The incoming EM radiation will have a wavevector $k_{light,x}$, as indicated in Figure 4.30. Changing the angle of incidence of the light θ alters the component of the wavevector parallel to the prism base $k_{light,x}$. When this component matches the real part of the surface plasmon wavevector, k_{sp}, then surface plasmons are excited. This condition requires that

$$k_{sp} = \frac{\omega}{c}\sqrt{\frac{\varepsilon_m \varepsilon_d}{\varepsilon_m + \varepsilon_d}} \qquad (4.38)$$

where ε_m and ε_d are the relative permittivities of the metal and the dielectric, respectively. Equation (4.38) describes a surface plasmon wave if the real part of ε_m

Figure 4.31 Surface plasmon resonance curve.
θ_c = critical angle; θ_{SPR} = surface plasmon resonance
(SPR) angle.

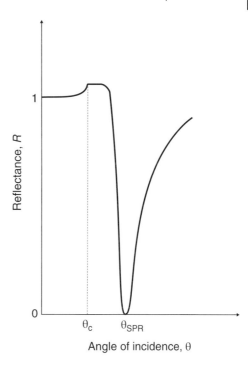

is negative and its absolute value is smaller than ε_d. At optical wavelengths, this condition is fulfilled by several metals; among these, silver and gold are the most commonly used.

A curve resulting from the measurement of the reflectance, R, against angle of incidence, θ, is shown in Figure 4.31. As the angle of incidence is increased past the critical angle, θ_c, all of the incidence light will be reflected ($R = 1$). When the component of the light's wavevector parallel to the interface is equal to that of the surface plasmons ($k_{light,x} = k_{sp}$), then all of the optical energy will couple into the SPPs and the reflected intensity will go to zero. This condition is known as *surface plasmon resonance* (SPR). The angle at which this occurs is shown as θ_{SPR} in Figure 4.30. The reflectance curve is often referred to as an SPR curve.

The reflectance versus angle relationship can be calculated using the Fresnel equations (Section 4.5.2). The reverse problem of finding optical parameters of a metal film from experimentally measured SPR curves has no analytical solution. However, it can be solved numerically by fitting the experimental SPR data to theory using a least squares technique.

The Kretschmann method described above is only one of the techniques available to enhance the momentum of the incident optical wave. In the Otto arrangement, the prism is separated by an air gap from a metal plate. The coupling takes place between the evanescent field in the gap and the plasmons on the metal surface. An advantage of the Otto approach is that there is no restriction on the thickness of the metal. Diffraction gratings and optical waveguides can also be used. Figure 4.32 contrasts the coupling to SPPs using the Kretschmann and diffraction grating approaches. In the former, Figure 4.32a, a prism or semi-cylinder may be used to change the momentum of the incident photons. The thin metal layer ($\approx 50\,\text{nm}$) can either be deposited directly onto the base of the semi-cylinder or deposited onto a glass slide, which is placed in contact with the semi-cylinder using a refractive index matching fluid. The use of a diffraction grating to couple light to the SPPs (Figure 4.32b) enables the plasmons to be generated on the same side of the metal coating as the incident light. The periodicity of the grating effectively enhances the momentum of the incoming photons. However, the reflected beam will consist of

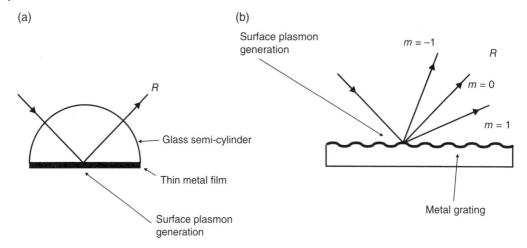

Figure 4.32 Methods for coupling light to surface plasmons. (a) Kretschmann configuration. (b) Grating coupling.

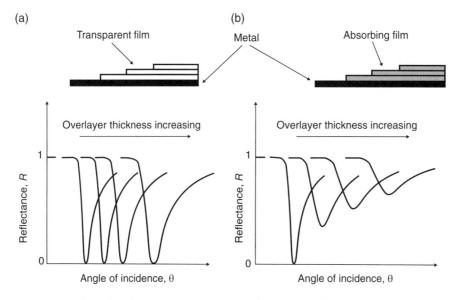

Figure 4.33 Effect of overlayers on SPR curve. The curves are shown for different thicknesses of overlayer. (a) Overlayers transparent to wavelength of incident radiation. (b) Overlayers absorbing at wavelength of incident radiation.

different *orders of diffraction*, different values of m shown in the figure. This results in the appearance of more than one minimum in the SPR curve.

A thin dielectric coating on top of the metal film will shift the surface plasmon dispersion curve to higher momentum. Consequently, the SPR will move to a higher angle. This can be conveniently demonstrated using Langmuir–Blodgett films. Figure 4.33 shows the effect of the multilayers on the SPR curves. In Figure 4.33a, the organic film is transparent to the incoming EM wave. The effect of increasing the number of layer in the LB array is to shift the SPR curves to higher angles; the minimum in R remains at zero, but the increasing film thickness produces a slight broadening of the resonance curves. The effects are somewhat different in the case of

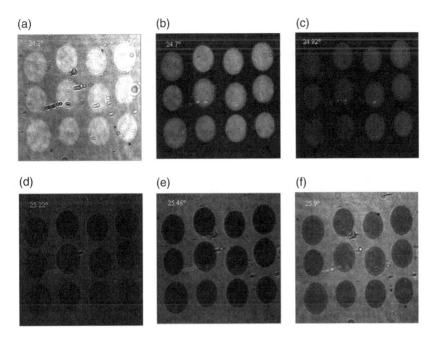

Figure 4.34 Surface plasmon microscopy images taken from a 3×4 matrix of identically functionalized sensor elements at various angles of incidence: (a) 24.2°; (b) 24.7°; (c) 24.92°; (d) 25.22°; (e) 25.46°; (f) 25.9°. *Source*: Reproduced from Zizlsperger and Knoll [12]. Copyright (1998), with kind permission of Springer Science and Business Media.

an absorbing organic film (Figure 4.33b). Here, the resonance depth is decreased and the resonance width is significantly increased as the film thickness increases. Surface plasmon resonance can be used to determine the thickness and refractive index of organic layers. The sensitivity of the method is high and changes in refractive index of the overlayer of about 10^{-5} may be monitored. In this respect, the technique compares favourably with ellipsometry (Section 4.5.3). In Chapter 10 (Section 10.4), the application of SPR to chemical sensing is described.

The shift of the SPR conditions with film thickness can be also used as the basis for a high contrast microscope. The reflected and scattered plasmonic light is converted by a lens to form an image of the interface on a television camera, for example. Only the areas at resonance appear dark. Variations in film thickness of as little as a few tenths of a nanometre are enough to generate sufficient contrast for an image. Figure 4.34 shows example of the SPR images from multispot, parallel on-line monitoring of interfacial binding reactions by surface plasmon microscopy [12]. The various photographs correspond to different incidence angles.

4.8 Photonic Crystals

The periodic arrangement of ions on a lattice gives rise to the energy band structure in semiconductors. The energy bands that result then control the motion of charge carriers through the crystal. Similarly, in a *photonic crystal*, the periodic arrangement of refractive index or dielectric constant variation controls how photons are able to move through the crystal. Photons react to the refractive index contrast in an analogous manner to the way electrons react when confronted with a periodic potential of ions. Each gives rise to a range of allowed energies and a band structure characterized by an energy gap or photonic band gap. A band gap forms when the electron wavelength is comparable to the inter-atomic spacing.

Photonic band gaps were first predicted in 1987 by Yablonovitch, at Bell Communications Research. A few years later, Yablonovitch and co-workers produced a photonic crystal using an array of 1 mm holes milled into a slab of material of refractive index 3.6. The resulting structure, which became known as Yablonovite, was found to prevent microwaves from propagating in any direction, i.e. it exhibited a three-dimensional photonic band gap. An estimate of the distance between voids is given by the wavelength of the light divided by the refractive index of the material. This relationship means that it is even more difficult to create photonic crystals in materials with high refractive index.

The simplest possible photonic crystal, shown in Figure 4.35, consists of alternating layers of material with different dielectric constants. Such a photonic crystal can form an *optical microcavity* and, as will be shown in Chapter 9 (Section 9.7), can be used to concentrate the output of light emitting devices. The traditional approach to an understanding of this system is to allow a plane wave to propagate through the material and to consider the multiple reflections that take place at each interface. Figure 4.36 shows how an optical band gap arises in the frequency (energy) versus wavevector diagram for the multilayer by changing the refractive index difference between the two layers [13]. The figure is shown in the reduced zone representation, as discussed for electrons in Chapter 3, Section 3.3.2. In Figure 4.36a, all of the layers in the multilayer structure have the same dielectric constant, so the medium is completely homogeneous. Figure 4.36b represents a structure with alternating dielectric constants of 12 and 13 (the value 13 is chosen, as it is approximately the value of the low frequency dielectric

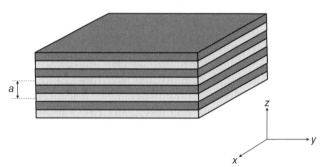

Figure 4.35 One-dimensional photonic crystal. The multilayer film consists of alternating layers with different dielectric constants, periodicity = a.

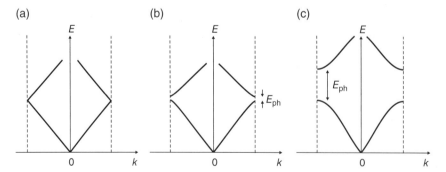

Figure 4.36 Photonic energy band structure showing energy E versus wavevector k for multilayer films, as depicted in Figure 4.35. Photonic band gap = E_{ph}. (a) Each layer has same dielectric constant $\varepsilon_r = 13$. (b) Layers alternate between $\varepsilon_r = 13$ and $\varepsilon_r = 12$. (c) Layers alternate between $\varepsilon_r = 13$ and $\varepsilon_r = 1$. *Source*: From Joannopoulos et al. [13]. Reprinted with permission of Princeton University Press.

constant for the inorganic semiconductor GaAs). Finally, Figure 4.36c is for a structure with a much higher dielectric constant ratio of 13 to 1 (air). It is evident that width of the photonic band gap, E_{ph}, increases as the difference between the dielectric constants increases. Most of the promising applications of photonic crystals exploit the location and width of the photonic band gaps. For example, a crystal with a band gap might make a very good, narrow-band filter, by rejecting all frequencies in the gap (i.e. these wavelengths would pass through the material). A resonant cavity, made from a photonic crystal, would have perfectly reflecting walls for frequencies in the gap.

Two- and three-dimensional photonic crystals can take on a large number of forms. Examples of such structures are shown in Figure 4.37. Figure 4.37a shows a square lattice of dielectric columns. For certain values of the column spacing, this crystal can have a photonic band gap in the x-y plane. Inside this gap, no extended states are permitted, and the incident ray is reflected. Although a multilayer film, such as depicted in Figure 4.35, only reflects light at normal incidence, this two-dimensional photonic crystal can reflect light from any direction in the plane. The optical analogue of an ordinary crystal is a three-dimensional photonic crystal, a dielectric that is periodic along three mutually orthogonal axes. A simple example is provided in Figure 4.37b, in which dielectric spheres are arranged in a lattice structure and surrounded by a different dielectric medium. This model is simply that of atoms making up solids, in which case the 'host' medium is air. The dielectric constants can also be reversed, with air bubbles embedded in a regular fashion in a dielectric material. Breaking the periodicity of the voids in the photonic crystal, either by enlarging or reducing the size of a few of the voids, introduces new energy levels within the photonic band gap. This is analogous to the creation of energy levels within the band gap by the addition of dopant atoms in semiconductor crystals. Examples of photonic crystals in organic thins films are described in Chapter 11, Section 11.3.3.

4.8.1 Subwavelength Optics

The nature of surface plasmons changes when they propagate on metal surfaces that are periodically textured on the scale of the wavelength of light. When the period of the nanostructure is one half that of the effective wavelength of the surface plasmon mode, scattering may lead to the formation of surface plasmon standing waves and the opening of a surface plasmon 'band gap' [14]. This is illustrated in Figure 4.38. There are two standing wave solutions, each with the same wavelength but of different frequencies (because of their different field and surface charge distributions). The upper frequency solution, ω_+ is of higher energy because of the greater

Figure 4.37 (a) A two-dimensional photonic crystal showing a square lattice of dielectric columns. (b) A three-dimensional photonic crystal consisting of a regular array of dielectric spheres.

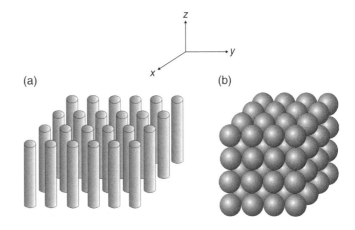

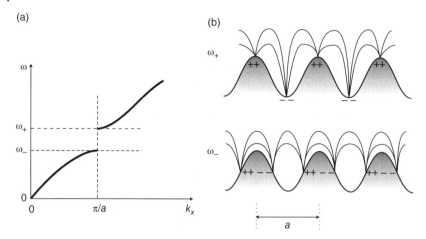

Figure 4.38 Periodic texturing of a surface can result in the formation of a surface plasmon band gap. (a) Frequency ω versus wavevector k_x dispersion curve in the x-direction showing band gap formation. (b) Electric field distributions for standing waves either side of the band gap, with frequencies ω_+ and ω_-. *Source*: Reprinted by permission from Macmillan Publishers Ltd: Barnes et al. [14] Copyright (2003).

distance between the surface charges and the greater distortion of the field. Surface plasmons with frequencies between ω_- and ω_+ cannot propagate, so this frequency interval is known as the *stop gap*. At the band edges, the density of surface plasmon states is high and there is a significant increase in the associated field enhancement.

Surface plasmon effects can give rise to enhanced transmission through periodic arrays of subwavelength (~100 nm) holes in optically thick metallic films [15]. Not only is the transmission much higher than expected from classic diffraction theory, it can be greater than the percentage area occupied by the holes, implying that even the light impinging on the metal between the holes can be transmitted. The entire periodic structure acts like an antenna in the optical regime. The transmission spectra of hole arrays display peaks that can be tuned by adjusting the period and symmetry. As the holes are too small to act as hollow waveguides, the light cannot get through directly. In fact, the interference pattern produced by the counter-propagating plasmons on the metal surface coincides with the positions of the holes and the amplitude of the field inside is boosted and light is transmitted.

One of the most intriguing developments in this area over recent years has been the development of materials with a negative index of refraction, i.e. light refracts in the opposite direction to normal as it enters a material. It has been suggested that if both the permittivity ε and permeability μ of a material were negative, then the refractive index, which determines the velocity of light within the material (Eqs. (4.2) and (4.3)) should also be negative [16]. A thin planar slab of such a 'meta' material could act as a perfect lens [17]. Such materials have been developed for microwave radiation, but it is more difficult to use the same approach for the optical regime. However, it has been shown that gold films patterned with small metal posts may have a negative permeability at visible light frequencies [18]. This led to the observation of optical impedance matching, i.e. the total suppression of reflection from an interface (Section 4.5.2) and paves the way towards components with negative refraction at visible frequencies.

Surface plasmon effects have been explored using a variety of other metallic nanostructures. These include chains of metal nanoparticles, nanostrips on a dielectric substrate, and nanogaps between metallic media and sharp metal wedges [19, 20]. Envisaged applications of such resonant plasmonic materials range from energy conversion (e.g. catalysis and chemical energy conversion) to uses in health and medicine (e.g. biosensing and photothermal therapy) [21].

Problems

4.1 Calculate the critical angle θ_c for an air/water interface. The refractive index of water is 1.33.

4.2 Light with a frequency of 5.0×10^{14} Hz travels in a polymer that has a refractive index of 2.0. What is the wavelength of the light while it is in the polymer and while it is in a vacuum?

4.3 The fraction of nonreflected radiation that is transmitted through a 10 mm thickness of a transparent plastic is 0.90. If the thickness is increased to 20 mm, what fraction of light will be transmitted?

4.4 A simple antireflection coating is required for a photographic lens that is most effective for green light of wavelength 550 nm. If the coating material has a refractive index of 1.38, how thick should it be?

4.5 Light falls from air ($n = 1.00$) perpendicularly on a transparent plate ($n = 1.72$). To the nearest percentage point, how much of the incident light is reflected? If the refractive index of the plate is decreased to $n = 1.33$, calculate the increase (in percent) in the transmitted light intensity.

References

1 Richardson, T., Roberts, G.G., Polywka, M.E.C., and Davies, S.G. (1988). Preparation and characterization of organotransition metal Langmuir-Blodgett films. *Thin Solid Films* 160: 231–239.

2 Friend, R.H., Bott, D.C., Bradley, D.D.C. et al. (1985). Electronic-properties of conjugated polymers. *Phil. Trans. R. Soc. London A* 314: 37–49.

3 Kuroda, H., Kunii, T., Hiroma, S., and Akamatu, H. (1967). Charge-transfer bands in crystal spectra of molecular complexes. *J. Mol. Spectrosc.* 22: 60–75.

4 Kuhn, H., Möbius, D., and Bücher, H. (1972). Spectroscopy of monolayer assemblies. In: *Techniques of Chemistry*, vol. 1Pt 3B (ed. A. Weissberger and B. Rossiter), p577–p702. New York: Wiley.

5 Bulović, V.G., Baldo, M.A., and Forrest, S.R. (2001). Excitons and energy transfer in doped luminescent molecular organic materials. In: *Organic Electronic Materials* (ed. R. Farchioni and G. Grosso), 391–441. Berlin: Springer.

6 Tanaka, J. (1963). The electronic spectra of aromatic molecular crystals. The crystal structure and spectra of perylene. *Bull. Chem. Soc. Jpn.* 36: 1237–1249.

7 Allen, S., Ryan, T.G., Hutchings, M.G. et al. (1993). Characterization of nonlinear optical Langmuir-Blodgett oligomers. In: *Organic Materials for Non-linear Optics III* (ed. G.J. Ashwell and D. Bloor), 50–60. London: Royal Society of Chemistry.

8 Glover, R.E. and Tinkham, M. (1957). Conductivity of superconducting films for photon energies between 0.3 and 40 kT_c. *Phys. Rev.* 108: 243–256.

9 Hu, L., Hecht, D.S., and Grüner, G. (2004). Percolation in transparent and conducting carbon nanotube networks. *Nano Lett.* 4: 2513–2517.

10 Saran, N., Parikh, K., Suh, D.-S. et al. (2004). Fabrication and characterization of thin films of single-walled carbon nanotube bundles on flexible plastic substrates. *J. Am. Chem. Soc.* 126: 4462–4463.

11 Swalen, J.G. (1986). Optical properties of Langmuir-Blodgett films. *J. Mol. Electron.* 2: 155–181.

12 Zizlsperger, M. and Knoll, W. (1998). Multispot parallel on-line monitoring of interfacial binding reactions by surface plasmon microscopy. *Progr. Colloid Polym. Sci.* 109: 244–253.

13 Joannopoulos, J.D., Meade, R.D., and Winn, J.D. (1995). *Photonic Crystals*, 2e. Princeton University Press: Princeton.

14 Barnes, W.L., Dereux, A., and Ebbesen, T.W. (2003). Surface plasmon subwavelength optics. *Nature* 424: 824–830.

15 Ebbesen, T.W., Lezec, H.J., Ghaemi, H.F. et al. (1998). Extraordinary optical transmission through sub-wavelength hole arrays. *Nature* 391: 667–669.

16 Pendry, J. (2003). Positively negative. *Nature* 423: 22–23.

17 Pendry, J.B. (2000). Negative diffraction makes a perfect lens. *Phys. Rev. Lett.* 85: 3966–3969.

18 Grigorenko, N., Geim, A.K., Gleeson, H.F. et al. (2005). Nanofabricated media with negative permeability at visible frequencies. *Nature* 438: 335–338.

19 Gramotnev, D.K. and Bozhevolnyl, S.I. (2010). Plasmonics beyond the diffraction limit. *Nat. Photonics* 4: 83–91.

20 Schuller, J.A., Barnard, E.S., Cai, W. et al. (2010). Plasmonics for extreme light concentration and manipulation. *Nat. Mater.* 9: 193–204.

21 Dionne, J.A., Baldi, A., Baum, B. et al. (2015). Localized fields, global impact: industrial applications of resonant plasmonic materials. *MRS Bull.* 40: 1138–1146.

Further Reading

Barford, W. (2005). *Electronic and Optical Properties of Conjugated Polymers*. Oxford: Oxford University Press.

Dressel, M. and Grüner, G. (2002). *Electrodynamics of Solids: Optical Properties of Electrons in Matter*. New York: Cambridge University Press.

Duffin, W.J. (1990). *Electricity and Magnetism*, 4e. London: McGraw-Hill.

Eisberg, R. and Resnick, R. (1985). *Quantum Physics of Atoms, Molecules, Solids, Nuclei, and Particles*, 2e. New York: Wiley.

Kitai, A. (ed.) (2008). *Luminescent Materials and Applications*. Chichester: Wiley.

Maier, S.A. (2007). *Plasmonics: Fundamentals and Applications*. New York: Springer.

Nabook, A. (2005). *Organic and Inorganic Nanostructures*. Boston: Artech House.

Raether, H. (1977). Surface plasma oscillations and their applications. *Phys. Thin Films* 9: 145–261.

Schwoerer, M. and Wolf, H.C. (2006). *Organic Molecular Solids*. Berlin: Wiley-VCH.

Wilson, J. and Hawkes, J. (1998). *Optoelectronics*. London: Prentice Hall.

5

Electroactive Organic Compounds

Fillet of a fenny snake,
Into the cauldron boil and bake

5.1 Introduction

This chapter concerns the nature and properties of some important groups of electroactive compounds used in organic electronics and molecular electronics research. The focus is on organic materials exhibiting interesting electrical and magnetic behaviour. Discussions of other important organic compounds are to be found elsewhere in this book. For example, materials with non-linear optical behaviour are introduced in Chapter 4, while photochromic compounds are described in Chapter 11. Liquid crystals and biological compounds are the subjects

Organic and Molecular Electronics: From Principles to Practice, Second Edition. Michael C. Petty.
© 2019 John Wiley & Sons Ltd. Published 2019 by John Wiley & Sons Ltd.
Companion website: www.wiley.com/go/petty/molecular-electronics2

of separate chapters (Chapters 8 and 12, respectively). Synthetic details on all the electroactive compounds are beyond the scope of this book and are covered in some detail elsewhere (for example, [1]). By way of an introduction, and particularly as an aid to the non-chemist, some notes on basic organic chemistry are first given.

5.2 Selected Topics in Chemistry

5.2.1 Moles and Molecules

An important principle stated in 1811 by the Italian chemist Amedeo Avogadro was that equal volumes of gases at the same temperature and pressure contain the same number of molecules, regardless of their chemical nature and physical properties. This idea was introduced in Chapter 2, Section 2.2.1. *Avogadro's number*, $N_A = 6.02 \times 10^{23}$ – the number of molecules of any gas present in a volume of 22.41 l, is the same for the lightest gas (hydrogen) as for a heavy gas such as carbon dioxide or bromine. It is one of the fundamental constants of chemistry and allows calculation of the amount of a pure substance, the mole (Chapter 2, Section 2.2.1).

One mole is the amount of substance of a system that contains as many elementary entities as there are atoms in 0.012 kg of the isotope carbon 12; it is often abbreviated simply as 'mol'. When the mole is used, the elementary entities must be stated and may be atoms, molecules, ions, electrons, other particles, or specified groups of such particles. A few books prefer to use the unit *kilogram-mole*, or *kmol*. When the kilogram-mole is used, Avogadro's constant becomes $6.02 \times 10^{26}\,kmol^{-1}$. For a material of molecular weight M and density ρ, the number of atoms per unit volume n may be calculated as follows:

$$n = \frac{N_A \rho}{M} \tag{5.1}$$

Because of its role as a scaling factor, Avogadro's number provides the link between a number of useful physical constants as we move between the atomic and macroscopic worlds. For example, it provides the relationship between the *universal gas constant R* and the Boltzmann constant k_B (Chapter 3, Section 3.2):

$$R = k_B N_A \tag{5.2}$$

5.2.2 Acids and Bases

In 1884, Svante Arrhenius suggested that salts such as NaCl dissociate when they dissolve in water to give electrically charged particles, which he called *ions*:

$$NaCl(s) \rightarrow Na^+(aq) + Cl^-(aq) \tag{5.3}$$

where (s) and (aq) refer to the solid and aqueous phases, respectively. Later, Arrhenius extended this theory by suggesting that *acids* are neutral compounds that ionize when they dissolve in water to give H^+ ions and a corresponding negative ion. According to the theory, hydrogen chloride is an acid because it ionizes when it dissolves in water to give hydrogen (H^+) *cations* and chloride (Cl^-) *anions*, according to the following scheme:

$$HCl(g) \rightarrow H^+(aq) + Cl^-(aq) \tag{5.4}$$

where (g) refers to the gaseous phase. In contrast, *bases*, such as NaOH, were considered to produce a hydroxide ion, OH⁻. Although a major advance in the understanding of these compounds, the Arrhenius theory ran into difficulties when compounds such as Na_2CO_3 were considered. These exhibited the properties of bases, but clearly could not ionize directly to yield a hydroxide ion. In the 1920s, the Arrhenius theory was independently developed by Thomas Lowry and Johannes Brønsted to a general and more powerful description of acids and bases. The Lowry–Brønsted (LB) definition of an acid is a species that tends to donate or lose a hydrogen ion, or *proton*. An LB base has a tendency to gain or accept a proton. In the LB model, HCl donates a proton to a water molecule (and is therefore an acid).

Under the LB definition, both acids and bases are related to the concentration of hydrogen ions present. Acids increase the concentration of hydrogen ions, while bases decrease the concentration of hydrogen ions (by accepting them). Hence, the acidity or basicity of something can be measured by its hydrogen ion concentration. The well-established *pH* scale for measuring acidity was introduced in 1909 by the Danish biochemist Sören Sörensen. This is defined as

$$pH = -\log\left[H^+\right] \tag{5.5}$$

The concentration is commonly abbreviated by using square brackets, thus [H⁺] = hydrogen ion concentration. When measuring pH, [H⁺] is in units of moles of H⁺ per litre of solution. For example, for pure water the experimentally determined value of [H⁺] is $1.0 \times 10^{-7}\,mol\,l^{-1}$ (or 100 nanomoles per litre). Thus water has a pH equal to 7. A change in [H⁺] by a factor of 2 will result in a pH change of 0.3. The pH scale ranges from 0–14. Substances with a pH between 0 and 7 are acids, while those with a pH greater than 7 and up to 14 are bases. In the middle, at pH = 7, are neutral substances, for example, pure water.

According to the LB theory, reactions between acids and bases always involve the transfer of an H⁺ ion from a proton donor to a proton acceptor. Acids can therefore be neutral molecules:

$$HCl + NH_3 \rightarrow NH_4^+ + Cl^- \tag{5.6}$$
$$\text{acid} \quad \text{base}$$

positive ions:

$$NH_4^+ + OH^- \rightarrow NH_3 + H_2O \tag{5.7}$$
$$\text{acid} \qquad \text{base}$$

and also negative ions:

$$H_2PO_4^- + H_2O \rightarrow HPO_4^{2-} + H_3O^+ \tag{5.8}$$
$$\text{acid} \qquad \text{base}$$

The LB theory therefore expands the number of potential acids. It also enables the prediction of whether a compound is an acid from its chemical formula.

Gilbert Lewis defined these two terms in even a less restrictive way and included even more examples: an acid was considered to be an electron acceptor, while a base was an electron donor. It may be more accurate to say that Lewis acids are substances which are electron-deficient (or low electron density) and Lewis bases are substances which are electron-rich (high electron density). This approach further expands the number of acids.

5.2.3 Ions

Many substances dissolve in water to produce ions. Water itself contains ions. At 25 °C, the density of water is about $1.0 \, g \, cm^{-3}$ (or $1.0 \, g \, ml^{-1}$). This corresponds to a concentration of water of about $55.4 \, mols \, l^{-1}$. As noted in the previous section, the concentration of the H^+ (or OH^-) ions formed by the dissociation of H_2O molecules at 25 °C is only $1.0 \times 10^{-7} \, mols \, l^{-1}$. Therefore, the ratio of the concentration of H^+ (or OH^-) ions to that of neutral H_2O molecules is about 2×10^{-9}. In other words, only about two parts per billion of the water molecules dissociate into ions at room temperature.

Because the proton is several orders of magnitude smaller than the water molecule, the charge on a particular H^+ ion is distributed over such a small volume that this H^+ ion is attracted towards any source of negative charge that exists in solution. Therefore, the instant that an H^+ ion is created in an aqueous solution, it bonds to a water molecule. In fact, each H^+ ion that an acid donates to water is bound to four neighbouring water molecules, so an appropriate formula for the entity produced when an acid loses a proton is $H(H_2O)_4{}^+$ or $H_9O_4{}^+$.

When an ion (positively or negatively charged) is inserted into water, it changes the structure of the hydrogen bond network. A water molecule tends to rotate (reorient) so that its dipole (polarized charge) faces the opposite charge of the ion. As the water molecules orient towards the ion, they break the hydrogen bonds to their nearest neighbours. The group of water molecules thus formed around an ion is called a *hydration shell*. The orientation of the molecules in the hydration shell results in a net charge on the outside of this shell, a charge of the same sign as that of the ion in the centre. The charge on the outside of the hydration shell tends to orient water molecules in the immediate vicinity, leading to a second hydration shell. A further highly diffuse region, the tertiary hydration shell, may well be present before distances are reached that are so far removed from the ion that the water molecules become essentially indistinguishable from those of bulk water. The result of forming hydration shells is to weaken the structure of the hydrogen bond network. This explains the observation that salt water has a lower freezing point than pure water: each ion in the liquid has a hydration shell of oriented water molecules around it, which prevents the water molecules from forming the hexagonal structure of ice.

The degree of hydration depends on a number of factors, in particular the ionic size and its associated charge density. Figure 5.1 shows a schematic diagram of the hydration sphere of a metal cation, M^{n+}, having a primary shell of six water molecules and a secondary shell of about 12 water molecules [2]. Such a situation is probably the case for trivalent metal ions such as Cr^{3+} and Rh^{3+}.

5.2.4 Solvents

If organic or molecular electronics are to prove useful technologies, the appropriate organic compounds must be able to be easily processed, i.e. they must be capable of being made into the shapes required for devices, probably in the form of thin films. The processing can take many forms but frequently requires the organic electroactive compounds to first be dissolved in a solvent to enable other steps, e.g. spin-coating or self-assembly, to proceed. The majority of chemical reactions are also undertaken in solution. Here, a solvent fulfils several functions. It solvates the reactants and reagents so that they dissolve. This facilitates collisions between the reactant(s) and reagents to transform the reactant(s) to product(s). The solvent also provides a means of temperature control, either to increase the energy of the colliding particles so that they will react more quickly, or to absorb heat that is generated during a reaction. This section briefly looks at the properties of some of the more common organic solvents.

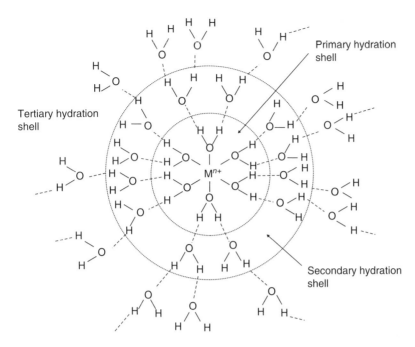

Figure 5.1 The localized structure of a hydrated metal cation in aqueous solution [2].

The rule of thumb for choosing a solvent is that 'like dissolves like'. For example, polar solvents will dissolve polar reactants. Generally, there are three measures of the polarity of a solvent: its dipole moment (Chapter 2, Section 2.3.5); its dielectric constant; and its miscibility with water. Molecules with large dipole moments and high dielectric constants are considered polar, whereas those with small dipole moments and small dielectric constants are classified as nonpolar. On a practical basis, solvents that are miscible with water are polar and those that are not are nonpolar.

Solvents can be conveniently categorized according to their polarity. A number of examples are shown in Table 5.1. There are three main groupings: polar *protic*; polar *aprotic*; and nonpolar. In this context, protic refers to a hydrogen atom attached to an electronegative atom (almost always an oxygen atom). Therefore, protic compounds can generally be represented by the formula ROH, where R is a chemical group. The dipole moment of polar protic solvents originates from the O—H bond. The large difference in electronegativities (Chapter 2, Section 2.2.4) of the oxygen and the hydrogen atom, combined with the small size of the hydrogen atom, justify separating molecules that contain an OH group from those polar compounds that do not. Examples of protic compounds from Table 5.1 are water (HOH), methanol (CH_3OH), and acetic acid (CH_3CO_2H).

The aprotic solvents do not contain an O—H bond, but nevertheless possess large dipole moments (and large dielectric constants). Typically, the dipole originates from a multiple bond between carbon and either oxygen or nitrogen. Most polar aprotic solvents contain a C=O double bond; a good example is acetone, $(CH_3)_2C$=O. Nonpolar compounds possess zero (i.e. benzene, C_6H_6) or low dipole moments (diethyl ether, CH_3CH_2OCH) and are not miscible with water.

The choice of solvent is known to affect the behaviour of certain organic electronic devices. For example, the efficiency of bulk heterojunction photovoltaic cells (Chapter 9, Section 9.7.1) can be influenced by the evolution of particular molecular morphologies during solvent evaporation.

Table 5.1 Properties of some common organic solvents.

Solvent	Structure	Boiling point (°C)	Dielectric Constant	Class Polar	Class Protic
Hexane	$CH_3(CH_2)_4CH_3$	69	2		
Benzene		80	2.3		
Diethyl ether	$CH_3CH_2OCH_2CH_3$	35	4.3		
Chloroform	$CHCl_3$	61	4.8		
Acetic acid	$H_3C-\overset{\overset{O}{\|\|}}{C}-OH$	118	6.1	*	*
Tetrahydrofuran (THF)		66	7.6		
Acetone	$H_3C\overset{\overset{O}{\|\|}}{\underset{}{C}}CH_3$	56	21	*	
1-Propanol	$CH_3CH_2CH_2-OH$	97	20	*	*
Ethanol	CH_3CH_2-OH	78	25	*	*
Methanol	CH_3-OH	65	33	*	*
Acetonitrile	$H_3C-C\equiv N$	82	38	*	
Dimethyl sulfoxide (DMSO)	$H_3C\overset{\overset{O}{\|\|}}{\underset{}{S}}CH_3$	189	47	*	
Water	$H-O-H$	100	78	*	*
Formamide	$H\overset{\overset{O}{\|\|}}{\underset{}{C}}NH_2$	210	111	*	*

Normally, the choice of low boiling point solvents is preferred when processing thin films (e.g. by spin-coating) as this leads to rapid drying. However, if the solvent evaporation occurs faster than the process of crystallization, then amorphous films can result. Higher boiling point solvents can be used to increase deliberately the film drying time. This so-called *solvent-annealing* can result in improved thin film morphology. Solvent vapour annealing allows macroscopic healing of molecular assemblies at surfaces to form suprastructures featuring a higher degree of order, and significant improvement of the performance of organic electronic devices.

In the case of films containing mixtures of different components, high boiling point solvents can lead to phase segregation. To circumvent this, mixed solvents can be used. For example, by changing the drying time of a mixed polymer film with mixtures of high and low boiling point solvents, control over the extent of phase demixing and crystallinity can be achieved.

5.2.5 Functional Groups

Functional groups are specific groups of atoms within organic molecules that are responsible for the characteristic chemical reactions of those molecules. The same functional group will undergo the same or similar chemical reaction(s) regardless of the size of the molecule it is a part of. Functional groups are attached to the carbon backbone of organic molecules. They determine the characteristics and chemical reactivity of molecules. Functional groups are far less stable than the carbon backbone and are likely to participate in chemical reactions. Common functional groups found in compounds used in organic electronics are depicted in Table 5.2.

5.2.6 Aromatic Compounds

In organic chemistry, hydrocarbons are divided into two classes: *aromatic* compounds and *aliphatic* compounds. Aromatic compounds contain an especially stable ring of atoms, such as benzene. Aliphatic compounds can be saturated, like hexane, or unsaturated, like hexene and hexyne. Open-chain compounds (whether straight or branched) contain no rings of any type, and are thus aliphatic.

Table 5.2 Chemical structure of common functional groups found in organic compounds.

Compound type	Functional Group	Comments
Alcohol	—C—O—H	Functional group is hydroxyl. Primary, secondary, and tertiary alcohols, depending on number of carbons to which C—OH carbon is bonded. Miscible in organic solvents. Can show either acidic or basic properties.
Aldehyde	C=O (with H)	Can be oxidized to carboxylic acid or reduced to primary alcohol.
Amine	—C—N	Primary, secondary, and tertiary compounds exist. The nitrogen atom has typically a lone pair of electrons. Amines act as strong bases.
Amide	—C—N (with O)	Derivatives of carboxylic acids.
Carboxylic acid	—C—O—H (with O)	Weak acids. The presence of electronegative groups (e.g. —OH or —Cl) next to carboxylic group increases the acidity.
Ester	—C—O—C (with O)	More water soluble than their parent hydrocarbons, but more hydrophobic and than either their parent alcohols or parent acids.
Ether	—C—O—C	Molecules cannot form hydrogen bonds among each other. More hydrophobic than esters or amides of comparable structure.
Ketone	C—C=O—C	More volatile than alcohols or carboxylic acids of similar molecular weight.
Thiol	—C—S—H	Formerly known as mercaptans. Bind strongly to skin proteins and to some metals.
Nitrile	—C≡N	Sometimes referred to as cyanide.
Epoxide	C—C (with O)	Also known as oxirane.

Aromatic compounds represent a large class of molecules that include benzene and compounds that resemble benzene in some of their chemical properties, e.g. toluene, naphthalene, and anthracene. These compounds contain at least one ring that consists of six carbon atoms, each joined to at least two other carbon atoms, and each joined to adjacent carbon atoms by one single and one double bond. The resulting hexagonal structure is characteristic of many aromatic compounds. Most aromatic compounds have a delocalized π-electron system of alternating single and double bonds and are planar structures. This conveys a particular chemical stability to these systems. The number of π-electrons is $(4n + 2)$, where n is an integral number. This is known as the *Hückel rule*. Thus the number of delocalized electrons in benzene (6), naphthalene (10), and anthracene (14) is consistent with their aromatic character. In contrast, cyclobutadiene, C_4H_4, is not aromatic, since the number of π-delocalized electrons is 4, which is not satisfied by any n integer value. The cyclobutadienide (2-) ion, however, is aromatic. The eight-membered cyclic compound with four alternating double bonds (cyclooctatetraene) is also not aromatic and shows reactivity similar to alkenes.

Aromatic molecules normally exhibit enhanced chemical stability, compared to similar non-aromatic molecules. The circulating π-electrons in an aromatic molecule generate significant local magnetic fields that can be detected by the *nuclear magnetic resonance* (NMR) technique. The bonds in an aromatic ring are less reactive than ordinary double bonds; aromatic compounds tend to undergo ionic substitution (e.g. replacement of a hydrogen bonded to the ring with some other group) rather than addition (which would involve breaking one of the resonant bonds in the ring).

The presence of the six-membered benzene ring is not essential for aromatic compounds. There are a large number of compounds containing heteroatoms (O, N, S) that are also aromatic. Some examples are shown in Table 5.3. The aromicity arises because the heteroatom is either involved in a double bond in the ring, or it can make use of a lone pair of electrons (Chapter 2, Section 2.4.1) to interact with the π-electrons to satisfy Hückel's rule. For example, pyridine (C_5H_5N) has a lone pair of electrons at the nitrogen atom that does not participate in

Table 5.3 Examples of aromatic molecules containing heteroatoms.

Molecule	Structure	Molecule	Structure
Pyridine		Indole	
Pyrazine		Purine	
Pyrimidine		Furan	
Thiophene		Oxazole	
Pyrrole		Thiazole	

Table 5.4 Structures of pyrene and corenene, two aromatic compounds that do not follow the Hückel rule.

Molecule	Structure
Pyrene	
Coronene	

the aromatic π-system. This makes pyridine a basic compound with chemical properties similar to tertiary amines. In the case of pyrrole (C_4H_5N), which has a very low basicity compared to pyridine, the ring nitrogen is connected to a hydrogen atom. The lone pair of electrons of the nitrogen atom then becomes delocalized in the aromatic ring. It should be noted that the purine molecule depicted in Table 5.3 is a precursor to the DNA bases adenine and guanine (Chapter 12, Section 12.3.1).

There are also some compounds that are aromatic but which do not follow Hückel's rule. Examples are the large macrocycles pyrene, $C_{16}H_{10}$, and coronene, $C_{24}H_{12}$, shown in Table 5.4. Such structures can be delocalized in 'subcycles', such that every atom participates in some number of resonance structures.

5.2.7 Material Purity

Ultra-high material purity is a fundamental requirement for ensuring high performance and reliability in all types of electronic devices. The relationships between the nature and concentration of impurities in inorganic semiconductor devices, such as those based on Si and GaAs, are well established. For example, substitutional lattice impurities such as phosphorus or boron can change the conductivity of a host silicon semiconductor by several orders of magnitude if introduced at only the parts per billion level. This is because each impurity atom disrupts the valence states of the neighbouring lattice atoms (Chapter 3, Section 3.3.5). In organic electronics, the types of impurities and their effect on performance may differ substantially from those affecting inorganic semiconductor devices; for example, no equivalent ordered lattice exists in most organic solids. Nonetheless, impurities can strongly influence the conductive properties of an organic material. Impurities do not necessarily form electrically active substitutional defects in organic solids, but may act as deep traps, extracting charge or acting as recombination sites within the thin film.

Organic electronic materials can generally be classified into two categories: 'small molecules', and polymers (although a third group – complex biological molecules – are making an impact in some research areas). Small-molecule materials have well-defined molecular weights, allowing for straightforward separation of the host from the impurities. One common means for accomplishing this is via thermal gradient sublimation, whereby the organic source material is heated in a vacuum furnace, and then allowed to condense downstream in a cooler region of the furnace. The low- and high-molecular-weight impurities each condense in a different temperature zone from the desired source material, making separation of these components possible. Using sublimation techniques, fractional impurity concentrations as low as 10^{-4} are potentially achievable, although it remains an important challenge to measure this quantity precisely, because of the complex role that impurities play in affecting the properties of the host material.

Because polymer chains in solution have a dispersity of molecular weights (Chapter 2, Section 2.6.1), there are few strategies for purifying the material based on molecular weight alone. Chromatography and other 'distillation' processes are commonly used to achieve the highest level of purity, although attaining $< 1\%$ impurity concentrations remains a challenge.

5.3 Conductive Polymers

The idea of an organic polymeric material being electrically conductive was introduced in Chapter 3, Section 3.4.1. The simplest conductive polymer is *trans*-polyacetylene, consisting of hydrocarbon chain with alternating single and double bonds. This results in a delocalized electron system, which allows charge carriers to move along the polymer backbone. *Trans*-polyacetylene is a semiconductor with an energy band gap of about 1.4 eV. One of the advantages of conjugated polymers is that, to some extent, it is possible to control and vary their properties through appropriate modifications to their structure. The local arrangement of the carbon atoms on the backbone can be altered, for instance by incorporating heteroatoms, or the hydrogen atoms can be replaced with other side groups. Many conductive polymers have been synthesized to provide particular electronic features (band gap, electron affinity).

The monomer repeat units are often based on five-membered or six-membered (benzene) carbon ring systems. These include polyphenylene, poly(*p*-phenelenevinylene), polypyrrole, polythiophene (and various other polythiophene derivatives), and polyaniline (PANi) [3]. The chemical structure of some of these materials is shown in Figure 5.2 (the structure of PANi has been previously given in Chapter 3, Section 3.4.1, Figure 3.34). Generally, the band gap of the conductive polymer decreases with the length of conjugation, in accordance with Eq. (3.59). Many conductive polymers have been developed so that they can be easily solution-processed and used as thin films in devices. This is accomplished by attaching alkyl chains, which confer solubility in organic solvents. These chains are positioned on the molecules so that they do not disrupt the delocalized electron system. Some examples are given in Chapter 9, Section 9.6.

Heteroatoms are often incorporated into the polymer backbone; polycarbonitrile is the simplest example. Starting with the *trans*-isomer of polyacetylene, polycarbonitrile can be obtained by replacing every second CH group with an N atom (Figure 5.2b). In Chapter 3, Section 3.4.1, it was noted that the occurrence of a bond-length alternation for *trans*-polyacetylene could be considered as a Peierls' distortion, i.e. a doubling of the unit cell that accompanies the creation of a band gap at the Fermi level. For polycarbonitrile, the presence of the N atoms leads by itself to a doubling of the unit cell and, therefore, one might not directly expect a C—N bond-length alternation. However, bond length alternation does occur and leads to an increase in the band gap at the Fermi energy. This particular polymer has the unusual property of having a sigma band very close to the Fermi level. This (occupied) band is formed by the nitrogen lone-pair orbitals.

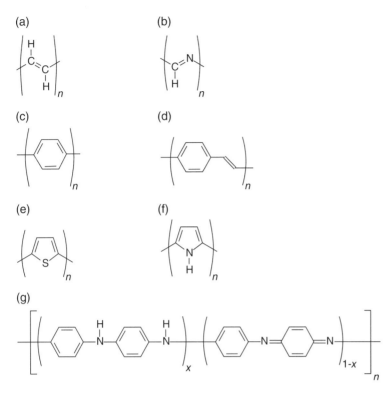

Figure 5.2 Chemical structures of some conductive polymers: (a) polyacetylene; (b) polycarbonitrile; (c) polypheneylene; (d) poly(p-phenylenevinylene); (e) polythiophene; (f) polypyrrole; and (g) polyaniline (PANi).

Poly(*p*-paraphenylenevinylene) (PPV) (depicted in Figure 5.2d) is one of the most studied conjugated polymers and was one of the first used in the fabrication of organic light emitting devices (Chapter 9, Section 9.6); the LUMO and HOMO configuration for this polymer have been described in Chapter 3, Section 3.4.1. Due to the larger unit cell compared with polyacetylene, there are many more bands [4]. The total width of the valence bands is somewhat larger than for polyacetylene, but the position of the Fermi level does not change significantly. The most important result is that both the highest occupied and lowest occupied bands are, as expected, of π symmetry and, in addition, have their major components on the *vinyl* linkages between the phenylene rings (the term vinyl is used generically to refer to a substitution at a carbon atom that is part of an alkene double bond). Therefore, when attempting to modify the electronic properties, the best approach is to replace the hydrogen atoms associated with the vinyl link by other side groups. Substitution of one of the hydrogens with the electron donor NH_2 leads to an overall upward shift of the bands in energy, whereas the electron acceptor CN produces an overall downward shift.

Instead of replacing the hydrogen atoms with other side groups, some of the backbone atoms may be substituted with others to modify the electronic properties, equivalent to passing from polyacetylene to polycabonitrile. Replacing one of the CH groups of the phenylene rings of PPV with an N atom produces poly(2,5-pyridine vinylene), PPyV. Conversely, replacing the vinyl linkages with either single N atoms or single NH groups results in PANi.

Polyaniline (Figure 5.2g) differs from many other conjugated polymers in a number of ways. The N heteroatom lies in the conjugation path along the polymer backbone. Due to steric hindrances, the backbone of this polymer is not planar, but the polymer instead forms helices.

Therefore, the separation into σ and π orbitals is only approximate. The lack of strict planarity has led to the proposal that local distortions, so-called ring-twist distortions, may occur. As previously noted in Chapter 3, PANi can exist in three oxidation states: fully reduced leucoemeraldine ($x = 1$ in Figure 5.2g), which has no quinoidimine units, oxidized emeraldine ($x = 0.5$), and doubly (or fully) oxidized pernigraniline ($x = 0$). Only the pernigraniline form is strictly conjugated. However, the emeraldine form has the property that upon protonation of the imine sites, i.e. placing the emeraldine base into an acidic solution of pH < 4, the conductivity increases by 11 orders of magnitude. The emeraldine form of PANi is also unique, in that both the base form and the salt (conductive) form are soluble and can be processed.

The charge carriers in PANi and the role of the lattice distortions are more complex than other nondegenerate ground state polymers. First, only protons are required to 'dope' emeraldine. The addition of a proton to imine nitrogen causes the number of π electrons on the backbone to remain constant; however, the conductivity is seen to increase by many orders of magnitude. Furthermore, *thermoelectric power* (a phenomenon exploited in the thermocouple) measurements reveal that, even in the highly conductive state, the charge carriers in the films have negative charge, i.e. electrons rather than holes. This seems strange as, upon protonation, holes are added to the backbone. An explanation is that protonation removes one of the imine nitrogens' lone pair electrons to form the N—H bond. The remaining unpaired electrons can then hop between the vacancies left at these sites, resulting in negative charge carriers. Polyaniline (or the family of aniline polymers) is thus an extremely complex system.

Compared with polyacetylene and polycarbonitrile, the PPV- and PANi-based polymers contain more or less aromatic rings as well as, in some cases, heteroatoms. A less complicated system, and one of the most intensively studied conjugated polymers, is polythiophene, shown in Figure 5.2e. In its simplest form, it consists of a planar zig-zag sequence of C_4H_2S units and of the two forms, shown in Figure 5.3. The aromatic structure, shown in Figure 5.3a, has a lower total energy than the *quinoid* one (Figure 5.3b). The aromatic structure has a larger band gap at the Fermi energy level than the quinoid structure, but both structures have π bands close to the Fermi level [4]. During the second half of the 1980s, scientists at the Bayer AG research laboratories in Germany developed a new polythiophene derivative, poly(3,4-ethylenedioxythiophene), having the backbone structure shown in Figure 5.4; the polymer is often abbreviated to PEDOT or PEDT [5, 6]. Initially, PEDOT was found to be an insoluble polymer, but is exhibited some interesting properties. For example, it possessed a high conductivity ($\sim 300 \, S \, cm^{-1}$), was almost transparent as thin oxidized films, and was highly stable in the oxidized state. The solubility problem was subsequently solved by using a water-soluble polyelectrolyte, poly(styrene sulfonic acid) PSS (Figure 5.4), to produce a water-soluble complex, PEDOT:PSS. The role of the PSS is two-fold: first, to provide a counterbalancing ion; second, to keep the PEDOT polymer segments dispersed in an aqueous medium. Figure 5.5 shows the optical transmission versus the surface, or sheet, resistance (in units of Ω per square) (Chapter 3, Section 3.2.2). Possible applications for the PEDOT:PSS complex include capacitors, antistatic coatings, printed circuit boards, and as a semi-transparent electrode in electroluminescent devices (Chapter 9, Section 9.6).

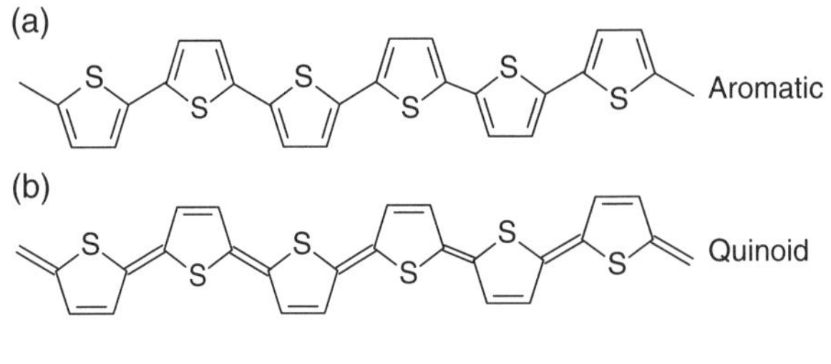

(a)

Aromatic

(b)

Quinoid

Figure 5.3 Aromatic and quinoid forms of polythiophene.

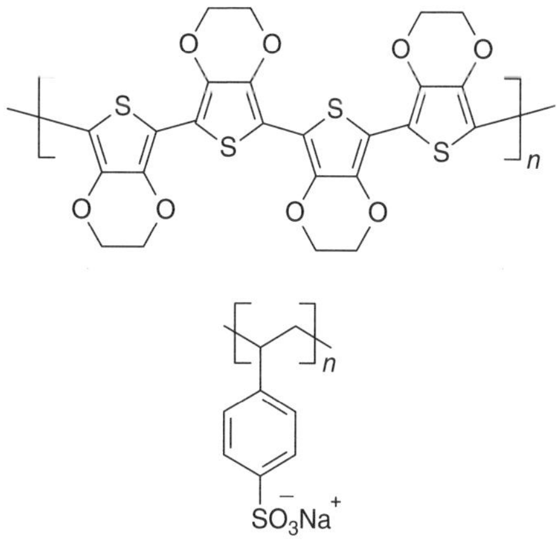

Figure 5.4 Top: chemical structure of poly(3,4-ethylenedioxythiophene) often abbreviated as PEDOT (or PEDT). Bottom: Polystyrene sulphonic acid, PSS. The PEDOT is frequently used as a complex with PSS.

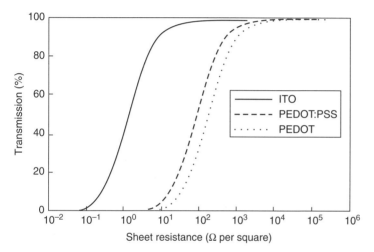

Figure 5.5 Transmission versus sheet resistivity for indium-tin-oxide (ITO) ($\sigma = 600 \mathrm{S\, cm}^{-1}$), a proprietary PEDOT:PSS formulation – Baytron® PHC ($\sigma = 450 \mathrm{S\, cm}^{-1}$), and in situ polymerized PEDOT ($\sigma = 600 \mathrm{S\, cm}^{-1}$). *Source:* From Kirchmeyer and Reuter [6]. Copyright (2005). Reproduced by permission of The Royal Society of Chemistry.

The optical and electrical behaviour of PEDOT:PSS is dependent on its electrochemical oxidation state. In its pristine state, the PEDOT component is partly oxidized with the PSS-counterions, ensuring the overall charge neutrality of the complex. However, under the application of an appropriate bias voltage, it is possible to oxidize further or reduce the PEDOT. Figure 5.6 shows how this can be achieved by the incorporation of PEDOT:PSS in a simple electrochemical cell: two thin adjacent PEDOT:PSS patterns are deposited onto solid substrate and in direct contact with a layer of a solid electrolyte [7]. When a voltage is applied to the adjacent PEDOT:PSS patterns, the PEDOT in the negatively biased pattern becomes reduced, while that in the positively biased pattern becomes further oxidized. The electrochemical reaction is described by

$$\mathrm{PEDOT : PSS + M^+ + e^- \Leftrightarrow PEDOT + M : PSS} \tag{5.9}$$

where $\mathrm{M^+}$ denotes positively charged metal ions in the electrolyte and $\mathrm{e^-}$ denotes the electrons in the PEDOT.

(a)

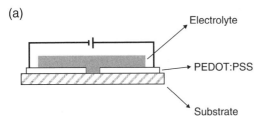

Electrolyte

PEDOT:PSS

Substrate

Figure 5.6 (a) Structure of electrochemical cell incorporating PEDOT:PSS. Application of a voltage between the left and right PEDOT patterns results in the oxidation and reduction of the polymer regions. (b) Optical absorption spectra of the oxidized and reduced PEDOT:PSS patterns. *Source: Reprinted from Andersson et al. [7]. Copyright (2002). Reproduced with permission from Wiley-VCH.*

(b)

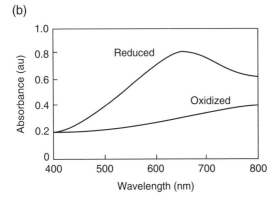

The above reaction is associated with a change of the electrical and optical properties of the PEDOT:PSS. In its oxidized state, PEDOT possesses a high concentration of free charge carriers and the optical properties are determined by transitions between bipolaronic states (Chapter 3, Section 3.4.3). This results in a low degree of optical absorption in the visible wavelength region (Figure 5.6b) and high electrical conductivity. In contrast, the reduced, neutral form of PEDOT exhibits semiconductive properties, characterized by a strong optical absorption in the visible wavelength region and low electrical conductivity. It has been suggested that the structure depicted in Figure 5.6 might be used as the basis for an electrochemical active matrix addressed display [7].

Atoms of sulphur or nitrogen possess s and p valence electrons, as in the case of carbon. The chemical bonds that they make in carbon-based polymers are therefore expected to be relatively similar. This situation may change when, instead, metal atoms with d valence electrons (Chapter 2, Section 2.2) are incorporated into the backbone. However, the presence of metal atoms does not necessarily destroy the conjugation along the polymer backbone [4].

5.4 Charge-Transfer Complexes

Charge-transfer compounds were introduced in Chapter 3, Section 3.4.1. The best known compounds are materials such as tetrathiafulvalene (TTF) and tetracyanoquinodimethane (TCNQ). The latter was first synthesized at the Dupont laboratories in 1960 and formed the basis of the first stable conductive organic materials. Wide ranges of other donor and acceptor complexes that exhibit semiconductive behaviour have been synthesized since this time. Some examples are shown in Figure 5.7. Many charge-transfer systems are arranged in separate donor and acceptor stacks (Chapter 3, Section 3.4.1, Figure 3.33), resulting in extensive electron delocalization. The classic complex TTF-TCNQ has room temperature conductivity of about $500\,\mathrm{S\,cm^{-1}}$ and, between temperatures of 298 and 54 K, shows metallic behaviour with

Figure 5.7 Examples of compounds that can form charge-transfer complexes. Electron donors are shown on the left: perylene, tetrathiafulvalene (TTF), tetramethyltetraselenafulvalene (TMTSF), and bis(ethylenedithio)-TTF (BEDT-TTF or ET). Electron acceptors and anions are shown on the right: tetracyanoquinodimethane (TCNQ), tetracyanonaphthalene (TNAP) and Ni(dmit)$_2$ (H$_2$dmit = 4,5-dimercapto-1,3-dithiole-2-thione).

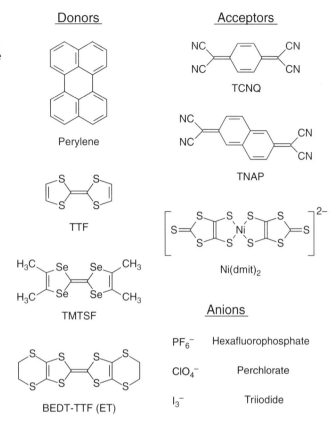

decreasing electrical resistance with decreasing temperature. Below 54 K, a Peierls transition opens up an energy gap in the band structure and the complex becomes semiconducting. This characteristic is depicted in Figure 5.8, together with the conductivity versus temperature data for other organic charge-transfer systems [9]. Organic superconductivity (Chapter 3, Section 3.4.4) was first observed in 1979 in salts of tetramethyltetraselenafulvalene (TMTSF) (Figure 5.7). For example, (TMTSF)$_2$$^+$ PF$_6$$^-$ has a transition temperature of 0.9 K under a pressure of about 9 kbar (which suppresses the Peierls transition); this compound also exhibits the Meissner effect (the exclusion of magnetic field from the material), an important test for superconductivity.

Several other salts of TMTSF are superconducting at temperatures less than 4 K; conductivity versus temperature data for (TMTSF)$_2$$^+$ ClO$_4$$^-$ are included in Figure 5.8. The molecule bis(ethylenedithio)tetrathiafulvalene (BEDT-TTF or ET) contains eight sulphur atoms and is an important donor. A family of electrically conductive salts has been synthesized with the stoichiometry (ET)$_2$X, where X is the charge-compensating monovalent anion. When the counter ion is Cu[N(CS)$_2$]Br, Figure 5.8 reveals a relatively high superconducting transition temperature of about 9 K. Such salts can possess little, or no, columnar stacking, and are characterized by short inter-stack S—S interactions. One of the highest superconducting transition temperatures reported for BEDT-TTF is 12.3 K at ambient pressure and 13.4 K at 8.2 GPa [10].

Metal complexes of the 1,2-dithiolene *ligand* (a ligand is an atom, ion, or functional group that is bonded to one or more central atoms or ions, usually metals generally through coordinate covalent bond) form the basis of a range of highly conductive materials and, in some cases, superconducting systems. The redox behaviour of these systems can be varied by changing the

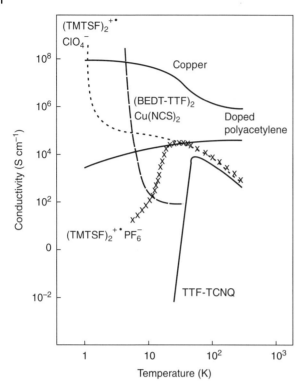

Figure 5.8 Conductivity versus temperature data at ambient pressure for a range of charge-transfer salts. The conductivity for copper is included for comparison. *Source:* From Petty M.C. et al. [8]. Reproduced with permission of Edward Arnold (Publishers) Ltd.

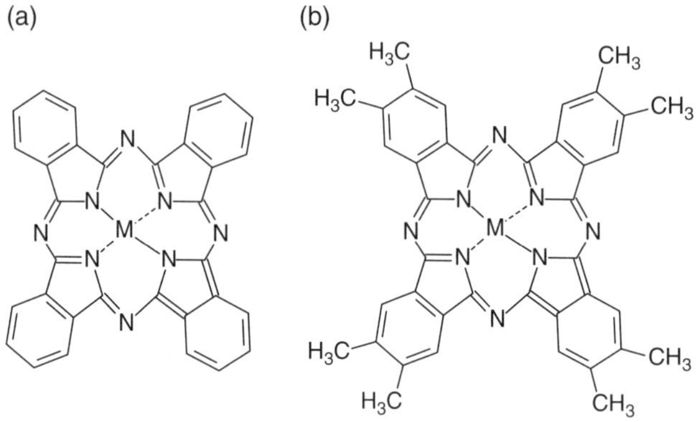

(a)   (b)

Figure 5.9 (a) Metal phthalocyanine. (b) Substituted phthalocyanine compound.

central metal atom. Figure 5.7 shows one of the most widely studied compounds, Ni(dmit)$_2$ (H$_2$dmit = 4,5-dimercapto-1,3-dithiole-2-thione).

Phthalocyanines, depicted in Figure 5.9, are examples of metallomacrocyclic compounds. These are flat ring-shaped molecules that become electrically conductive on oxidation; for example, on doping with iodine or bromine. Figure 5.9a shows the structure of a metallic phthalocyanine; a variety of transition metals are found to co-ordinate into the centre of the molecule, e.g. M = Fe, Co, Ni, Cu, Zn, and Pt. In many of the charge-transfer systems, the metal retains a +2 oxidation state and does not play a significant role in the conduction process. The charge carriers are associated with the delocalized π orbitals on the macrocyclic ligand. Phthalocyanines are highly stable compounds and many of these can be heated to several 100 °C before they sublime. This makes them suitable for processing into thin films by physical vapour deposition (thermal evaporation or molecular beam epitaxy – Chapter 7, Section 7.2.2). The phthalocyanine molecule can also be substituted using, for example, methyl groups (Figure 5.9b) to make the material soluble in common organic solvents.

Highly conductive organic solids differ from inorganic metals in several respects. First, the overlap between adjacent units is highly anisotropic, leading to the formation of one-dimensional bands rather than the three-dimensional bands of conventional metals. Second, the weak intermolecular forces lead to a high probability of defects even in regular one-dimensional structures, and these severely limit the mean free path of charge carriers in otherwise ideal bands. And finally, the molecules in an organic solid are separated by distances of the order of van der Waals contacts, so that some shortening of these intermolecular spacings is possible without severe consequences for electron repulsions – as would be encountered for species which started at separations already significantly shorter than van der Waals distances. This favours Peierls distortions at low temperatures.

5.5 Graphene, Fullerenes, and Nanotubes

5.5.1 Graphene

Important electroactive compounds that may find application in organic and molecular electronics are based on forms of carbon. Graphite consists of vast carbon sheets, which are stacked one on top of another like a sheaf of papers. In pure graphite, these layers are about 0.335 nm apart, but intercalating various molecules can separate them further. The bonding between the carbon atoms in the planes is mainly sp^2 hybridizations consisting of a network of single and double bonds. Weak interactions between the delocalized electron orbitals hold adjacent sheets together. *Graphene* is the name given to a flat monolayer of carbon atoms tightly packed into a two-dimensional honeycomb lattice and is a basic building block for graphitic materials of all other dimensionalities. A graphene sheet is depicted in Figure 5.10a. This can be wrapped into 0-d *fullerenes*, rolled into 1-d *nanotubes*, or stacked into 3-d graphite.

Graphene was isolated in 2004 by two researchers at The University of Manchester, UK, Andre Geim and Konstantin Novoselov. One Friday, the two scientists removed some flakes from a lump of bulk graphite with sticky tape (mechanical exfoliation). They noticed some flakes were thinner than others. By separating the graphite fragments repeatedly they managed to create flakes that were just one atom thick [11, 12]. Six years after their ground-breaking isolation of graphene, Geim and Novoselov were awarded the 2010 Nobel Prize for Physics. Graphene is the thinnest material ever made. It is 100 times stronger than steel, a better electrical conductor and heat conductor than copper, flexible, and optically transparent.

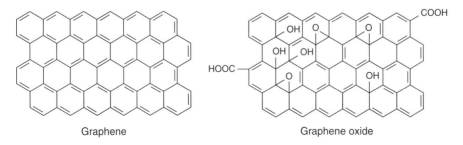

Figure 5.10 Chemical structures of graphene (left) and a single layer of graphene oxide (graphite oxide) (right).

The electronic properties of graphene depend very much on the number of layers. Architectures consisting of more than 10 graphene layers can be considered as a graphite thin film, since these essentially exhibit the electronic properties of graphite. Graphene (single layers and bilayers) is best described as a zero band gap semiconductor. In intrinsic (undoped) graphene each carbon atom contributes one electron completely filling the valence band and leaving the conduction band empty. As such, the Fermi level (Chapter 3, Section 3.2.4) is situated precisely at the energies where the conduction and valence bands meet. These are known as the *Dirac or charge neutrality points*. The carrier mobility in graphene remains high in both electrically and chemically doped devices. This translates into ballistic transport (Chapter 3, Section 3.2.1) on the submicrometre scale. Room temperature carrier mobilities in excess of $200\,000\,cm^2\,V^{-1}\,s^{-1}$ have been reported for graphene devices [13]. This figure should be contrasted to the values given in Table 3.1 (Chapter 3) for inorganic and organic semiconductors. Unfortunately, the lack of an intrinsic band gap in graphene is a fundamental obstacle to its use as the channel layer in a field effect transistor (Chapter 9, Section 9.4). This is further discussed in Chapter 11, Section 11.10.

Many other applications for graphene have been proposed or are under development, in areas including biological engineering, filtration, lightweight/strong composite materials, photovoltaics, and energy storage. Graphene also meets the electrical and optical requirements for transparent conductive coatings (e.g. Figure 5.5). This is discussed in Chapter 9, Section 9.6.

There are a number of important derivatives of graphite and graphene. *Graphite oxide* (formerly called *graphitic oxide* or *graphitic acid*) is a compound of carbon, oxygen, and hydrogen in variable ratios, obtained by treating graphite with strong oxidizers. The maximally oxidized bulk product is a yellow solid with C : O ratio between 2.0 and 2.9. In contrast to hydrophobic graphite, graphite oxide is a highly hydrophilic layered material and can easily be exfoliated by sonication in water, yielding stable dispersions that consist mostly of single-layer sheets. These monomolecular sheets are also referred to as *graphene oxide* by analogy to graphene. The structure and properties of graphite (or graphene) oxide depend on particular synthesis method and degree of oxidation. The material typically preserves the layer structure of the parent graphite, but the layers are buckled and the interlayer spacing is about two times larger (~0.7 nm) than that of graphite. Chemical reduction to *reduced graphene oxide* removes a significant amount of oxygen. While graphene oxide is an insulating material, *reduced graphene oxide* is conductive (but exhibits strongly reduced conductivity as compared to graphene).

The atomic structures of graphene oxide and reduced graphene oxide remain a matter of debate, due to a rather random functionalization of each layer and compositional variations depending on the method of preparation. Figure 5.10b shows a structure for a single layer of graphene oxide that includes many of the features observed by different research groups. The most prevalent chemical groups are hydroxyl and epoxide, with carboxylic acid groups at the sheet edges (Table 5.2); sulphur impurity is often found, for example in a form of organosulfate groups. Graphene oxide and reduced graphene oxide lend themselves to covalent functionalization due to the presence of defects in the graphene lattice that act as sited for reactivity. The resulting chemically modified graphenes could potentially be much more adaptable for a lot of applications.

Graphene's many superior properties justify its nickname of 'miracle material' [13]. However, it is unclear whether graphene or its derivatives will lead to disruptive technologies (Chapter 1, Section 1.5). The intense interest in graphene (15 000 scientific papers were published in 2014) has also led to significant work in related two-dimensional materials, such as molybdenum disulphide and phosphorene (a double layer of phosphorus atoms) [14].

Figure 5.11 Structure of C_{60}.

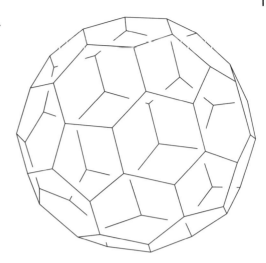

5.5.2 Fullerenes

Under certain conditions, carbon forms regular clusters of 60, 70, 84, etc. atoms. A C_{60} cluster, shown in Figure 5.11, is composed of 20 hexagons and 12 pentagons and resembles a football. The diameter of the ball is about 1 nm. As with graphite, each carbon atom in C_{60} is bonded to three other carbon atoms. Thus, C_{60} can be considered as a rolled up layer of a single graphene sheet. The term *buckministerfullerene* was given originally to the C_{60} molecule because of the resemblance to the geodesic domes designed and built by Richard Buckminster Fuller. However, this term (or fullerene or buckyball) is generally used to describe C_{60} and related compounds. For example, a molecule with the formula C_{70} can be formed by inserting an extra ring of hexagons around the equator of the sphere, producing an elongated shell more like a rugby ball.

The electronic structure of C_{60} is unique for a π-bonded hydrocarbon, in that the molecule is a strong electron acceptor with an electron affinity of 2.65 eV. This is a consequence of its geometric structure, which influences the electronic energy levels in two simple ways. First, as noted above, the structure may be regarded as 20 six-membered rings with 12 five-membered rings. Conjugated five-membered rings always lead to higher electron affinity as a result of the aromatic stability associated with the $C_5H_5^-$ (cyclopentadienyl) anion. The non-planarity of the structure also means that the π-electrons are no longer pure p in character. The slight pyramidalization of each carbon atom (referring to the downward deflection of the three atoms surrounding each carbon from the plane in which they would all lie in the graphite structure) induces a small re-hybridization in which some s-character is introduced into the π-orbitals. The molecular orbital energy diagram for C_{60} reveals that the molecule might be expected to accept at least 6 electrons, and possibly up to 12. Figure 5.12 shows six separate reversible electrochemical redox waves for C_{60} in solution [15].

In the solid state, the C_{60} molecules form a crystal lattice with a face-centered cubic structure and a molecular spacing of 1 nm (Chapter 2, Section 2.5.3). Alkali atoms can easily fit into the empty spaces in the fcc lattice. Consequently, when C_{60} crystals are heated with potassium, the metal vapour diffuses into the fullerene lattice to form the compound K_3C_{60}. Figure 5.13 shows the location of the alkali metal atoms in the fullerene lattice, where they occupy the two vacant tetrahedral sites (shown by the open spheres) and a larger octahedral site (shaded spheres) per C_{60} molecule [16]. In the tetrahedral site, the potassium ion has four surrounding C_{60} molecules and in the octahedral site, there are six surrounding fullerene molecules.

(a)

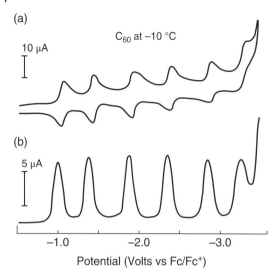

Figure 5.12 Reduction of C_{60} in CH_3CN-toluene solution at $-10\,°C$ using (a) cyclic voltammetry at a $100\,mV\,s^{-1}$ scan rate and (b) differential pulse voltammetry ($50\,mV$ pulse width, $300\,ms$ period, $25\,mV\,s^{-1}$ scan rate). Fc = ferrocene. *Source:* Reprinted with permission from Xie et al. [15]. Copyright (1992) American Chemical Society.

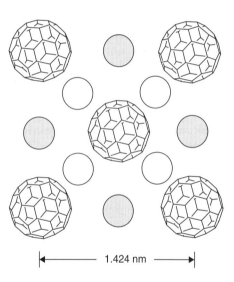

Figure 5.13 Structure of K_3C_{60}. The open and shaded spheres represent the potassium ions at tetrahedral and octahedral sites, respectively [16].

The molecule C_{60} is an insulator, but when it becomes doped with an alkali metal it becomes electrically conducting. In the case of K_3C_{60}, the potassium atoms become ionized to form K^+ and their electrons are associated with the C_{60}, which becomes a C_{60}^{3-} triply negative ion. Each C_{60} molecule has three electrons that are loosely bonded to the fullerene molecule and can move through the lattice. The K_3C_{60} has a superconducting transition temperature of $18\,K$. Higher transition temperatures are found for other alkali atoms, e.g. $33\,K$ at ambient pressure and $38\,K$ at $0.8\,GPa$ for Cs_3C_{60} [10, 17]. The transition temperature increases with the radius of the dopant alkali metal ion.

Larger fullerenes such as C_{70}, C_{76}, C_{80}, and C_{84} have been found. Some smaller fullerenes also exist. For example, a solid phase of C_{22} has been identified in which the lattice consists of C_{20} molecules bonded together by an intermediate carbon atom. When suitably doped, such smaller fullerenes may exhibit higher superconducting transition temperatures than the C_{60} materials.

The existence of the carbon fullerenes has stimulated some discussion about similar clusters of other atoms, such as silicon or nitrogen. Theory has shown that the N_{20} cluster should be stable. This material has also been predicted to be a powerful explosive! However, it has yet to be synthesized.

5.5.3 Carbon Nanotubes

In addition to the spherical-shaped fullerenes, it is possible to synthesize tubular variations – carbon nanotubes. Such tubes are comprised of graphene sheets, curled into a cylinder. Each tube may contain several cylinders nested inside each other. The tubes are capped at the end by cones or faceted hemispheres. Because of their very small diameters (down to around 0.7 nm) carbon nanotubes are prototype one-dimensional nanostructures.

Figure 5.14 shows how the unit vectors $\mathbf{a_1}$ and $\mathbf{a_2}$ generate the hexagonal graphene lattice (these unit vectors are not orthogonal in the hexagonal lattice, Chapter 2, Section 2.5.3). If a (= 0.142 nm) is the carbon–carbon bond length, then $|\mathbf{a_1}| = a(\sqrt{3}, 0)$ and $|\mathbf{a_2}| = a(\sqrt{3}/2, 3/2)$. An important feature of a carbon nanotube is the orientation of the six-membered carbon ring in the honeycomb lattice relative to the axis of the nanotube. In cutting a rectangular sheet, shown as the shaded region in Figure 5.14, a circumference vector $\mathbf{C} = n\mathbf{a_1} + m\mathbf{a_2}$ is defined. The integers n and m denote the number of unit vectors along the crystallographic axes. The direction of the axis of the nanotube is shown by the vector $\mathbf{T}$ in the figure (in a direction orthogonal to $\mathbf{C}$). The radius of the resulting nanotube R is then given by

$$R = \frac{C}{2\pi} = \left(\frac{\sqrt{3}}{2\pi}\right) a\sqrt{n^2 + m^2 + nm} \tag{5.10}$$

The primary classification of a carbon nanotube is as either being chiral or achiral (Chapter 2, Section 2.4.2). An achiral nanotube is one whose mirror image has an identical structure to the original. There are only two cases of achiral nanotubes: armchair and zig-zag (these names arise from the shape of the cross-sectional ring). When the circumference vector lies along one of the two basis vectors, then nanotube is said to be of the zig-zag type, for which $m = 0$. For an armchair nanotube, the circumference vector is along the direction exactly between the two

Figure 5.14 Graphene sheet showing the unit vectors $\mathbf{a_1}$ and $\mathbf{a_2}$ of the two-dimensional unit cell. In cutting a rectangular sheet, shown as the shaded region, a circumference vector $\mathbf{C} = n\mathbf{a_1} + m\mathbf{a_2}$ is defined. The direction of the axis of the nanotube is shown by the vector $\mathbf{T}$. Zig-zag and armchair structures are defined by the depicted directions of $\mathbf{C}$.

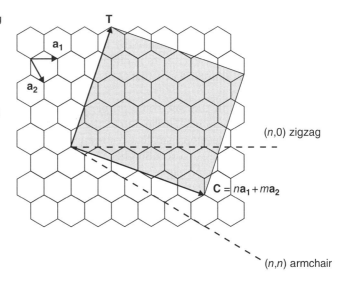

(a) (b) (c)

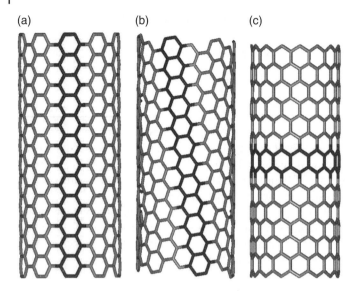

Figure 5.15 Three classes of single-wall (carbon) nanotube (SWNT) (a) (10,10) armchair SWNT, (b) (12,7) chiral SWNT, and (c) (15,0) zigzag SWNT.

basis vectors; in this case $n = m$. All other (n, m) indexes correspond to chiral nanotubes. Three examples of single-wall (carbon) nanotubes (SWNTs) are shown in Figure 5.15.

The electronic structure of an SWNT is either metallic or semiconducting, depending on its diameter and chirality. Each carbon atom in the hexagonal lattice of the graphene sheet possesses six electrons. The inner 1s orbital contains two electrons, while three electrons in $2sp^2$ hybridized orbitals form three bonds in the plane of the graphene sheet. This leaves the final electron in a 2p orbital, perpendicular to the graphene sheet and to the nanotube surface. A delocalized π-electron network is therefore formed across the nanotube surface, responsible for its electronic properties. The band structure of a carbon nanotube can be derived from that of graphene, which is *semi-metal*, with the valence and conduction bands meeting a several points in the Brillouin zone (Chapter 3, Section 3.3.2). A full analysis leads to the following condition for carbon nanotubes to be metallic [18]:

$$|n - m| = 3I \tag{5.11}$$

where I is an integer. Nanotubes, for which this condition does not hold, are semiconducting. Furthermore, it can be shown that the band gap of semiconducting nanotubes decreases inversely with an increase in the tube radius, as depicted in Figure 5.16 for zig-zag nanotubes. The relationship is approximately that the band gap $E_g = 0.45/R$ eV, where R is the radius of the tube in nanometres.

Some deviations in the electronic properties of nanotubes (from the simple π-electron network model of graphene) arise due to the curvature of the tube. As a result, nanotubes satisfying Eq. (5.11) develop a small curvature-induced band gap and hence are semi-metallic. Armchair nanotubes are an exception because of their special symmetry, and remain metallic for all diameters. The band gap of semi-metallic nanotubes is small and varies inversely as the square of the nanotube diameter. For example, a semi-metallic nanotube with a diameter of 1 nm has a band gap of about 40 meV [18].

Nanotubes are found in a variety of forms and shapes other than the single wall tube. Bundles of SWNTs are frequently observed. The individual tubes in the bundle are attracted to their nearest neighbours via van der Waals interactions (Chapter 2, Section 2.3.5), with typical

Figure 5.16 Band gap versus radius for ziq-zaq carbon nanotubes. *Source:* Reprinted from Anantram and Léonard [18]. Copyright (2006), with permission from IOP Publishing Ltd.

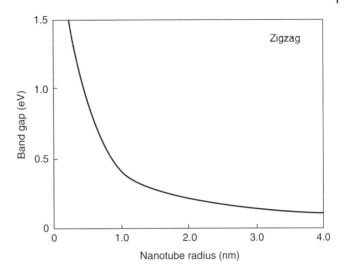

distances between the tubes being comparable to the interplanar spacings in graphite, 0.31 nm. Multiwall nanotubes (noted above) consist of SWNTs nested inside one another, like a Russian doll. Carbon nanotubes also occur in more interesting shapes such as junctions between nanotubes of different chiralities and three terminal junctions. Such junctions are atomically precise in that each carbon atom is bonded primarily to its three nearest neighbours and there are no dangling bonds.

At low temperature, a single-wall carbon nanotube is a quantum wire in which the electrons in the wire move without being scattered (ballistic transport – Chapter 3, Section 3.2.1). Carbon nanotubes hold promise as interconnects in both silicon nanoelectronics and molecular electronics applications, because of their low resistance and strong mechanical properties. Carbon nanotubes can also be doped either by electron donors or electron acceptors. After reaction with the host materials, the dopants are intercalated in the intershell spaces of the multiwalled nanotubes and, in the case of single-walled nanotubes, either in between the individual tubes or inside the tubes. The remarkable electronic behaviour of carbon nanotubes is discussed separately in Chapter 11, Section 11.10. In addition to their unique electronic properties, single-wall carbon nanotubes are also some of the strongest materials that are known, exhibiting very high *tensile strengths*, around 45×10^9 Pa (20 times stronger than steel) and large values of Young's modulus, $1.3–1.8 \times 10^9$ Pa (almost 10 times that of steel). When carbon nanotubes are bent, they buckle like straws but do not break, and can be straightened back without damage. This results from the sp^2 bonds, which can rehybridize as they are bent. Multiwall nanotubes also possess excellent mechanical properties, but they are not quite as good in this respect as their single-walled counterparts.

The above confirms carbon's uniqueness as an electronic material. It can be a good conductor in the form of graphite, an insulator in the form of diamond, or a flexible polymer (conductive or insulating) when reacted with hydrogen and other species. Carbon differs from other group IV elements, such as Si and Ge, which exhibit sp^3 hybridization. Carbon does not have any inner atomic orbitals except for the spherical 1s orbital, and the absence of nearby inner orbitals facilitates hybridizations involving only the valence (outer) s and p orbitals. The fact that sp and sp^2 hybridizations do not readily occur in Si and Ge might be related to the absence of 'organic materials' made from these elements.

5.6 Piezoelectricity, Pyroelectricity, and Ferroelectricity

5.6.1 Basic Principles

Piezoelectricity is the ability of certain crystals (e.g. quartz) to generate a voltage in response to applied mechanical stress. The word is derived from the Greek *piezein*, which means to squeeze or press. The piezoelectric effect is reversible; piezoelectric crystals, subject to an externally applied voltage, can change shape by a small amount. The deformation, about 0.1% of the original dimension in certain ceramic materials, may be only of the order of nanometers, but nevertheless finds useful applications such as the generation and detection of sound, high voltages and high (MHz) frequencies, and ultrafine manipulation of optical assemblies [19, 20].

On a microscopic scale, piezoelectricity results from a nonuniform charge distribution within the unit cell of a crystal. When the crystal is mechanically deformed, the positive and negative charge centres are displaced by differing amounts. Although the overall crystal remains electrically neutral, the difference in charge centre displacements results in an electric polarization. Only certain classes of crystals (Chapter 2, Section 2.5.3), those without a centre of symmetry, can be piezoelectric. This is illustrated by the two-dimensional unit cells shown in Figures 5.17 and 5.18, which comprise of positively and negatively charged ions [20]. In Figure 5.17, the cubic unit cell has a centre of symmetry. When unstressed, the centres of mass of the negative and positive ions coincide with the centre of the unit cell. Therefore, there is no net dipole moment or polarization, P (Chapter 2, Section 2.3.5, P is the dipole moment per unit volume). This situation does not change if a force is applied, as depicted in Figure 5.17b.

If the unit cell has no centre of symmetry, the two centres of mass become displaced on application of a force. The example shown in Figure 5.18 is for a hexagonal unit cell. The direction of the induced polarization depends on the direction of the applied stress. For Figure 5.18b, a polarization is produced in the same direction as the applied stress. When the same unit cell is stressed along a different direction, as shown in Figure 5.18c, there is no polarization in this direction, because there is no net displacement of the centres of mass in the direction of the force. However, the direction of the force will produce a polarization in an orthogonal direction. Generally, an applied stress in one direction in a piezoelectric crystal can give rise to a polarization in other crystal directions. If T_j is the applied mechanical stress along some j direction ($j = x, y, z$) and P_i is the induced polarization along some i direction, then the two are linearly related by

$$P_i = d_{ij}T_j \tag{5.12}$$

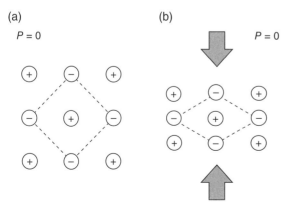

(a)
$P = 0$

(b)
$P = 0$

Figure 5.17 Schematic representation of a cubic unit cell. (a) In the absence of an applied force, the centre of mass of the positive ions coincides with that of the negative ions and there is no resulting dipole moment or polarization. (b) Under an applied force, the situation does not change. *Source:* Reproduced from Kasap [20]. Copyright (1997), with permission from McGraw-Hill Education.

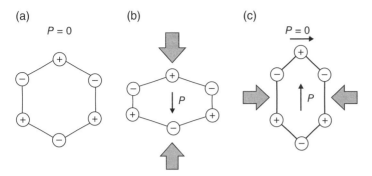

Figure 5.18 The hexagonal unit cell depicted has no centre of symmetry. (a) In the absence of an applied force, the centres of mass for positive and negative ions coincide. (b) Under an applied force, there is a displacement in the centres of mass for the positive and negative ions, resulting in a polarization in the direction of the applied force. (c) When the force is in a different direction, the polarization does not occur in the direction of the force. *Source:* Reproduced from Kasap [20]. Copyright (1997), with permission from McGraw-Hill Education.

where d_{ij} are called the *piezoelectric coefficients*. In fact, d_{ij} is a tensor quantity. Such quantities have been noted earlier, i.e. the case of permittivity, Chapter 4, Section 4.3.1. Formally, d_{ij} is a *third rank tensor* (an array of $3 \times 3 \times 3$ coefficients), as it relates a vector quantity (polarization) to a second-rank tensor quantity (stress). In the converse piezoelectric effect, the application of an electric field will produce a strain (change in dimensions) of the crystal, or

$$S_j = d_{ij}E_i \qquad (5.13)$$

where S_j is the strain along the j direction and E_i is the electric field along i. The coefficients in Eqs (5.12) and (5.13) are the same.

Of the 32 crystal classes, 21 are non-centrosymmetric (not having a centre of symmetry) and of these, 20 exhibit direct piezoelectricity (the 21st class is a particular cubic class in which the various symmetry elements combine to produce non-polar behaviour). Ten of these are polar (i.e. spontaneously polarize) having a dipole in their unit cell, and exhibit *pyroelectricity*. If this dipole can be reversed by the application of an electric field, the material is said to be *ferroelectric*. Figure 5.19 illustrates the relationship between the crystal structure and the phenomena of piezoelectricity, pyroelectricity, and ferroelectricity.

Figure 5.19 Relationship between the crystal symmetry and the properties of piezoelectricity, pyroelectricity, and ferroelectricity.

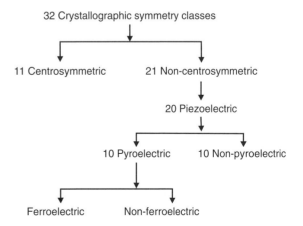

If a pyroelectric material is heated, a change in the material polarization is produced. This can be measured as a voltage developed across the crystal. The *pyroelectric coefficient p* is given by the rate of change of polarization with temperature

$$p = \frac{dP}{dT} \tag{5.14}$$

Very small temperature changes ($\sim 10^{-6}$ °C) can give rise to voltages that are easily measurable. Because pyroelectric materials are also piezoelectric, they may also develop an additional polarization when heated due to any temperature-induced stresses (this will depend on if and how the material is mechanically constrained, e.g. in the form a thin film of material on a substrate with a different thermal expansion coefficient). This is referred to as a secondary pyroelectric effect.

Ferroelectric crystals are permanently polarized, even in the absence of an applied field. This effect disappears above a certain temperature, the *Curie temperature*, T_C, and a *paraelectric* phase is formed. There are two main types of ferroelectrics: displacive and order–disorder. The effect in barium titanate, $BaTiO_3$, a typical ferroelectric of the displacive type, is due to a the Ti ion being displaced from equilibrium slightly, the force from the local electric fields due to the ions in the crystal increases faster than the elastic restoring forces. This leads to an asymmetrical shift in the equilibrium ion positions and hence to a permanent dipole moment. In an order–disorder ferroelectric, there is a dipole moment in each unit cell, but above T_C these are pointing in random directions. Upon lowering the temperature and going through the Curie point, the dipoles order, all pointing in the same direction within a region or domain.

Ferroelectric crystals often show several Curie points and domain structure hysteresis, much as do ferromagnetic crystals (Section 5.7). By analogy to magnetic core memory, this hysteresis can be used to store information (see below) in memory elements based on ferroelectric capacitors. The nature of the phase transition in some ferroelectric crystals is still not well understood. Ferroelectrics often have very large dielectric constants, and thus are often found in capacitors. They also often have unusually large nonlinear optical coefficients. The term *electret*, introduced by Oliver Heaviside to suggest the electrical equivalent of a magnet, is often used to describe a ferroelectric material.

5.6.2 Organic Piezoelectric, Pyroelectric, and Ferroelectric Compounds

Generally, inorganic piezoelectric and pyroelectric materials, as shown in Table 5.5, possess higher values of d and p coefficients than their organic counterparts. However, piezoelectric and pyroelectric polymeric sensors and actuators offer the advantage of processing flexibility because they are lightweight, tough, readily manufactured into large areas, and can be cut and

Table 5.5 Piezoelectric and pyroelectric properties of common materials [19–21].

Material	T_C (°C)	ε_r	d (pC N^{-1})	p (μC m^{-2} K^{-1})
$BaTiO_3$, barium titanate	120	1900	190	200
TGS, triglycine sulphate	49	43–50	25	280–350
PZT, lead zirconate titanate $PbTi_{1-x}Zr_xO_3$	200–400	1000–2000	250–500	60–500

formed into complex shapes. Polymers also exhibit high strength and high impact resistance. Other notable features of polymers are their low dielectric constants, low elastic stiffnesses, and low densities, which result in a high voltage sensitivity (an important sensor characteristic). Furthermore, these materials possess low acoustic and mechanical impedances, essential for medical and underwater applications. Polymers also typically have high dielectric breakdowns and high operating field strengths, which means that they can withstand much higher driving fields than ceramics.

An interesting crystalline organic ferroelectric compound is triglycine sulphate, $(NH_2CH_2COOH)_3H_2SO_4$, usually abbreviated to TGS (Table 5.5). This is an order–disorder ferroelectric that is known to exhibit a second-order phase transition (Chapter 2, Section 2.5.2). Above the Curie temperature, $T_C = 49\,°C$, the crystal is centrosymmetric with monoclinic symmetry. The glycine group NH_2CH_2COOH (glycine is the only amino acid that is not chiral – Chapter 12, Section 12.2.1) commonly crystallizes in two different forms. One is a structure in which the two carbon atoms and the two oxygen atoms are approximately coplanar, while the nitrogen is significantly displaced out of this plane; the other is a structure in which all the carbon, nitrogen, and oxygen atoms are close to planar. In TGS, two of the glycine groups (denoted II and III) are quasi-planar, while the other (I) is non-planar. Chemically, two protons from the H_2SO_4 group are more properly associated with the glycines. Glycine I is more correctly a glycinium ion; the other proton can be thought of as an O_{II}—H—O_{III} proton and resonates between glycine II and glycine III, at least in the paraelectric phase above T_C. All three glycines groups participate in the polarization reversal on application of an electric field, but the main reversible dipole is that associated with glycine I.

Table 5.6 compares the piezoelectric and pyroelectric properties of some polymeric compounds, along with data on the glass transition temperatures T_g and melting points T_m (NB: for many polymeric materials, the Curie temperature is greater than the melting point). Probably the best-known organic piezoelectric material is the polymer poly(vinylidene difluoride) (PVDF). The crystal structure of this material has been described in Chapter 2, Section 2.6.3. This material is a semicrystalline polymer with a monomer unit (CH_2CF_2). The dipole moment is 7.0×10^{-30} C m perpendicular to the chain direction. The pyroelectric β-phase of PVDF is obtained from the parent α-phase by a process of stretching and then poling, in which a high electric field is applied to align the dipoles. If all the monomer dipoles were aligned along the chain direction, a maximum microscopic polarization of about $100\,mC\,m^{-2}$ can be obtained.

Table 5.6 Piezoelectric and pyroelectric properties of polymers [19–21].

Polymer	Mer unit	T_g (°C)	T_m (°C)	ε_r	d (pC N^{-1})	p (μC m^{-2} K^{-1})
PVC	H H | | —C–C— | | H Cl	83	212	3.5	0.7	1
PVDF	F H | | —C–C— | | F H	−35	175	12	28	40
Nylon-11	H O | || —N–C+(CH$_2$)$_{10}$	68	195	3.7	0.3	5

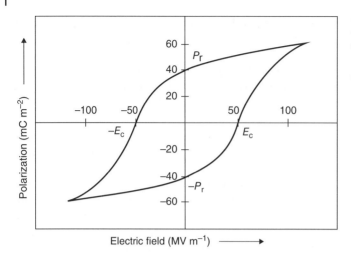

Figure 5.20 Typical polarization versus electric field hysteresis loop for poly(vinylidene difluoride) (PVDF). P_r = remanent polarization; E_c = coercive field.

However, semi-crystalline PVDF is approximately 50% crystalline and the observed polarization is around one half of this maximum value [19, 21].

PVDF is also a pyroelectric and ferroelectric polymer containing polar crystals, in which the direction of polarization can be reversed by the application of an electric field. The material can therefore be taken through a *hysteresis loop*, shown in Figure 5.20 in the form of a plot of the polarization as a function of the applied electric field [19]. The applied electric field required to reduce the polarization to zero, E_c, is called the *coercive field*, while the polarization retained by the material with no applied field, P_r, is the *remanent* or *residual polarization*.

Copolymers of PVDF with trifluoroethylene (TrFE) and tetrafluoroethylene (TFE) have also been shown to exhibit strong piezoelectric, pyroelectric, and ferroelectric effects. An attractive morphological feature of the comonomers is that they force the polymer into an all-*trans* conformation that has a polar crystalline phase. This eliminates the need for mechanical stretching to yield a polar phase. Poly(VDF-TrFE) crystallizes to a much greater extent than PVDF (up to 90% crystalline), yielding a higher remanent polarization, lower coercive field, and much sharper hysteresis loops. TrFE also extends the operational temperature by about 20°, to close to 100 °C. Conversely, copolymers with TFE have been shown to exhibit a lower degree of crystallinity and a suppressed melting temperature as compared to the PVDF homopolymer.

A low level of piezoelectricity is also observed in polyamides (also known as nylons). The monomer unit of odd-numbered nylons consists of even numbers of methylene groups and one amide group with a large dipole moment. Polyamides crystallize in all-*trans* conformations and are packed so as to maximize hydrogen bonding between adjacent amine and carbonyl groups. The amide dipoles align synergistically for the odd-numbered monomer, resulting in a net dipole moment. The unit dipole density is dependant on the number of methylene groups present and the polarization increases with decreasing number of methylene groups. Other polymer materials exhibiting piezoelectric and pyroelectric effects include polyurea, polytrifluoroethylene, and copolymers of vinylidene cyanide and vinyl acetate [19]. However, the electroactive responses of these polymers are inferior to those of PVDF and its copolymer poly(VDF-TrFE). Pyroelectric polymers can also act as a host matrix for other pyroelectric compounds, such as TGS [22].

The phenomena of piezoelectricity and pyroelectricity are also found throughout the natural world. For example, piezoelectricity effects were found in keratin (Chapter 12, Section 12.2.2) in 1941 [21]. When a bundle of hair was immersed in liquid air, a potential of a few volts was generated between the tip and the root. When pressure was applied on the cross-section of the bundle, a voltage was generated. Subsequently, piezoelectricity has been observed in a wide range of other biopolymers including collagen, polypeptides such as polymethylglutamate and polybenzyl ʟ glutamate, oriented films of DNA, poly-lactic acid, and chitin [20]. Since most natural biopolymers possess ᴅ symmetry (Chapter 2, Section 2.4.2), they exhibit 'shear' piezoelectricity. A shear stress in the plane of polarization produces electric displacement perpendicular to the plane of the applied stress, as depicted in Figure 5.18c. The piezoelectric constants of biopolymers are small relative to synthetic polymers (Table 5.6), ranging in value from $0.01\,pC\,N^{-1}$ for DNA to $2.5\,pC\,N^{-1}$ for collagen. Currently, the physiological significance of piezoelectricity in many biopolymers is not well understood, but it is believed that such electromechanical phenomena may have a distinct role in biochemical processes. For example, it is known that electric polarization in bone influences bone growth.

As noted above, the bistable polarization of ferroelectrics makes them candidates for binary memory applications in the same way as the bistable magnetization of ferromagnetics. The memory is non-volatile (Chapter 11, Section 11.6) and does not require a holding voltage. To record information, the polarization may be reversed or re-oriented by application of a field greater than the coercive field. Commercial applications for this type of memory are discussed in Chapter 11, Section 11.6.3.

The topic of ferroelectric behaviour in liquid crystals is covered separately in Chapter 8, Section 8.5. It should be noted that many organic ferroelectric materials possess high second-order nonlinear optical properties and can be used for second-harmonic generation and electro-optic switching. This is discussed in Chapter 4, Section 4.3.2.

5.7 Magnetic Materials

5.7.1 Basic Principles

In Chapter 4, Section 4.3, the definition of an electric dipole moment was given. Magnetism has its origin in the *atomic magnetic dipole*. Magnetic dipoles, or magnetic moments, result on the atomic scale from the two kinds of movement of electrons. The first is the orbital motion of the electron around the nucleus. The second, much stronger, source of electronic magnetic moment is due to the electron spin (although current quantum mechanical theory suggests that electrons neither physically spin, nor orbit the nucleus).

Consider a current loop carrying a current I, as depicted in Figure 5.21a. The area enclosed by the current loop (enclosed within a single plane) is A. The magnetic dipole moment, $\mathbf{m_m}$, is then defined:

$$\mathbf{m_m} = I\,A\,\mathbf{u_n} \tag{5.15}$$

where $\mathbf{u_n}$ is a unit vector perpendicular to the plane of A. The direction of this unit vector is such that looking along it, the (conventional) current circulates clockwise. When a magnetic

Figure 5.21 (a) Magnetic dipole moment $\mathbf{m_m}$. (b) In a magnetic field, a magnetic dipole moment experiences a torque.

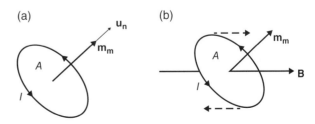

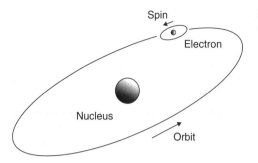

Figure 5.22 Motion of an electron around the nucleus of an atom.

moment is placed in a magnetic field, it experiences a torque that tries to rotate the magnetic moment to align with the axis of the magnetic field, Figure 5.21b.

An orbiting electron, as illustrated in Figure 5.22, behaves much like a current loop. The circulating electron therefore produces its own *orbital magnetic moment*, $\mathbf{m_{orb}}$. There is also a *spin magnetic moment*, $\mathbf{m_{spin}}$, associated with it, due to the electron itself spinning on its own axis. The overall magnetic moment of an electron consists of $\mathbf{m_{orb}}$ and $\mathbf{m_{spin}}$. However, as these are both vector quantities, they cannot be added numerically. The total magnetic moment of the atom $\mathbf{m_{atom}}$ depends on the orbital motions of all the electrons. However, only unfilled subshells contribute to the overall magnetic moment of an atom (the magnetic moments of pairs of electrons in closed subshells cancel with each other).

When a magnetic field $\mathbf{B}$ is applied to a material, the combined effect of the individual atomic magnets produced is that the material develops a net magnetic moment along the field and becomes magnetized. The *magnetization*, $\mathbf{M}$, is a vector quantity and is defined (in a similar way to polarization) as the magnetic dipole moment per unit volume. The magnetizing field (also called the applied field or the magnetic field intensity) is given the symbol $\mathbf{H}$, and is the total field that would be present if the field were applied to a vacuum. The magnetic field, $\mathbf{B}$ (sometimes called the magnetic induction) is the total flux of magnetic field lines through a unit cross-sectional area of the material, considering both lines of force from the applied field and from the magnetization of the material. $\mathbf{B}$, $\mathbf{H}$, and $\mathbf{M}$ are related by

$$\mathbf{B} = \mu_0 \left(\mathbf{H} + \mathbf{M} \right) \tag{5.16}$$

and

$$\mathbf{B} = \hat{\mu}_r \mu_0 \mathbf{H} \tag{5.17}$$

where $\hat{\mu}_r$ is the relative permeability, formally a tensor quantity since both $\mathbf{B}$ and $\mathbf{H}$ are vectors. The constant μ_0 is the *permeability of free space* ($4\pi \times 10^{-7}\,\mathrm{H\,m^{-1}}$), which is the ratio of $\mathbf{B}/\mathbf{H}$ measured in a vacuum. Equation (5.16) can be contrasted to Eq. (4.8) in Chapter 4, which relates the electric displacement $\mathbf{D}$ to the polarization $\mathbf{P}$. The SI units of magnetic field $\mathbf{B}$ are Tesla, while those of the magnetizing field $\mathbf{H}$ are $\mathrm{A\,m^{-1}}$. The latter is the *cause* (e.g. depending only on the external conduction currents in, say, a coil), while $\mathbf{B}$ is the *effect* and depends on the magnetization of the material.

In the electrical world, the polarization $\mathbf{P}$ produced in a material is related to the applied electric field $\mathbf{E}$ (Chapter 4, Eq. (4.9)). The equivalent expression in magnetism involves the magnetizing field $\mathbf{H}$:

$$\mathbf{M} = \hat{\chi}_m \mathbf{H} \tag{5.18}$$

where $\hat{\chi}_m (= \hat{\mu}_r - 1)$ is the *magnetic susceptibility* – a tensor quantity.

Table 5.7 Units for magnetic properties. Multiply a number in the CGS system by the conversion factor to change it to SI units. In the CGS system, $4\pi M$ is usually quoted as it has units of Gauss and is numerically equivalent to B and H.

Magnetic quantity	Symbol	CGS Unit	Conversion Factor	SI Unit
Magnetic dipole moment	m_m	emu	10^{-3}	$A\,m^2$
Magnetic field	B	Gauss	10^{-4}	Tesla
Magnetizing field; magnetic field intensity	H	Oersted	$10^3/4\pi$	$A\,m^{-1}$
Magnetization	M	$emu\,cm^{-3}$	10^3	$A\,m^{-1}$
Magnetization	$4\pi M$	Gauss		
Permeability	μ	dimensionless	$4\pi \times 10^{-7}$	$H\,m^{-1}$
Magnetic susceptibility	χ_m	$emu\,cm^{-3}\,Oersted^{-1}$	4π	dimensionless

Magnetic units are frequently the cause of confusion. While the SI (Système International d'Unités) system is generally preferred by engineers, CGS and Gaussian units are widespread [the defining equation here is $B = H + 4\pi M$, rather than that given in Eq. (5.16)]. Table 5.7 shows the conversion factors between these systems for common magnetic units.

All materials can be classified in terms of their magnetic behaviour falling into one of five categories, depending on their bulk magnetic susceptibility. The two most common types of magnetism are *diamagnetism* and *paramagnetism*, which account for the magnetic properties of most of the periodic table of elements at room temperature, as shown in Figure 5.23. These elements are usually referred to as non-magnetic, whereas those which are referred to as magnetic are actually classified as *ferromagnetic*. The only other type of magnetism observed in pure elements at room temperature is *antiferromagnetism*. Finally, magnetic materials can also be classified as *ferrimagnetic*, although this is not observed in any pure element but can only be found in compounds, such as the mixed oxides, known as ferrites, from which ferrimagnetism derives its name. The origins of the different forms of magnetism, which are contrasted in Table 5.8, are outlined in the following sections.

Diamagnetism

In a diamagnetic material, the atoms have no net magnetic moment when there is no applied field. Under the influence of an applied field, the spinning electrons precess and this motion, which is a type of electric current, produces a magnetization in the opposite direction to that of the applied field (the susceptibility is negative). All materials have a diamagnetic effect; however, it is often the case that the diamagnetic effect is masked by the larger paramagnetic or ferromagnetic term. The value of susceptibility is independent of temperature. As noted in Chapter 3, Section 3.4.4, below its critical temperature, a superconductor is a perfectly diamagnetic substance ($\chi_m = -1$).

Paramagnetism

There are several theories of paramagnetism, which are valid for specific types of material. In the *Langevin model*, each atom has a magnetic moment, which is randomly oriented as a result of thermal energy. The application of a magnetic field creates a slight alignment of these moments and hence a low magnetization in the same direction as the applied field. As the temperature increases, the thermal agitation will increase and it will become harder to align

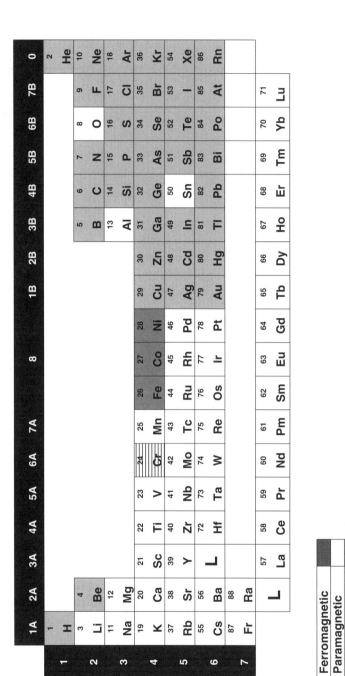

Figure 5.23 The periodic table showing magnetic properties of the elements.

Table 5.8 Summary of different types of magnetic behaviour.

Type of magnetism	Susceptibility	Atomic arrangement	Example
Diamagnetism	Small and negative.	Atoms have no magnetic moment.	Au, Ag, Cu, many polymers. Atoms have closed shells.
Paramagnetism	Small and positive.	Atoms have randomly oriented magnetic moments.	Gaseous and liquid oxygen. Alkali and transition metals.
Ferromagnetism	Large and positive.	Atoms have parallel aligned magnetic moments.	Fe, Co, Ni.
Antiferromagnetism	Small and positive.	Opposite and equal magnetic moments on two different sublattices.	Salts and oxides of transition metals, e.g. MnO, NiO.
Ferrimagnetism	Large and positive.	Magnetic moments on different sublattices do not cancel.	Ferrites.

the atomic magnetic moments. As a consequence, the susceptibility decreases. This behaviour is described by the *Curie law*, which relates magnetic susceptibility to the temperature:

$$\chi_m = \frac{C}{T} \tag{5.19}$$

where C is a constant, in units of temperature, known as the *Curie temperature* (or Curie constant). Materials, which obey this law, are substances in which the magnetic moments are localized at the atomic or ionic sites and where there is no interaction between neighbouring magnetic moments. The hydrated salts of the transition metals, e.g. $CuSO_4 \cdot 5H_2O$, are examples of this type of behaviour because the transition metal ions, which have a magnetic moment, are surrounded by a number of non-magnetic ions / atoms. These prevent interaction between neighbouring magnetic moments.

The *Pauli model of paramagnetism* is applicable to materials where the electrons interact to form a conduction band; this is valid for most paramagnetic metals. In this model, the conduction electrons are considered essentially to be free and under an applied field. An imbalance between electrons with opposite spin is set up leading to a low magnetization in the same direction as the applied field. The susceptibility is independent of temperature, although the electronic band structure may be affected, which will then have an effect on the susceptibility.

Ferromagnetism

Ferromagnetism is only possible when atoms are arranged in a lattice and the atomic magnetic moments can interact to align parallel to each other. This effect is explained in classical theory by the presence of a molecular field within the ferromagnetic material, which was first postulated by Weiss in 1907. This field is sufficient to magnetize the material to saturation. In quantum mechanics, the *Heisenberg model of ferromagnetism* describes the parallel alignment of magnetic moments in terms of an exchange interaction between neighbouring moments.

On the macroscopic level, ferromagnetic materials contain billions of individual spins, coupled in such a way that the individual moments will respond together to an external magnetic field. Weiss postulated the presence of *magnetic domains* within the material, which are regions where the atomic magnetic moments are aligned. By forming such domains, the magnetic energy is minimized. In the absence of an external field, the orientation of the moments of the domains will be random and the magnetization of the sample will be zero. The transition from one domain to its neighbour occurs through a region where the local magnetic moments are rapidly varying, called *Bloch walls*, illustrated in Figure 5.24. Exchange forces between neighbouring atomic spins tends to keep the moments aligned, and therefore to keep the Bloch walls as thick as possible. However, the magnetic moments within the walls will possess relatively

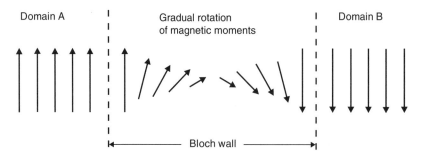

Figure 5.24 A Bloch wall separating two magnetic domains with opposite magnetic polarizations. The magnetic moments change direction gradually within the Bloch wall.

high energy as they are unaligned with the 'easy' directions of magnetization in the two domains. This *anisotropy energy* favours a thin Bloch wall. Consequently, the actual wall thickness is a compromise between the magnetic anisotropy and the exchange energy. In unmagnetized polycrystalline samples, each crystal grain will possess domains, as shown in Figure 5.25. Very small grains may constitute a single domain but, in general, the majority of grains will have many domains.

When a ferromagnetic sample is magnetized, the domains that are favourably aligned with the applied field will grow at the expense of the other domains. Eventually, all the individual moments will be parallel to each other and the magnetization reaches its saturation value. If the field is decreased, the formation of domains will not be reversible. Thus, the magnetization at zero field will not be zero. The finite value of magnetization in zero field (apart from the initial magnetization) is called the *remanent magnetization M_r* (similar to the remanent polarization of a ferroelectric material – Section 5.6.1). In order to demagnetize the sample, it is necessary to apply a negative magnetic field, called the *coercive field H_c* (again, an analogy may be made with ferroelectric materials). A ferromagnetic material will therefore exhibit a hysteresis loop, such as that shown in Figure 5.26, where the magnetization M is shown as a function of the

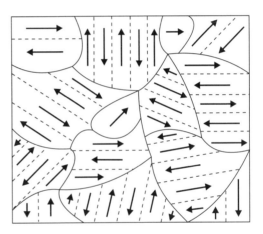

Figure 5.25 Schematic representation of magnetic domains in an unmagnetized polycrystalline material.

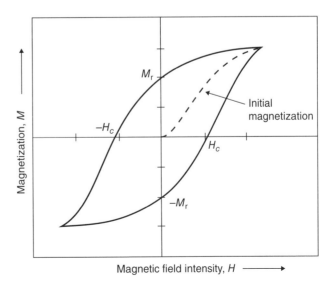

Figure 5.26 Typical magnetization versus magnetic field intensity curve for a ferromagnetic material. H_c = coercive field; M_r = remanent magnetization.

applied magnetic field H. The movement of the domains determines how the material responds to a magnetic field. As a consequence, the susceptibility is a function of applied magnetic field.

In the periodic table of elements, only Fe, Co, and Ni are ferromagnetic at and above room temperature. As ferromagnetic materials are heated, the thermal agitation of the atoms results in a decrease in the degree of alignment of the atomic magnetic moments and hence in the saturation magnetization. The thermal agitation finally becomes so great that the material becomes paramagnetic; the temperature of this transition is the Curie temperature, T_C (Fe, $T_C = 770\,°C$; Co, $T_C = 1131\,°C$; and Ni, $T_C = 358\,°C$). Above T_C, the susceptibility varies according to the *Curie–Weiss law*. The Curie law, described above for paramagnetism, is a special case of this more general law, which incorporates a temperature constant θ and derives from Weiss theory. This model incorporates the interaction between magnetic moments:

$$\chi_m = \frac{C}{T-\theta} \tag{5.20}$$

In the above equation, θ can either be positive, negative, or zero. Clearly, when $\theta = 0$, then the Curie–Weiss law equates to the Curie law (Eq. (5.19)). When θ is non-zero, then there is an interaction between neighbouring magnetic moments and the material is only paramagnetic above a certain transition temperature. If θ is positive, then the material is ferromagnetic below the transition temperature, the Curie temperature, T_C (compare this with T_C for a ferroelectric material described in Section 5.6.1 above). It is important to note that Eq. (5.20) is only valid when the material is in a paramagnetic state. It is also not valid for many metals, as the electrons contributing to the magnetic moment are not localized. However, the law does apply to some metals, e.g. the rare-earths elements, where the 4f electrons, which create the magnetic moment, are closely bound.

Antiferromagnetism

In the periodic table, the only element exhibiting antiferromagnetism at room temperature is chromium. Antiferromagnetic materials are very similar to ferromagnetic materials, but the exchange interaction between neighbouring atoms leads to the anti-parallel alignment of the atomic magnetic moments. The magnetic field cancels out and the material appears to behave in the same way as a paramagnetic material. Like ferromagnetic materials, these materials become paramagnetic above a transition temperature, the *Néel temperature*, T_N (Cr, $T_N = 37\,°C$) [θ is negative in Eq. (5.20)–however the value of θ does not relate to T_N].

Ferrimagnetism

Ferrimagnetism is only observed in compounds, which have more complex crystal structures than pure elements. Within these materials, the exchange interactions lead to parallel alignment of atoms in some of the crystal sites and anti-parallel alignment of others. The material breaks down into magnetic domains, just like a ferromagnetic material and the magnetic behaviour is also very similar, although ferrimagnetic materials usually have lower saturation magnetizations.

5.7.2 Organic Magnets

Molecular magnets are systems where a permanent magnetization and magnetic hysteresis can be achieved (although usually at low temperatures) not through a three-dimensional magnetic ordering, but as a purely single molecule phenomenon. Many organometallic charge-transfer salts exhibit ferromagnetic effects. A well-known example is decamethyl-ferrocene (DMeFc) radical cation (Chapter 3, Section 3.4.1)/tetracyanoethylene (TCNE) radical anion complex, studied in extensively by Miller, Epstein, and colleagues in the 1980s, and shown in Figure 5.26 [23]. This was the first ferromagnetic system with electron spins residing in p orbitals (in contrast to the more familiar ferromagnets with spins solely residing in d or f orbitals) that exhibited hysteresis and did not have an extended network bonding in one, two, or three dimensions. However, the Curie temperature was somewhat low, at 4.8K. The driving force behind the ferromagnetism is the coupling between $[DMeFc]^{+\bullet}$ and $[TCNE]^{-\bullet}$, both within the stack and between out-of-registry stacks, providing the full three-dimensional coupling of spins required for bulk ferromagnetism. Organic-based magnets now include many diverse examples of materials exhibiting magnetic ordering, including p-orbital-based organic nitroxides, p- or d-orbital-based mixed organic radicals/organometallic or inorganic co-ordination systems [24–26]. Electron-transfer salts of TCNE and manganese porphyrin compounds form a large family of ferrimagnets ($T_C < 28K$), demonstrating that metallomacrocycles are viable components of organic magnets [24, 25]. The reaction of bis(benzene)vanadium with TCNE in solution or by chemical vapour deposition forms $V[TCNE]_x$, a ferrimagnet that magnetically orders above room temperature (127 °C) [24]. The material is a magnetic semiconductor as it has a room temperature electrical conductivity of $0.01\,S\,cm^{-1}$. Several studies indicate that electrons in the valence and conduction bands of the $V[TCNE]_x$ films are spin polarized, which suggest that spintronic devices might be achievable (Figure 5.27).

The magnetic behaviour of electron-doped C_{60} compounds was first reported in 1991. Buckminsterfullerene has no intrinsic magnetic moment. For a magnetic moment to exist, an electron must be transferred to C_{60} from a donor molecule. The fullerene-based charge-transfer salt, TDAE-C_{60}, has a Curie temperature of 16K. A further example is a cobaltocene-doped fullerene derivative, which has a T_C of 19K. In this case, only the fullerene molecules have magnetic spins, and so are solely responsible for the observed magnetism. Pristine C_{60} is a van der Waals crystal, which can be converted to covalently-bonded crystalline phases by compression. Depending on the treatment, the molecules interconnect to form one-, two-, or three-dimensional polymers. It has been suggested that the two-dimensional polymerized highly-oriented rhombohedral C_{60} (Rh−C_{60}), which resembles highly oriented pyrolytic graphite, is ferromagnetic [27]. However, the origin of the ferromagnetism, which persisted up to 500K, is unclear. Carbon nanotubes may be filled with ferromagnetic materials such as iron nanoparticles [28]. The magnetic response of these hybrid structures is dominated by that of the ferromagnetic filling, since the graphitic carbon of the nanotube walls is weakly diamagnetic.

Reports of magnetic ordering in organic materials needs to be treated with some caution as, in many instances, there are serious purity issues and the compounds are not well defined. There have been a number of false reports on 'organic magnets' that included trace amounts of

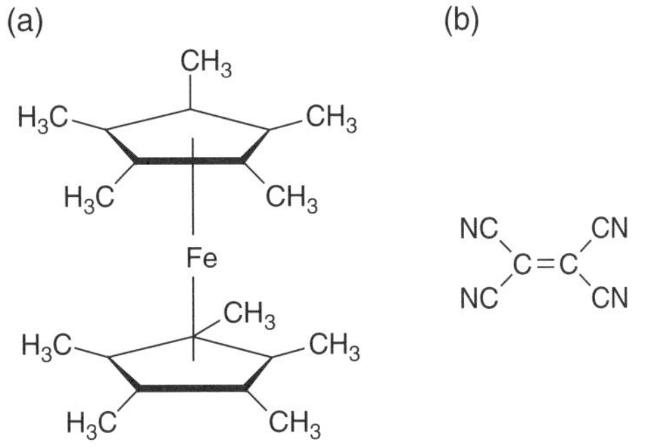

Figure 5.27 (a) Decamethylferrocene. (b) Tetracyanoethylene (TCNE).

ferromagnetic material. Due to the high density of iron, molecular magnetic materials are generally inferior to conventional magnets on a 'magnet per unit weight or unit volume' basis [29]. However, for more specialist applications, such as data storage, factors such as the coercive field (i.e. magnetizing field required to totally demagnetize the sample) become important and, in this respect, molecular systems are interesting. Molecular magnets may also be candidates for quantum computing qubits (Chapter 11, Section 11.14).

Problems

5.1 (a) How many moles are in 25 g of water?
   (b) How many molecules are in 25 g of water?
   (c) What is the pH of a solution of 25 g HCl in 1.5 l of water?
   (d) Which of the following the most polar: methanol (CH_3OH) or diethyl ether ((CH_3)2O)?

5.2 The paramagnetic susceptibility of PANi doped with HCl is 1.6×10^{-4} at 290 K. Suggest a value for the susceptibility at 100 K? State clearly any assumptions that are made. If the measured susceptibility of the polymer is 2.6×10^{-4} at 100 K, suggest a reason for the discrepancy in your theoretical value.

5.3 A device that converts heat directly into electricity may be produced by heat cycling a ferroelectric material through its Curie temperature. One particular arrangement utilizes the ferroelectric in the form of a parallel plate capacitor. The capacitor is charged from a voltage source, heated to above its Curie temperature, and then discharged through a load resistor.

   Such an arrangement is based on barium titanate and operates between 110 and 140 °C. The volume of the material is 10^{-3} m^3 and the low temperature displacement is 0.25 C m^{-2}. Calculate the work output and the efficiency of this energy conversion system.

   (Constants for BaTiO$_3$: density 6×10^3 kg m^{-3}, specific heat $= 500$ J kg^{-1} K^{-1}, relative permittivity at 110 °C = 4500, at 140 °C = 2500).

5.4 Download a diagram of a graphene sheet and print three copies, and cut along the edges of the outermost hexagons. Starting from an origin carbon atom in the top left hexagon, label the corresponding carbons in the other hexagons (0,0), (1,0), (2,0), ..., (1,1), ..., (6,6). By rolling up the 2-dimensional sheets, create the following nanotubes: zig-zag (7,0), armchair (6,6), and chiral (9,2). Calculate the radius of each tube.

5.5 A pyroelectric material in the form of a parallel plate capacitor is heated at a uniform rate. Derive an expression for the current that is produced (Eq. (11.1) in Chapter 11)). Hence, calculate the current generated in nA as a 1.25 cm^2 area sample of a PVDF polymer (pyroelectric coefficient $= 40$ μC m^{-2} K^{-1}) is heated at a rate of 1 K s^{-1}.

References

1 Tour, J.M. (2003). *Molecular Electronics*. Singapore: World Scientific.
2 Richens, D.T. (1997). *The Chemistry of Aqua Ions*. Chichester: Wiley.
3 Feast, W.J., Tsibouklis, J., Pouwer, K.L. et al. (1996). Synthesis, processing and material properties of conjugated polymers. *Polymer* 37: 5017–5047.

4 Springborg, M., Schmidt, K., Meider, H., and De Maria, L. (2001). Theoretical studies of electronic properties of conjugated polymers. In: *Organic Electronic Materials* (ed. R. Farchioni and G. Grosso), 39–87. Berlin: Springer.

5 Groenendaal, L., Jonas, F., Freitag, D. et al. (2000). Poly(3,4-ethylenedioxythiophene) and its derivatives: past, present, and future. *Adv. Mater.* 12: 481–494.

6 Kirchmeyer, S. and Reuter, K. (2005). Scientific importance, properties and growing applications of poly(3,4-ethylenedioxythiophene). *J. Mater. Chem.* 15: 2077–2088.

7 Andersson, P., Nilsson, D., Svensson, P.-O. et al. (2002). Active matrix displays based on all-organic electrochemical smart pixels printed on paper. *Adv. Mater.* 14: 1460–1464.

8 Petty, M.C., Bryce, M.R., and Bloor, D. (eds.) (1995). *An Introduction to Molecular Electronics*. London: Edward Arnold.

9 Bryce, M.R. (1995). Conductive charge-transfer complexes. In: *An Introduction to Molecular Electronics* (ed. M.C. Petty, M.R. Bryce and D. Bloor), 168–184. London: Edward Arnold.

10 Saito, G. and Yoshida, Y. (2011). Organic superconductors. *Chem. Rec.* 11: 124–145.

11 Geim, A.K. and Novoselov, K.S. (2007). The rise of graphene. *Nat. Mater.* 6: 183–191.

12 Novoselov, K.S., Geim, A.K., Morozov, S.V. et al. (2004). Electric field effect in atomically thin carbon films. *Science* 306: 666–669.

13 Novoselov, K.S., Fal'ko, V.I., Colombo, L. et al. (2012). A roadmap for graphene. *Nature* 490: 192–200.

14 Service RF (2015). Beyond graphene. *Science* 348: 490–492.

15 Xie, Q., Pérez-Cordero, E., and Echegoyen, L. (1992). Electrochemical detection of C_{60}^{6-} and C_{70}^{6-}: enhanced stability of fullerides in solution. *J. Am. Chem. Soc.* 114: 3978–3980.

16 Stephens, P.W., Mihaly, L., Lee, P.L. et al. (1991). Structure of single-phase superconducting K_3C_{60}. *Nature* 351: 632–634.

17 Ganin, A.Y., Takabayashi, Y., Khimyak, Y.Z. et al. (2008). Bulk superconductivity at 38 K in a molecular system. *Nat. Mater.* 7: 367–371.

18 Anantram, M.P. and Léonard, F. (2006). Physics of carbon nanotube electronic devices. *Rep. Prog. Phys.* 69: 507–561.

19 Das-Gupta, D.K. (1995). Piezoelectric and pyroelectric materials. In: *An Introduction to Molecular Electronics* (ed. M.C. Petty, M.R. Bryce and D. Bloor), 47–71. London: Edward Arnold.

20 Kasap, S.O. (1997). *Principles of Electrical Engineering Materials and Devices*. Boston: McGraw-Hill.

21 Harrison, J.S. and Ounaies, Z. (2001). *Piezoelectric Polymers* NASA ICASE Report No. 2001-43. Washington, DC: NASA.

22 Petty, M., Tsibouklis, J., Petty, M.C., and Feast, W.J. (1993). A TGS/70:30 VDF:TrFE copolymer composite: thin film formation and pyroelectric properties. *Ferroelectrics* 150: 267–278.

23 Miller, J.S., Epstein, A.J., and Reiff, W.M. (1988). Ferromagnetic molecular charge-transfer complexes. *Chem. Rev.* 88: 201–220.

24 Miller, J.S. (2014). Organic- and molecule-based magnets. *Mater. Today* 17: 224–235.

25 Miller, J.S. and Epstein, A.J. (1998). Tetracyanoethylene-based organic magnets. *Chem. Commun.* 1319–1325.

26 Mroziñsli, J. (2005). New trends in molecular magnetism. *Coord. Chem. Rev.* 249: 2534–2548.

27 Makarova, T.L., Sundqvist, B., Höhne, R. et al. (2001). Magnetic carbon. *Nature* 413: 716–718.

28 Boi, F.S., Mpumtjoy, G., and Baxendale, M. (2013). Boundary layer chemical vapor synthesis of self-organized radial filled-carbon-nanotube structures. *Carbon* 64: 516–526.

29 Bushby, R.J. and Paillaud, J.-L. (1995). Molecular magnets. In: *An Introduction to Molecular Electronics* (ed. M.C. Petty, M.R. Bryce and D. Bloor), 72–91. London: Edward Arnold.

Further Reading

Atkins, P. and de Paula, J. (2002). *Physical Chemistry*, 7e. Oxford: Oxford University Press.

Bardosova, M. and Wagner, T. (eds.) (2015). *Nanomaterials and Nanoarchitectures*. Dordrecht: Springer.

Barford, W. (2005). *Electronic and Optical Properties of Conjugated Polymers*. Oxford: Oxford University Press.

Brand, O., Fedder, G.K., Hierold, C. et al. (eds.) (2008). Carbon nanotube devices. In: *Advanced Micro & Nanosystems*, vol. 8. Weinheim: Wiley-VCH.

Bredas, J.-L. and Marder, S.R. (eds.) (2016). *The WSPC Reference on Organic Electronics: Basic Concepts*, vol. 1. World Scientific.

Burgess, J. (1988). *Ions in Solution: Basic Principles of Chemical Interactions*. Chichester: Ellis Horwood.

D'Souza, F. and Kadish, K.M. (eds.) (2016). *Handbook of Carbon Nano Materials*. World Scientific Series on Carbon Nanoscience, vol. 7 and 8. Singapore: World Scientific.

Farchioni, R. and Grosso, G. (eds.) (2001). *Organic Electronic Materials*. Berlin: Springer.

Ferraro, J.R. and Williams, J.M. (1987). *Introduction to Synthetic Electrical Conductors*. Orlando: Academic Press.

Gatteschi, D., Sessoli, R., and Villain, J. (2006). *Molecular Nanomagnets*. Oxford: Oxford University Press.

Jones, R.A.L. (2004). *Soft Machines*. Oxford: Oxford University Press.

Katsnelson, M.I. (2012). *Graphene: Carbon in Two Dimensions*. Cambridge: Cambridge University Press.

Kroto, H.W. and Walton, D.R.M. (eds.) (1993). *The Fullerenes*. Cambridge: Cambridge University Press.

Leznoff, C.C. and Lever, A.B.P. (eds.) (1989). *Phthalocyanines: Properties and Applications*. New York: VCH.

Lines, M.E. and Glass, A.M. (1977). *Principles and Applications of Ferroelectric and Related Materials*. Oxford: Clarendon Press.

Ludwigs, S. (ed.) (2014). *P3HT Revisited – From Molecular Scale to Solar Cell Devices*. Heidelberg: Springer.

Miller, J.S. and Drillon, M. (eds.) (2001). *Magnetism: Molecules to Materials: Molecule-based Materials*, vol. 2. Weinheim: Wiley-VCH.

Müller, T.J.J. and Bunz, U.H.F. (eds.) (2006). *Functional Organic Materials*. Weinheim: Wiley-VCH.

Nishinaga, T. (ed.) (2016). *Organic Redox Systems: Synthesis, Properties, and Applications*. Hoboken: Wiley.

Poole, C.P. Jr. and Owens, F.J. (2003). *Introduction to Nanotechnology*. Hoboken: Wiley-Interscience.

Saito, R., Dresselhaus, G., and Dresselhaus, M.S. (1998). *Physical Properties of Carbon Nanotubes*. London: Imperial College Press.

Salaneck, W.R., Lundström, I., and Rånby, B. (eds.) (1993). *Conjugated Polymers and Related Materials*. Oxford: Oxford University Press.

Schwoerer, M. and Wolf, H.C. (2006). *Organic Molecular Solids*. Berlin: Wiley-VCH.

Streitwieser, A. Jr. and Heathcock, C.H. (1985). *Introduction to Organic Chemistry*, 3e. New York: Macmillan.

Wallace, G.G., Spinks, G.M., Kane-Maguire, L.A.P., and Teasdale, P.R. (2002). *Conductive Electroactive Polymers: Intelligent Materials Systems*. Boca Raton, FL: CRC Press.

Warner, J.H., Schäffel, F., Bachmatiul, A., and Rümmeli, M.H. (2013). *Graphene: Fundamentals and Emergent Applications*. Waltham: Elsevier.

Whatmore, R.W. (1986). Pyroelectric devices and materials. *Rep. Prog. Phys.* 49: 1335–1386.

Wolf, E.L. (2013). *Graphene: A New Paradigm in Condensed Matter*. Oxford: Oxford University Press.

Wright, J.D. (1995). *Molecular Crystals*. Cambridge: Cambridge University Press.

6

Tools for Molecular Electronics

Make instruments to plague us

6.1   Introduction

This chapter gives an overview of some of the techniques that can be used to provide information on the physical structure of organic compounds. The list is by no means exhaustive, but illustrates the physical principles behind the more popular methods. In most instances, the materials are in the form of thin films (e.g. suitable for device fabrication). A detailed understanding of and ultimate control over the morphology of these thin layers is essential if the compounds are to be exploited technologically.

The physical properties that underpin the operation of electronic devices ultimately depend on the chemical bonding and crystalline nature of the material. One crystalline form of the same substance (e.g. diamond) can exhibit markedly different properties from another (e.g. graphite). The influence of impurities can be just as dramatic. In Chapter 3, the intrinsic electrical conductivity was shown to depend on the nature and organization of the atoms and molecules that constitute the material (Band Theory). Chemical impurities, either intentionally

Organic and Molecular Electronics: From Principles to Practice, Second Edition. Michael C. Petty.
© 2019 John Wiley & Sons Ltd. Published 2019 by John Wiley & Sons Ltd.
Companion website: www.wiley.com/go/petty/molecular-electronics2

or unintentionally added, may ultimately determine the electrical conductivity of a particular semiconductor. It has taken many years (over 60, and work is continuing!) to understand and control the properties of just one important electronic material – silicon. Sadly, researchers of organic and molecular electronics often ignore the lessons learned from inorganic semi-conductor technology and there are many reports in the literature of irreproducible and poor-quality devices based on impure (or even unknown) materials. This has not helped in establishing the credibility of the subject.

Although there are many simple characterization tools (e.g. the humble microscope) that can provide a useful insight into the morphology of organic thin films, many of the techniques that need to be used are sophisticated, and usually very expensive. Part of the reason is that highly sensitive instrumentation is required for the study of nanoscale amounts of material. With progress down the Moore's Law curve, new equipment has needed to be developed. The surface analytical methods noted in Section 6.7 and the scanning microscopies described in Section 6.8 are good examples.

The characterization of organic materials can also be challenging. In contrast to their inorganic counterparts, the materials are usually soft and fragile. In a high vacuum environment under bombardment by an electron beam, organic substances can be easily damaged. The results from such a physical investigation may therefore not be representative of the material in its natural environment, a particular problem for biological scientists. Consequently, elaborate and careful sample preparation usually accompanies the physical study of organic materials.

6.2 Direct Imaging

6.2.1 Optical Microscopy

The observation of a material under an optical microscope can provide a simple insight into its morphology. Unfortunately, organic compounds, particularly in the form of thin films, do not generally provide high contrast images and therefore special techniques have been devised for their study. Two examples are *phase contrast microscopy* and *confocal microscopy*; these are described below. A further technique, *scanning near-field optical microscopy* (SNOM or NSOM), is also introduced. This has much in common with the scanning probe microscopy methods discussed later in this chapter (Section 6.8).

Phase contrast microscopy was developed in the early twentieth century by Frits Zernike (who was awarded the Nobel Prize for Physics in 1953). In an optical microscope, light from the object being imaged passes through the centre of the lens as well as the periphery. If the light passing through the edge of the objective lens is retarded by a half wavelength and the light at the centre is not retarded, then the light rays will be out of phase by a half wavelength and they will cancel each other out when the objective lens brings the image into focus. A reduction in brightness of the object is then observed (the degree of reduction in brightness will depend on the refractive index of the object). The phase shift is introduced by rings etched accurately onto glass plates so that they introduce the required degree of phase shift when inserted into the optical path of the microscope. When in use, this technique allows the phase of the light passing through the object to be inferred from the intensity of the image produced by the microscope. Phase contrast is only useful on specimens that are colourless and transparent and difficult to distinguish from their surroundings. The method is, therefore, suited to the study of organic, particularly biological, specimens.

Confocal microscopy also offers several advantages over conventional optical microscopy, including controllable depth of field, the elimination of image degrading out-of-focus information, and the ability to collect serial optical sections from thick specimens. The key

Figure 6.1 A confocal optical microscopy system. The small pinhole enables data collection from a thin optical plane within the sample. Points that are outside the plane of focus will have a different secondary focal plane and thus most of the light is deflected (dashed lines).

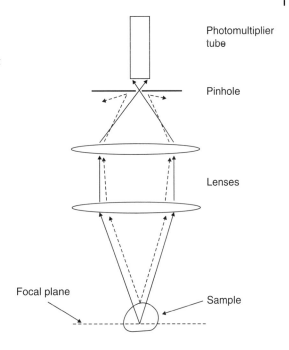

Photomultiplier tube

Pinhole

Lenses

Focal plane

Sample

to the confocal approach is the use of *spatial filtering* to eliminate out-of-focus light or flare in specimens that are thicker than the plane of focus. Current instruments are highly evolved from the earliest versions, but the principle of confocal imaging, which was patented by Marvin Minsky in the 1950s, is employed in all modern confocal microscopes; a schematic diagram is shown in Figure 6.1. In a conventional wide-field microscope, the entire specimen is bathed in light from a mercury or xenon source, and the image can be viewed directly by eye or projected onto an image capture device or photographic film. The method of image formation in a confocal microscope is fundamentally different. Illumination is achieved by scanning one or more focused beams of light, usually from a laser or arc-discharge source, across the specimen. This point of illumination is brought to focus in the specimen by the objective lens, and laterally scanned under computer control. The sequences of points of light from the specimen are detected by a photomultiplier tube through a pinhole (or in some cases, a slit) and the output is built into an image and displayed by the computer. Although unstained specimens can be viewed using light reflected back from the specimen, they usually are labelled with one or more fluorescent probes.

In conventional optical microscopy, or *far-field optical microscopy*, the standard arrangement consists of a system of lenses that focus light from the sample into a virtual, magnified image. In the early 1870s, Ernst Abbé formulated a rigorous criterion for being able to resolve two objects in a light microscope; this is the basis of *diffraction-limited optics*:

$$d > \frac{\lambda}{2\theta} \tag{6.1}$$

where d = the distance between the two objects, λ = the wavelength of the incident light, and 2θ = the angle through which the light is collected. According to this equation, the best resolution achievable with optical light is about 200 nm.

SNOM or NSOM is an imaging technique used to obtain resolution beyond the Abbé diffraction limit. The operational principle, as shown in Figure 6.2, involves illuminating a specimen through a sub-wavelength sized aperture, whilst keeping the specimen within the near-field

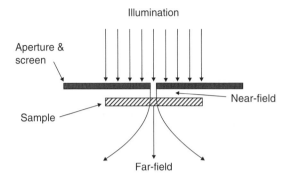

Illumination

Aperture & screen

Sample

Near-field

Far-field

Figure 6.2 Schematic diagram showing the principles of scanning near-field optical microscopy (SNOM or NSOM).

regime of the source. The first suggestion that this was possible came from E.H. Synge in 1928. Although not strictly feasible with the technology of the time, his technical criteria form an accurate basis for a super-resolution optical microscope. If the aperture-to-sample separation is kept roughly less than half the diameter of the aperture, the source does not have the opportunity to diffract before it interacts with the sample, and the resolution of the system is determined by the aperture diameter as opposed to the wavelength of light used. In essence, the SNOM uses the evanescent field (Chapter 4, Section 4.7.1) that can be confined on structures much smaller than the wavelength; this field does not propagate into the far-field. A common way to define the aperture and confine the optical field is to stretch an optical fibre, so as to obtain a conical tip, and to coat it with a thin, but opaque metal layer, usually aluminium. The end of the fibre is uncoated, and is therefore a small pinhole through which an evanescent light wave can pass. The diameter of the hole is often 50–100 nm. An image is built up by raster-scanning the aperture across the sample and recording the optical response of the specimen through a conventional far-field microscope objective. An optical resolution of < 50 nm can be achieved using SNOM.

6.2.2 Electron Microscopy

Electrons are generated in an electron microscope by thermionic emission from a metal filament and accelerated through a potential. For an accelerating voltage of 100 kV, the electron wavelength is 3.7×10^{-3} nm. Atomic resolution should therefore be achievable.

The main interactions taking place when the electron beam is incident on matter are shown in Figure 6.3. Unscattered and diffracted electrons form the basis for conventional *transmission electron microscopy* (TEM) and diffraction. Diffracted electrons will form a pattern that can be transformed directly into an image by a magnetic lens. Either the diffraction pattern or the image can be projected onto a viewing screen. Electron diffraction is discussed later in Section 6.5.

Low-energy (< 50 eV) secondary electrons emitted from the surface of the sample can be used for *scanning electron microscopy* (SEM); the experimental set-up is depicted in Figure 6.4. A beam of electrons is condensed by a condenser lens, and then focused to a very fine point on the sample by the objective lens. The beam can be concentrated in a small probe (≈ 2 nm in diameter), which may be deflected across the sample in a raster fashion using scanning coils (which create a magnetic field). Secondary electrons can be detected above the sample and an image showing the intensity of secondary electrons emitted from different parts of the sample can be built up.

A related technique involves the collection of electrons transmitted through the specimen during scanning. These can be used to produce a scanning transmission electron microscope

Figure 6.3 Interaction of an electron beam with a sample.

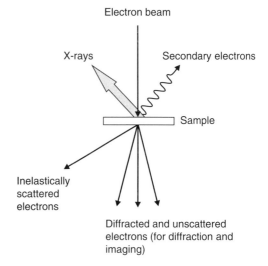

Figure 6.4 Schematic diagram of a scanning electron microscope.

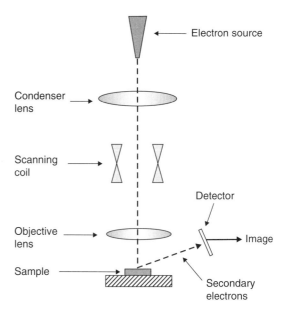

(STEM) image, with the advantage, compared with TEM, that radiation damage is reduced because the beam is not stationary. Emitted X-rays are characteristic of the elements in the sample and may be used for elemental analysis.

In the electron microscope, the electrons are scattered by the atomic potentials of the atoms in the sample. Generally, the scattering increases with atomic number. In contrast to X-rays and neutrons, the scattering of electrons by matter is very strong and diffraction is even feasible with gaseous samples.

Transmission electron microscopy studies of thin organic films, such as those produced by self-assembly or the Langmuir–Blodgett (LB) technique (Chapter 7, Section 7.3), require the layers either to be removed from the substrate or to be deposited onto amorphous supports. A relatively simple sample preparation technique for LB layers involves the deposition of the film onto a previously anodically oxidized aluminium substrate. The organic film on its

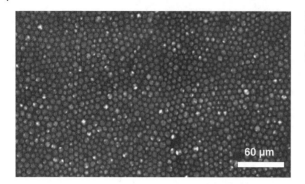

Figure 6.5 Transmission electron micrograph of Au nanoparticles deposited on a carbon-coated microscope grid using the LB technique [1].

alumina support is then removed by etching the aluminium layer in a mercuric chloride solution. Alternatively, LB films may be floated from glass substrates in highly diluted hydrofluoric acid and picked up on a carbon-coated copper microscope grid, or transferred to carbon-coated grids by raising the latter through the condensed floating monolayer. Figure 6.5 shows an example of a transmission electron micrograph of an LB film of gold nanoparticles deposited on a carbon-coated grid [1]. The nanoparticles possess a well ordered close-packed arrangement with an average size of about 8 nm.

Reflection measurements are less demanding on sample preparation: thin films of organic materials can be deposited onto a variety of substrates and then rapidly studied in the electron microscope.

6.3 X-Ray Reflection

When electromagnetic energy is incident on the surface of a material, some of it will be reflected specularly, so that the angle of incidence is equal to the angle of reflection (well known for visible light, Chapter 4, Section 4.5). For crystalline materials, constructive and destructive interference between the radiation reflected from successive crystal planes will occur when the wavelength λ of the incident radiation is of the same order as the lattice spacing (tenths of a nanometre), as shown in Figure 6.6. Bragg's law provides the condition for maxima in the reflected radiation:

$$n\lambda = 2d_{hkl}\sin\theta \qquad (6.2)$$

where θ is the angle of incidence, conventionally measured from the plane of reflection in X-ray crystallography (in contrast to geometrical optics), d_{hkl} is the interplanar spacing (Chapter 2,

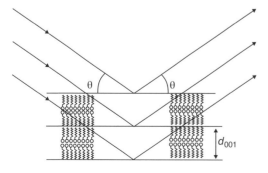

Figure 6.6 Bragg reflection from LB film planes with spacing d_{001}.

Section 2.5.6), and the integer n is known as the *order of the reflection*. Clearly, high-order (n = 2, 3, etc.) scattering from planes d_{hkl} is indistinguishable from first-order scattering from planes $(d_{hkl})/2$, $(d_{hkl})/3$, etc.

A small correction is required to Eq. (6.2) to take into account the fact that the refractive index for X-rays will be slightly less than unity. Bragg's law then becomes

$$n\lambda = 2d \sin\left(1 - \frac{\delta}{\sin^2\theta}\right) \qquad (6.3)$$

where δ is related to the X-ray refractive index by $n = 1 - \delta$. Typically δ is of the order 10^{-6}. (NB: a refractive index less than unity may be thought to be impossible, as it leads to an EM wave velocity greater than the speed of light (Chapter 4, Section 4.3). However the refractive index is related to the *phase velocity* of the radiation (Chapter 3, Section 3.3.1)).

The simplest interplanar spacing in an LB multilayer film is d_{001}, which for a fatty acid salt (e.g. cadmium octadecanoate) is approximately 5.0 nm. Using Eq. 6.2 with an X-ray wavelength of 0.154 nm, gives $\theta \approx 1°$ for the first-order Bragg peak. Therefore, low-angle measurements are necessary for X-ray studies of LB and similar films of long-chain compounds.

Figure 6.7 shows X-ray diffraction data (both experimental and theoretical) from a 43-layer fatty acid salt film [2]. As predicted, the first-order Bragg reflection is observed for an angle of incidence close to one degree. In a Y-type LB film, the d_{001} spacing is equal to the distance between the polar planes, i.e. the thickness of two monomolecular layers. The monolayer thicknesses for fatty acid salts, obtained from X-ray experiments, are plotted as a function of the number of carbon atoms in the molecule in Figure 6.8 [3]. The X-ray data are in close agreement with those calculated from the lengths of the molecules, inferring that the hydrocarbon chains in transferred monolayers are oriented almost at right angles to the substrate surface. For Y-type LB films of other amphiphilic materials (including simple fatty acids), the X-ray d-spacing is often less than twice the molecular length, suggesting some tilting or interdigitation (or both) of the molecules. Unfortunately, the d-spacing value alone cannot be used to distinguish between these situations.

Figure 6.7 X-ray diffraction data from a 43-layer LB film of perdeuterated manganese octadecanoate on a silicon substrate. Experimental values are shown as points. The solid curve is based on calculation and is displaced from the data points. *Source*: Reprinted with permission from Nicklow et al. [2]. Copyright (1981), American Physical Society.

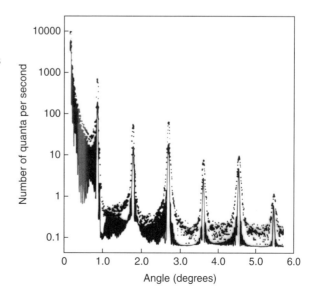

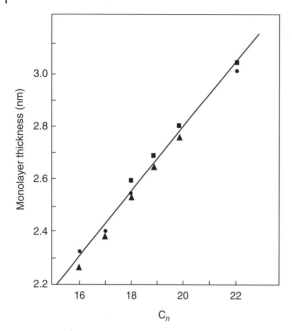

Figure 6.8 Monolayer thickness, obtained from X-ray diffraction experiments, versus number of carbon atoms in the molecule for salts of long-chain fatty acids. *Source*: Reprinted with permission from Petty [3].

X-rays are scattered by their interaction with atomic electrons and interference takes place between X-rays scattered from different parts of an atom. The scattering power or scattering factor decreases with increasing scattering angle 2θ, resulting in a decrease in the intensity of the Bragg peaks. However, it is evident that the height of the third Bragg reflection in Figure 6.7 is greater than that of the second. This is because the intensity of a particular Bragg peak is not only related to the scattering power of each atom in the lattice but also to the position of the atom in the unit cell. In summing the individual waves to give the resultant diffracted beam, both the amplitude and phase of each wave scattered by the individual atoms are important. The intensity of the scattered radiation I_{hkl} from a set of planes $\{hkl\}$ may be written as

$$I_{hkl} \propto F_{hkl}^2 \tag{6.4}$$

where F_{hkl} is called the *structure factor*.

The structure factor of the unit cell depends upon the constituent atoms and their individual scattering factors. For a particular d-spacing, the only variable in the cell structure is the nature of the atom. Since F depends directly on the effective number of scattering electrons per atom, a large F value is given by a high atomic number. Thus organic materials containing heavy metals are generally superior for X-ray diffraction experiments.

The angular width of the Bragg peaks at half-height, Δ, is inversely proportional to the length of the ordered region L according to the Scherrer equation [4]:

$$\Delta = \frac{\lambda}{L\cos\theta} \tag{6.5}$$

The parameter L can be interpreted either as an apparent size of the crystalline grain, assuming a grain structure with perfectly ordered domains, or, in the case of an imperfect crystalline order, L is related to the correlation length of the order.

6.3.1 Electron Density Profile

If the structure factors F_{hjk} for a complete set of X-ray reflections are known, the electron density ρ at any position xyz in the unit cell may be calculated using

$$\rho(xyz) = \frac{1}{V} \sum_h \sum_k \sum_l F_{hkl} \cos 2\pi (hx + ky + lz) \tag{6.6}$$

where V is the unit cell volume. This is an example of a *Fourier summation*; the electron density profile is the Fourier transform of the X-ray diffraction pattern. To calculate $\rho(xyz)$, values of F_{hkl} are needed. Unfortunately, only F_{hkl}^2 can be measured. (The + or − signs of F_{hkl} correspond to phases of 0° and 180°.) It is therefore possible to obtain F_{hkl}, but not the sign of the structure factor. This dilemma is known as the *phase problem*.

For a set of data containing m structure factors, each of which may be positive or negative, there are 2^m possible combinations of phases; e.g. for 300 reflections there are 2×10^{90} possibilities! Clearly, the solution of the phase problem by trial-and-error methods is not very practicable! A physical criterion is often used to evaluate the electron density profile of multilayer films perpendicular to the substrate. For example, the optimization of a plateau in the electron density profile (representing the region of constant electron density corresponding to the alkyl chain) can be used to simplify these computations in many LB film systems.

6.3.2 Kiessig Fringes

The observation of X-ray interferences, first described by Kiessig, is one of the most precise methods to measure the thickness of a thin film. Specular reflection of X-rays by a flat surface is observed at glancing angles close to the critical angle θ_c (Chapter 4, Section 4.5) for total reflection. For all materials, θ_c is a few tenths of a degree. In a thin film system, each interface of two layers with different refractive indices gives rise to a specularly reflected X-ray beam. With monochromatic X-radiation, interference maxima and minima can be observed near the critical angle. Kiessig showed that is was possible to determine the thickness of the film from the angular spacing of these intensity maxima. If deviation of the X-ray refractive index from unity is neglected, the fringe maxima θ_m are related to the total thickness t by

$$\sin \theta_m = m \frac{\lambda}{2t} \tag{6.7}$$

where m is an integer. The film thickness can therefore be obtained from the slope of the $\sin \theta_m$ versus m plot. Figure 6.9 reveals a series of Kiessig fringes obtained for layer-by-layer assemblies of the polyelectrolytes [5]. There are no Bragg peaks, as might be expected for a multilayer assembly, suggesting a poor degree of layer-by-layer ordering. The number of Kiessig fringes increases with the increase in the number of polyelectrolyte bilayers.

6.3.3 In-plane Measurements

X-ray diffraction can also be used to obtain information about the in-plane structure of multilayer molecular assemblies. However, since most substrates supporting the films (e.g. glass slides) will almost completely attenuate an X-ray beam, diffraction in transmission either requires the organic layer to be removed from the substrates or special substrate materials to be used. The technique of Prakash et al. [6] consists of tilting the plane of the multilayer to small angles with respect to the horizontal (with the incident and diffracted X-ray beams in the horizontal plane). This allows the observation of (11l) diffraction peaks.

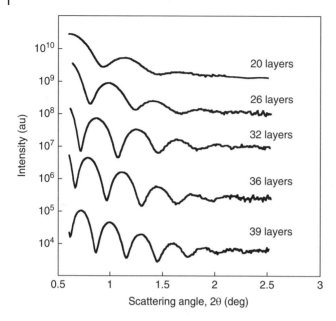

Figure 6.9 X-ray diffraction data for different numbers of layers of alternating anionic and cationic polyelectrolyte films deposited by the layer-by-layer technique. Kiessig interference fringes are evident. *Source*: Reprinted from Lvov [5, p. 133], Copyright (1999), with permission from Taylor & Francis Group LLC.

6.4 Neutron Reflection

The use of a neutron beam gives rise to interference effects in a similar way to an X-ray beam. The major difference between the two types of radiation lies in the factors governing the intensity of diffraction. For X-rays, this depends on the electron density variation across the layers. In the case of neutrons, it is the variation of nuclear scattering length density that determines the Bragg intensity. Because neutron scattering is a nuclear property, it may vary considerably from one element to the next and is different for different isotopes, e.g. hydrogen and deuterium. Neutron absorption is usually negligible and the interference effects that cause X-ray scattering to diminish with increasing angle are absent with neutrons and the scattering is isotropic.

6.5 Electron Diffraction

Electron diffraction also provides useful structural information on organic films. To obtain the structure normal to the film plane, the electron beam impinges at grazing angles and the reflected beam is observed. For fatty acid type LB films, electron diffraction experiments reveal the packing of the C_2H_4 subcells in the aliphatic chains. Consequently, two levels of organization must be considered for long-chain molecules in layered structures. These are illustrated in Figure 6.10.

Besides the nature of the subcell, the packing of the individual molecules in the multilayer will define the main cell. The X-ray diffraction data shown in Figure 6.7 reveal the packing associated with the main cell. In the solid state, long-chain fatty acid type compounds exhibit polymorphism, i.e. they can exist in a variety of crystallographic states, depending on the packing arrangements of the main and subcells. In all, there are 10 different packing arrangements for long-chain fatty acids, depending on the relative displacements of neighbouring chains [7]. These can conveniently be distinguished by the Miller indices of the interface plane between

Figure 6.10 Main cell and subcells for fatty acid multilayers.

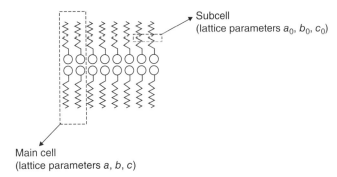

Subcell
(lattice parameters a_0, b_0, c_0)

Main cell
(lattice parameters a, b, c)

Figure 6.11 Transmission electron diffraction (TED) pattern for 22-tricosenoic acid LB film. *Source*: Reprinted with permission from Peterson and Russell. [8]. Copyright (1984), with permission from Taylor & Francis Group LLC.

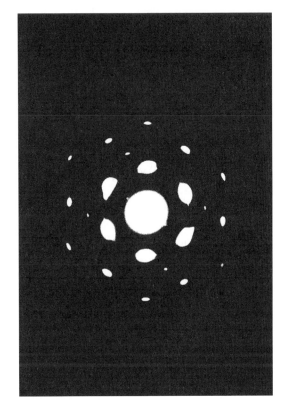

layers of molecules and the type of subcell symmetry. There are three possible close-packed structures with similar packing densities: orthorhombic (R), monoclinic (M), and triclinic (T) (Chapter 2, Section 2.5.4). In the T and M arrangements, the zig-zags in the chain are parallel to one another. However, this is not the case for the R structure.

The transmission geometry is used to obtain in-plane structural information. As noted above, an advantage of the electron beam compared with X-rays is that the interaction of electrons with matter is much stronger. Consequently, a diffraction pattern can be obtained with a beam of smaller diameter. Both transmission electron diffraction (TED) and reflection high energy electron diffraction (RHEED) techniques may be used. Figure 6.11 shows a TED pattern for a LB film built up from 22-tricosenoic acid [8]. The diffraction pattern can be indexed as arising from the orthorhombic R(001) packing of the subcells. In-plane X-ray diffraction data (Section 6.3.3) can provide similar information on the packing of the subcells in LB arrays of molecules [9].

If the LB multilayer structure were a simple crystalline system, the above identification system would allow only a few possible orientations for the molecular chains. In practice, it is found that the tilt angle of the alkyl chains in fatty acid materials (e.g. 22-tricosenoic acid) varies continuously with the deposition pressure; the greater the pressure, the smaller the angle of tilt from the substrate normal [10, 11]. The structure of these LB layers of tilted molecules may therefore be quite complex, consisting of regions of crystallinity in which grains are inclined to the substrate with a distribution of tilt azimuths and permeated by holes [12].

The angle of inclination to the shadow edge of the line of the spots in RHEED patterns of fatty acid LB films also varies with the deposition pressure [10, 11]. Typical data are shown in Figure 6.12 [10, 12]. As the deposition pressure is increased, the molecular tilt elevation also increases (i.e. the molecules become more upright). However, this is a continuous process and is not what would be expected from the molecules taking up one of a few fixed orientations.

6.6 Infrared Spectroscopy

Infrared (IR) spectroscopy (wavelength in the range 1–100 μm) probes the vibrational features of an organic molecule. A particular bond must have a permanent dipole moment associated with it to interact with the infrared radiation.

The compression and extension of a bond can be likened to the behaviour of a spring and this analogy may be taken further by assuming that the bond, like a simple spring, obeys Hooke's law. For a diatomic molecule (e.g. HCl), the vibrational energy levels E_v are given by

$$E_v = h\nu\left(\upsilon + \frac{1}{2}\right) \tag{6.8}$$

where h is Planck's constant and the vibrational quantum number $\upsilon = 0,1,2$.

In IR spectroscopy, it is usual for the energy of the electromagnetic radiation to be quoted in terms of a *wavenumber*, $\bar{\nu}$. This is the reciprocal of the wavelength, in centimetres, i.e.

$$\bar{\nu} = \frac{1}{\lambda}\,\mathrm{cm}^{-1} \tag{6.9}$$

The wavenumber expresses the number of waves or cycles contained in each centimetre length of the radiation and is a useful concept in spectroscopy.

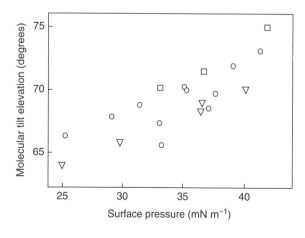

Figure 6.12 Molecular tilt (measured from substrate plane) versus deposition surface pressure for 22-tricosenoic acid LB films. □ deposition pH 7; ○ pH 3. *Source*: Reprinted from Peterson et al. [10]. Copyright (1998), with permission from Elsevier. ∇ pH 7. Data from Barnes and Sambles [12].

The number of vibrational modes associated with polyatomic molecules can be very large. An N-atomic molecule has $3N$-5 normal modes of vibration if it is linear and $3N$-6 if it is nonlinear. A normal mode is one in which all the nuclei undergo harmonic motion, have the same frequency of oscillation and move in phase (but with different amplitudes). The form of these vibrations may be obtained from knowledge of the bond lengths and angles and of the bond-stretching and angle-bending force constants. Consider, for example, a molecule of water. This is nonlinear and triatomic, with three allowed normal vibrational modes, as depicted in Figure 6.13. Each motion is described as stretching or bending, depending on the nature of the change in molecular shape. Furthermore, the motions are designated either symmetric or antisymmetric.

Although a normal mode of vibration involves movement of all the atoms in a molecule, there are circumstances in which movement is almost localized in one part of the molecule. If the vibration involves the stretching or bending of a terminal-XY group, where X is heavy compared to Y (e.g. an OH group in a fatty acid), the corresponding vibration wavenumbers are almost independent of the rest of the molecule to which -XY is attached. A typical wavenumber for the XY stretching vibration may therefore be referred to. For example, the (OH) stretching frequency is normally in the region of $3600\,\mathrm{cm}^{-1}$. Many group vibrations occur in the region 1500–$3700\,\mathrm{cm}^{-1}$; stretching and bending vibrations of some well-known groups are listed in Table 6.1.

Not all parts of a molecule are characterized by group vibrations. Many normal modes involve strong coupling between stretching or bending motions of atoms in a straight chain, a

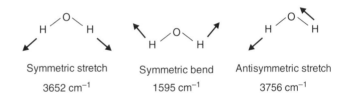

Figure 6.13 Three fundamental vibrations of the water molecule.

Symmetric stretch
$3652\,\mathrm{cm}^{-1}$

Symmetric bend
$1595\,\mathrm{cm}^{-1}$

Antisymmetric stretch
$3756\,\mathrm{cm}^{-1}$

Table 6.1 Characteristic stretching and bending frequencies of molecular groups.

Group	Approximate wavenumber (cm^{-1})
—OH	3600
—NH$_2$	3400
=CH$_2$	3030
—CH$_3$	2960 (antisym. stretch)
	2870 (sym. stretch)
	1460 (antisym. bend)
	1375 (sym. bend)
—CH$_2$—	2920 (antisym. stretch)
	2850 (sym. stretch)
	1470 (bend)
—C≡C—	2220
>C=O	1750–1600
>C=C<	1650

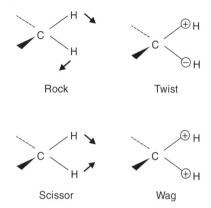

Figure 6.14 Rocking, twisting, scissoring, wagging vibrations in a CH_2 group. The + and –signify movement in and out of the plane of the diagram; the arrows indicate movement in this plane.

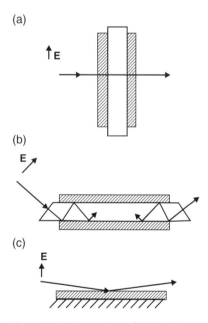

Figure 6.15 Comparison of (a) simple transmission, (b) attenuated total reflection (ATR), and (c) reflection absorption infrared spectroscopy (RAIRS) sampling techniques.

branched chain or a ring. Such vibrations are called skeletal vibrations and tend to be specific to a particular molecule. For this reason, the region where skeletal vibrations mostly occur, from about $1400\,cm^{-1}$ to low wavenumbers, is sometimes called the *fingerprint region*.

In addition to the description of group vibrations as stretch and bend (or deformation) the terms rock, twist, scissor, wag, torsion, ring breathing, and inversion (umbrella) are used frequently. Some of these are illustrated for the CH_2 group in Figure 6.14.

Only the component of radiation polarized in the direction of the transition dipole moment (Chapter 4, Section 4.4.1) will induce a change from one vibrational energy level to another. The absorption intensity of the radiation may be written as

$$I \propto \left| \mathbf{p} \cdot \mathbf{E} \right|^2 \tag{6.10}$$

where $\mathbf{p}$ is the transition dipole moment and $\mathbf{E}$ is the electric field vector. If $\mathbf{E}$ is parallel to the transition dipole moment, the probability of absorption is high, whereas if $\mathbf{E}$ is perpendicular to it no radiation is absorbed.

The absorption or reflection intensities resulting from the interaction of IR radiation with monolayer samples are very low, a direct result of the relatively small numbers of molecules being sampled. In a transmission experiment with normal incidence of the IR beam, the electric field vector is oriented parallel to the layer plane. In this geometry (Figure 6.15a), the projection of the transition moments on the layer plane is probed. An increase in surface sensitivity may be obtained by using a method based on *attenuated total reflection* (ATR) (Figure 6.15b). In this technique, the organic compound is deposited onto either side of an IR-transmitting crystal (e.g. silicon or germanium). The radiation is incident at an angle greater than the critical angle and undergoes multiple reflections inside the crystal. On each reflection, the evanescent field of the IR beam penetrates the organic material under study and may be absorbed by it (the evanescent field decays over micrometre distances – Chapter 4, Section 4.7.1). Both the simple transmission and ATR experiments may be used with either polarized or unpolarized radiation.

Reflection absorption infrared spectroscopy, RAIRS, requires IR radiation to be incident at a grazing angle (85°–88°) to a metal surface, onto which an organic layer has been deposited (Figure 6.15c). Incident s-polarized radiation (Chapter 4, Section 4.2) undergoes a phase shift of 180° on reflection from the metal surface and so the electric field components of the incident and reflected radiation cancel at the metal/LB film interface. In contrast, the incident and

Figure 6.16 Infrared spectra of 21 LB layers of cadmium docosanoate on a silicon ATR crystal. (a) ATR mode; (b) RAIRS mode. *Source*: Reprinted with permission from Davies and Yarwood [13]. Copyright (1989) American Chemical Society.

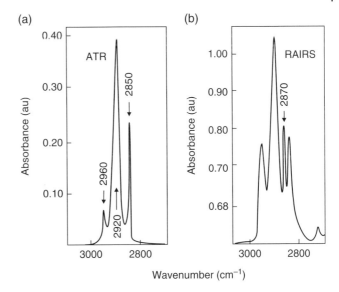

reflected components of p-polarized IR radiation differ by only about 90° at grazing incidence. This is the origin of the *surface selection rule*, which results in the very useful ability to distinguish vibrations that possess a transition dipole moment with a large component perpendicular to the surface.

The orientation of the molecules in self-assembled or LB films may be investigated using different polarizations of incident radiation or by comparing ATR and RAIRS measurements. Figure 6.16 contrasts the ATR spectrum with an approximate RAIRS spectrum (taken in ATR mode with a metal overlayer on the organic film) in the CH stretching region for 21-layers of cadmium docosenoate [13]; both spectra were recorded using p-polarized radiation. The bands that are observed are $2960\,cm^{-1}$ (antisymmetric CH_3 stretch), $2920\,cm^{-1}$ (antisymmetric CH_2 stretch), $2870\,cm^{-1}$ (symmetric CH_3 stretch), and $2850\,cm^{-1}$ (symmetric CH_2 stretch). The symmetric and antisymmetric CH_2 stretches are both intense in the ATR spectrum. In the RAIRS mode, the CH_3 bands increase in intensity. The transition dipole moments of the CH_2 vibrations are perpendicular to the long axis of the fatty acid molecule, whereas the symmetric and antisymmetric CH_3 stretches have components along the chains axis. Therefore, the experimental data shown in Figure 6.16 are consistent with the long axes of the fatty acid molecules being almost perpendicular to the substrate plane.

Infrared spectroscopy may also be used to monitor chemical and structural changes occurring in multilayer films. For example, Figure 6.17 shows the results of an IR study of LB layers of a derivative of the charge-transfer compound tetrathiafulvalene (TTF) (Chapter 3, Section 3.4.1) before and after doping with iodine vapour [14]. The molecules of this compound, known as hexadecanolyl tetrathiafulvalene (HDTTF), transfer to the substrate with their long axes roughly perpendicular to the surface. As depicted in Figure 6.17, the C=O stretching vibration, at $1650\,cm^{-1}$, has its transition dipole oriented perpendicular to the long axis of the molecule, while the C=C stretching mode of the fulvalene ring, at approximately $1530\,cm^{-1}$, is parallel to this direction. The intensity of these bands reflects strongly the chemistry of the initial doping (oxidation) and subsequent loss of I_2 to produce an organic conductor. Immediately after doping, the C=O stretching band moves to a higher frequency (curve b in Figure 6.17) because of electronic changes associated with the oxidation of the TTF derivative HD tetrathiafulvalene to produce a radical cation HDTTF•+. The C=C band ($1530\,cm^{-1}$) almost

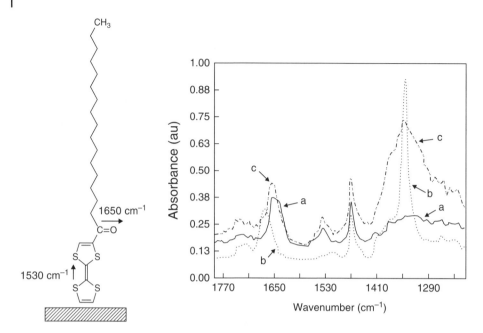

Figure 6.17 ATR infrared spectra of HDTTF LB film as a function of doping with Br_2 vapour. (a) Before doping. (b) Immediately after doping. (c) One hour after doping. The orientation of the HDTTF molecule on the substrate is shown on the left. *Source*: Reprinted with permission from Dhindsa et al. [14]. Copyright (1990) American Chemical Society.

disappears, showing that the radical cation carries a largely C—C single bond. The band reappears when iodine molecules leave the film to form a mixed valence complex, with some fulvalene rings in their ground (not charge-transfer) state. The charge-transfer in such systems often leads to vibronic excitation of originally IR-forbidden vibrations. Figure 6.17 shows a very intense band near $1350\,cm^{-1}$ immediately after doping with I_2. This arises from the coupling of the motion of electrons with vibrational modes in the fulvalene ring, referred to as electron-molecule vibronic coupling.

6.6.1 Raman Scattering

When electromagnetic energy falls on an atomic or molecular sample, it may be absorbed if the energy of the radiation corresponds to the separation of two energy levels in the atoms or molecules. If it does not, the radiation will be either transmitted or scattered. Of the scattered radiation, most is of unchanged wavelength λ and is called *Rayleigh scattering*. The intensity I_s is related to the wavelength by

$$I_s \propto \lambda^4 \tag{6.11}$$

However, a small amount of the scattered radiation is at a slightly increased or decreased wavelength. This is *Raman scattering*; the radiation scattered with an energy lower than the incident beam is referred to as *Stokes' radiation*, while that at higher frequency is called *anti-Stokes' radiation*. To be Raman active, a molecular rotation or vibration must cause some change in a component of the molecular polarizability. The process can be easily understood in the context of the simplified energy-level diagram shown in Figure 6.18. Incident light of energy $h\nu_0$ impinges on a molecule, originally in its ground vibrational level, causing a momentary

Figure 6.18 Energy level diagram illustrating the Raman scattering process.

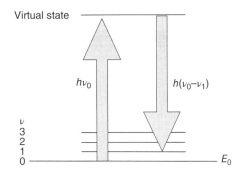

excitation to what is termed a *virtual state*. During this brief interaction, the molecule from the light acquires a quantum of vibrational energy. In this case, the scattered light is Stokes' radiation with energy, $h(\nu_0 - \nu_n)$, less than the original light by an amount equal to that necessary for excitation of a particular vibrational mode, ν_n, of the molecule, and the molecule is left vibrationally excited.

Raman scattering is an inherently weak process, so an enhancement mechanism is usually needed to organic monolayer samples. An increase in the signal can usually be achieved by depositing films onto the surfaces of noble metals such as Au or Ag or by the interaction with surface plasmons. These methods are generally referred to as *surface-enhanced Raman spectroscopy*. For materials with electronic transitions in the visible, the resonance Raman effect can be used to enhance certain bands. The Raman process may be used as a basis for a microscopy technique. The mode of operation is similar to a fluorescence microscope. Images are recorded with a sensitive (e.g. charge-coupled device (CCD)) camera in light that has been scattered in a particular Raman band by using optical filters.

6.7 Surface Analytical Techniques

Surface analytical techniques such as *Auger electron spectroscopy* (AES), *X-ray photoelectron spectroscopy* (XPS), and *secondary-ion mass spectrometry* (SIMS), may be used to provide useful information about the structure and/or chemical composition of the surfaces of organic materials. These methods are contrasted in Table 6.2.

Auger electron spectroscopy is a two-step process and is depicted schematically in Figure 6.19. First, a high-energy photon ejects an electron from the core orbital of an atom, E_A, in the

Table 6.2 Important surface analytical methods.

Technique	Acronym	Probe Beam	Depth probed	Sample Beam	Comments
Auger electron spectroscopy	AES	Low-energy (1–10 keV) electrons	0.5–10 nm	Electron energy (20–1000 eV)	Very high surface sensitivity
X-ray photoelectron spectroscopy	XPS	Low-energy X-rays	0.5–5 nm	Photo-electrons	Little surface damage; chemical composition can be determined
Secondary-ion mass spectrometry	SIMS	Pulsed ion (Ar^+) beam (few keV)	2 nm – 100 μm	Secondary ions	Identification of chemical compounds

(a)   (b)

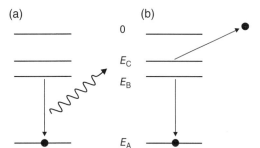

Figure 6.19 Possible processes occurring following the removal of an inner core electron by an incident high-energy electron. An electron from a higher energy immediately fills this vacancy. (a) X-ray photon emission occurs as the vacancy is filled. (b) Auger effect – the energy is transferred to a second electron, which is subsequently emitted.

material under investigation. An electron from a higher energy, E_B, immediately fills this vacancy, so an X-ray photon emission can occur (Figure 6.19a). However, the energy may also be transferred to a second electron, an Auger electron, releasing it from one of the higher energy orbitals, E_C, as shown in Figure 6.19b. The energies of these Auger electrons are low (20–1000 eV) so that, although they may be generated from as far within the sample as the original electron beam penetrates, only those produced within the first few atomic layers of the surface can escape. Therefore, the technique has immense surface sensitivity.

In the XPS experiment, the sample surface is irradiated by a source of low-energy X-rays. Photoionization takes place in the sample, producing photoelectrons of a characteristic energy distribution. Because X-rays do not normally cause appreciable surface damage, XPS is usually preferred as an analytical technique for organic materials. Analysis of the kinetic energy of the photoelectrons permits the elemental composition of the surface layers (up to 5 nm) to be determined quantitatively. Information concerning the bonding environments of the elements may also be obtained.

Secondary ion mass spectrometry involves bombarding a sample surface with a pulsed primary ion beam (usually Ar^+) with energy of a few keV. This results in the emission of both positively and negatively charged secondary ions from the uppermost surface layers (a few nm). The identification of chemical compounds is possible by the detection of either molecular ions with masses of up to 10^4 atomic mass units or by the detection of characteristic fragments.

The relative position of a particular atom with respect to its nearest neighbours can sometimes be determined with great precision from the *extended X-ray absorption fine structure* (EXAFS). Above an X-ray absorption edge, the absorption by the atom is slightly affected by the waves being scattered by neighbouring atoms. This is manifested by weak oscillations in the absorption as a function of frequency for a considerable range above the edge. Just above the absorption edge (around 5 eV beyond the absorption threshold) there are stronger oscillations, called *X-ray absorption near-edge structure* (XANES), which derive from the same process of scattering from neighbouring atoms. *Near edge X-ray absorption fine structure* (NEXAFS) is another method using synchrotron radiation that may be used to obtain orientation information on the molecules in thin organic films.

6.8 Scanning Probe Microscopies

Gerd Binnig and Heinrich Rohrer introduced the *scanning tunnelling microscope* (STM) in 1981. For this, they were awarded one half of the Nobel Prize for Physics in 1986. (The other half went to Ernst Ruska for the design of the first electron microscope.) Like *atomic force microscopy* (AFM–invented in 1986 by Binnig, Quate, and Gerber), the STM technique may be used to provide direct images of metal surfaces with nanometre resolution. There are now very many other related methods (see below), the instrumentation of which can all be classified as *scanning probe microscopes* (SPMs). These devices work by measuring a local property – such as height, optical absorption, or magnetism – with a probe or tip placed very close to the sample. The small probe-sample separation (on the order of the instrument's resolution) makes it possible to take measurements over a small area. To acquire an image, the microscope raster-scans the probe over the sample while measuring the local property in question.

Scanning tunnelling microscopy measures the current flowing between the tip and the surface of the sample. The method may provide lateral and vertical resolutions of less than 0.3 and 0.02 nm, respectively. Furthermore, the electron energies are usually less than 3 eV, thus avoiding the degradation of an organic thin film, which can be a problem with other imaging methods.

Whereas STM records the overlap of the local electron density of states between a tip and a surface (or its modulation by adsorbate molecules in the gap), AFM measures the interatomic forces between a cantilevered spring tip and the sample surface. The image contrast in AFM is achieved by probing the elastic response of the molecules to the force exerted by the scanning tip. Figure 6.20 shows the principles of the technique. The microscope, which essentially consists of a tip attached to a cantilever (Figure 6.20a), can operate in three different modes: contact, noncontact, and tapping. Figure 6.20b shows the dependence on the interaction force on the distance between the tip and the surface (Chapter 2, Section 2.3.1). At a relatively large distance from the surface, the force of attraction between the tip and the surface dominates, while the force of repulsion is the most significant at very small distances.

In the contact mode, the tip is brought into close contact with the sample surface, so that the force between the tip and the sample becomes repulsive. The resulting deflection by the cantilever is then measured. For organic materials, one problem with this mode of operation is that is can easily scratch the soft surface of the sample. If the tip is moved away from the surface by a small distance (5–10 nm), the tip will be attracted to the surface, causing the cantilever to bend towards it. This is the noncontact mode of operation. However, since the force of attraction is much weaker than that of repulsion, the resolution is normally lower than for

Figure 6.20 (a) Cantilever interaction with a surface in atomic force microscopy (AFM). (b) Force versus separation between the AFM tip and the substrate.

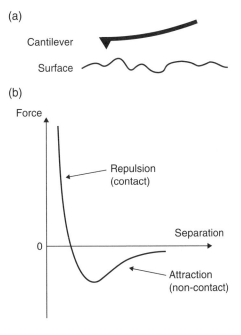

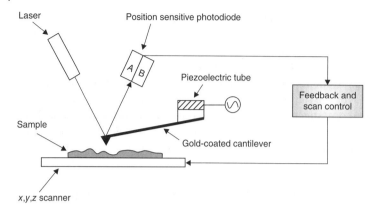

Figure 6.21 Schematic diagram of an atomic force microscope.

the contact mode of operation. The tapping mode combines the advantages of the contact and noncontact techniques. The cantilever is oscillated at a frequency in the range 100–500 kHz, with amplitude of approximately 20 nm, so that the tip is just touching (tapping) the surface. The resolution of this tapping mode is almost the same as that of the contact mode, but it is faster and much less damaging.

Figure 6.21 shows a schematic diagram of an AFM. The instrument can generally measure the vertical deflection of the cantilever with picometre resolution. To achieve this, most AFMs use an optical lever, a device that achieves resolution comparable to an interferometer, while remaining inexpensive and easy to use. It works by reflecting a laser beam off the cantilever. Angular deflection of the cantilever causes a two-fold larger angular deflection of the laser beam. The reflected laser beam then strikes a position-sensitive photodetector, consisting of two side-by-side photodiodes. The difference between the two photodiode signals indicates the position of the laser spot on the detector and thus the angular deflection of the cantilever. Because the cantilever-to-detector distance generally measures thousands of times the length of the cantilever, the optical lever greatly magnifies motions of the tip.

Most scanned-probe microscopes use piezoceramic tubes (usually made of ceramic materials) to position the tip, as these combine a simple one-piece construction with high stability and large scan range. Four electrodes cover the outer surface of the tube, while a single electrode covers the inner surface. Application of voltages to one or more of the electrodes causes the tube to bend or stretch, moving the sample in three dimensions. If the tip were scanned at constant height, there would be a risk that the tip would collide with the surface, causing damage. Hence, in most cases, a feedback mechanism is employed to adjust the tip-to-sample distance to keep the force between the tip and the sample constant.

Early experiments using STM on organic samples exhibited poor reproducibility. Artefacts such as steps, domain walls, or superstructures, due to multiple tip effects and *Moiré patterns*, made the images difficult to interpret. A major problem with thin film samples, such as self-assembled films, is that these possess highly insulating regions (due to the hydrocarbon tails) through which the tunnelling current must pass. Such difficulties are circumvented using AFM. Figure 6.22 shows an AFM image of the surface of a 12-layer 22-tricosenoic acid LB film; lines of individual molecules are evident at the magnification shown [15].

Other measurements can be made using modifications of the SPM. These include variations in surface microfriction with a *lateral force microscope* (LFM), orientation of magnetic domains with a *magnetic force microscope* (MFM), and differences in elastic modulii on the micro-scale with a *force modulation microscope* (FMM). *Scanning capacitance microscopy* (SCM) is sensitive

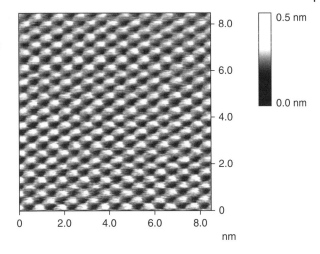

Figure 6.22 Atomic force microscope image of a 12-layer 22-tricosenoic acid LB film deposited onto silicon. Unfiltered data obtained with a Digital Instruments Nanoscope III. *Source*: Reprinted with permission Evanson et al. [15]. Copyright (1996) American Chemical Society.

to electrical charges that develop on the surface and on the probe. This is most useful for determining the level of doping on semiconductor chips. These current microscopes can detect electronic charges of attofarads (10^{-18}).

A further adaptation of the SPM has been developed to probe differences in chemical forces across a surface at the molecular scale. This technique has been called the *chemical force microscope* (CFM). The AFM and STM can also be used to do electrochemistry on the microscale.

6.9 Film Thickness Measurements

Film thickness is an essential parameter in the characterization of thin organic films. Many different measurement techniques have been used and some of the most popular are listed in Table 6.3. However, some of these methods do not provide an independent measure of film thickness; other physical parameters must be first determined. For example, some optical techniques (e.g. surface plasmon resonance, Chapter 4, Section 4.7.2) measure the optical thickness (or optical path length), which is equal to the metric thickness multiplied by the refractive index. Figure 6.23 shows a schematic diagram of an ellipsometer. A range of such instruments is available commercially and is used extensively for thin film (e.g. silicon dioxide) thickness measurements in the semiconductor industry. However, as noted in Chapter 4, Section 4.5.3, the method needed to obtain accurate values of film thickness is complex. The problem of

Table 6.3 Summary of common methods to determine the thickness of thin organic films.

Technique	Comments
X-ray diffraction	Provides d_{001} lattice spacing. Section 6.3.
Ellipsometry	Commercial instruments available; sample refractive index should be known for high accuracy. Chapter 4, Section 4.5.3.
Surface plasmon resonance	Sample refractive index should be known. Chapter 4, Section 4.7.
Capacitance versus number of layers	Provides dielectric thickness. Requires insulating samples. Possibility of damage to organic layer during metallisation. Chapter 7, Section 7.3.1.
Mechanical probe	Commercial instruments available. Provides metric thickness directly. Organic film may be damaged. Not suitable for monolayer sensitivity.

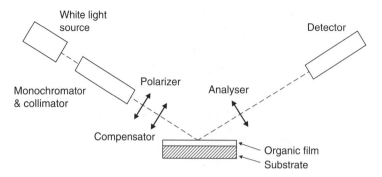

Figure 6.23 Schematic diagram of an ellipsometer.

obtaining values of film thickness and refractive index from the ellipsometric parameters ψ and Δ involves a least squares routine, in which theoretical values of ψ and Δ are calculated and the subsequent error function is minimized. However, the solutions are not unique (the same values of ψ and Δ correspond to different refractive index and thickness combinations) and some experience is needed for accurate results.

A further problem with ellipsometric measurements on ultra-thin organic films concerns the nature of the thin film system under investigation. The organic layer is invariably deposited onto a solid surface: this may be a metal (or metallized glass microscope slides) or semiconductor. For such substrates, there is almost certainly a surface ('oxide') layer between the bulk substrate and the organic film. This may result from exposing an evaporated metal film or a freshly etched semiconductor to the atmosphere. It may also be augmented during the organic film deposition process (e.g. for LB deposition, the substrate is lowered into and out of the aqueous subphase). Such an interfacial film may be several nm in thickness and can introduce considerable errors if its presence is ignored. The usual solution is to modify the optical constants of the underlying substrate (which are used in the computer iteration) by taking measurements on an uncoated portion of the substrate.

Electrical measurements based on measuring the capacitance of a metal/LB film/metal structure as a function of the number of monolayers (Chapter 7, Section 7.3.1) yield the dielectric thickness (metric thickness ÷ relative permittivity). Direct measurement of the thickness of an LB film is conveniently accomplished using a mechanical probe; however, care must be taken to avoid damage to the relatively soft organic layer.

Problems

6.1 Derive an expression for the (de Broglie) wavelength of an electron in terms of the potential difference V (in volts) through which it has been accelerated. Consequently, calculate the wavelength of an electron accelerated by 100 kV in an electron microscope. Compare this to the wavelength of a 2 eV photon.

6.2 The Raman effect can be summarized by the following equation:

$$\Delta v = \overline{v_0} - \overline{v_s}$$

where $\overline{v_0}$ is the wavenumber of incident light and $\overline{v_s}$ is the wavenumber of scattered light. Explain the Stokes' shift behaviour in terms of this equation with respect to the energy state of a molecule excited by a laser. How does this differ from Rayleigh scattering and anti-Stokes' behaviour?

6.3 A beam of X-rays of wavelength 0.071 nm is diffracted by (110) plane of a face-centred cubic crystal with lattice constant of 0.28 nm. Find the glancing angle for the second-order diffraction.

6.4 The wavenumbers (in cm^{-1}) corresponding to the main absorption bands for three C_2H_8O isomers are given in the table below.

Spectrum 1	Spectrum 2	Spectrum 3
3136	3336	2981
3121	3078	2940
3050	2919	2883
2984	2866	1716
2934	1662	1365
2907	1036	1170
2883	896	
1641		
1614		
1218		
1072		
966		
814		

The structures of the isomers are as follows:

2-butanone   ethyl vinyl ether   2-methyl-2-propen-1-ol

Associate each spectrum with the structure of each isomer.

6.5 In a low-angle X-ray study of a thin organic film, Kiessig fringes are observed with a series of consecutive maxima corresponding to $\sin \theta$ at 0.0035, 0.0040, 0.0045, 0.0050. If the X-ray wavelength is 0.154 nm, calculate the thickness of the film in nm.

References

1 Paul, S., Pearson, C., Molloy, A. et al. (2003). Langmuir-Blodgett film deposition of metallic nanoparticles and their application to electronic memory structures. *Nano Lett.* 3: 533–536.
2 Nicklow, R.M., Pomerantz, M., and Segmüller, A. (1981). Neutron diffraction from small numbers of Langmuir-Blodgett monolayers of manganese stearate. *Phys. Rev. B* 23: 1081–1087.
3 Petty, M.C. (1996). *An Introduction to Langmuir-Blodgett Films.* Cambridge: Cambridge University Press.

4 Leuthe, A. and Riegler, H. (1992). Thermal behaviour of Langmuir-Blodgett films. II. X-ray and polarized reflection microscopy studies on coexisting polymorphism, thermal annealing and epitaxial layer growth of behenic acid multilayers. *J. Phys. D. Appl. Phys.* 25: 1786–1797.

5 Lvov, Y. (2000). Electrostatic layer-by-layer of proteins and polyions. In: *Protein Architecture, Interfacing Molecular Assemblies and Immobilization Biotechnology* (ed. Y. Lvov and H. Mohwald), 125–167. New York: Marcel Dekker.

6 Prakash, M., Ketterson, J.B., and Dutta, P. (1985). Study of in-plane structure in lead-fatty acid LB films using X-ray diffraction. *Thin Solid Films* 134: 1–4.

7 Kitaigorodskii, A.I. (1961). *Organic Chemical Crystallography*. New York: Consultants Bureau.

8 Peterson, I.R. and Russell, G.J. (1984). An electron diffraction study of ω-tricosenoic acid Langmuir-Blodgett films. *Philos. Mag. A* 49: 463–473.

9 Kumar, N.P., Major, S., Vitta, S. et al. (2002). Molecular packing in cadmium and zinc arachidate LB multilayers. *Colloids Surf. A* 198-200: 75–81.

10 Peterson, I.R., Russell, G.J., Earls, J.D., and Girling, I.R. (1988). Surface pressure dependence of molecular tilt in Langmuir-Blodgett films of 22-tricosenoic acid. *Thin Solid Films* 161: 325–331.

11 Robinson, I., Sambles, J.R., and Peterson, I.R. (1989). A reflection high energy electron diffraction analysis of the orientation of the monoclinic subcell of 22-tricosenoic acid Langmuir-Blodgett bilayers as a function of the deposition pressure. *Thin Solid Films* 172: 149–158.

12 Barnes, W.L. and Sambles, J.R. (1987). Surface pressure effects on Langmuir-Blodgett multilayers of 22-tricosenoic acid. *Surf. Sci.* 187: 144–152.

13 Davies, G.H. and Yarwood, J. (1989). Infrared intensity enhancement for Langmuir-Blodgett monolayers using thick metal overlayers. *Langmuir* 5: 229–232.

14 Dhindsa, A.S., Bryce, M.R., Ancelin, H. et al. (1990). Infrared spectroscopic studies on the structure and ordering of hexadecanolytetrathiafulvalene conducting Langmuir-Blodgett multilayers. *Langmuir* 6: 1680–1682.

15 Evanson, S.A., Badyal, J.P.S., Pearson, C., and Petty, M.C. (1996). Variation in intermolecular spacing with dipping pressure for arachidic acid LB films. *J. Phys. Chem.* 100: 11672–11674.

Further Reading

Bellamy, L.J. (1986). *The Infrared Spectra of Complex Molecules*, 3e. London: Chapman & Hall.

Cheetham, A.K. (1988). Diffraction methods. In: *Solid State Chemistry Techniques* (ed. A.K. Cheetham and P. Day), 39–51. Oxford: Oxford Scientific Publications.

Feigin, L.A., Lvov, Y.M., and Troitsky, V.I. (1989). X-ray and electron diffraction study of Langmuir-Blodgett films. *Sov. Sci. Rev. Phys.* 11: 285–377.

Goins, B.A. and Phillips, W.T. (eds.) (2011). *Nanoimaging*. Singapore: Pan Stanford.

Hollas, J.M. (1992). *Modern Spectroscopy*, 2e. Chichester: Wiley.

Novotny, L. and Hecht, B. (2012). *Principles of Nano-Optics*, 2e. Cambridge: Cambridge University Press.

Roberts, G.G. (ed.) (1990). *Langmuir-Blodgett Films*. New York: Plenum Press.

Russell, G.J. (1982). Practical reflection electron diffraction. *Prog. Cryst. Growth Charact.* 5: 291–321.

Tredgold, R.H. (1994). *Order in Thin Organic Films*. Cambridge: Cambridge University Press.

Ulman, A. (1991). *Ultrathin Organic Films*. San Diego, CA: Academic Press.

An, U. (ed.) (1995). *Characterization of Organic Thin Films*. Boston: Butterworth-Heinemann.

Woodruff, D.R. and Delchar, T.A. (1994). *Modern Techniques of Surface Science*, 2e. Cambridge: Cambridge University Press.

7

Thin Film Processing and Device Fabrication

When I consider everything that grows

7.1   Introduction

For most current (e.g. organic light-emitting displays) and future (e.g. plastic transistors) applications of organic and molecular electronics, the materials are required in the form of thin films (in this context, 'thin' is generally taken to mean $1–10\,\mu m$). This presents a considerable challenge to materials scientists, as organic compounds, in their bulk form, can be fragile and difficult to handle. The physical properties of a thin film may be quite different from those of the bulk, particularly if the film thickness is very small. This behaviour can be related to the morphology of the layer, which is determined by the processes that occur during the film formation.

Organic and Molecular Electronics: From Principles to Practice, Second Edition. Michael C. Petty.
© 2019 John Wiley & Sons Ltd. Published 2019 by John Wiley & Sons Ltd.
Companion website: www.wiley.com/go/petty/molecular-electronics2

In this chapter, an overview of the more popular methods that may be used to fabricate thin layers of organic compounds is presented. A distinction will be drawn between established deposition technologies (in many cases developed for use with inorganic materials) and those techniques that actually allow molecular-scale architectures to be assembled. Each deposition method differs in complexity and may be more suited to provide films in a particular thickness range. Specific types of organic compound are necessary for certain processes, e.g. self-assembly exploits the attraction between certain chemical groups. Some methods are 'wet' (spinning), while others are inherently 'dry' (plasma deposition). There are also implications for the degree of order and contamination levels in the deposited film.

Film thickness uniformity may also be an important parameter. For instance, thin film optical interference filters can demand a thickness uniformity of ±1%. A related issue is conformal coverage, which refers to the ability to coat both the vertical and horizontal surfaces of substrates. The problem arises primarily in the fabrication of integrated circuits as the semi-conductor contacts and device interconnection metallization are needed to cover complex topography, with micro-steps, grooves, and raised stripes. Film coverage will not be uniform when geometrical shadowing effects cause unequal deposition on the top and sidewalls of steps.

In all forms of thin film growth, it is important to know whether the deposition is *epitaxial*, i.e. is the orientation of the atoms/molecules in the thin film the same as that of the atoms/molecules in the underlying substrate? Molecular beam epitaxy (MBE) achieves this and allows the growth of extremely pure crystalline films. Finally, important considerations for commercialization are factors such as the cost of the thin film process, the time taken, and/or whether the method can easily be adapted to continuous coating.

7.2 Established Deposition Methods

7.2.1 Spin-coating

Spin-coating is exploited extensively by the microelectronics industry for depositing layers of photoresist films, generally polymers such as polyimides, onto silicon wafers. The various steps involved are illustrated in Figure 7.1. A quantity of a polymer solution is first placed on the semiconductor wafer, which is then rotated at a fixed speed of several thousand revolutions per minute (or the solution can be applied while the wafer is slowly rotating). The resist solution flows radially outwards, reducing the fluid layer thickness. Evaporation of the solvent results in a film of uniform thickness.

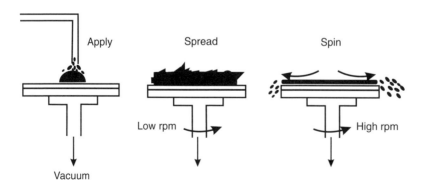

Figure 7.1 Schematic diagram of spin-coating.

A quantity of solution is first applied to the substrate surface; this may be pre-treated with an adhesion promoter to improve wetting. The initial volume of fluid dispensed onto the rotating disk and the rate of fluid delivery have a negligible effect on the final film thickness. In contrast, the resist viscosity (dependent on the concentration of the starting solution) and final film speed are both important process parameters. An increase in angular velocity decreases the film thickness; an inverse power-law relationship usually holds for the thickness dependence on the final spin speed. For a given speed, the film thickness decreases rapidly at first, but then slows over longer times. A simple theory predicts the following relationship between the thickness of the spun film, d, the viscosity coefficient of the solution, η, its density, ρ, the angular velocity of the spinning, ω, and the spinning time, t [1]:

$$d = \left[\frac{\eta}{4\pi\rho\omega^2} \right]^{1/2} t^{-1/2} \tag{7.1}$$

More sophisticated models have been developed to allow for changes in the resist resulting from solvent evaporation and the non-Newtonian character (a non-Newtonian fluid is one that does not follow Newton's Law of Viscosity) of the rheological behaviour of the photoresist [2–4]. These work reasonably well for practical concentrations of spinning solutions. However, very dilute solutions often lead to pinholes in the final film, and failure of the mathematical predictions.

An important practical consideration is the formation of an *edge bead*. Due to the increased friction with air at the periphery of the substrate, the fluid in this region dries first, causing the remaining resist to flow over the step and dry, leading to the formation of the bead. Studies have shown that the edge bead width is inversely dependent on the spin speed [5]. The bead can be removed by bevelling the edges of the substrate. A further issue occurs when the processing constrains the substrate to rectangular dimensions (e.g. flat panel displays) [5, 6]. A waveform pattern can develop in the corners of the substrate, where radial uniformity is lost. An air barrier plate, positioned above the spinning substrate, can be used to minimize this effect. This creates a partially saturated atmosphere above the substrate, retarding evaporation and slowing down the formation of the solid film in the substrate corners. The *Bernoulli effect* is also significant for the coating of rectangular substrates. This is a result of the leading edge of the substrate in addition to the contact angle of the edge bead creating aerofoil. This effect can be reduced using a recessed or carrier chuck on which the substrate is mounted [6].

Organic compounds that have been successfully deposited by spin-coating include electrically insulating polymers, such as polyvinylidene fluoride, polymers, and dyes developed for electroluminescent displays, and certain phthalocyanine materials. Although spin-coating is expected to produce films in which individual molecules are relatively disordered, this is not always the case. For instance, organized phthalocyanine layers have been deposited [7]. A study using X-ray diffraction revealed a crystalline order similar to that observed for Langmuir–Blodgett films (Section 7.3.1) of analogous compounds. The way in which this order is achieved is not fully understood, but may result from the centrifugal forces acting upon the individual molecules during spinning. In other cases, order can be achieved by a post-deposition treatment. Application of heat and an electric field normal to film plane (the process of poling, Chapter 2, Section 2.6.3) can be used to align the C–F dipoles in poly(vinylidene difluoride) films. The result is a polar film possessing both piezoelectric and pyroelectric properties (Chapter 5, Section 5.7). A similar method can be used to induce second-order nonlinear optical behaviour (e.g. second-harmonic generation, Chapter 4, Section 4.3.2) in films of appropriate molecules.

7.2.2 Physical Vapour Deposition

Thermal Evaporation

Solid materials vaporize when heated to sufficiently high temperatures; this process may proceed through the liquid phase. A thin film is then obtained by the condensation of the vapour onto a colder substrate. This method has traditionally been used for the deposition of films of inorganic materials, such as metals and their alloys. However, the technique is now being exploited for the formation of layers of low molecular weight organic compounds. The rate of evaporation Γ (in $\mathrm{kg\,m^{-2}\,s^{-1}}$) from a surface is given by the Langmuir expression [8]:

$$\Gamma = P\left(\frac{M}{2\pi RT}\right)^{1/2} \tag{7.2}$$

where P is the vapour pressure (in $\mathrm{N\,m^{-2}}$) of the material at temperature T (in K), M is the molecular weight, and R is the gas constant.

Because of collisions with ambient gas atoms, a fraction of the vapour atoms will be scattered. For a straight-line path between the evaporating material (source) and the substrate, it is necessary to use low pressures ($< 10^{-4}$ mbar), where the mean free path of the gas atoms λ (Chapter 3, Section 3.2.1) is much greater than the source–substrate distance. From kinetic theory, λ is given by

$$\lambda = \frac{k_B T}{P\pi d^2 \sqrt{2}} \tag{7.3}$$

where d is the diameter of the molecules. The mean free path is therefore inversely proportional to gas pressure. For the common gases, He, H_2, O_2, N_2, Ar, CH_4, CO_2, and H_2O, the effective cross-sectional diameters range from 0.2 nm (for He) to 0.5 nm (for H_2O). This gives average mean free path values of about 10 cm at a pressure of 10^{-3} mbar and about 10^4 cm at 10^{-6} mbar [8]. Therefore, at low pressures a shadow mask, as depicted in Figure 7.2, can be used

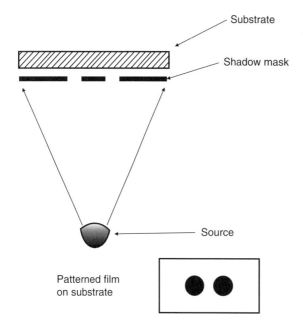

Substrate

Shadow mask

Source

Patterned film on substrate

Figure 7.2 Use of a shadow mask to pattern an evaporated thin film.

Figure 7.3 Vacuum evaporation system. Evaporating atoms or molecules traverse the space between the source and the substrate at reduced pressure.

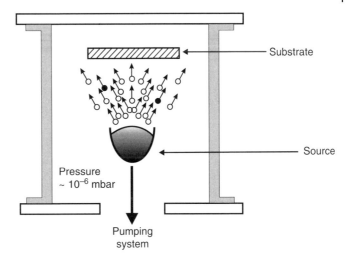

Substrate

Source

Pressure
~ 10^{-6} mbar

Pumping
system

immediately in front of the substrate to define patterns. The low pressure also prevents contamination of the source material (e.g. by oxidation).

Figure 7.3 shows a schematic diagram of a typical evaporation arrangement. The system chamber, which can be made out of glass or metal, is evacuated to a pressure of 10^{-4}–10^{-6} mbar, normally with two types of vacuum pump operating in series (e.g. a rotary and diffusion pump). It is common to introduce a shutter (either mechanically or electronically operated) between the source and the substrate. This can be used to prevent the material that evaporates initially from the source surface (which may be contaminated) from depositing onto the substrate. A film thickness sensor is usually included in the evaporation chamber. This is based on an oscillating quartz crystal. As the evaporating species condenses on this crystal, its vibration frequency is altered. The change of frequency can then be used as an indication of the film thickness. (This technique is also exploited in certain chemical sensors, Chapter 10, Section 10.4.5.)

Although commonly thought of as a single process, the deposition of thin films by thermal evaporation consists of several distinguishable steps:

1) transformation of the source material, in solid or liquid form, into the gaseous state;
2) vapour molecules traversing the space between the source and the substrate; and
3) condensation of the vapour upon arrival on the substrate.

The first step, governed by Eq. (7.2), requires the conversion of thermal to mechanical energy, and is achieved by a variety of physical methods. Resistive heating has been used to deposit low molecular weight dyes, charge-transfer salts, and large macromolecules, such as the phthalocyanines. Typical evaporation rates are 1–10 nm min^{-1}. Other techniques include: arc evaporation; radio frequency (RF) heating; or heating by electron bombardment. Deposition of polymer films by laser ablation is an area that offers some promise [9]. Laser pulsed methods have been used successfully for polyethylene, polycarbonate, polyimide, and poly(methyl methacrylate). Typical film growth rates are 0.02–0.1 nm per laser pulse.

Materials that dissociate in the vapour phase may provide solid films with a stoichiometry (i.e. composition) that differs from that of the source. Therefore, special techniques have been devised. One approach is to use the method of *flash evaporation*. Rapid evaporation is achieved by continuously dropping fine particles of the materials onto a hot surface. Although fractionation occurs during the evaporation of each particle (the more volatile component evaporating

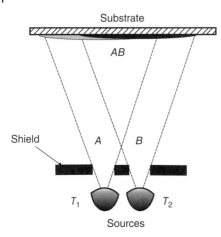

Substrate

AB

Shield

A   B

T_1 T_2

Sources

Figure 7.4 Two-source evaporation arrangement. Two independent sources, held at temperatures T_1 and T_2, are used to evaporate independently materials *A* and *B*. The two vapour fluxes combine at the substrate.

first), at any time there will be several particles at different stages of fractionation. Consequently, the vapour phase will possess a similar composition to that of the source material. A further technique is to evaporate from two or more sources and control the flux from each to obtain a vapour with the required composition (Figure 7.4). This has been used effectively to deposit thin films of doped organic charge-transfer salts [10]: one source is the charge-transfer salt, e.g. tetrathiafulvalene (TTF), while the other is the dopant, e.g. iodine. Laser co-ablation techniques can be used to deposit films of metal-polymer composites [9].

Whatever method is used to heat the material, the vapour will be in an atomic or molecular form when it arrives at the substrate. Figure 7.5 illustrates the various stages that occur during the formation of a solid film on the surface of the substrate [11]. The surface of the substrate is covered with a large number of adsorption sites and a molecule becomes bound to one of these with a characteristic energy. The adsorbed atoms do not remain stationary (Figure 7.5a) but may re-evaporate, requiring energy equal to the adsorption energy, or migrate or hop to an adjacent adsorption site (Figure 7.5b). Other processes are collision and recombination (Figure 7.5c). Each group of atoms eventually reaches a size that is more likely to grow than to decay. The formation of such stable islands of material is known as nucleation (Figure 7.5d). Individual islands continue to grow by the addition of more single hopping atoms (Figures 7.5e and f).

Eventually, the islands coalesce with neighbours to form an interlinking network (Figure 7.5g). The uncoated areas gradually diminish until a single continuous film is formed (Figure 7.5h). This may not occur until a great deal of material has been deposited. For example, an evaporated film of gold on glass at room temperature becomes continuous at an average thickness of about 30 nm, or 100 atoms thick!

The microstructure of an evaporated thin film depends on the evaporation rate, substrate temperature, and on the chemical and physical nature of the substrate surface. The size of the grains in a polycrystalline film will generally be larger for high source and substrate temperatures. However, if the kinetic energy of the incoming molecules is too high, the surface mobility of the adsorbed species is reduced because the vapour molecules will penetrate the condensed film. The effect of substrate temperature on grain size is greater for relatively thick films. For a given material-substrate combination and under a fixed set of deposition conditions, the grain size increases initially as the film thickness increases. Beyond a certain thickness, the grain size remains constant, suggesting that coherent growth with the underlying grains does not go on forever. Figure 7.6 shows an atomic force microscope (AFM) image (see Chapter 6, Section 6.8)

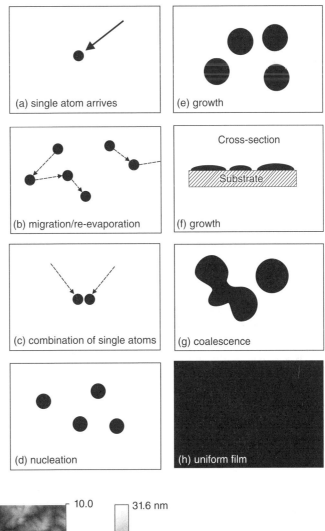

Figure 7.5 Stages of growth of an evaporated thin film. *Source:* From Leaver and Chapman [11]. Copyright (1971), Taylor & Francis.

(a) single atom arrives

(b) migration/re-evaporation

(c) combination of single atoms

(d) nucleation

(e) growth

Cross-section

Substrate

(f) growth

(g) coalescence

(h) uniform film

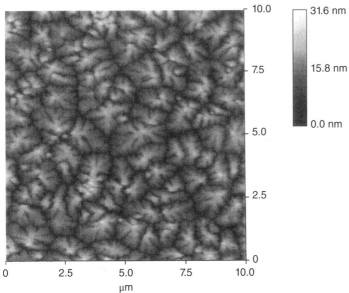

Figure 7.6 Atomic force microscope (AFM) image of an evaporated film of pentacene, thickness ~50 nm [12]. *Source:* Reprinted with permission from Dan Kolb.

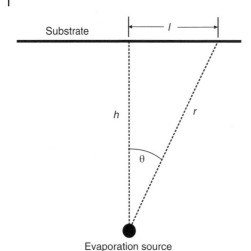

Substrate

$\leftarrow l \rightarrow$

h

r

θ

Evaporation source

Figure 7.7 Evaporation geometry from a point thermal evaporation source onto a plane substrate.

of an evaporated film of the organic semiconductor pentacene on a glass substrate [12]. The film thickness is approximately 50 nm. A granular structure, with grain sizes of the order of 100 nanometres, is evident.

The physical nature of evaporated films can be changed by post-deposition heat treatment (annealing). This is usually associated with a change in the crystallinity of the film. For example, following thermal evaporation, the dc in-plane conductivity of thin films of the electron donor bis(ethylenedithio) tetrathiafulvalene (BEDT-TTF) can be increased by many orders of magnitude by doping with iodine and then annealing at 60 °C [13].

One important aspect, often neglected, is the thickness distribution in evaporated films. This will depend on the geometry of the evaporation system. For a point source and plane substrate, parallel to the plane of the emitting surface, depicted in Figure 7.7, the rate of deposition varies as $(\cos \theta)/r^2$, where r is the radial distance of the substrate from the source and θ is the angle between the radial vector and the normal to the substrate. The following expression can be derived for the film thickness d as a function of l, the distance along the substrate measured from its central position [8]:

$$\frac{d}{d_0} = \left[1 + \left(\frac{l}{h}\right)^2\right]^{-3/2} \tag{7.4}$$

where d_0 is the film thickness at the central position on the substrate (where the vapour condenses normally) and h is the normal distance of the point source to the substrate. Equation (7.4) reveals that the film thickness at a distance along the substrate surface equal to h will be about 35% of its maximum value, measured in the centre of the substrate. Using a large area source may reduce this effect.

Molecular Beam Epitaxy

Molecular beam epitaxy is, in principle, similar to the method of the vacuum evaporation described in the previous section [14]. However, an ultrahigh vacuum ($<10^{-9}$ mbar) is required to eliminate the scattering by residual gas molecules. The technique consists of directing controlled 'beams' of the required molecules towards a heated substrate. The experimental arrangement is shown in Figure 7.8. Multiple sources, called *Knudsen cells* (for source materials *A*, *B*, and *C* in Figure 7.8) can be shuttered (usually computer-controlled) and used to create a

Figure 7.8 The use of Knudsen cells for molecular beam epitaxy (MBE). The three cells shown are used to provide molecular beams of materials *A*, *B*, and *C*.

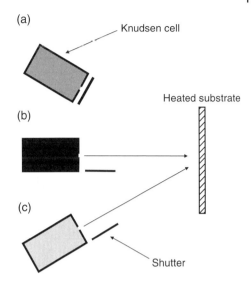

superlattice structure on the substrate, with precise control of the molecular composition, orientation, and packing in two dimensions. Each Knudsen cell encloses an evaporating surface that is large compared to the orifice. The diameter of the orifice must be about one-tenth or less of the mean free path of the gas molecules, and the wall around the orifice must be vanishingly thin so that gas molecules leaving the enclosure are not scattered, adsorbed, and desorbed by the orifice wall.

Some of the first trial observations of real-time epitaxial growth in organic molecular systems have involved the growth of phthalocyanine monolayers on cleaved surfaces of MoS_2 and on highly oriented pyrolyzed graphite and alkali halides. In-situ reflection high-energy diffraction (RHEED – Chapter 6, Section 6.5) can be used to monitor the actual film growth. Multiple quantum well structures based on organic molecular materials, such as perylene and naphthalene derivatives, and copper phthalocyanine can be fabricated using MBE. The method has also been used successfully to deposit electrically conductive organic semiconductors for the fabrication of organic field effect transistors.

Molecular beam epitaxy is a very slow process. The typical rate of growth is a single monolayer per second, or approximately one micrometre per hour. However, this permits abrupt changes in the material composition during film deposition.

Sputtering

Sputtering is based on the momentum exchange of accelerated ions incident on a target of source material [8]. At very low kinetic energies (<5 eV) the interaction is confined to the outermost surface layer of the target material. Ions, such as Ar and Xe, rather than neutral atoms, are used for bombardment as these can be accelerated to any desired kinetic energy with applied electric fields.

A source of ions is provided by a glow discharge created by an electric field between two electrodes in a gas at low pressure. Figure 7.9 shows a schematic diagram of a simple sputtering system. The material to be sputtered, the target, is the cathode. The vacuum chamber is evacuated and then filled with the inert gas at a low pressure. A DC potential difference of several kilovolts is applied to the electrodes, which causes the gas to become ionized and form a plasma. The positive gas ions are then accelerated by the electric field so that they arrive at the cathode with considerable energy and sputter the target atoms. Some secondary

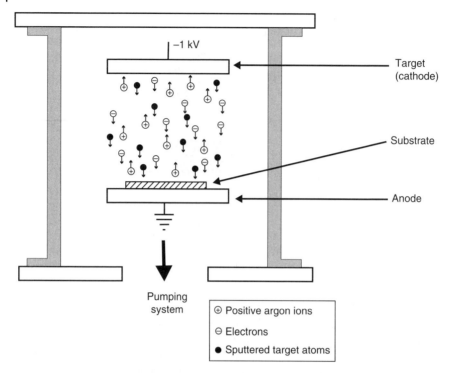

Figure 7.9 Sputtering system. A target, held at a negative potential, is bombarded with positively charged ions.

electrons are also produced at the cathode and these accelerate towards the anode and help to maintain the plasma.

Effective sputtering is possible only when both the number of ions and their energy are large and controllable. As the gas pressure is increased, the discharge current increases and the number of ions increases (approximately proportional to the pressure squared). There is however an upper limit to the pressure, since ejected atoms suffer more collisions and are prevented from reaching the anode. Optimum pressures for glow discharge sputtering are in the range 25–75 mbar.

There are several variations in the basic sputtering arrangement described above. If the substrate is held at a negative potential, it will be subjected to steady ion bombardment, which effectively 'cleans' the film of adsorbed gases otherwise trapped in as impurities. This is called bias sputtering. A magnetic field may also be applied to improve the ionization efficiency (by increasing the path length of the ionizing electrons). For insulating source materials, DC sputtering is inappropriate because of the accumulation of positive surface charges on the target; RF sputtering is therefore used.

The principal advantage of sputtering is that almost any material can be deposited. Since no heating is required, materials that are difficult to melt (and therefore to evaporate) are easily sputtered, as are compounds that would dissociate in an evaporation source. The method has been used for some organic polymers, e.g. polytetrafluoroethylene (PTFE). However, a relatively large amount of the material is needed as a target. This is not always practical for state-of-the-art organic compounds, as these may only be available in very small quantities.

Sputtering takes place from the whole of the target surface, and this means a uniformly thick deposit over larger areas can be obtained, in contrast to thermal evaporation. Furthermore, the sputtering rate is proportional to the current flowing between the electrodes, and hence the process is more controllable than evaporation.

7.2.3 Chemical Vapour Deposition

Chemical vapour deposition (CVD) is a process in which one or more gaseous species react on a solid surface and one or more of the reaction products is a solid phase material [8, 15]. The CVD process is based on the decomposition and/or radical generation of chemical species by stimulating vapour with heat, plasma (discharge) or light (laser). The method is used in the microelectronics industry for the fabrication of inorganic semiconducting and insulating films. Chemical reactions that are commonly exploited include: pyrolysis (thermal decomposition); oxidation; reduction; disproportionation; and various transfer reactions. During the pyrolysis of silane (SiH_4), the molecules strike a hot surface and decompose into Si and H_2, with the latter going back into the gas phase. The silicon left behind can build up as a solid film. Pyrolysis can be used to produce layers of insulating polymer films. For example, solid poly-*p*-xylene films have been formed by pyrolysis of *p*-xylene at 600°°C [14]. Polymers can also be formed by the high temperature electron bombardment of monomers. This approach has been used to form polymer films of siloxane, styrene, butadiene, and methacrylate [1, 16].

For plasma-enhanced CVD (PECVD), glow discharge plasmas are sustained within chambers where simultaneous CVD reactions occur [17, 18]. An RF field normally excites the discharge. One example is the formation of polystyrene films for use as the dielectric in a nuclear battery [1, 17]. The surface polymer may be simultaneously removed by reaction with the residual gases (ablation); adding gases that promote one or other effect may control the balance between ablation and polymerization. For instance, CF_4 is an effective plasma etch material under normal discharge conditions; however, if a small amount of hydrogen is added, a polymer is deposited.

Laser, or more generally, optical chemical processing involves the use of monochromatic photons to enhance and control reactions at substrates [19]. Two reactions may be involved during laser-assisted deposition. In the pyrolytic mechanism, the laser heats the substrate to decompose gases above it and to enhance rates of chemical reactions. Pyrolytic deposition requires substrates that melt above the temperatures necessary for gas decomposition. Photolytic processes, on the other hand, involve direct dissociation of molecules by energetic photons.

Metalorganic CVD (MOCVD) is a further common technique for device growth. For the formation of the inorganic semiconducting material gallium aluminium arsenide, different gases containing the molecular species of interest (in this case, Ga, Al, and As) pass through a reactor that incorporates a heated substrate. The composition of the deposited material is controlled by the respective composition of the gases and the temperature of the substrate. The gases pass through the chamber by forced convection; these then participate in a chemical reaction, which encourages the desired growth material to precipitate onto the substrate. The following equation shows the basic reaction between the organometallic compound trimethylgallium and arsine to form gallium arsenide and methane:

$$\left(CH_3\right)_3 Ga + AsH_3 \rightarrow GaAs + 3CH_4 \tag{7.5}$$

This reaction takes place at about 700°C and epitaxial growth of high-quality GaAs layers can be obtained. The addition of a trimethylaluminium source allows the alloy AlGaAs to be grown. The flexibility of gas mixing arrangements permits the growth of multiple thin layers similar to those discussed for MBE. The gas flow rates, reactor pressure, and reactor temperature are important parameters in determining the composition and quality of material deposited using MOCVD.

Atomic layer deposition (ALD) is a subclass of CVD [20]. The majority of ALD reactions use two precursor chemicals, which react with the surface of a material one at a time in a sequential, self-limiting, manner. Through the repeated exposure to separate precursors, a thin film is slowly deposited. In contrast to CVD, the precursors are never present simultaneously in the reactor. The advantage of ALD is the precise thickness control at the monolayer level. The deposition of Al_2O_3 has been developed as a model ALD system, usually performed using trimethylaluminium and water. Single-element materials, such as metal and semiconductors, can be deposited using plasma or radical-enhanced ALD. Low temperature ALD enables deposition on thermally sensitive materials, such as organic polymers. This approach may be useful to functionalize a polymer surface, to create unique inorganic/organic polymer composites, or to deposit gas diffusion barriers on polymers. While ALD processes have been developed for a wide range of inorganic materials, similar self-limiting surface reactions can be used for the growth of organic polymers. This film growth process is more correctly described as *molecular layer deposition* (MLD), because a molecular fragment is deposited during each ALD cycle. MLD has been developed for materials such as polyamides and polyimides.

7.2.4 Electrochemical Methods

Electrochemical deposition, or electroplating, has been known for at least 100 years. Approximately half of the 70 or so metals can be electrodeposited, either singly or as alloys. The equipment required consists of an anode and a cathode immersed in a suitable electrolyte. Metal is deposited onto the cathode and the relationship between the weight of the material deposited and the various parameters can be expressed by Faraday's first and second laws of electrolysis. These state:

1) the weight of the deposit is proportional to the amount of electric charge that is passed; and
2) the weight of material deposited by the same charge is proportional to the *equivalent weight E* (defined by the atomic weight of an element or radical divided by the valence it assumes in compounds), provided that there are no other electrolysis reactions occurring that consume a fraction of the current.

Expressed as an equation, the weight deposited m is given by

$$m = q\alpha E = q\alpha \frac{M}{z} = \frac{1}{F}\frac{qM}{z} \tag{7.6}$$

where q is the total charge passed through the electrolyte, M is the molar mass, z is the valence number of the substance as an ion in solution (electrons per ion), and α is known as the current efficiency, i.e. the fraction of the total current associated with deposition of the material concerned. In Eq. (7.6), F is *Faraday's constant*, a common unit of charge in electrochemistry. This is equal to the charge of 6.02×10^{26} electrons (or one kilomole – see Chapter 2, Section 2.2.1), i.e.

$$F = N_A e \tag{7.7}$$

Therefore $F = 9.65 \times 10^7$ C kilomole^{-1}. A charge of one Faraday passing through an electrochemical cell will liberate 1 kg equivalent of a substance at an electrode. A deposition speed of about $1\,\text{nm}\,\text{s}^{-1}$ for an electrode current density of $100\,\text{A}\,\text{m}^{-2}$ is typical for an electrochemical process.

Anodization is a particular kind of electrolysis. In a copper-electroplating bath, the copper anode goes into the electrolytic solution as a positive ion. With some metals this does not happen, but instead the surface of the anode slowly oxidizes by reaction with water in the electrolyte; the increasing resistance of the growing anodic oxide layer causes a continuous decrease in the electrolytic current. This anodization process is limited to a few metals, such as aluminium and tantalum, which can be oxidized to form excellent thin film capacitors. Many polymeric films can be prepared by the anodic oxidation of suitable monomer species, such as pyrrole, thiophene, and aniline [21]. The full details of electropolymerization are not fully understood. In the very first step, the neutral monomer is oxidized to a radical cation. It must therefore have an oxidation potential that is accessible via a suitable solvent-electrolyte system and should react more quickly with other monomers than with other nucleophiles in the electrolyte solution.

Electroless deposition does not require electrodes or an external energy source. Ions in the plating solution are simply converted to neutral atoms by a reducing agent in the presence of a catalyst. The process is confined to particular chemical reactions, i.e. each depositing layer of the film must be catalysed by the preceding layer. In some cases, such as silver, electroless deposition begins and proceeds readily, but most surfaces have to be specifically treated ('sensitized') before a film will deposit. A general method for the electroless metallization of polyamide has been described [22]. Here, the polymer itself is used as either an oxidizing or reducing agent to deposit a metal species from solution. The equipment required for electroless deposition is very simple and both conductive and non-conductive substrates can be used. Furthermore, as there are no requirements for uniform current density, irregular objects and inaccessible surfaces can be coated.

7.2.5 Inkjet Printing

The need to combine large area coatings with device patterning has resulted in the development of direct-write fabrication methods, such as inkjet printing [23–25]. A number of different inkjet technologies have been developed. Continuous inkjet deposition is based on inducing an electric charge in the liquid by ejecting a jet of ink from an orifice through a region with an external electric field. The drops can then be deflected to a substrate or to a reservoir for recycling. In the case of drop-on-demand printers, droplet ejection can be achieved with thermal (bubble-jet) and piezoelectric modes of operation, both shown schematically in Figure 7.10. The majority of published literature on inkjet printing as a tool for manufacturing organic devices has been the result of using piezoelectric actuated printers. This technology is favoured primarily because it applies no thermal load to the organic 'inks'. The combination of solution-processable emissive polymers with inkjet printing offers some promise in the development of low-cost, high-resolution displays (Chapter 9, Section 9.6). The technique has also been applied to the manufacture of all-polymer transistor circuits (Chapter 9, Section 9.4).

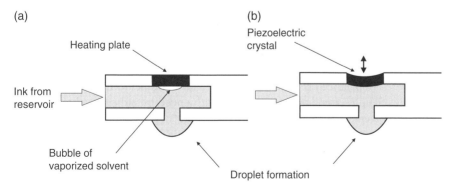

Figure 7.10 Inkjet print heads: (a) thermal (bubble-jet) operation; (b) piezoelectric operation.

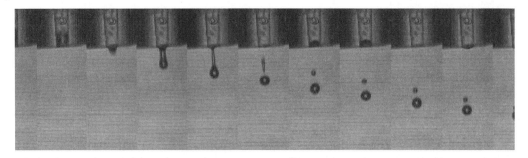

Figure 7.11 Ejection of drops from print head [26]. Nozzle size = 50 μm. *Source:* Reprinted with permission from David Morris.

The sizes of nozzles used by inkjet printers are in the range 20–70 μm, giving droplet sizes 4–180 pl. However, some printers can produce 1 pL droplets. Figure 7.11 shows a sequence of photographs taken during the formation and ejection of a droplet from a piezoelectric inkjet head [26]. The formation of the satellite drop (which eventually re-combines with the main droplet) is a typical phenomenon. The physical properties of the ink must be matched to the performance of a specific printer; the ink viscosity and surface tension are both important parameters. The viscosity must be low enough to allow for rapid refilling (a typical ink has a viscosity of about 2 cP) and the surface tension (typically 40 mN m^{-1}) must be sufficiently high to hold the ink in the nozzle without dripping. The clogging of the nozzle with partly dried ink is a major problem for this method of thin film deposition. Low-volatility, water-miscible liquids, such as ethylene glycol, are added at 10–20% to prevent this.

When the droplet strikes the substrate, it may be adsorbed. But this will depend on the nature of the surface. For example, overhead transparency sheets used for conventional printing are non-absorbing, and have to be coated with a hydrophilic layer or may be patterned to achieve a rough surface. 'Phase-change' inks can be used on hard surfaces [24]. These materials freeze rapidly on deposition and are cold-rolled to produce a strongly bonded layer. The rate of expulsion of the droplets from the nozzle and the relative movement of the inkjet head and substrate will determine if sequential drops coalesce on the surface.

A common feature of inkjet-printed droplets, particularly those printed on non-absorbing substrates, is the so-called *coffee-ring* effect. This occurs as a result of the higher evaporation rate at the solvent's pinned contact line compared to its centre [27]. The thinning in the middle of the droplet is a consequence of solvent being transported to the boundary by a refilling flow from the centre, as the contact line cannot retract. However, the situation is likely to be more complex. For example, *Marangoni flow*, which results from the presence of a surface tension gradient, can redistribute material back to the centre of the droplet [28]. Figure 7.12 shows a proposed schematic diagram for the process of droplet drying for inkjet-printed films of triisopropylsilylethynyl pentacene [29]. To obtain a well-ordered crystalline structure from a drying droplet, it is essential that there is sufficient evaporation of solvent at the contact line for nucleation to occur, and there is a balanced recirculation follow in the droplet to transport and self-assemble the molecules. The recirculation flow can be influenced by judicious choice of the solvent (or mixture of solvents). Such approaches to thin film deposition can lead to single crystal films, which can be used to yield high performance thin-film transistors (Chapter 9, Section 9.4) [30]. The morphology of very thin films of inkjet-printed conductive polymers can also be exploited in chemical sensing devices (Chapter 10, Section 10.4.3) [31].

Figure 7.13 shows the effect of depositing droplets of a PEDOT:PSS (Chapter 5, Figure 5.3) onto or near to other droplets [26]. When successive droplets are deposited so that part of the

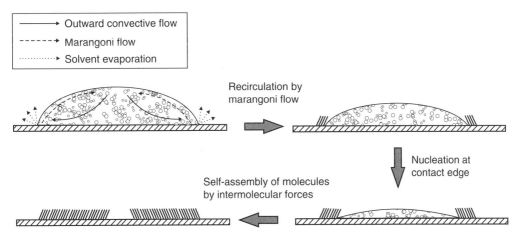

Figure 7.12 Schematic diagram of the possible mechanism of the kinetically driven development of self-aligned TIPS/PEN crystals. *Source:* Reprinted from Lim et al. [29]. Copyright (2008), with permission from Wiley-VCH.

wet droplet overlaps with a previous, dry droplet, the wet droplet interacts with it in such a way as to cover a greater area than if there had been no overlap. The printing of thin lines is limited by the size of the inkjet drop, which depends ultimately on jetting conditions such as the ink, the surface energy of the substrate, and the nozzle diameter. The resolution of the printing process can be improved by using various surface pattering techniques [23]. This is illustrated in Figure 7.14. The substrate surface is pretreated to form hydrophilic and hydrophobic regions. This can be accomplished with a variety of methods, including laser printing, deposition of self-assembled monolayers using a rubber stamp (see Section 7.4.3), or ultraviolet light exposure. Water-based ink droplets are then confined to the hydrophilic pattern by the surrounding hydrophobic regions.

7.2.6 Spray-coating

Spray-coating is a generic term that covers a multitude of thin film deposition processes (some industrial spray coating process can provide a coating up to several millimetres in thickness). In comparison with some of the methods described above, such as spin-coating or ink-jet printing, spray-coating provides advantages such as low equipment cost and good thickness uniformity over a large area. The technique enables direct patterning of the material, so wastage is minimized and the processing speed is increased. A typical system involves vaporization of a solution and transport of the resulting mist (aerosol) onto a substrate where it forms a thin solid film. In some forms of spray-coating, the starting solution is a precursor solution; for example, in *chemical spray pyrolysis*, the aerosol reacts with a heated substrate to form a thin film of the desired material [32, 33].

Typical spray-coating equipment, as depicted in Figure 7.15, consists of an atomizer to generate the aerosol, solution reservoir, and substrate/heater. Formation of thin films by spray pyrolysis can be divided into three parts: atomization of the solution; transportation of the resultant aerosol onto the substrate, and subsequent deposition/decomposition onto the substrate. The atomization procedure is the first step in the spray deposition. The aim is to generate droplets from a spray solution and direct them, with some initial velocity, towards the substrate surface. The process normally uses an air blast, ultrasonic, or electrostatic techniques. The atomizers differ in resulting droplet size, rate of atomization, and the initial velocity of the

(a)

z:1.1µm

Y:163.0µm

X:163.0µm

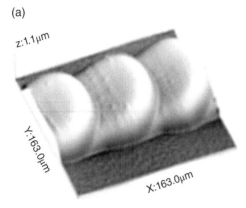

(b)

Z:846.9nm

Y:163.0µm

X:163.0µm

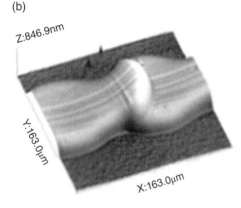

(c)

Z:650.3nm

Y:163.0µm

X:163.0µm

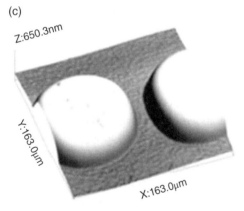

Figure 7.13 Inkjet-printed dots of PEDOT: PSS on a substrate [26]. (a) 60 µm line pitch. (b) 100 µm line pitch. (c) 120 µm line pitch. *Source:* Reprinted with permission from David Morris.

droplets. It has been shown that the size of the generated droplet is not related to any fluid property of the precursor solution and depends solely on the fluid charge density. After the droplet leaves the atomizer, it travels through the ambient with an initial velocity determined by the atomizer. In the aerosol form, the droplets are transported with the intention of as many droplets as possible reaching the surface. As the droplets move through the ambient, there are four forces simultaneously acting on them, which can result in both physical and chemical changes. The forces are gravitational, electrical (the droplets may be charged), thermophoretic (a retarding force causing the droplets to decrease significantly their velocity as they approach a heated substrate), and the Stokes' force (due to drag experienced by the droplet due to air resistance). The droplets also experience evaporation during their flight, which will influence their trajectory. On arrival at the substrate, the droplets will impact with the substrate and may decompose (in the case of a pyrolysis reaction). The reactant molecules can re-evaporate, and migrate over the substrate surface, in some respects in a similar manner to the formation processes in a thermally evaporated film (Figure 7.5). Chemical spray pyrolysis has traditionally been used to deposit layers of inorganic semiconductors (e.g., CdS, ZnO, SnO_2). However, spraying is increasingly being applied for the formation of thin layer of organic compounds, such as PEDOT:PSS (Chapter 5, Figure 5.4), particularly for applications in large area photovoltaic cells (Chapter 9, Section 9.7) [31].

7.2.7 Sol–Gel Processing

Sol–gel processing involves the suspension of a solid in a liquid (the *sol*), the removal of the liquid (the *gel*), and finally the densification of the solid, e.g. by sintering [34, 35]. The method is used to form highly dense films of ceramics and glasses. Thin layers can be prepared using much lower temperatures than would be needed in the 'conventional' processing of the same material.

The sol–gel technique covers a wide range of processing of colloidal and macromolecular systems. The starting point is the formation of a stable colloid (Chapter 2, Section 2.7). This can be achieved by breaking a material down from a macroscopic to a microscopic state, perhaps by mechanical means such as milling. Alternatively, chemical processes may be used to

Figure 7.14 Use of surface treatments to define patterns for ink-jet printing. (a) Pretreatment to form hydrophilic and hydrophobic regions. (b) Water-based ink droplet confined to the hydrophilic pattern by the surrounding hydrophobic regions.

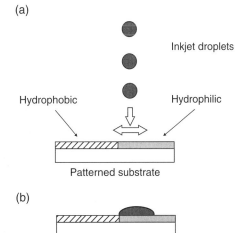

(a)

Inkjet droplets

Hydrophobic   Hydrophilic

Patterned substrate

(b)

Figure 7.15 Schematic diagram of a spray pyrolysis thin film coating system.

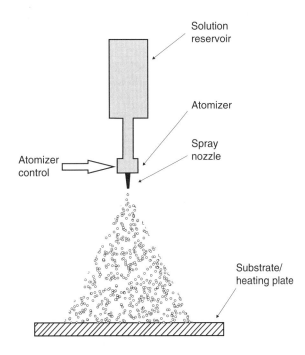

Solution reservoir

Atomizer

Spray nozzle

Atomizer control

Substrate/ heating plate

build up appropriately sized particles from smaller ones. The chemistry of the sol–gel process is largely based on an alkoxide solution route. Alkoxides are traditional organometallic precursors for silica, alumina, titania, and other metal oxides. The sol–gel reaction, driven by a catalyst, starts with the hydrolysis of alkoxides in a water-alcohol mixed solution, followed by poly-condensation reactions.

The crucial second stage is the conversion of the sol to a gel. In the latter state, the particles have cross-linked and formed a structure that is capable of immobilizing the remaining solvent. The sol–gel reaction is often performed in the dip-coating regime, in which the solid substrate is slowly pulled out of the solution containing the required chemicals. This reaction, known as *gelation*, takes place in a thin layer wetting the substrate.

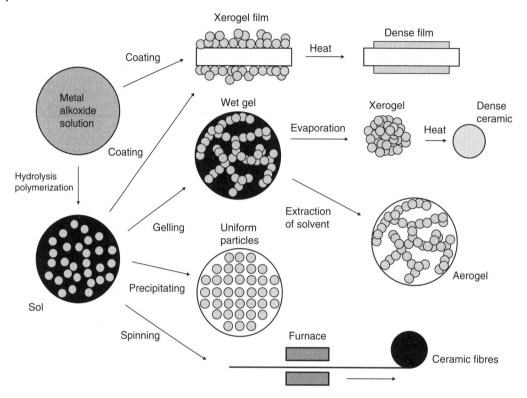

Figure 7.16 Sol–gel technologies and products. *Source:* Reproduced from Brinker and Scherer [35]. Copyright (1990), with permission from Academic Press.

Drying by evaporation under normal conditions gives rise to capillary pressure that causes shrinkage of the gel network. The resulting dried gel, called a *xerogel* ('xero' means dry), is often reduced in volume by a factor of five to ten compared to the original wet gel. Most gels are amorphous, even after drying, but many recrystallize when heated. Heating a xerogel to high temperatures may form a relatively dense product.

If the wet gel is placed in an autoclave and dried under *supercritical* conditions, there is no interface between the liquid and vapour, so there is no capillary pressure and relatively little shrinkage. The process is called *supercritical* (or *hypercritical*) *drying* and the product is called an *aerogel*. This may be extremely porous, made mostly of air. The classification of sol-gel technology and products is given in Figure 7.16. Applications of sol–gel technology to organic electronics include the development of inorganic gate insulators for thin film transistors [36], the fabrication of nanocomposites incorporating reduced graphene oxide for vapour sensing [37], and the manufacture of polymer-inorganic hybrid films [38, 39].

7.2.8   Other Techniques

Other traditional methods of thin film formation include: dip-coating; painting; and screen printing. Most of these techniques are relatively easy to perform and require a minimum of equipment. Polymer films of materials, such as polypropylene, polystyrene, and poly(vinyl chloride) (PVC), can be obtained by the technique of direct isothermal immersion of a substrate into a suitable solution of the polymer (e.g. PVC in cyclohexanone). Material will deposit on the immersed substrate until equilibrium is reached between the deposition and re-solution rates. Satisfactory films can be also obtained by solution casting – allowing the evaporation of a polymer-containing solution placed on a substrate (e.g. polystyrene in chloroform). Both these methods will give relatively thick films.

Screen printing (sometimes called silk screen printing) offers a further inexpensive method of preparation of films. The process consists of dispersing a paste (the ink) of material onto a mesh-type screen, on which a desired pattern may be defined photolithographically. The substrate is placed a short distance beneath the screen. A flexible wiper (squeegee) then moves across the screen surface, deflecting it vertically and bringing it into contact with the substrate. This forces the paste through the open mesh areas. The substrate is allowed to stand at ambient temperature for some time to enable the paste to coalesce to form a coherent film.

7.3 Molecular Architectures

The techniques outlined in the following sections are generally restricted to the deposition of organic materials. These methods can, in theory, provide monolayer and multilayer structures with a high degree of order of the constituent molecules. However, meticulous attention to experimental detail is required. Diagrams of such thin film architectures, showing the molecules aligned like soldiers on parade, are commonplace; however, these can be very deceptive.

7.3.1 Langmuir–Blodgett Technique

A method that allows the manipulation of materials on the nanometre scale is the Langmuir–Blodgett technique [40–43]. Langmuir–Blodgett (LB) films are prepared by first depositing a small quantity of an *amphiphilic* compound (i.e. one incorporating both polar and nonpolar groups) dissolved in a volatile solvent, onto the surface of a carefully purified *subphase*, usually water. Figure 7.17 shows the chemical structure and molecular dimensions of stearic acid

Figure 7.17 Chemical structure for *n*-octadecanoic acid (stearic acid). The approximate geometrical shape and dimensions of the molecule are shown on the right.

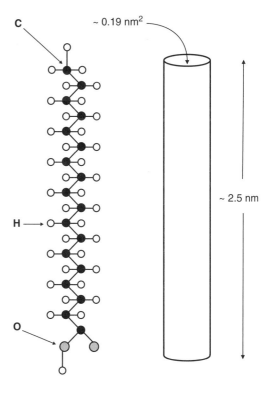

(*n*-octadecanoic acid), one of the classical materials used for LB film formation. The hydrocarbon chain forms the non-polar, hydrophobic part of the molecule, whereas the polar carboxylic acid —COOH group confers some water solubility.

When the solvent has evaporated, the organic molecules may be carefully compressed to form a floating two-dimensional solid. During this process, the floating film will undergo a multiplicity of phase transformations. These are similar to three-dimensional gases, liquids, and solids, but perhaps the closest analogy is with the mesophases shown by liquid crystals (Chapter 8, Section 8.2). Phase changes may be readily identified by monitoring the surface pressure, Π, as a function of the area occupied by the film, a. This is the two-dimensional equivalent of the pressure versus volume isotherm for a gas/liquid/solid. Figure 7.18 shows such a plot for a hypothetical long-chain organic monolayer material (e.g. a long-chain fatty acid). This diagram is not meant to represent that observed for any particular substance, but displays most of the features observed for long-chain compounds.

In the 'gaseous' state (G in Figure 7.18), the molecules are far enough apart on the water surface that they exert little force on one another. As the surface area of the monolayer is reduced, the hydrocarbon chains will begin to interact. The 'liquid' state that is formed is generally called the expanded monolayer phase (E). The hydrocarbon chains of the molecules in such a film are in a random, rather than a regular orientation, with their polar groups in contact with the subphase. As the molecular area is progressively reduced, condensed (C) phases may appear. There may be more than one of these and their emergence can be accompanied by constant pressure regions of the isotherm, as observed in the cases of a gas condensing to a liquid and a liquid solidifying. These regions will be associated with enthalpy changes (Chapter 2, Section 2.5.2) in the monolayer. In the condensed monolayer states, the molecules are closely packed and are oriented with the hydrocarbon chain pointing away from the water surface. The area per molecule in such a state will be similar to the cross-sectional area of the hydrocarbon chain, i.e. about $0.19\,\mathrm{nm}^2\,\mathrm{molecule}^{-1}$.

If the surface pressure versus area measurements are undertaken at several temperatures, and the points corresponding to the same phase transitions are plotted on a pressure versus

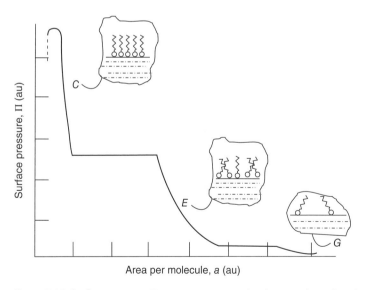

Figure 7.18 Surface pressure, Π, versus area per molecule, a, isotherm for a long-chain organic compound. The surface pressure and area are in arbitrary units (au). *Source:* Reprinted from Petty [40]. Copyright (1996), Cambridge University Press.

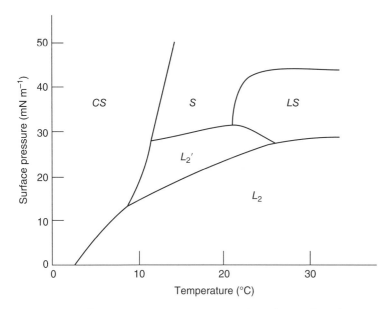

Figure 7.19 Surface pressure versus temperature phase diagram for *n*-docosanoic acid. The different phases (see text for details) are depicted as *CS, S, LS, L₂*, and *L₂'. Source:* From Peterson [44]. Reproduced by permission of Research Studies Press Ltd.

temperature graph, the resulting diagram will show the range of temperatures and pressures over which the various phases exist. Such a phase diagram for *n*-docosanoic acid is given in Figure 7.19 [44]. The nomenclature of the various phases follows the method adopted by Harkins [45]. The condensed phases include L_2, L_2' (liquid condensed), *LS* (super-liquid), *S* (solid), and *CS* (close-packed solid). The use of the term 'liquid' for some of these monolayer states simply reflects the historical assumptions that the phases were liquid-like. However, it is now known that all the condensed phases have well-defined in-plane structures and exhibit X-ray diffraction peaks (Chapter 6, Section 6.3). The monolayer characteristics of many organic materials may be found in the comprehensive handbook edited by Mingotaud et al. [46].

The two most common methods for monitoring the surface pressure are the Langmuir balance and the Wilhelmy plate. Both have similar sensitivities ($10^{-3}\,\mathrm{mN\,m^{-1}}$), but the use of the Wilhelmy plate technique is perhaps the more popular. Suspending a plate from a sensitive balance in the monolayer provides an absolute measurement of Π. Figure 7.20 shows the experimental arrangement. The forces acting on the plate are due to gravity and surface tension downwards and buoyancy, due to displaced water, upwards. For a rectangular plate of dimensions l, w, and t and of material of density ρ_W immersed to a depth h in a liquid of density ρ_L, the net downward force F is given by

$$F = \rho_W\,g\,lwt + 2\gamma\big(t+w\big)\cos\theta - \rho_L\,g\,twh \tag{7.8}$$

where γ is the surface tension of the liquid, θ is the contact angle on the solid plate, and g is the acceleration due to gravity. The usual procedure is to choose a plate that is completely wetted by the liquid (i.e. $\theta = 0$) and measure the change in F for a stationary plate. This change in force ΔF is then related to the change in surface tension $\Delta\gamma$ by

$$\Delta\gamma = \frac{\Delta F}{2\big(t+w\big)} \tag{7.9}$$

(a)

(b)

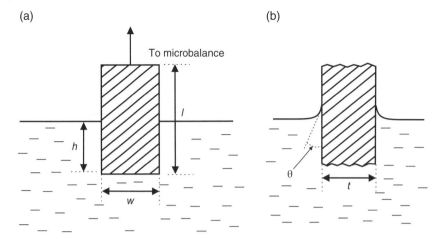

Figure 7.20 A Wilhelmy plate: (a) front view; (b) side view.

If the plate is thin enough, so that $t << w$:

$$\Delta\gamma = \frac{\Delta F}{2w} \tag{7.10}$$

The Langmuir balance is a differential technique. A clean water surface is separated from the monolayer-covered area by a partition (usually a movable PTFE float connected to a conventional balance) and the force acting on this partition is measured [41].

Figure 7.21 shows the commonest form of LB film deposition. The substrate is hydrophilic and the first monolayer is transferred, like a carpet, as the substrate is raised through the water. The substrate may therefore be placed in the subphase before the monolayer is spread. Subsequently, a monolayer is deposited on each traversal of the monolayer/air interface. As shown, these stack in a head-to-head and tail-to-tail pattern; this deposition mode is called the *Y-type*. Although this is the most frequently encountered situation, instances in which the floating monolayer is only transferred to the substrate as it is being inserted into the subphase, or only as it is being removed, are observed. These deposition modes are called *X-type* (monolayer transfer on the downstroke only) and *Z-type* (transfer on the upstroke only) and are illustrated in Figure 7.22. Mixed deposition modes are sometimes encountered and, for some materials, the deposition type can change as the LB film is built-up.

Langmuir–Blodgett films of long-chain fatty acids are often prepared by adding divalent cations to the subphase to improve the deposition characteristics of the monolayer material. The floating layer will be a mixture of the fatty acid and the fatty acid salt. The salt concentration in the monolayer will depend on the pH. In a subphase containing, say, cadmium chloride, the following reaction will take place:

$$2C_nH_{2n+1}COOH + CdCl_2 \Leftrightarrow \left(C_nH_{2n+1}CO_2\right)_2^- Cd^{2+} + 2HCl \tag{7.11}$$

The salt formation is favoured by a high subphase pH. In the specific case of a floating monolayer of *n*-eicosanoic acid (arachidic acid) on a water subphase at room temperature, containing a cadmium salt in a concentration of 10^{-4} M and having a pH value of 5.7, a monolayer comprising about 50% cadmium eicosanoate and 50% eicosanoic acid will be formed [41].

Figure 7.21 Y-type Langmuir–Blodgett film deposition. Film transfer on both the upward and downward movements of the substrate.

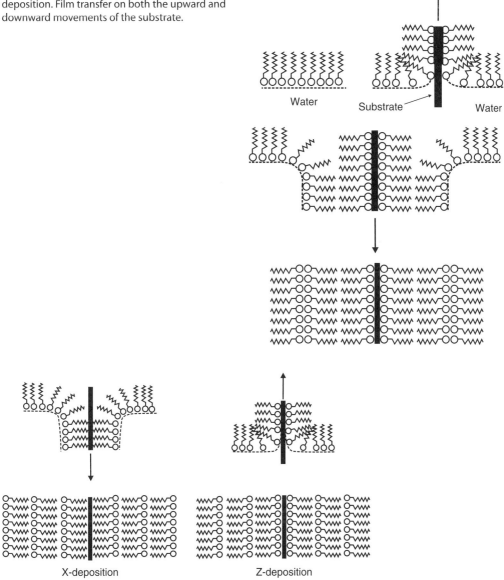

Figure 7.22 X-type and Z-type Langmuir–Blodgett film deposition.

Film transfer is characterized by measurement of the *deposition ratio*, τ (also called the *transfer ratio*). This is the decrease in the area occupied by the monolayer (held at constant pressure) on the water surface, divided by the coated area of the solid substrate, i.e.

$$\tau = \frac{A_L}{A_S} \tag{7.12}$$

where A_L is the decrease in the area occupied by the monolayer on the water surface and A_S is the coated area of the solid substrate. If asymmetric substrates are used (e.g. a glass slide

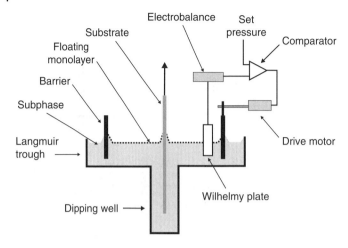

Figure 7.23 Schematic diagram of equipment for Langmuir–Blodgett film deposition.

metallized on one surface), then it is unlikely that τ will be identical for both surfaces. Transfer ratios significantly outside the range 0.95–1.05 suggest poor film homogeneity. The accurate measurement of τ for alternate-layer films (see later in this section) can present a particular problem as, in most alternate-layer deposition systems, the substrate holder is passed through the floating monolayer [40].

Many different approaches have been made to the design of equipment for the deposition of LB films. Figure 7.23 shows a schematic diagram of a simple experimental arrangement. The subphase container – the *Langmuir trough* – is made from PTFE and a PTFE-coated glass fibre barrier, which can be moved using a low-geared electric motor, defines a working area. The barrier motor is coupled to a sensitive electronic balance that continuously monitors, using a Wilhelmy plate, the surface pressure of the monolayer. Using a feedback arrangement, this pressure can be maintained at a predetermined value. The physical dimensions of the Langmuir trough arrangement are not critical (systems with dimensions ranging from centimetres to metres have been demonstrated) and are governed by the size of the substrate used.

The diagrams shown in Figures 7.21 and 7.22 are simple cartoons. These may not represent accurately the real arrangements of molecules on a solid surface. For LB films of *fatty acid salts*, X-ray diffraction experiments show that the long axes of the molecules in the LB film are indeed orthogonal to the substrate plane, as shown in most 'molecular stick' drawings [40]. The molecular arrangement in the deposited film is similar to that of the *LS, S,* and *CS* condensed monolayer phases on the water surface. However, the long molecular axes in many simple *fatty acid* LB layers are inclined at an angle to the substrate normal, the tilt angle depending on the precise deposition conditions. For example, the molecular tilt elevation of fatty acid films deposited from the L_2' phase can vary with the deposition pressure (Chapter 6, Figure 6.12). This tilt elevation may also change from layer to layer.

During LB film deposition, the first monomolecular layer will be transferred onto a solid substrate of a different material. This is an example of *heterogeneous crystal growth* [47]. For subsequent monolayers, when transferring onto an existing film, the deposition will be *homogeneous*. It is likely that the chemical and physical structure of the first monolayer will be different to that of subsequent layers. There is now much evidence to support this. For example, infrared investigations of fatty acid films show that the first monolayer possesses a hexagonal packing of the CH_2 subcells in the alkyl chain, with the chain axis oriented perpendicular to the substrate surface [40]. As the film thickness increases, a transition occurs to a structure in which the subcells have an orthorhombic packing and the main cells are packed in a monoclinic crystallographic form, with the axes of the alkyl chains inclined at an angle of 20–30° to the substrate normal.

For fatty acid salt materials deposited onto metallic substrates, there may be an ion exchange between the fatty acid salt and the thin oxide layer on the substrate surface [40]; e.g. when a calcium stearate film is transferred onto an aluminium plate, a layer of aluminium stearate is formed. Consequently, a strong chemical bond can anchor the polar head of the first LB layer to the substrate surface. Under certain conditions, epitaxial deposition is observed for LB films. For fatty acids, evidence from electron diffraction shows that each monolayer has the same local orientation of its molecular lattice as that of the underlying monolayer; however, this does not mean that translational order also extends from layer to layer [48]. Langmuir–Blodgett films of fatty acids generally consist of domains with in-plane dimensions ranging from several 100 micrometres to a few tenths of a millimetre. If the long axes of the molecules are tilted with respect to the substrate normal, then the films will be birefringent (Chapter 4, Section 4.3.1) and the domain structure can be seen by observation under a polarizing microscope. Figure 7.24 shows such an image of a 170-layer LB film of 22-tricosenoic acid deposited onto an etched silicon substrate [49]. Most of the area depicted in the photograph was coated initially with a bilayer of *n*-octadecanoic (stearic) acid. However, the region in the top left corner of the photograph was not coated, providing clear evidence that the 22-tricosenoic acid had developed the domain features of the underlying *n*-octadecanoic acid bilayer.

The reproducibility of the LB deposition process is monitored by measuring a suitable physical characteristic of the organic film as a function of time. Measurements of the film thickness, its optical density, or the frequency shift of a quartz crystal oscillator are all straightforward. A further useful method, particularly if the multilayer structure is to be used for some electrical application, is to sandwich the multilayer film between two metal electrodes and to monitor the capacitance of this structure as a function of the number of dipping cycles. Since capacitance varies as the reciprocal of thickness, a plot of reciprocal capacitance of the metal/ LB film/metal structure versus the number of dipping cycles should be a straight line. Figure 7.25 shows such a plot for cadmium stearate layers sandwiched between evaporated aluminium electrodes [50]. The linear form of the graph confirms the reproducibility of monolayer capacitance and therefore of the film deposition for one monolayer to the next. The slope of the straight line is related to the permittivity (Chapter 3, Section 3.6.1) and thickness of each monolayer, while the intercept on the ordinate yields similar information about the surface (e.g. oxide) layers on the metal electrodes.

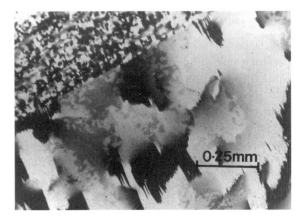

Figure 7.24 Photomicrograph of 22-tricosenoic acid (170 bilayers) deposited onto single crystal silicon. The substrate, apart from that shown in the top left-hand corner, was initially covered with a bilayer of *n*-octadecanoic acid. *Source:* From Peterson and Russell [49]. Copyright Society of Chemical Industry. Reproduced with permission from John Wiley & Sons, Ltd., on behalf of the SCI.

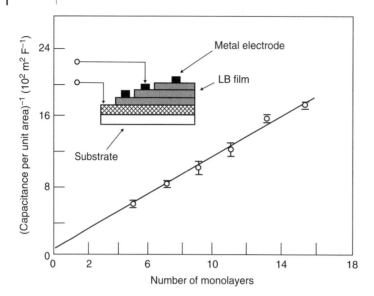

Figure 7.25 Reciprocal capacitance per unit area versus number of monolayers for Au/cadmium stearate/Al structures. The inset shows the structure of the metal/LB film/metal capacitors [50]. *Source:* Reprinted with permission from John Batey.

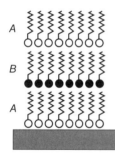

Figure 7.26 Schematic diagram of an alternate-layer Langmuir–Blodgett film built up from monolayers of compound *A* and monolayers of compound *B*.

It is possible to build up films containing more than one type of monomolecular layer. In the simplest case, condensed floating monolayers of two different amphiphilic materials are confined (using mechanical barriers) to different regions of the water surface. By lowering the solid substrate through the first layer of, say material *A*, and raising it up through the other, material *B*, alternate layers of structure *ABABAB* ... may be built-up, as shown in Figure 7.26. This facilitates the fabrication of organic superlattices with precisely defined symmetry properties. Such molecular assemblies can exhibit pyroelectric, piezoelectric, and nonlinear optical phenomena (Chapter 11, Section 11.3).

The LB technique may also be combined with solid-state chemistry methods to produce novel molecular architectures. Figure 7.27 shows a method to produce a network of conductive polypyrrole in a fatty acid matrix [51]. First, monolayers of the iron salt of a long-chain fatty acid (e.g. ferric palmitate) are assembled on an appropriate substrate. The multilayer film is then exposed to saturated HCl vapour at room temperature for several minutes. During this process, a chemical reaction transforms the fatty acid salt into layers of ferric chloride separated by layers of fatty acid. In the third and final step, the film is exposed to pyrrole vapour and a reaction occurs between the pyrrole and the ferric chloride, producing polypyrrole within the multilayer assembly.

As described above, long-chain fatty acid materials are often deposited as a mixture of a fatty acid and a salt. Following deposition, the free acid may be removed from the film by soaking the LB layers in a suitable solvent. This *skeletonization* process reduces the refractive index of the multilayer structure, making it suitable for use as an antireflection coating for glass.

It should be noted that the LB process is very different to many other thin film techniques. During film deposition, a monomolecular layer will be transferred, like a carpet, from the surface

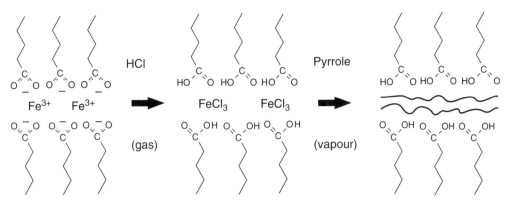

Figure 7.27 Idealized solid-state reactions of ferric stearate multilayers with HCl and pyrrole vapour to form a conductive polymer film within the fatty acid matrix [51].

of the subphase to the substrate. The resulting LB film is therefore continuous at one monolayer coverage, in contrast to vacuum evaporation (Figure 7.5).

Many biological molecules form condensed monolayers on a water surface. Phospholipids, chlorophyll a (the green pigment in higher plants), and vitamins A, E, and K are all examples. Biochemists and biophysicists have also been long aware that monomolecular films bear a close resemblance to naturally occurring biological membranes (Chapter 12, Section 12.6), and many revealing experiments may be undertaken with floating and transferred layers of biological compounds. The structurally similar proteins avidin (a toxic protein found in egg white) and streptavidin and have been the model system for many protein binding studies. Each avidin molecule tightly binds biotin, a member of the B-vitamin family (found in egg yolk), at four symmetrically located sites. Streptavidin in an aqueous subphase will bind to a biotin lipid at the air/water interface, as shown in Figure 7.28. The resulting complex forms two-dimensional

Figure 7.28 Illustration of the specific binding of streptavidin (in solution) to a biotin-derivatized lipid at the air/water interface.

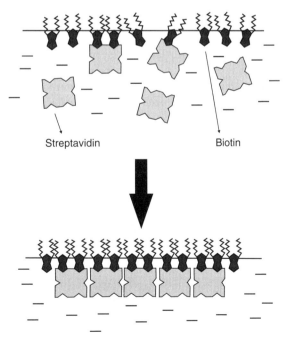

Streptavidin   Biotin

crystalline domains. The vertical dipping LB process is not the only way to transfer a floating molecular film to a solid substrate or to build up multilayer films. Other methods are based on touching one edge of a hydrophilic substrate with the monolayer-covered subphase or lowering the substrate horizontally, so that it contacts the ends of the floating molecules, the *Langmuir-Schaefer* technique [40]. The latter approach is useful for the transfer of highly rigid monolayers to solid supports.

7.3.2 Chemical Self-Assembly

Self-assembly is a much simpler process than that of LB deposition. Monomolecular layers are formed by the immersion of an appropriate substrate into a solution of the organic material, as illustrated in Figure 7.29 [43]. The best-known examples of self-assembled systems are organosilicon compounds on hydroxylated surfaces (SiO_2, Al_2O_3, glass, etc.) and alkanethiols on gold, silver, and copper. However, other combinations include: dialkyl sulfides or disulfides on gold; alcohols and amines on platinum; carboxylic acids on aluminium oxide and silver; and phosphonic acids on Al_2O_3 and indium tin oxide.

The self-assembly process is determined predominantly by strong interactions between the head group of the self-assembling molecule and the substrate, resulting in a chemical bond between the head group and a specific surface site. This can be a covalent Si—O bond for alkyltrichlorosilanes on hydroxylated surfaces; a covalent, but slightly polar, Au—S bond in the case of alkanethiols on gold; or an ionic $—CO_2^-Ag^+$ bond for carboxylic acids on AgO/Ag. Secondary considerations are the interactions between the alkyl chains and those between the terminal functionalities, a methyl (CH_3) group in the case of a simple alkyl chain.

Chemical means can be used to build up multilayer organic films. A method pioneered by Sagiv is based on the successive absorption and reaction of appropriate molecules [52]. As shown in Figure 7.30, the headgroups react with the substrate to give a permanent chemical attachment and each subsequent layer is chemically bonded to the one before, in a very similar way to that used in systems for supported synthesis of proteins.

7.3.3 Electrostatic Layer-by-Layer Deposition

In the electrostatic layer-by-layer (LbL) method, the ionic attraction between opposite charges is the driving force for the multilayer build-up [53]. In contrast to the Sagiv technique described in the previous section, which requires a reaction yield of 100% to maintain surface functional density in each layer, no covalent bonds need to be formed. The starting materials are usually *polyelectrolytes* (PEs) [54]. This term refers to polymer systems consisting of a macroion, i.e. a macromolecule carrying covalently bound anionic or cationic groups, and a low-molecular weight counterion for electroneutrality. Such materials possess a wide range of

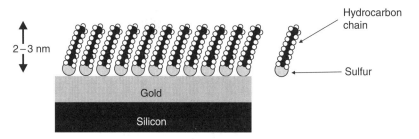

Figure 7.29 Self-assembled monolayer film of an alkane-thiol on an Au-coated substrate.

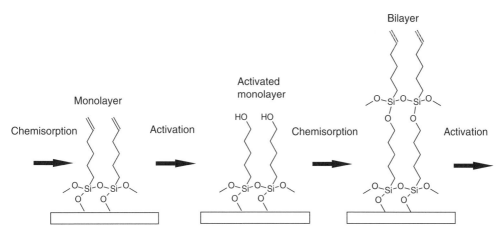

Figure 7.30 Preparation of a chemically attached polymeric multilayer. *Source:* Reprinted from Netzer et al. [52]. Copyright (1983), with permission from Elsevier.

molecular and supramolecular structures and a strong dependence of their properties, in solution or as a dispersion, on the surrounding medium. The interest in polyelectrolytes stems mainly from their ability to form organized structures in, or from, solution.

The PE family can be divided into three main sub-groups, shown in Figure 7.31. The first group includes all the strong polyelectrolytes. Such polyions possess a well-defined and constant charge in solution, e.g. sodium poly(styrene sulfonate), PSS (Figure 7.31a), or poly(diallyldimethylammoium chloride), poly-DADMAC (Figure 7.31b). These materials are dissociated into macroion and counterion in aqueous solutions over the entire pH range of 0 to 14. All weak polyelectrolytes belong to the second class. These charged polymers are able to form a polyion-counterion system in solution only over a limited range of pH (being dissociated outside this range), e.g. polyethyleneimine (PEI) (Figure 7.31c), or poly(ethylene-*co*-maleic acid) (PMAE) (Figure 7.31d). Finally, polyampholytes carry both anionic and cationic groups that are activated in alkaline or acid media, respectively, e.g. proteins or copolymers (Figure 7.31e). Therefore, by varying the pH of the solution in which they are immersed, it is possible to reverse the sign of their electrostatic charge.

One of the most interesting properties is the ability of PEs to dissolve in water (aqueous media are generally the most common environments) even if, as for polystyrene, they possess a hydrophobic backbone. The physical chemistry and processing characteristics of PEs are dependent on their behaviour in a solution or dispersion. These materials combine properties derived from long-chain molecules with those that result from charge interactions.

The electrostatic LbL deposition equipment simply consists of the process steps shown in Figure 7.32. A suitably charged substrate is immersed sequentially in the polyanion and poly-cation solutions; an intermediate washing step is included to remove excess polyelectrolyte. A schematic molecular illustration of the processing sequence is shown in Figure 7.33. The solid substrate with a positively charged planar surface is placed in the solution containing the anionic polyelectrolyte and a monolayer of the polyanion is adsorbed (Figure 7.33a). Since the adsorption is carried out at relatively high concentrations of the polyelectrolyte, most ionic groups remain exposed to the interface with the solution and the surface charge is reversed. After rinsing in pure water, the substrate is immersed in the solution containing the cationic polyelectrolyte. Again, a monolayer is adsorbed but now the original surface charge is restored (Figure 7.33b), resulting in the formation of multilayer assemblies of both polymers (Figure 7.33c).

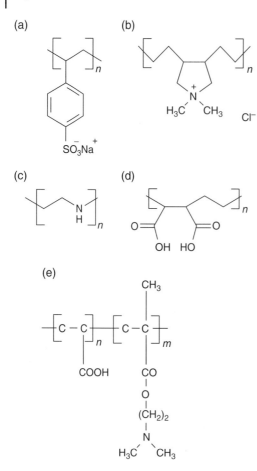

The range of substrates that can be used as templates for the multilayer self-assembly is wide, with little restriction on their type or shape. The only feature required by the surface of the substrate is for it to possess an electric charge. This can be achieved in different ways. For example, it is possible to use freshly cleaved mica, a glass or quartz slide, or a silicon wafer covered by a layer of cationic polyethylenimine. Good quality surfaces are obtainable by amino silanization procedures, by plasma treatment, as well as by sonification in an appropriate solution or by the deposition of a charged amphiphile monolayer using the LB method. Freshly prepared metal substrates, usually slightly negatively charged, can also be used. After a few polyelectrolyte adsorption cycles, the top surface will develop a strong electric charge. The choice of substrate is not limited to planar surfaces; microcapsules, colloids, tubules, or biological cells are all common templates.

The advantages of LbL deposition over other thin film deposition methods include: (i) a high versatility (being applicable to almost every solvent accessible surface); (ii) a wide range of substances may be deposited (not just polymers, but colloids, proteins, DNA, and inorganic compounds [53]); and (iii) tailoring of surface properties and therefore of the interaction between the assembled object and its environment (e.g. surfaces specially designed for bio-compatibility, corrosion protection, anti-static coatings, sensing, surfaces with specific adhesion, or wetting properties, improved surface conductivity or ion transport).

Figure 7.31 Types of polyelectrolyte: (a) sodium poly(styrene sulphonate) (PSS); (b) poly(diallyldimethylammoium chloride) (poly-DADMAC); (c) polyethyleneimine (PEI); (d) poly(ethylene-*co*-maleic acid) (PMAE); (e) a copolymer formed from acrylic acid and dimethylaminoethyl-methacrylate. (a), (b) are examples of strong polyelectrolytes; (c), (d) are weak polyelectrolytes; and (e) is a polyampholyte.

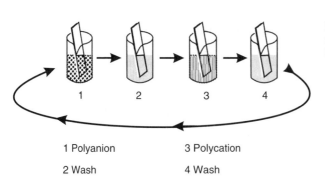

1 Polyanion

2 Wash

3 Polycation

4 Wash

Figure 7.32 Layer-by-layer (LbL) electrostatic deposition in a series of beakers.

(a)

(b)

(c)

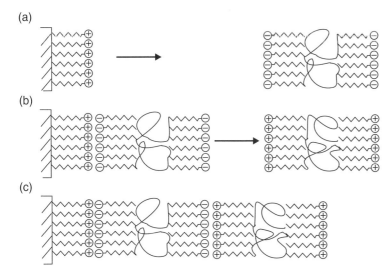

Figure 7.33 Build-up of multilayer assemblies by consecutive adsorption of anionic and cationic polyelectrolytes [52].

Figure 7.34 Atomic microscope scan in air using contact mode of polyethyleneimine/poly(styrene sulphonate) layer-by-layer (PEI/PSS LbL) film, where the electrostatic self-assembly has only been partial. A group of polymer coils is visible in the upper half of the scan. *Source:* Reprinted from Palumbo et al. [55]. Copyright (2005), with permission from Elsevier.

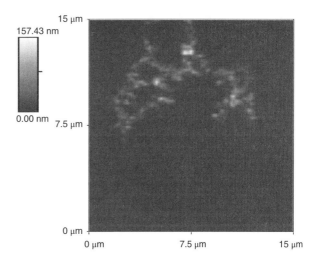

Many studies reveal that the LbL assembly process exhibits a non-linear film growth in its early stages. During the first several minutes of deposition, the charged macromolecules tend to adsorb preferentially on selected defect sites of the oppositely charged support (scratches, micro-particles, and edges) and form islands composed of polymer coils, as shown by the atomic force microscopy image in Figure 7.34 [55]. As the number of deposition steps in increased, the islands coalesce until a complete coverage of the underlying support is achieved. In this respect, LbL deposition shows some similarities to the nucleation and growth of thermally evaporated thin films (Figure 7.5).

Weak and strong polyelectrolytes can generate LbL thin films with different surface morphologies. Such behaviour can be attributed to their different charge density in solution. For example, polymeric chains can assume the form of a coil (strong polyelectrolytes) or a globule (weak polyelectrolytes at low charge density). Consequently, during the process of adsorption, the former would produce a more uniform surface, while the latter would generate a rougher

(a)

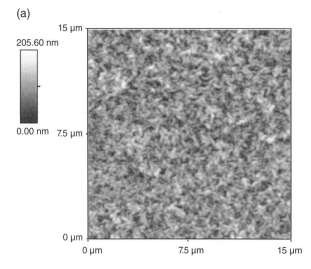

Figure 7.35 Atomic force microscopy scans in air using contact mode of layer-by-layer (LbL) films. (a) Three bilayers of polyethyleneimine/poly(styrene sulphonate)/poly(ethylene-*co*-maleic acid) (PEI/PMAE). (b) Three bilayers of PEI/ poly(styrene sulphonate) (PSS). *Source:* Reprinted from Palumbo et al. [55]. Copyright (2005), with permission from Elsevier.

(b)

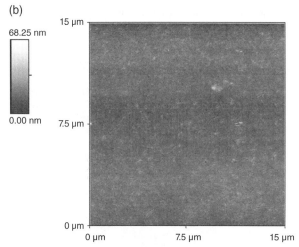

and more porous surface. Atomic force microscope images of two different LbL architectures are shown in Figure 7.35 [55]. The image in Figure 7.35a is for three bilayers of PEI/ PMAE, whereas that in Figure 7.35b depicts the morphology of a three bilayer film of PEI/ PSS (the chemical structures have been given in Figure 7.31). The much smoother surface for the latter structure is probably related to the strong polyelectrolyte character of the PSS. In solution, the chains of this polymer will be rigid. In contrast, PMAE is a relatively weak polyacid and, in solution, lower repulsion forces along their length will influence its chains. The molecular conformation in solution will more globular as the chains will not be as rigid as those of the PSS.

A related, but alternative, approach to the electrostatic LbL assembly outlined above uses LbL adsorption driven by hydrogen-bonding interactions [56]. This has been accomplished with polyvinylpyrrolidone, poly(vinyl alcohol), polyacrylamide, and poly(ethylene oxide). In the case of polyaniline, comparisons with films assembled via the electrostatic mechanism, using sulfonated polystyrene, indicate that the non-ionic polymers adsorb onto polyaniline with a greater density of loops and tails and form highly interpenetrated bilayers with high polyaniline content.

7.4 Micro- and Nanofabrication

In this section, some methods that may prove useful for defining nanometre-scale patterns in organic thin films are described. First, the well-established principles of photolithography are outlined.

7.4.1 Photolithography

Planar microelectronic components are patterned using photolithography. A surface is covered with a light-sensitive photoresist and then exposed to ultraviolet light through a mask. Either the exposed photoresist (positive resist) or the unexposed regions (negative resist) can be washed away to leave positive or negative image of the mask on the surface. Positive resists are typically workable to finer feature sizes and are more generally used. Figure 7.36 illustrates the processing steps that are needed (from a–h) to define a pattern on a silicon substrate using a positive resist. A thin oxide layer is first grown on the silicon in a furnace (Figure 7.36b). Subsequently, the oxide is coated with the resist layer (e.g. by spin-coating) and the resist is exposed through a contact mask to UV radiation (Figure 7.36d). After removing the exposed areas of photoresist, the silicon oxide defined by the patterned mask is etched away (Figure 7.36g). The resist is dissolved in a solvent, and the resulting pattern in the silicon dioxide

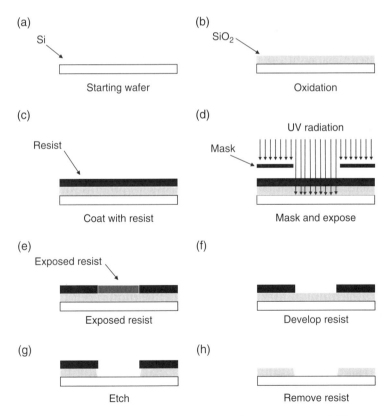

Figure 7.36 Schematic diagram of the process of photolithography. (a) Starting wafer. (b) Oxidation. (c) Coat with resist. (d) Mask and expose. (e) Exposed resist. (f) Develop resist. (g) Etch. (h) Remove resist.

(Figure 7.36h) can be used as a window to diffuse or implant dopant atoms in the silicon in the predefined regions.

The lateral resolution of diffraction-limited optics (Chapter 6, Section 6.2.1) can be expressed by

$$\text{minimum feature size} = \frac{k_1 \times \text{wavelength}}{\text{numerical aperture}} \tag{7.13}$$

where

$$\text{numerical aperture} = \frac{1}{f\text{ number}} = \frac{\text{aperture diameter}}{\text{focal length}}$$

and k_1 is a system-dependent constant in the range 0.5–0.8.

A reasonable depth of focus is also needed to expose wafers with a particular topography with height variations patterned in previous lithography steps. This is given by

$$\text{depth of focus} = \frac{k_2 \times \text{wavelength}}{\left(\text{numerical aperture}\right)^2} \tag{7.14}$$

k_2 is a further constant depending on the coherence of the source and on the resist.

To define ever-smaller patterns, radiation of a lower wavelength can be used. The mercury light source (365 nm line) used in the early days of photolithography has been replaced by a krypton fluoride excimer laser radiating at 248 nm (deep ultraviolet). This improves both the resolution and depth of focus and enables features down to 250 nm to be defined. Further improvements are possible utilizing an argon fluoride excimer laser operating at 193 nm and excimer lasers have been demonstrated to 126 nm. Additional resolution enhancement techniques are increasingly being used to provide the 10–20 nm features in current (2018) integrated circuits. Examples are the use of *immersion lithography*, in which the conventional air gap between the final lens and the wafer surface is replaced by a liquid, mask illumination improvements and the introduction of phase-shifting masks. A surface plasmon interference lithography method has enabled one-dimensional and two-dimensional patterns with a half-pitch (half the distance between identical features) resolution of 14.6 nm to be generated in a 25 nm-thick photoresist layer under 193 nm illumination [57].

7.4.2 Nanometre Pattern Definition

To extend the ongoing technologies to wavelengths in the ultraviolet and soft X-ray range (collectively known as *extreme ultraviolet*, XUV) below 100 nm, some effort is being directed at the use of reduction optics. This uses reflecting optics and masks with special multilayer coatings. The preferred wavelength is 13 nm, just below the absorption edge of silicon. The reflective coatings consist of a stack of coherently reflecting bilayer systems of Si and Mo. The XUV radiation requires optical elements fabricated with a surface accuracy and roughness in the sub-nanometre range. This specification is beyond values achieved so far, yet is being rapidly approached.

X-ray lithography involves the use of even shorter wavelengths of about 1 nm. For this wavelength, no suitable optics is available and the imaging scheme reduces to a one-to-one shadow printing process. The source of X-rays is a *synchrotron* (a very expensive capital investment).

One synchrotron can have several exposure beam lines. X-ray mask fabrication is a non-trivial task and has not yet been successfully commercialized.

Electron-beam lithography is a method widely used in research and is the primary technique for mask fabrication. However, serial writing is too slow for integrated circuit manufacture. The *SCALPEL* (SCattering with Angular Limitation Projection Electron-beam Lithography) circumvents the serial limitations. The ultimate resolution of e-beam lithography is not determined by the wavelength (in contrast to optical lithography), as wavelengths of around 10^{-13} m (smaller than atomic dimensions) are achievable. An e-beam can be focused to a spot size of approximately 1 nm and the resolution is then determined by the resist properties. *Ion-beam projection lithography* utilizes a beam of hydrogen or helium ions, which are focused electrostatically.

Structures of less than 10 nm may also be fabricated using techniques of shadowing and edge-step deposition [1]. Substrate steps with a square profile are first formed by ion-beam etching, as shown in Figure 7.37. Ion-etching the film-coated substrate at an angle then produces wires of triangular cross section, so that the wire is formed in the shadow of the step. Metal wires as narrow as 30 nm and as long as 0.5 mm have been fabricated in this way.

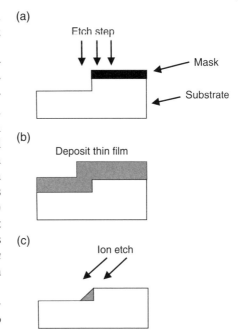

Figure 7.37 Fabrication process for the production of fine wires by the step-edge technique. (a) Etch step. (b) Deposition of thin film. (c) Ion etch.

Conventional lithography has enabled the development of micro- and nanoscale electronic devices. However, it has limitations, including high cost, limited resolution, and poor compatibility with unconventional (e.g. organic) materials that may be soft, nonplanar, or difficult to process. Some of the methods that address these issues are described below.

7.4.3 Nanoimprint Lithography

Nanoimprint lithography (NIL) is a simple process with low cost, high throughput, and high resolution [58]. There are several different types of NIL, including *thermoplastic NIL*, *photo NIL*, and *resist-free NIL*. In a standard thermoplastic NIL process, a thin layer of a thermoplastic polymer (the imprint resist) is spin coated onto a substrate. Then a master mould, with predefined topological features, is brought into contact with the thin film of resist and these are pressed together under certain pressures and subsequently heated. Above the polymer's glass transition temperature (Chapter 2, Section 2.6.3), the pattern on the master is transferred to the softened polymer. After cooling, the master is separated from the sample, which retains the master pattern. A pattern transfer process (e.g. reactive ion etching) is finally used to transfer the pattern in the resist to the underlying substrate. Cold-welding between two metal surfaces can also be used to transfer patterns without heating.

In photo NIL, a photo (UV) curable liquid resist is applied to a substrate and the mould is usually made of a transparent material such as fused silica. After the mould and resist are pressed together, the resist is cured in the UV light and becomes solid. Resist-free NIL does not require an additional etching step to transfer patterns from imprint resist to the substrate layer. An example of pattern generation using a rubber stamp, so-called *soft-lithography*, is

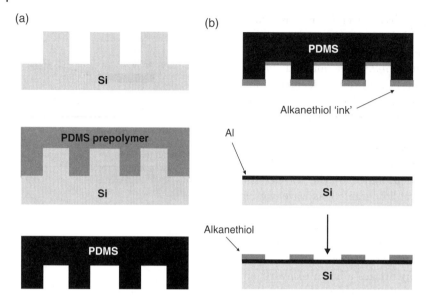

Figure 7.38 Soft lithography method to apply a pattern to a surface using a rubber stamp. (a) A pattern-transfer element is formed by pouring a liquid polymer onto a master made from silicon to form an elastomer. (b) Replica to be used as a stamp to transfer a chemical ink to a surface. *Source:* Reprinted from Brittain et al. [59]. Copyright (1998), with permission from IPO Publishing Limited.

shown in Figure 7.38 [59]. A pattern-transfer element is formed by pouring a liquid polymer, such as polydimethylsiloxane (PDMS), onto a master made from silicon. The polymer is allowed to cure to form an elastomer, which can then be removed from the master. This replica can be used as a stamp to transfer a chemical ink, such as a solution of an alkanethiol, to a surface.

Chemical approaches to the deposition of ultra-thin organic films also offer some control over the composition and structure of the surface. For example, a substrate is first patterned with gold and aluminium strips using conventional photolithography [60]. The aluminium oxidizes spontaneously in air and provides a layer of aluminium oxide at the solid/vapour interface; in contrast, the gold remains clean. Two adsorbates, L_1 and L_2, are chosen so that they adsorb strongly and selectively on gold and alumina. In the example, L_1 is an alkanethiol and L_2 is a fluorine-labelled carboxylic acid. Exposure of the substrate to a solution containing L_1 and L_2 results in replicating the gold pattern with a self-assembled monolayer derived from the alkane-thiol and the aluminium pattern with a self-assembled monolayer derived from the carboxylic acid.

A relatively high throughput method for lithographically processing one-dimensional nanowires is *on-wire lithography* [61]. This procedure combines template-directed synthesis of nanowires with electrochemical deposition and wet-chemical etching and allows routine fabrication of face-to-face disk arrays and gap structures in the range five to several 100 nanometres.

7.4.4 Scanning Probe Manipulation

The tip of a scanning tunnelling or an AFM (Chapter 6, Section 6.8) may be used to manipulate atoms and molecules on a surface. Careful control of the tip can allow patterns to be drawn in an organic film. Scanning microscopy can also be used to reposition molecules, such as the fullerene C_{60}, on surfaces and to break up an individual molecule [62]. A wide range of scanning probe lithographic techniques is available to pattern self-assembled monolayers [63].

Figure 7.39 IBM company logo written using a scanning tunnelling microscope with Xe atoms on a (110) Ni surface. Each of the letters is 5 nm, from top to bottom. *Source:* Reprinted from Eigler and Schweizer [64]. Copyright (1990), with permission from Macmillan Publishers Ltd.

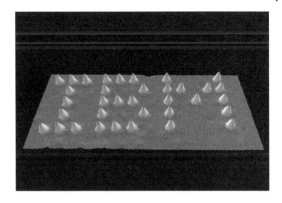

Figure 7.39 shows the IBM company logo written using a scanning tunnelling microscope with Xe atoms on a (110) Ni surface [64]. This experiment was undertaken in an ultra-high vacuum system and the entire chamber housing the microscope was cooled to 4 K. Each of the letters is 5 nm, from top to bottom.

The photolithographic process may also be used to create nanometre-scale patterns in self-assembled monolayers by utilizing a UV laser coupled to a scanning near-field optical microscope (SNOM – Chapter 6, Section 6.2.1) as a light source [65–67]. The alkanethiolates that are formed by the adsorption of alkanethiols on a gold surface (Section 7.3.2) can be oxidized by UV light in the presence of air to alkylsulfonates. The reaction is

$$Au\text{-}SR\left(CH_2\right)_n X + 3/2 O_2 \rightarrow Au^+ + X\left(CH_2\right)_n SO_3 - \tag{7.15}$$

The process is depicted schematically in Figure 7.40. The coupling of the UV source to the SNOM allows features smaller than the diffraction limit to be fabricated. The alkylsulfonates are weakly bound to the underlying gold surface and may be displaced by immersing the substrate into a solution of a contrasting thiol. Figure 7.41 shows a lateral force microscopy (LFM; Chapter 6, Section 6.8) image that has been created using this technique [67]. The lines represent $C_{15}CH_3$ written into C_2COOH self-assembled monolayers on polycrystalline Au, with a grain size of approximately 20 nm. Parallel lines as small as 20 nm, with a separation of about 20 nm, are evident. The contrast between regions of different terminal group functionality is very clear, and lines written close together can be differentiated. The resolution achieved ($\lambda/12$) is significantly beyond the diffraction limit and rivals the performance of electron beam lithography

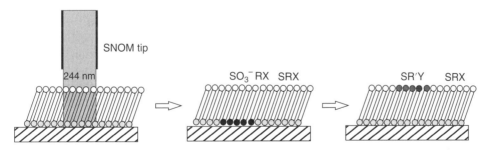

Figure 7.40 Schematic diagrams showing the selective replacement of a self-assembled layer of an X-terminated thiol following oxidation using a UV source and subsequent exposure to a solution of a contrasting Y-terminated thiol. The use of the scanning near-field optical microscope (SNOM) tip allows high resolution structures to be defined.

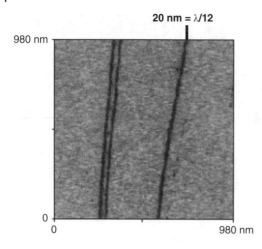

20 nm = λ/12

980 nm

0

0   980 nm

Figure 7.41 Lateral force microscopy image showing lines of $C_{15}CH_3$ written into a C_2COOH self-assembled monolayer on polycrystalline gold (grain size ≈ 20 nm). *Source:* Reprinted from Sun and Leggett [67]. Copyright (2004) American Chemical Society.

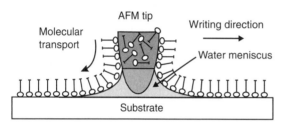

AFM tip

Writing direction

Molecular transport

Water meniscus

Substrate

Figure 7.42 Dip-pen nanolithography.

for materials of this type. Furthermore, the method is an ambient technique compatible, in principle, with operation in a fluid medium.

7.4.5 Dip-Pen Nanolithography

A further approach that has been developed at Northwestern University is called *dip-pen nanolithography* (DPN) [68–70]. This technique, illustrated in Figure 7.42, is able to deliver organic molecules in a positive printing mode. An AFM tip is used to 'write' alkanethiols on a gold thin film in a manner analogous to that of a fountain pen. Molecules flow from the AFM tip to a solid substrate (the 'paper') via capillary transport, making DPN a potentially useful tool for assembling nanoscale devices.

The chemisorption of the 'ink' is the driving force that moves the organic molecules from the AFM tip through the water to the substrate, as the tip is scanned across this surface. Adjusting the scan rate and relative humidity can control line widths. The basic DPN method has some similarity to LB film deposition, described above in Section 7.3.1. For example, if the DPN experimental arrangement shown in Figure 7.42 is turned by 90°, a 'nano-LB trough' (similar to those that use a moving subphase to compress the monolayer film) is evident. As the amphiphilic molecules flow down the AFM tip, supported by a thin layer of water, their surface pressure will rise. The resulting condensed monomolecular film is then transferred to a solid substrate. However, for DPN, the 'substrate' is stationary while the 'nano-trough' moves in relation to it.

Developments of DPN have included an overwriting capability that allows one nanostructure to be generated and the areas surrounding that nanostructure to be filled with a

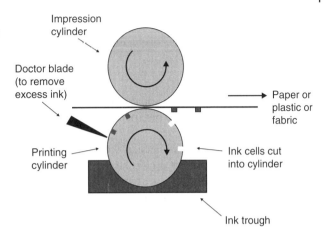

Figure 7.43 Schematic diagram of gravure printing where the ink is held in a recess etched or cut into the printing roller.

second type of ink. Perhaps the greatest limitation in using scanning probe methodologies for ultrahigh resolution nanolithography over large areas derives from the serial nature of most techniques. However, the use of multiple scanning-probe microscopy cantilever probes as pens has been shown to increase patterning speeds.

7.4.6 Gravure Printing

For the large-scale production of organic electronic devices, methods such as roll-to-roll processing will become important. In the *gravure* (or *rotogravure*) process, depicted schematically in Figure 7.43, a pattern or image is engraved onto a metal cylinder, usually with a diamond-tipped or laser etching machine. This image is composed of small recessed cells (or dots) that act as tiny wells. Their depth and size controls the amount of ink that gets transferred to the substrate. A *doctor blade* removes the excess ink. Such a process can be easily adapted as a continuous deposition process to fabricate organic devices [71]. Starting with a roll of plastic (or, perhaps, fabric) substrate material, the organic materials forming the device active regions are first deposited from the liquid or vapour phase, and then a metal 'strike' layer is deposited. The film is then passed between rollers with the embossed pattern representing the ultimate electrode scheme required on their surfaces; these rollers directly pattern the surface. In the final step, the strike layer metal is removed by dry etching.

7.4.7 Other Methods

Conventional photolithography is a key technology that has underpinned the progress of modern electronics. However, it has many limitations including the need for a mask, focusing optics, light sensitive materials, and expensive equipment. There is also poor compatibility with unconventional materials that may be soft and nonplanar. Some of the alternatives to photolithography have been described in the sections above. Other methods that have been proposed exploit self-folding (*origami*) [72]. Such processes use patterning and material strategies to convert planar substrates into three-dimensional shapes in response to external stimuli. For example, the use of strategic cuts in materials followed by folding (*kirigami*) allows otherwise rigid materials to extend significantly, both in-plane and out-of-plane. This has enabled the formation of stretchable electrodes for nonplanar devices [73] and photovoltaic cells with integrated solar tracking [74].

Problems

7.1 The depth of focus for an optical lithographic system is 0.62 μm for an operating wavelength of 248 nm. What will be the depth of focus if the wavelength is reduced to 193 nm?

7.2 For an inkjet printer, calculate the size of the individual droplets to produce an output on a substrate of 300 dots per inch. Estimate the number of atoms in each droplet.

7.3 A thin film of a low molecular weight organic compound is formed by thermal evaporation. The substrate is a circular disc with a diameter of 5 cm and is positioned directly above the evaporation source such that the distance of the source to the centre of the substrate is 10 cm. Calculate the expected percentage decrease in the organic film thickness between the centre and edge of the substrate.

7.4 Suggest methods that might be used to form thin layers of the following organic compounds: tris(8-hydroxyquinolene)aluminium (Alq_3), poly(3,4-ethylene dioxythiophene) doped with polystyrene sulfonated acid (PEDOT:PSS), copper phthalocyanine, cadmium tricosanoate, 1-octdecanethiol, polyaniline.

7.5 What physical phenomenon might be used to measure the thickness of a single Langmuir–Blodgett (LB) layer of a long-chain fatty acid: (a) nuclear magnetic resonance (NMR), surface plasmons (SP), or electron spin resonance (ESR)?

In a process to deposit a thin organic film using the LB technique, 55 μl of a long-chain fatty acid material (molecular weight 364) are spread onto the surface of purified water from a solution in chloroform of concentration $1.1\,g\,l^{-1}$. When the solvent has evaporated, the resulting surface film is compressed. The surface pressure begins to rise rapidly when the floating film occupies an area of $0.20\,nm^2$ molecule^{-1} (condensed phase). What surface area of film (in cm^2, to 3 significant figures) does this correspond to? If the same volume of a different fatty acid is used, the area corresponding to the condensed phase is measured as $132\,cm^2$. Calculate the molecular weight of this compound (to 3 significant figures).

References

1 Brodie, I. and Muray, J.J. (1992). *The Physics of Micro/Nano-Fabrication*. New York: Plenum Press.

2 Flack, W.W., Soong, D.S., Bell, A.T., and Hess, D.W. (1984). A mathematical model for spin-coating of polymer resists. *J. Appl. Phys.* 56: 1199–1206.

3 Bornside, D.E., Macosko, C.W., and Scriven, L.E. (1987). On the modelling of spin-coating. *J. Imaging Technol.* 13: 122–130.

4 Bornside, D.E., Macosko, C.W., and Scriven, L.E. (1989). Spin-coating: one-dimensional model. *J. Appl. Phys.* 66: 5185–5193.

5 Carcano, G., Ceriani, M., and Soglio, F. (1993). Spin coating with high viscosity photoresist on square substrates – applications in the thin film hybrid microwave integrated circuit field. *Microelectron. Int.* 10: 12–20.

6 Luurtsema GA. (1997). *Spin Coating for Rectangular Substrates*. MSc Thesis. Berkeley: University of California.

7 Critchley, S.M., Willis, M.R., Cook, M.J. et al. (1992). Deposition of ordered phthalocyanine films by spin-coating. *J. Mater. Chem.* 2: 157–159.

8 Glang, R. (1970). Vacuum evaporation. In: *Handbook of Thin Film Technology* (ed. L.I. Maissel and R. Glang), 1-3–1-330. New York: McGraw-Hill.

9 Chrisey, D.B. and Hubler, G.K. (eds.) (1994). *Pulsed Laser Deposition of Thin Films*. New York: Wiley-Interscience.

10 Breen, J.J., Tolman, J.S., and G W Flynn, G.W. (1993). Scanning tunnelling microscopy studies of vapour deposited films of tetrathiafulvalene with iodine. *Appl. Phys. Lett.* 62: 1074–1076.

11 Leaver, K.D. and Chapman, B.N. (1971). *Thin Films*. London: Wykeham.

12 Kolb D. 2009. Organic transistors based on pentacene and dibenzothiophene derivatives. PhD thesis. Durham: University of Durham.

13 Kilitziraki, M., Moore, A.J., Petty, M.C., and Bryce, M.R. (1998). Evaporated thin films of tetrathiafulvalene derivatives and their charge-transfer complexes. *Thin Solid Films* 335: 209–213.

14 Hara, M. and Sasabe, H. (1995). Organic molecular beam epitaxy. In: *An Introduction to Molecular Electronics* (ed. M.C. Petty, M.R. Bryce and D. Bloor), 243–260. London: Edward Arnold.

15 Sherman, A. (1987). *Chemical Vapour Deposition for Microelectronics*. Noyes: Park Ridge.

16 Chopra, K.L. (1969). *Thin Film Phenomena*. New York: McGraw-Hill.

17 Yasuda, H. (1985). *Plasma Polymerization*. New York: Academic Press.

18 Lucovsky, G., Ibbotson, D.E., and Hess, D.W. (eds.) (1990). *Characterization of Plasma-Enhanced CVD Processes*. Pittsburg: Materials Research Society.

19 Eden, J.G. (1992). *Photochemical Vapour Deposition*. New York: Wiley-Interscience.

20 George, S.M. (2010). Atomic layer deposition: an overview. *Chem. Rev.* 110: 111–131.

21 Heinze, J. (1991). Electrochemistry of conducting polymers. *Synth. Met.* 41–43: 2805–2823.

22 Haushalter, R.C. and Krause, L.J. (1983). Electroless metallization of organic polymers using the polymer as a redox reagent: reaction of polyamide with zintl anions. *Thin Solid Films* 102: 161–171.

23 Sirringhaus, H. and Shimoda, T. (eds.) (2003). Inkjet printing of functional materials. *MRS Bull* 28: 802–803.

24 Calvert, P. (2001). Inkjet printing for materials and devices. *Chem. Mater.* 13: 3299–3305.

25 Sing, M., Haverinen, H.M., Dhagat, P., and Jabbour, G.E. (2010). Inkjet printing – process and applications. *Adv. Mater.* 22: 673–685.

26 Morris DJ. (2004). *Electrical and spectral characterization of inkjet printed poly(3,4 ethylenedioxythiophene): poly(4-styrenesulfonate). PhD Thesis*. Bangor: University of Wales.

27 Deegan, R.D., Bakajin, O., Dupont, T.F. et al. (1997). Capillary flow as the cause of ring stains from dried liquid drops. *Nature* 389: 827–829.

28 Hu, H. and Larson, R. (2006). Marangoni effect reverses coffee-ring depositions. *J. Phys. Chem.* 110: 7090–7094.

29 Lim, J.A., Lee, W.H., Lee, H.S. et al. (2008). Self-organization of ink-jet-printed triisopropylsilylethynyl pentacene via evaporation-induced flows in a drying droplet. *Adv. Funct. Mater.* 18: 229–234.

30 Minemawari, H., Yamada, T., Matsui, H. et al. (2011). Inkjet printing of single-crystal films. *Nature* 475: 364–367.

31 Mabrook, M.F., Pearson, C., and Petty, M.C. (2005). An inkjet-printed chemical fuse. *App. Phys. Lett.* 86: 013507.

32 Eslamian, M. (2014). Spray-on thin film PV solar cells: advances, potentials and challenges. *Coatings* 4: 60–84.

33 Perednis, D. and Gauckler, L.J. (2005). Thin film deposition using spray pyrolysis. *J. Electroceramics* 14: 103–111.

34 Jones, R.W. (1989). *Fundamental Principles of Sol-Gel Technology*. London: Institute of Metals.

35 Brinker, C.J. and Scherer, G.W. (1990). *Sol-Gel Science: The Physics and Chemistry of Sol-Gel Processing*. San Diego, CA: Academic Press.

36 Sung, S., Park, S., Lee, W.J. et al. (2015). Low-voltage flexible organic electronics based on high-performance sol gel titanium dioxide dielectric. *ACS Appl. Mater. Interfaces* 7: 7456–7461.

37 Guo DM, Cai PJ, Sun J, He WN, Wu XH, et al. (2016). Reduced-graphene-oxide/metal-oxide p-n heterojunction aerogels as efficient 3D sensing frameworks for phenol detection. *Carbon* 99: 571–578.

38 Ogoshi, T. and Chujo, Y. (2005). Organic-inorganic polymer hybrids prepared by the sol-gel method. *Compos. Interfaces* 11: 539–566.

39 Morales-Acosta, M.D., Alvarado-Beltran, C.G., Quevedo-Lopez, M.A. et al. (2013). Adjustable structural, optical and dielectric characteristics in sol-gel PMMA-SiO$_2$ hybrid films. *J. Non-Cryst. Solids* 362: 124–135.

40 Petty, M.C. (1996). *An Introduction to Langmuir-Blodgett Films*. Cambridge: Cambridge University Press.

41 Roberts, G.G. (ed.) (1990). *Langmuir-Blodgett Films*. New York: Plenum Press.

42 Tredgold, R.H. (1994). *Order in Thin Organic Films*. Cambridge: Cambridge University Press.

43 Ulman, A. (1991). *An Introduction to Organic Thin Films*. San Diego, CA: Academic Press.

44 Peterson, I.R. (1992). Langmuir-Blodgett films. In: *Molecular Electronics* (ed. G.J. Ashwell), 117–206. Taunton: Research Studies Press.

45 Harkins, W.D. (1952). *Physical Chemistry of Surface Films*. New York: Reinhold.

46 Mingotaud, A.-F., Mingotaud, C., and Patterson, L.K. (eds.) (1993). *Handbook of Monolayers*. Orlando, FL: Academic Press.

47 Neugebauer, C.A. (1970). Condensation, nucleation and growth of thin films. In: *Handbook of Thin Film Technology* (ed. L.I. Maissel and R. Glang), 8-3–8-44. New York: McGraw-Hill.

48 Peterson, I.R. (1990). Langmuir-Blodgett films. *J. Phys. D Appl. Phys.* 23: 379–395.

49 Peterson, I.R. and Russell, G.J. (1985). Deposition mechanisms in Langmuir-Blodgett films. *Br. Polym. J.* 17: 364–367.

50 Batey J. (1983). *Electroluminescent MIS Structures Incorporating Langmuir-Blodgett Films. PhD Thesis*. Durham: University of Durham.

51 Rosner, R.B. and Rubner, M.F. (1994). Solid state polymerization of pyrrole within a Langmuir–Blodgett film of ferric stearate. *Chem. Mater.* 6: 581–586.

52 Netzer, L., Iscovici, R., and Sagiv, J. (1983). Adsorbed monolayers versus Langmuir–Blodgett monolayers - why and how? I: from monolayer to multilayer, by adsorption. *Thin Solid Films* 99: 235–241.

53 Decher, G. and Schlenoff, J.B. (eds.) (2003). *Multilayer Thin Films: Sequential Assembly of Nanocomposite Materials*. Weinheim: Wiley-VCH.

54 Dautzenberg, H., Jaeger, W., Kötz, J. et al. (1994). *Polyelectrolytes: Formation, Characterization and Application*. Munich: Hanser.

55 Palumbo, M., Pearson, C., and Petty, M.C. (2005). Atomic force microscope characterization of poly(ethyleneimine) / poly(ethylene-*co*-maleic acid) and poly(ethyleneimine) / poly(styrene sulfonate) multilayers. *Thin Solid Films* 483: 114–121.

56 Stockton, W.B. and Rubner, M.F. (1997). Molecular-level processing of conjugated polymers. 4. Layer-by-layer manipulation of polyaniline via hydrogen-bonding interactions. *Macromolecules* 30: 2717–2725.

57 Dong, J., Liu, J., Kang, G. et al. (2014). Pushing the resolution of photolithography down to 15 nm by surface plasmon interference. *Sci. Rep.* 4: 5618.

58 Chou, S.Y., Krauss, P.R., and Renstrom, P.J. (1996). Imprint lithography with 25-nanometer resolution. *Science* 272: 85–87.

59 Brittain, S., Paul, K., Zhao, X.-M., and Whitesides, G. (1998). Soft lithography and microfabrication. *Phys. World* 11 (5): 31–36.

60 Whitesides, G.M. and Laibinis, P.E. (1990). Wet chemical approaches to the characterization of organic surfaces: self-assembled monolayers, wetting, and the physical-organic chemistry of the solid-liquid interface. *Langmuir* 6: 87–96.

61 Qin, L., Park, S., Huang, L., and Mirkin, C.A. (2005). On-wire lithography. *Science* 309: 113–115.

62 Gimzewski, J. (1998). Molecules, nanophysics and nanoelectronics. *Phys. World* 11 (6): 29–33.

63 Krämer, S., Fuierer, R.R., and Gorman, C.B. (2003). Scanning probe lithography using self-assembled monolayers. *Chem. Rev.* 103: 4367–4418.

64 Eigler, D.M. and Schweizer, E.K. (1990). Positioning single atoms with a scanning tunnelling microscope. *Nature* 344: 524–526.

65 Sun, S., Chong, K.S.L., and Leggett, G.J. (2002). Nanoscale molecular patterns fabricated by scanning near-field optical lithography. *J. Am. Chem. Soc.* 124: 2414–2415.

66 Sun, S., Chong, K.S.L., and Leggett, G.J. (2005). Photopatterning of self-assembled monolayers at 244 nm and applications to the fabrication of functional microstructures and nanostructures. *Nanotechnology* 16: 1798–1808.

67 Sun, S. and Leggett, G.J. (2004). Matching the resolution of electron beam lithography by scanning near-field photolithography. *Nano Lett.* 4: 1381–1384.

68 Piner, R.D., Zhu, J., Xu, F. et al. (1999). 'Dip-pen' nanolithography. *Science* 283: 661–663.

69 Lee, S.W., Sanedrin, R.G., Oh, B.-K., and Mirkin, C.A. (2005). Nanostructured polyelectrolyte multilayer organic thin films generated via parallel dip-pen nanolithography. *Adv. Mater.* 17: 2749–2753.

70 Salaita, K., Wang, Y., and Mirkin, C.A. (2007). Applications of dip-pen nanolithography. *Nat. Nanotechnol.* 2: 145–155.

71 Forrest, S.R. (2004). The path to ubiquitous low-cost organic electronic applications on plastic. *Nature* 428: 911–918.

72 Kang, S.H. and Dickey, M.D. (2016). Patterning via self-organization and self-folding: beyond conventional lithography. *MRS Bull.* 41: 93–96.

73 Cho, Y., Shin, J.-H., Costa, A. et al. (2014). Engineering the shape and structure of materials by fractal cut. *Proc. Natl. Acad. Sci. U.S.A.* 111: 17390–17395.

74 Lamoureux, A., Lee, K., Shlian, M. et al. (2015). Dynamic kirigami structures for integrated solar tracking. *Nat. Commun* 6: 8092.

Further Reading

Adamson, A.W. (1990). *Physical Chemistry of Surfaces*, 5e. Chichester: Wiley-Interscience.

Chapman, B. (1980). *Glow Discharge Processes: Sputtering and Plasma Etching*. New York: Wiley.

Greene, J.E. (2015). Tracing the 4000 year history of organic thin films: from monolayers on liquids to multilayers on solids. *App. Phys. Rev.* 2: 011101.

Hodes, G. (ed.) (2001). *Electrochemistry of Nanomaterials*. Weinheim: Wiley-VCH.

Maissel, L.I. and Glang, R. (eds.) (1970). *Handbook of Thin Film Technology*. New York: McGraw-Hill.

McShane, M.J. and Lvov, Y.M. (2004). Layer-by-layer electrostatic self-assembly. In: *Dekker Encyclopaedia of Nanoscience and Technology* (ed. J. Schwartz and C. Contescu), 1–20. New York: Marcel Dekker.

Mitsuishi, M., Matsui, J., and Miyashita, T. (2006). Functional organized molecular assemblies based on polymer nano-sheets. *Polym. J.* 38: 877–896.

Nabook, A. (2005). *Organic and Inorganic Nanostructures*. Boston: Artech House.

Paunovic, M. (1998). *Fundamentals of Electrochemical Deposition*. New York: Wiley.

Pierre, A.C. (1998). *Introduction to Sol–Gel Processing*. Boston: Kluwer.

Talham, D.R. (2004). Conducting and magnetic Langmuir–Blodgett films. *Chem. Rev.* 104: 5479–5501.

Wallace, G.G., Spinks, G.M., Kane-Maguire, L.A.P., and Teasdale, P.R. (2002). *Conductive Electroactive Polymers: Intelligent Materials Systems*. Boca Raton, FL: CRC Press.

8

Liquid Crystals and Devices

But, soft! what nymphs are these?

8.1   Introduction

The melting of a solid is normally a sharp transition. As the material is heated, the arrangement of the atoms or molecules abruptly changes from that of the three-dimensional order of the solid to the zero order associated with the liquid, as described in Chapter 2, Section 2.5. However, it is not uncommon for an organic solid to pass through intermediate phases as it heated from a solid to a liquid, depicted in Figure 8.1. The Austrian botanist, Friedrich Reinitzer, is often credited with discovering this new phase of matter, the *liquid crystal phase*, in 1888. Reinitzer observed that, on heating a sample of cholesteryl benzoate, the solid crystal changed into a hazy liquid. As he increased the temperature further, the material changed into a clear, transparent liquid. The compound seemed to possess two melting points!

Organic and Molecular Electronics: From Principles to Practice, Second Edition. Michael C. Petty.
© 2019 John Wiley & Sons Ltd. Published 2019 by John Wiley & Sons Ltd.
Companion website: www.wiley.com/go/petty/molecular-electronics2

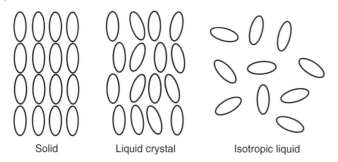

Figure 8.1 Representation of solid, liquid, and liquid crystal phases of rod-shaped molecules.

Solid   Liquid crystal   Isotropic liquid

8.2 Liquid Crystal Phases

Perhaps one in every few hundred organic compounds exhibits liquid crystalline behaviour. These phases are known as *mesophases*. When such a state is formed, the translational order of the solid phase may be lost, but the orientational order remains. Sometimes a mesophase may display translational order but no orientational order; this is termed a *plastic crystal*. A very wide range of liquid crystal phases is now known. These are identified by the degree of short-range translational order and by the shape of the molecules.

8.2.1 Thermotropic Liquid Crystals

Liquid crystals compounds can be *thermotropic* or *lyotropic*. In the former, a phase transition into a liquid crystal occurs as the temperature is changed. In contrast, lyotropic liquid crystals exhibit phase transitions as a function of concentration. Thermotropic liquid crystals can be further subdivided into two main classes: high molecular weight compounds, including both main chain and side chain polymers; and low molecular weight materials, which encompass *calamitic* (rod-like molecules) and *discotic* (disc-like molecules). The classes and their relationships to one another are shown in Figure 8.2. The following section provides an overview of these liquid crystalline mesophases.

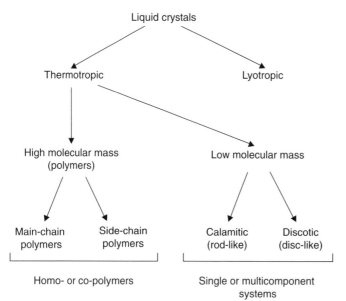

Figure 8.2 Liquid crystal categories.

Nematic Phases

The *nematic* (from the Greek word for thread) phase is the least ordered liquid crystal phase and is exploited extensively in electro-optic applications. This phase has no long-range translational order and only orientational order. There is only one nematic phase and, on heating, this will eventually become an isotropic liquid. A schematic diagram, showing the arrangement of rod-shaped molecules in a nematic phase, is given in Figure 8.3. The molecules in the liquid crystal are free to move about in much the same fashion as a liquid; as they do so they tend to remain orientated in a certain direction. The direction of preferred orientation is called the *director* of the liquid crystal. Each molecule is oriented at some angle θ about the director. The degree of orientational order is then given by the *order parameter S*, which is defined as.

$$S = \frac{1}{2}\langle 3\cos^2\theta - 1 \rangle \tag{8.1}$$

where the angular brackets denote a statistical average. A value of $S = 1$ indicates perfect orientational order, whereas no orientational order results in $S = 0$. Typical values for the order parameter of a liquid crystal range between 0.3 and 0.9, with the exact value a function of temperature, as a result of kinetic molecular motion. Liquid crystals are anisotropic materials, and the physical properties of the system vary with the average alignment of the director. If the alignment is large, the material is highly anisotropic. Similarly, if the alignment is small, the material is almost isotropic. Figure 8.4 shows an example of a nematic liquid crystal material based on a cyanobiphenyl group – the compound 4′-*n*-pentyl-4-cyanobiphenyl. The CN group provides the molecule with a strong electric dipole moment (Chapter 2, Section 2.3.5). The transitions from solid to nematic to liquid take place reversibly on heating and cooling.

Smectic Phases

Smectic (from the Greek word for soap) phases are usually formed by thermotropic liquid crystals at lower temperature than the nematic phase. Besides the orientational order, smectic phases possess one-dimensional translational ordering into layers. The smectic phases can be further subdivided and, at present, 12 different types have been identified. These are designated S_A, S_B, etc., up to S_K; there are two smectic S_B phases – the crystal *B* and *hexatic* S_B phases [1]. Some of these mesophases (crystal *B*, S_E, S_G, S_H, and S_J) have very long-range correlation of position over many layers and are more similar to crystalline solids.

The S_A phase is the least ordered of the thermotropic smectic phases. The molecules are arranged in disordered layers, each layer having a liquid-like freedom of motion of its constituent molecules in two dimensions, with the director perpendicular to the layer planes. By contrast, in the S_C phase, the molecules are tilted from this direction by about 35°. This tilt is correlated between molecules within each layer and from one layer to another. In the hexatic S_B phase, there is again ordering of orientationally aligned molecules into layers. The molecules are arranged in an hexagonal array, but the translational order is short-range only. The orientation of the hexagonal net

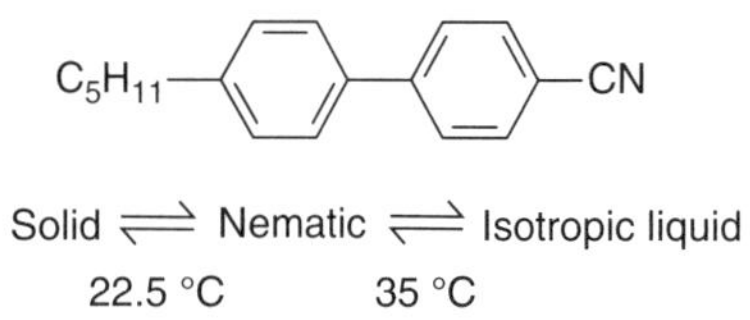

Figure 8.3 Arrangement of rod-shaped molecules in a liquid crystalline phase. The long axis of each molecule makes an angle θ with the director **n**.

$$C_5H_{11}\!-\!\!\bigcirc\!\!-\!\!\bigcirc\!\!-\!CN$$

Solid $\rightleftharpoons$ Nematic $\rightleftharpoons$ Isotropic liquid
  22.5 °C   35 °C

Figure 8.4 Example of an alkylcyanobiphenyl molecule, 4′-*n*-pentyl-4-cyanobiphenyl, that exhibits a nematic liquid crystal phase.

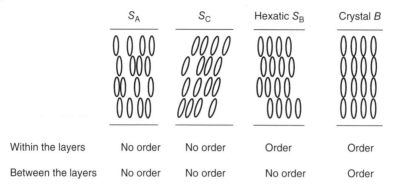

S_A	S_C	Hexatic S_B	Crystal B
Within the layers			
No order	No order	Order	Order
Between the layers			
No order	No order	No order	Order

Figure 8.5 Different smectic liquid crystalline arrangements of rod-shaped molecules. S_C is a tilted version of S_A.

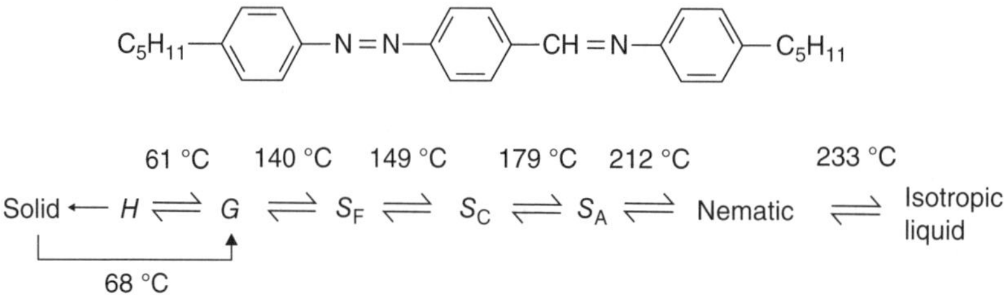

$$C_5H_{11}\text{—}\bigcirc\text{—}N=N\text{—}\bigcirc\text{—}CH=N\text{—}\bigcirc\text{—}C_5H_{11}$$

61 °C   140 °C   149 °C   179 °C   212 °C   233 °C

Solid ⟵ H ⇌ G ⇌ S_F ⇌ S_C ⇌ S_A ⇌ Nematic ⇌ Isotropic liquid

68 °C

Figure 8.6 Multiphase liquid crystal behaviour exhibited by the organic molecule depicted.

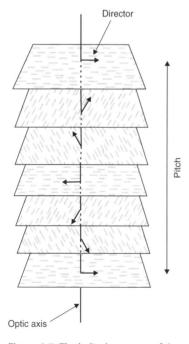

Figure 8.7 The helical structure of the cholesteric (chiral nematic) phase. The director direction (shown by an arrow) gradually rotates from one layer to the next, providing a unique helical structure.

is, however, maintained over a long range and unlike the ordering of molecular positions, is correlated between layers. Two tilted variations of the S_B phase exist in which the tilt direction is constrained to point either towards one face of the hexagonal lattice (S_F) or towards one apex (S_I). The S_A, S_C, hexatic S_B, and crystal B arrangements of rod-shaped molecules are contrasted in Figure 8.5.

A material may show several liquid crystalline phases as it is heated from the solid state to an isotropic liquid. Figure 8.6 shows an example. Most of the phase transitions are observed on both heating and cooling, although the crystal H phase is only seen on cooling the isotropic liquid below the melting point of the compound.

Chiral Phases

The final distinct type of liquid crystalline mesophase is the *cholesteric* or *chiral nematic*. The molecules in such a phase are optically active (Chapter 2, Section 2.4.2) and the optical chiral centre results in a unique helical structure in which the director gradually rotates from one plane of molecules to the next. An important characteristic of the cholesteric mesophase is the pitch. This is defined as the distance it takes for the director to rotate one full turn in the helix, as illustrated in Figure 8.7. A by-product of the helical structure of the chiral nematic phase is its ability to reflect selectively light of wavelengths equal to

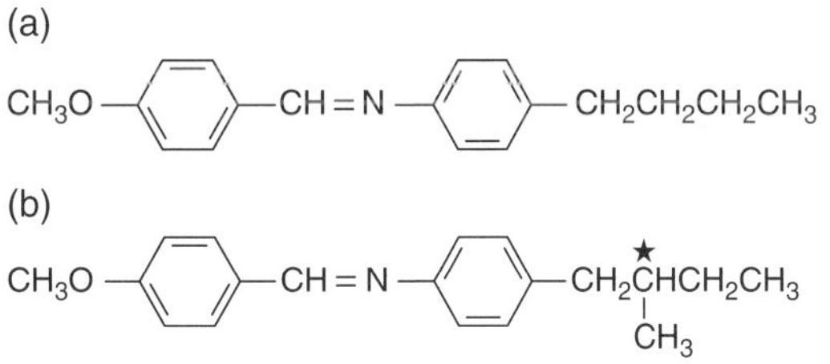

Figure 8.8 (a) An organic compound, 4-methoxylbenzylidene-4′-butylaniline, that exhibits a nematic liquid crystal phase. (b) A similar compound, now incorporating a chiral centre, which shows chiral nematic (cholesteric) behaviour. The chiral centre is denoted by an asterisk.

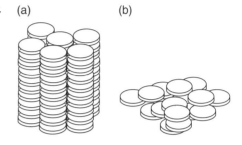

Figure 8.9 Discotic liquid crystal phases. (a) Columnar. (b) Nematic.

the pitch length, so that a colour will be reflected when the pitch is equal to the corresponding wavelength of light in the visible spectrum. An increase in temperature tightens the pitch. This effect is exploited in the liquid crystal thermometer, which displays the temperature of its environment by the reflected colour. Mixtures of various types of these liquid crystals are often used to create sensors with different responses to temperature change. Adjusting the chemical composition can also control the wavelength of the reflected light, since cholesterics can either consist of exclusively chiral molecules or of nematic molecules with a chiral dopant dispersed throughout. In this case, the dopant concentration is used to adjust the chirality and thus the pitch.

The two molecules depicted in Figure 8.8 illustrate the importance of the chiral centre for cholesteric behaviour. The compound shown in Figure 8.8a, 4-methoxylbenzylidene-4′-butyl-aniline, transforms from crystalline to nematic liquid crystal at 20 °C, and from nematic to an isotropic liquid at 74 °C. In contrast, the very similar molecule shown in Figure 8.8b forms a cholesteric liquid crystal. The chiral centre is marked with a *.

Discotic Phases

As their name suggests, discotic liquid crystals consists of disc-shaped molecules. Structurally, most of the mesophases fall into two distinct categories: the *columnar* and the nematic, contrasted in Figure 8.9. The columnar phase, in its simplest form shown in Figure 8.9a, consists of discs stacked one on top of the other to form liquid-like columns, while the nematic phase, Figure 8.9b, has an orientationally ordered arrangement of the discs without any long-range translational order. Two examples of the types of compounds forming discotic liquid crystal phases are shown in Figure 8.10. The molecules possess a fairly rigid, planar centre with hydrocarbon chains emanating in all directions. Such features are common in discotic liquid crystals.

8.2.2 Lyotropic Liquid Crystals

In contrast to thermotropic mesophases, lyotropic liquid crystal transitions occur with the influence of solvents, and not by a change in temperature. Lyotropic mesogens are typically amphiphilic, meaning that they are composed of both *lyophilic* (solvent-attracting) and

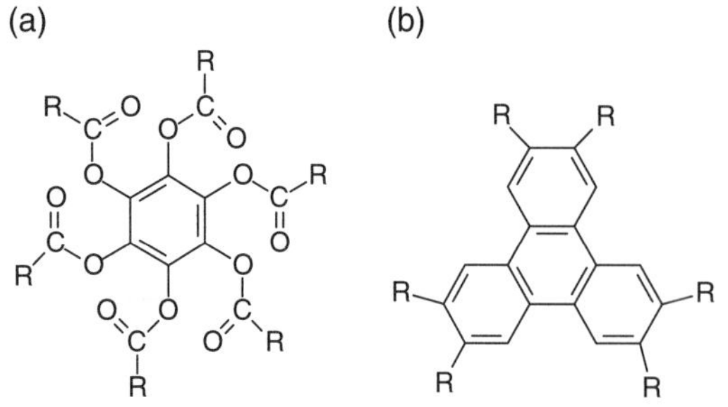

(a) (b)

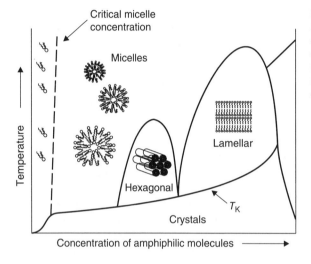

Figure 8.10 Examples of discotic liquid crystalline compounds. (a) Hexa-substituted benzene. (b) Triphenylene.

Figure 8.11 Phase diagram of a typical lyotropic liquid crystal. The nearly vertical dashed line on the left shows the minimum concentration for micelle formation. T_K is the Kraft temperature. Various liquid crystal phases occur in the region close to the 100% concentration axis.

lyophobic (solvent-repelling) parts (in the case of water being the solvent, the terms hydrophilic and hydrophobic are used – Chapter 7, Section 7.3.1). This causes them to form into *micellar* structures (Chapter 12, Section 12.2.5) in the presence of a solvent, since the lyophobic ends will stay together as the lyophilic ends extend outward towards the solution. As the concentration of the solution is increased and the solution is cooled, the micelles increase in size and eventually coalesce. This separates the newly formed liquid crystalline state from the solvent.

The molecules that make up lyotropic liquid crystals are *surfactants*, similar to compounds that form LB films (Chapter 7, Section 7.3.1) or biological membranes (Chapter 12, Section 12.6). Not all surfactants, however, form lyotropic liquid crystals. When dissolved in high enough concentrations, the molecules arrange themselves so that the polar heads are in contact with a polar solvent and/or the nonpolar tails are in contact with a nonpolar solvent.

Figure 8.11 shows a generic phase diagram of a typical lyotropic compound, illustrating the changes in structure that occur as the concentration of amphiphilic molecules increases. In very dilute solutions (far left of the phase diagram), the individual molecules are dispersed in the solvent, which is assumed to be water for the purposes of this discussion. However, as the concentration of the molecules in solution increases, they take on different micelle structures. These are dictated by the packing efficiencies of different shaped objects. For example, the molecules can begin to arrange themselves in hollow spheres, rods, and sheets. The concentration at which micelles form in solution, called the *critical micelle concentration*, is shown as the nearly vertical dashed line in Figure 8.11. If the temperature is too low, the molecules tend to form rigid crystalline structures. The temperature above which crystals do not form, but liquid crystalline structures do develop, is called the *Kraft* temperature – T_K in Figure 8.11. Micelles come in various sizes, but the smallest ones have a diameter about twice as long as the length of a hydrocarbon chain with all-*trans* bonds. As the weight concentration of the amphiphile increases, the micelles become increasingly able to dissolve nonpolar substances. When this occurs, the micelles become large and swollen. If they reach a sufficiently large size, the solution becomes cloudy and becomes an emulsion (Chapter 2, Section 2.7). At lower concentrations, the swollen micelles are not large enough to interfere with light, but they are still extremely stable and exist in equilibrium. This is an example of a microemulsion.

Rod-shaped micelles often form into hexagonal arrays made out of six rods grouped around a central micelle, depicted in Figure 8.11. This hexagonal arrangement offers more efficient packing than can be achieved by using spheres (closely-packed spheres can fill no more than 74% of space, but closely-packed cylinders can fill up to 91% of space). Hexagonal liquid crystals generally exist in solutions that are 40–70% amphiphile. The liquid crystals may come apart if too much water or salt is added to the solution, but many varieties can absorb oil by expanding the diameter of the rod-shaped micelle.

At even higher concentrations, the molecules form another liquid crystalline phase – the lyotropic liquid crystal bilayer (such sheet-like structures can fill up 100% of space). The molecular arrangement is similar to that of smectic liquid crystals in the thermotropic category. Because the sheet-like layers can slide easily past each other, this phase is less viscous than the hexagonal phase, at least in the direction of the sliding, despite its lower water content. Other behaviour can occur when the situation is something other than a simple water solution. If the molecules are placed on the surface of water without actually being dissolved in it, they form a monolayer in which the polar heads are in contact with the water and the hydrophobic tails point into the air (Chapter 7, Section 7.3.1).

If the concentration by weight of amphiphilic molecules is higher than that of water, the molecules form a sort of matrix with water droplets scattered inside, in contact with the polar heads. In the case of molecules dissolved in a nonpolar solvent, their behaviour is similar to that when dissolved in water, except that now the nonpolar tails are in contact with the solvent and the polar heads are isolated in the centres of the micelles and bilayers. If the solution contains both water and a higher concentration of nonpolar solvent, similar *inverse micelles* form with water droplets trapped inside the micelle and nonpolar solvent on the outside (Chapter 12, Section 12.2.5). Finally, if weaker amphiphilic molecules and simple salts are dissolved together in water, they form lyotropic nematic phases.

One very common example of lyotropic liquid crystalline behaviour is household soap. Soaps work better than pure water at removing dirt and grease, because the nonpolar insides of the micelles are capable of dissolving nonpolar substances that will not dissolve in water. (This also works in reverse if the solvent is nonpolar and some of the substance to be removed is polar.) Soaps also help water dissolve more dirt and grease because the molecules tend to remain at the surface, hydrocarbon tails away from the water, thus lowering the surface tension of the water and allowing more material to enter it and be dissolved.

8.3 Liquid Crystal Polymers

Polymer liquid crystals are a class of materials that combine the properties of polymers with those of liquid crystals. Such hybrids show the same mesophases that are characteristic of ordinary liquid crystals, yet retain many of the useful and versatile properties of polymers.

In order for normally flexible polymers to display liquid crystal characteristics, the mesogens must be incorporated into their chains. The placement of the mesogens plays a large role in determining the type of polymer liquid crystal that is formed. *Main-chain polymer liquid*

(a)

(b)

Backbone

Spacer

Mesogen

(c)

Polymer

Mesogen

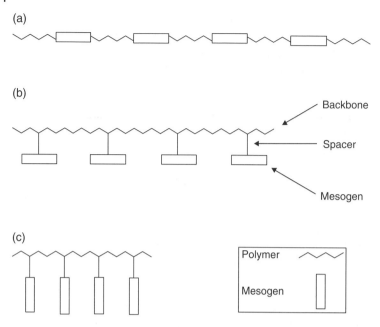

Figure 8.12 Examples of polymer liquid crystals: (a) represents a main chain polymer liquid crystal and (b) and (c) are examples of side chain liquid crystals.

crystals are formed when the mesogens are themselves part of the main chain of a polymer. In contrast, *side-chain polymer liquid crystals* are formed when the mesogens are connected as side chains to the polymer by a flexible 'bridge', called the *spacer*. Figure 8.12 shows examples of main chain, (a), and side chain, (b) and (c), polymer liquid crystals.

Main-chain polymer liquid crystals are formed when rigid elements are incorporated into the backbone of normally flexible polymers. These inflexible regions along the chain allow the polymer to orient in a manner similar to ordinary liquid crystals, and thus display liquid crystal characteristics. There are two distinct groups of materials, differentiated by the manner in which the rigid regions are formed. The first group is characterized by stiff, rod-like monomers; these monomers are typically made up of several aromatic rings. In the second and more prevalent group of main chain polymer liquid crystals, a spacer group separates the mesogens. Decoupling of the mesogens provides for independent movement of the molecules, thereby facilitating proper alignment. Generally, the mesogenic units are made up of two or more aromatic rings, which give the necessary restriction on movement to allow the polymer to display liquid crystal properties. The stiffness necessary for liquid crystallinity results from restrictions on rotation caused by *steric hindrance* (arising from the fact that each atom within a molecule occupies a certain amount of space) and resonance (Chapter 2, Section 2.4.3).

Another characteristic of the mesogen is its *axial ratio*. This is defined as the length of the molecule divided by its diameter. Experimental results have indicated that these molecules must be at least three times long as they are wide. Otherwise, the molecules are not sufficiently 'rod-like' to display the characteristics of liquid crystals.

Side-chain polymer liquid crystals have three major components: the backbone, the spacer, and the mesogen (Figure 8.12b and c). The versatility of such compounds arises because these structures can be varied in a large number of ways. For example, the nature of the backbone can be crucial in determining if the polymer shows liquid crystal behaviour. Polymers with rigid backbones normally have high glass transition temperatures, and thus liquid crystal behaviour is often difficult to observe. Perhaps the most important part of a side-chain polymer liquid crystal is the mesogen. It is the alignment of these groups that gives rise to the liquid crystal behaviour. Figure 8.13 shows the chemical structure of a side-chain liquid crystalline polymer based on a polysiloxane backbone. The spacer unit comprises of methylene units, while the mesogen is composed of aromatic rings. Like their main-chain counterparts, mesogens attached as side groups on the backbone of side-chain polymer liquid crystals are able to orient, because the spacer allows for independent movement. Even though the polymer may be in a tangled conformation, orientation of the mesogens is still possible because of the decoupling action of the spacer. The spacer length has a profound effect on the temperature and type of phase transitions. Usually, the glass transition temperature decreases with increasing spacer length. Short spacers tend to lead to nematic phases, while longer spacers lead to smectic phases.

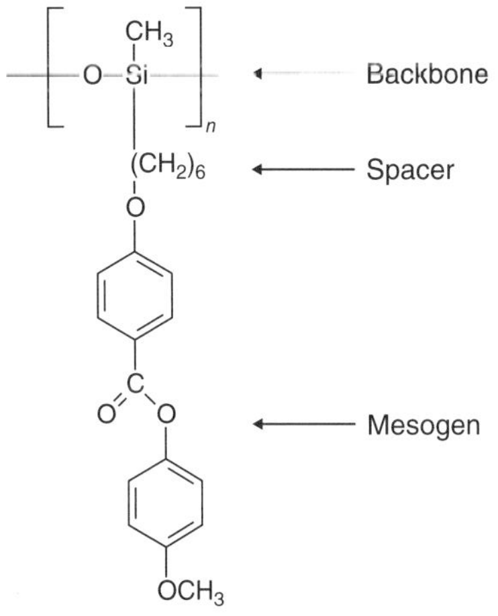

Figure 8.13 Molecular structure of a side-chain polymer liquid crystal based on a polysiloxane backbone.

8.4   Display Devices

In the 1930s, it was suggested that display devices could be based on liquid crystals, where their optical behaviour would be controlled by temperature. However, the early liquid crystals were unstable with small temperature changes, and their behaviour was too erratic for practical use. The breakthrough finally came 40 years later; in the chemistry laboratories of the University of Hull, UK, where George Gray and his colleagues succeeded in making the first really stable liquid crystals. These were the alkylcyanobiphenyls (Figure 8.4), which were cheap as well as stable. Furthermore, their optical properties could be controlled precisely using an electric field rather than heat.

8.4.1   Birefringence

Due to their anisotropic nature, liquid crystals are birefringent (Chapter 4, Section 4.3.1). They possess two indices of refraction, one for light polarized parallel to the director, $n_\parallel$, and the other, $n_\perp$, for light polarized perpendicular to the director. The former gives rise to the extraordinary ray, while the latter is responsible for the ordinary ray propagating through the sample. The birefringence is characterized by the difference in the refractive indices for the ordinary and extraordinary rays. Figure 8.14 shows the temperature dependence of the birefringence of a nematic liquid crystal. At the nematic-liquid transition around 54 °C, the two refractive indices merge into one value, n_i, associated with the isotropic liquid phase.

If an isotropic, transparent material is placed between two polarizers, oriented at 90° to each other (known as *crossed polarizers*), no light will normally emerge because the light from the first polarizer is completely absorbed by the second. Insertion of an isotropic material does not change this situation, because the polarization of light is unchanged as it travels through an isotropic material. The polarized light propagating through a liquid crystal may be considered to comprise of two components – light polarized along the director and light polarized perpendicular to the director. As the radiation propagates through the material, these two polarizations become out of phase and emerge from the material as elliptically polarized light. The electric

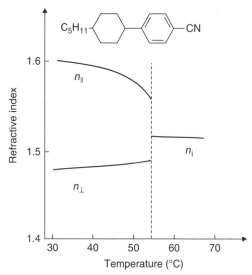

Figure 8.14 Temperature dependence of the birefringence of the liquid crystal molecule shown. The upper curve shows the extraordinary refractive index $n_{\parallel}$ while the lower curve gives the ordinary refractive index $n_{\perp}$. At the nematic liquid transition of around 54 °C, both refractive indices merge into that of the isotropic liquid n_i. *Source:* Reprinted from Zorn and Wu [4]. Copyright (2005), with permission from Wiley-VCH.

vector of such light is constantly rotating during each cycle of the electromagnetic (EM) radiation. Therefore, it will be parallel to the polarization axis of the second polarizer twice during each cycle and some light will emerge from it. The introduction of a liquid crystal between two polarizers will therefore cause the field of view to appear bright (unless the incident polarized light has its polarization direction either parallel or perpendicular to the director). Devices that exploit this effect are called *phase retarders* and are found in many optical applications. If the thickness of the birefringent material is carefully adjusted to cause a 90° change in the phase difference, light that is linearly polarized upon entering the material emerges as circularly polarized light. Similarly, incident circularly polarized light emerges as linearly polarized light.

8.4.2 Freedericksz Transition

The response of liquid crystal molecules to an electric field is widely used in display applications. The ability of the director to align along an external field is caused by the electric nature of the molecules. Permanent electric dipoles result when one end of a molecule has a net positive charge while the other end has a net negative charge (Chapter 2, Section 2.3.5). When an external electric field is applied to the liquid crystal, as depicted in Figure 8.15, the dipole molecules attempt to orient themselves along the direction of the field. This tendency is, of course, opposed by thermal motion.

Even if a molecule does not form a permanent dipole, it can still be influenced by an electric field. In some cases, the field produces slight re-arrangement of electrons and protons in molecules such that an induced electric dipole results. While not as strong as permanent dipoles, orientation with the external field still occurs. The effects of magnetic fields on liquid crystal molecules are analogous to electric fields. When a magnetic field is applied, the molecules will tend to align with or against the field. Liquid crystals are very sensitive to magnetic fields, achieving complete alignment of the director for magnetic fields of relatively low strength.

In the absence of an external field, the director of a liquid crystal is free to point in any direction. It is possible, however, to force the director to point in a specific direction by introducing an external

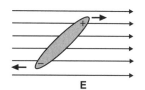

Figure 8.15 Effect of applied electric field on a dipolar molecule.

Figure 8.16 (a) Planar (or homogeneous) alignment and (b) homeotropic alignment of liquid crystal molecules between parallel plates.

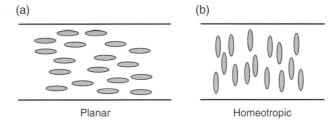

influence other than an electric or magnetic field. For example, when a thin polymer coating, usually a polyimide, is spread on a glass substrate and rubbed in a single direction with a cloth, it is observed that the liquid crystal molecules in contact with that surface align with the rubbing direction. The mechanism for this is believed to be an epitaxial growth of the liquid crystal layers on the partially aligned polymer chains in the near surface layers of the polyimide.

If mesogenic materials are confined between closely spaced plates with rubbed surfaces (as described above) and oriented with rubbing directions parallel, the entire liquid crystal sample can be oriented in a *planar* or *homogeneous texture*, as shown in Figure 8.16a. Mesogens can also be oriented normal to a surface with the use of appropriate surface layers (e.g. trichlorosilanes incorporating a long aliphatic chain) or in the presence of an electric field applied normal to the surface, giving rise to the *homeotropic texture*, as illustrated in Figure 8.16b.

The competition between orientation produced by surface anchoring and by electric field effects is exploited in many liquid crystal devices. Consider the case in which liquid crystal molecules are aligned parallel to the surface and an electric field is applied perpendicular to the cell. At first, as the electric field increases in magnitude, no change in alignment occurs. However, at a threshold magnitude of electric field, deformation occurs where the director changes its orientation from one molecule to the next. The occurrence of such a change from an aligned to a deformed state is called a *Freedericksz transition* and can also be produced by the application of a magnetic field of sufficient strength. An important aspect of the Freedericksz transition is that typical values for the threshold field (electric or magnetic) are quite modest, approximately $4 \times 10^4 \, \mathrm{V \, m^{-1}}$ for the electric field (which can be achieved by the application of 1 V to a sample 25 μm in thickness) and about 0.2 T in the case of the magnetic field. This is the result of the relative freedom that liquid crystal molecules possess as they diffuse throughout the sample. The Freedericksz transition is fundamental to the operation of many liquid crystal displays, because the director orientation (and thus the properties) can be controlled easily by the application of external fields. Liquid crystal displays require an AC drive voltage. Prolonged DC operation may cause electrochemical reactions inside the displays which cause significantly reduced life.

8.4.3 Twisted Nematic Display

The simplest liquid crystal display device is the *twisted nematic display*, developed in the early 1970s. Figure 8.17 shows a schematic diagram of its construction. The liquid crystal is sandwiched between two glass plates, which have each been coated with a transparent conductive coating such as indium tin oxide. These form the conductive electrodes across which a voltage may be applied. Polymer spacer beads maintain a uniform gap between the sheets of glass and the edges are sealed with an epoxy resin.

In a twisted nematic display, alignment layers are positioned with their rubbing directions perpendicular to each other and polarizers are applied to match the orientation of the alignment layers. The arrangement is shown in Figure 8.18. With no electric field applied between

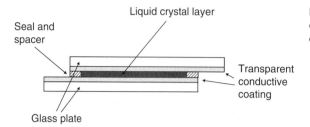

Seal and spacer

Liquid crystal layer

Glass plate

Transparent conductive coating

Figure 8.17 Schematic diagram of the construction of a twisted nematic liquid crystal cell.

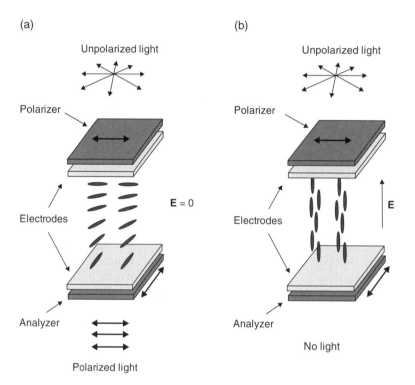

(a)

Unpolarized light

Polarizer

Electrodes

$E = 0$

Analyzer

Polarized light

(b)

Unpolarized light

Polarizer

Electrodes

E

Analyzer

No light

Figure 8.18 Twisted nematic cell. (a) With no applied field. (b) Electric field **E** applied perpendicular to the cell plates.

the electrodes, there will be no light transmitted, as the twist in the alignment of the liquid crystal molecules will rotate the plane of polarization of the incoming polarized light by 90°. However, when a voltage above the threshold value for the Freedericksz is applied to the cell, there will be a torque on the liquid crystal molecules, tending to align them parallel to the field. Although the molecules immediately adjacent to the glass surfaces will retain their parallel order (this feature is not shown in Figure 8.18), the majority of the liquid crystal molecules will align with the electric field. The polarization direction of the light crossing the cell will now only be rotated very slightly and the cell will transmit almost no light.

A twisted nematic polymer liquid crystal cell can also be used to make energy efficient displays. However, the response times of these devices can be relatively slow. Where a fast response is not needed, a laser may be used to melt selectively portions of the display into the liquid crystal phase. Applying a field across it, just as in an ordinary twisted nematic liquid crystal cell, then chooses the orientation of the cell. When the polymer cools down and hardens into a glass, the mesogens will be locked in that configuration and the field can be turned off.

For a display to function, it must have a light source. A twisted nematic display can be operated in either a reflection or a transmission mode. In the former, a reflective layer forms the base layer of the cell and the applied voltage controls the amount of ambient light that is reflected back to the viewer. The liquid crystal displays found in most calculators and watches are of this type. These displays are not very bright because the light must pass through multiple polarizers, which severely reduce its intensity, in addition to the various layers of the display that are only semi-transparent. Therefore, a more intense source is employed in the form of a *back lighting* system. Lights (e.g. incandescent bulbs or a solid state source) mounted behind and at the edges of the display replace the reflected ambient light. This results in brighter displays for two reasons: the light does not have to pass through the display and therefore does not lose intensity; and the lighting system can be made more intense than ambient light. Although back lighting has the disadvantage of being very power intensive, it is found widely in the more complex displays such as laptop computer screens.

8.4.4   Passive and Active Addressing

The information on a twisted nematic device can be displayed either as a series of dots or in the form of a *seven-segment array* (common in clocks and calculators). In the latter case, the seven segments can be used to show numeric and alphabetical characters (by choosing which segments to turn on). *Addressing* is the process by which individual display elements, known as *pixels*, are turned on and off in order to create an image. There are two main types of addressing, direct and multiplexing. Direct addressing is convenient for displays where there are only a few elements that have to be activated, such as the seven-segment display. With direct addressing, each pixel in the display has its own drive circuit. A microprocessor must individually apply a voltage to each element.

For a large laptop computer display with $768 \times 1024 \approx 8 \times 10^5$ pixels (*XGA* or eXtended Graphics Array – an IBM display standard introduced in 1990), it is impossible to realize such a large number of drive circuits. The individual pixels are therefore addressed as a matrix through rows and columns. For example, for a 100×100 matrix of pixels, 10^4 drivers are required for direct addressing in contrast to only 200 for multiplex addressing. This *passive matrix display* is addressed by a set of multiplexed transparent electrodes, perpendicular to one another, above and below the liquid crystal layer in a row and column formation (similar to the cross-bar arrangement used for memory devices and described in Chapter 11, Section 11.4). A passive pixel is addressed when there is a sufficient voltage across it to cause the liquid crystal molecules to align parallel to the electric field. A display can have more than one pixel on at any one time because of the response time of the liquid crystal material. When addressed, a pixel has a short turn-on time during which the liquid crystal molecules align in such a way as to make the pixel opaque. When the voltage is removed, the pixel behaves in a similar fashion to the discharge of a capacitor, slowly turning off as charge dissipates and the molecules return to their undeformed orientation. Because of this response time, a display can scan across the matrix of pixels, turning on the appropriate ones to form an image. As long as the time to scan the entire matrix is shorter than the turn-off time, a multiple pixel image can be displayed.

In the case of *active matrix displays*, the addressing takes place completely behind the liquid crystal film. The front surface of the display is coated with a continuous electrode, while the rear surface electrode is patterned into individual pixels. As shown in Figure 8.19, a transistor (a thin film transistor) acts as a switch for each pixel. Electrical contacts to the transistor are made using a set of narrow multiplexed electrodes (gate lines and source lines) running along the gaps between pixels. A pixel is addressed by applying a voltage to a gate line which switches the transistor on and allows charge from the source line to flow to the rear electrode. A voltage

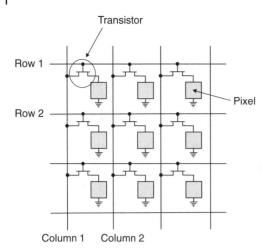

Row 1

Transistor

Row 2

Pixel

Column 1   Column 2

Figure 8.19 Active matrix display diagram. Each liquid crystal cell, or pixel, is turned on or off by a transistor.

is thereby established across the pixel, which turns on. An image is created in a similar fashion to the passive display, as the addressing circuitry scans across the matrix. An active matrix display does not suffer from many of the limitations of the passive display. It can be viewed at an angle of up to 45° and can have a contrast of 40 : 1, meaning that the brightness of an 'on' pixel is 40 times greater than an 'off' pixel. It does, however, require a more intense back lighting system, because the transistors and the gate and source lines are not very transparent.

8.4.5 Full-colour Displays

To achieve a colour display, it is first necessary to have a device that is black in one state and white in the other. In a white display, all wavelengths pass through and therefore, all wavelengths can be manipulated to create the desired colour. To get full colour, each individual pixel is divided into three subpixels: red, green, and blue. So, for each full colour pixel, three distinct pixels are used. These subpixels are formed by using colour filters that only allow certain wavelengths to pass through them, while absorbing the rest. With a combination of red, blue, and green subpixels of various intensities, a pixel can be made to produce any number of different colours. This is analogous to a colour cathode ray tube, such as used in a television or computer monitor, in which different phosphors glow red, green, or blue when excited by an electron beam. The number of colours that can be made by mixing red, green, and blue subpixels depends on the number of distinct *grayscales* (intensities) that can be achieved by the display.

8.4.6 Super-twisted Nematic Display

The simple twisted nematic display has a number of drawbacks for high density information display devices. For example, the difference between the 'on' and 'off' voltages in displays with many rows and columns can be very small. For this reason, the simple twisted nematic device is impractical for large information displays with conventional addressing Schemes. A significant improvement may be achieved using a *super-twisted nematic display*. Here, the alignment layers are placed with their rubbing directions at a variety of angles to one another to set up a twist from 180 to 270°, and the polarizers are not applied parallel to the alignment layers. Figure 8.20 shows the calculated dependence of the mid-cell tilt angle on the applied voltage [2]. For small twist angles, the molecules all remain approximately parallel to the glass plates of the cell until the threshold voltage is exceeded; the tilt then rises slowly as the applied voltage

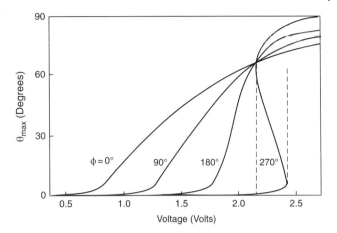

Figure 8.20 Theoretical dependence of the mid-cell angle θ_{max} of a twisted nematic liquid crystal cell on the applied voltage. The different curves represent different total twist angles ϕ. *Source:* Reprinted from Raynes [2]. Copyright (1986) with permission from Taylor & Francis/Gordon & Breach (US).

increases. As the twist increases, the voltage dependence of the mid-cell angle on voltage becomes much steeper, leading to improved multiplexing capabilities. Beyond about 240° a bistable situation even occurs (where the tilt increases with decreasing voltage). However, one disadvantage of the super-twisted display arises from the wavelength dependency of the transmitted light. The result is that most displays of this type are coloured, e.g. a pale blue or pale yellow/green background. As noted in the previous section, black and white operation, i.e. no colour dependency, is a necessary condition if a display is to achieve true full-colour operation. The best, but most expensive, solution is to use optical compensation in the form of a double super-twisted nematic cell. Here, two layers of a super-twisted nematic are used, but with opposite twists. In the 'off' state, the second layer compensates the phase shift resulting from the first layer. This pixel appears black. The second super-twisted nematic layer does not affect the 'on' state, and white light emerges. Since the two layers consist of the same liquid crystal material, the behaviour is constant over the entire temperature range.

8.5 Ferroelectric Liquid Crystals

The requirements for a material to exhibit ferroelectricity are that it must possess a permanent polarization and that the polarization direction can be changed using an applied electric field (Chapter 5, Section 5.6). The smectic *C* phase of liquid crystals first appeared in literature in 1933. However, it was not until 1974 that it was realized that the phase ought to be ferroelectric. This was discovered by Robert Meyer, who later demonstrated it in a synthesized chiral smectic C material DOBAMBC, *p*-decyloxybenzylidene-*p*'-amino-2-methylbutylcinnamate, the chemical structure of which is shown in Figure 8.21.

In the S_A phase, Figure 8.5, the molecules are upright, and since there is no head-to-tail ordering, there is no polarization normal to the layers. Moreover, even if the molecules themselves are chiral, there is equal probability of their assuming any orientation about their long

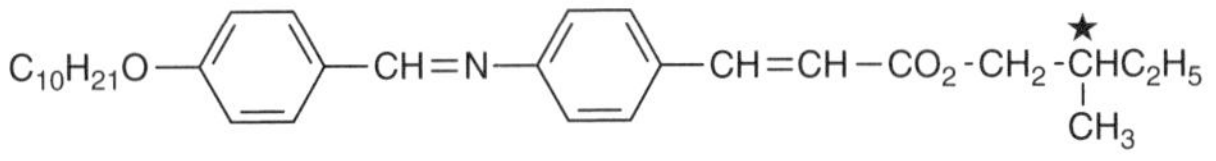

$$C_{10}H_{21}O-\bigcirc-CH=N-\bigcirc-CH=CH-CO_2-CH_2-\overset{\star}{\underset{CH_3}{C}}HC_2H_5$$

Figure 8.21 The structure of DOBAMBC, *p*-decyloxybenzylidene-*p*'-amino-2-methylbutylcinnamate, a ferroelectric liquid crystal. The chiral centre is marked with an asterisk.

axes. Hence, the transverse component of the dipole moment is averaged out and there is also no net polarization parallel to the layers.

In the chiral S_C phase (S^*_C), the molecules are arranged as for an ordinary S_C phase However, if the molecules possess a lateral dipole, then this will lie parallel to the layer plane but perpendicular to the tilt plane. Thus, in the S^*_C phase, each layer is spontaneously polarized. Since the structure has a twist about the layer normal, the polarization direction will rotate from one layer to the next, resulting in an average dipole of zero. A bulk S^*_C sample, free to develop its helical structure, will therefore not show ferroelectric behaviour, since the spontaneous polarization will average to zero over one pitch (since polarization vectors go around an entire circle and cancel each other out). This is often referred to as the *helielectric phase*.

In 1980, Clark and Lagerwall proposed a way to suppress the helix and developed the surface stabilized ferroelectric liquid crystal [3]. The helix is constrained by using a cell gap that is less than the helical pitch. The smectic layers are oriented approximately perpendicular to the glass bounding plate. Interaction forces between the liquid crystal and the bounding plates unwind the intrinsic helix. Symmetry arguments show that this boundary condition also causes the molecular orientation for each layer to be the same and the material exhibits ferroelectric behaviour. The director likes to lie in the plane of the bounding plates. Because of this condition and the fact that the director is constrained to be at a certain angle from the normal to the layer (i.e. to lie on the intersection of a cone and the bounding plate), there are two stable states, as shown in Figure 8.22. The polarization vector, therefore, must be normal to the bounding plates and its two states are in opposite directions. The values of the spontaneous polarization in these materials are quite small, usually between 10 and $1000\,nC\,cm^{-2}$, i.e. one or two orders of magnitude less than that for a solid inorganic ferroelectric such as KH_2PO_4. The switching times for ferroelectric liquid crystal devices can be just a few tens of microseconds, about 1000 times faster than the twisted nematic device. This is because (a) the ferroelectric liquid crystal possesses ordered permanent electric dipoles and (b) the electric field forces both the on and the off transitions.

The bistability offered by surface stabilized ferroelectric liquid crystal devices makes them ideal where low energy consumption is important for still images. This is because additional power is not required once the image is created. The switching time is still fast enough to provide high frame rates required for video and also allows full colour on each pixel, giving a higher quality on a display of a given size. A further advantage of ferroelectric liquid crystal displays is the small pixel size, made possible by the thin cells. With a size of approximately $5\,\mu m$, the pixels can be made a factor of ten smaller than those in nematic displays. This is significantly better than required for computer displays, e.g. in an XGA 768×1024 display of 14 in. diagonal, the subpixel representing a single colour is $90\,\mu m$ wide [4]. Ferroelectric displays are therefore very suitable for microdisplay applications, e.g. projection or head-mounted displays.

Perhaps the main drawback of the liquid crystal display is that the molecular orientation may be lost by mechanical stress during the use of the display. Experience with an ordinary nematic display reveals that if the display is gently pressed, it will lose its function momentarily because the twisted structure is distorted. This may lead to unwanted colours or even a completely black display. However, within a short time the display will again function correctly. Unfortunately,

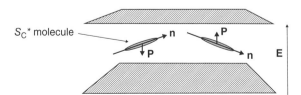

Figure 8.22 Surface-stabilized ferroelectric liquid crystal. Between closely spaced plates only two orientations are possible, with the directors **n** oriented into or out of the page. The corresponding polarizations **P** are down and up, respectively.

this is not true for a ferroelectric display. As soon as regions, or domains, are formed which do not have a correct orientation, they will remain stable.

In the late 1980s, a different arrangement of the molecules in their layer plane was discovered. The phase is known as the *antiferroelectric liquid crystal* (AFLC) phase. This phase occurs in some materials at a temperature below the ferroelectric liquid crystal (FLC) phase. These materials, like FLCs, are chiral and possess a spontaneous polarization. The difference is that, in the AFLC phase, the director is tilted in the opposite direction in alternate layers (see antiferromagnetic behaviour in Chapter 5, Section 5.7). For AFLCs, the pitch is the distance for the director to precess 180° instead of 360° as for FLCs – due to the opposite tilt in adjacent layers, the director also has gone around half of the cone. As for the FLC, the AFLC helix must be unwound through a boundary constraint for the material to be used in displays.

8.6   Polymer-dispersed Liquid Crystals

Polymer-dispersed liquid crystals represent a class of materials that may have applications ranging from switchable windows to projection displays. They consist of liquid crystal droplets that are dispersed in a solid polymer matrix. The liquid crystal director inside each droplet adopts one of two configurations, depending on whether the liquid crystal prefers to align parallel or perpendicular to the polymer surface; the two arrangements are illustrated in Figure 8.23. In the case of parallel alignment, two points where the director is undefined occur; these are called *disclinations*, by analogy with dislocations in crystals (Chapter 2, Section 2.5.7). Perpendicular alignment causes only one disclination to form at the centre of the droplet.

For the case of parallel alignment, a random orientation is expected from the directors in the liquid crystal droplets. Light passing through each droplet will travel at one velocity for light polarized parallel to the director (determined by the extraordinary refractive index $n_{\parallel}$) and at a different velocity for light polarized perpendicular to the director (determined by the ordinary refractive index $n_{\perp}$). The solid polymer is isotropic, possessing a single refractive index n_s, which, in general, will be different from either refractive index of the liquid crystal. This situation is depicted in Figure 8.24a. The difference in refractive indices between the polymer and the liquid crystal results in a structure that scatters light very effectively.

If an electric field is now applied to the polymer-dispersed liquid crystal, the directors will try to orient with the field, as shown in Figure 8.24b. Light passing through the cell now has its electric field nearly perpendicular to the director, so if $n_{\perp}$, the ordinary refractive index of the liquid crystal, is made equal to the polymer index n_s, little reflection occurs at the boundary of each droplet (Chapter 4, Section 4.5.2). The polymer-dispersed liquid crystal then appears clear.

Figure 8.23 (a) Polymer-dispersed liquid crystal with droplets of liquid crystal distributed throughout the polymer. (b) Two possible configurations for the directors, shown as lines, with respect to the polymer surface: top, parallel alignment; bottom, perpendicular alignment.

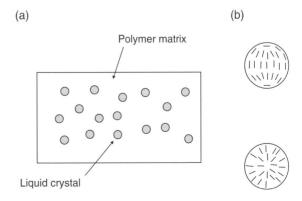

(a)

(b)

Polymer matrix

Liquid crystal

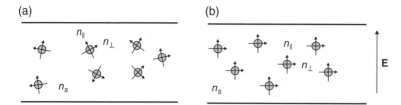

Figure 8.24 Reorientation of the director within a liquid crystal droplet in a polymer-dispersed liquid crystal. $n_∥$ and $n_⊥$ are the extraordinary and ordinary refractive indices of the liquid crystal and n_s is the refractive index of the polymer. (a) With no applied electric field, the directors within the individual liquid crystal droplets are unaligned and incident light is scattered. (b) Application of an electric field aligns the directors and the cell appears clear if $n_⊥$ and n_s are equal.

It is relatively easy to produce polymer-dispersed liquid crystals in large areas. As a consequence, their cost is low. Applications for these materials range from switchable windows to displays. The main drawback to their widespread application originates from incomplete alignment of the liquid crystal molecules under an applied electric field. Figure 8.23 reveals that the liquid crystal molecules close to the droplet surfaces tend to orient parallel to the surface and not always to the field. This results in a residual scattering that is manifest by cloudiness in window applications.

8.7 Liquid Crystal Lenses

The ability of liquid crystals to control the phase of a light beam, as well as its intensity, has led to their exploitation in electrically switchable optical elements such as lenses. A conventional fixed lens can be thought of as a device that produces a phase shift across the incident beam, as a function of distance. It achieves this by a medium of constant refractive index (usually glass) with a variable thickness across the lens. A liquid crystal lens can produce the same result by keeping the physical thickness constant and by tuning the refractive index.

Most liquid crystal lenses are made up of a number of individual pixels, as in the case of displays. These can be in the form of a rectangular grid, or as concentric rings, in order to realize a circular lens. The desired shape of the wavefront is then approximated by controlling the phase of the individual pixels, illustrated in Figure 8.25. Alternatively, modal addressing may be used [5]. This removes the need for individual pixels and exploits the fact that a liquid crystal cell can be modelled as a distributed electrical circuit, similar to a transmission line. By using a relatively high resistance transparent top electrode, and applying a voltage to via low resistance contact electrode surrounding it, the voltage across the liquid crystal cell, and thereby the

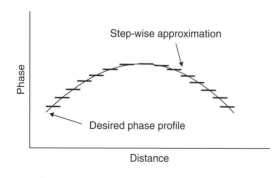

Figure 8.25 Approximation of a continuous phase profile (full line) with zonal (pixelated) liquid crystal elements.

(a)　　　　　　　　(b)　　　　　　　　(c)

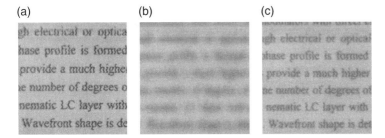

Figure 8.26 Example of a liquid crystal lens in operation. (a) Lens is off and the optical imaging system is adjusted to provide a good focus. (b) Optical system is mechanically adjusted to defocus the image. (c) Liquid crystal lens is turned on to refocus the image. *Source:* Reprinted from Love and Maumov [6]. Copyright (2001), with permission of Taylor & Francis Ltd. http://informaworld.com

phase profile, will vary with position. Such lenses are simple to make and can produce a smooth change in focal length. The lens will respond to an applied ac field. The frequency of this field defines the magnitude of the voltage across the liquid crystal, which in turn shapes the electro-optic response. Figure 8.26 shows the effect of focusing such a lens [6]. A simple imaging system is set up using fixed lenses to focus some text onto a camera (Figure 8.26a). The liquid crystal lens is placed in front of the system, but turned off. One of the fixed lenses is then displaced to defocus the image (Figure 8.26b) and this is subsequently corrected using the liquid crystal lens (Figure 8.26c).

Smart lenses based on hydrogels (Chapter 2, Section 2.7) that adapt to the environment have also been demonstrated [7]. For example, the focal length of a lens can be made to change as the result of a biological reaction.

8.8　Other Application Areas

The most common application of liquid crystal technology is liquid crystal displays, which have been discussed in some detail above. Other applications noted above have included liquid crystal thermometers, phase retarders, electro-optic switches, and lenses. Side-chain polymer liquid crystals exhibit good properties for applications in optically nonlinear devices, including optical waveguides and electro-optic modulators in poled polymeric slab waveguides. Other devices that may be fabricated from polymer liquid crystals include optically-addressed spatial light modulators, tuneable notch filters, optical amplifiers, and laser beam deflectors. The properties of ferroelectric chiral smectic *C* phases make this material useful for films with applications in nonlinear optics.

One area that is being explored is optical imaging and recording. In this technology, a liquid crystal cell is placed between two layers of photoconductor. Light is applied to the photoconductor, which increases its conductivity. This causes an electric field to develop in the liquid crystal corresponding to the intensity of the light. The electric pattern can be transmitted by an electrode, which enables the image to be recorded.

Liquid crystals have been put to many other uses. They are used for non-destructive mechanical testing of materials under stress. This technique is also used for the visualization of *RF* (radio frequency) waves in waveguides. They are exploited in medical applications where, for example, the transient pressure transmitted by a foot walking on the ground is measured. Low molar mass liquid crystals have applications including erasable optical disks, full colour 'electronic slides' for computer-aided drawing (CAD), and light modulators for colour electronic imaging.

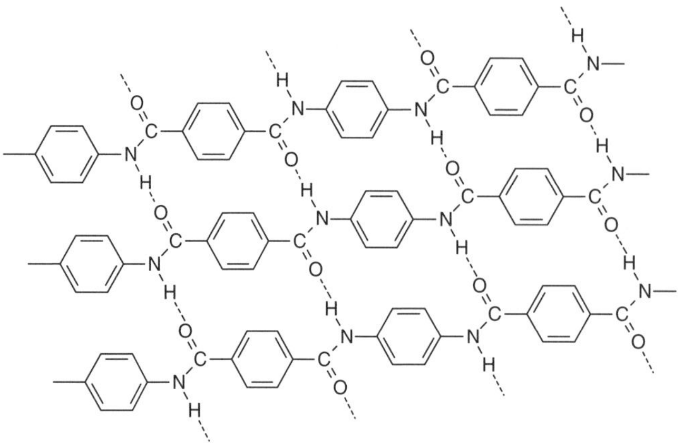

Figure 8.27 Chemical structure of Kevlar, poly(*p*-phenylene terephthalamide). Hydrogen bonding is depicted by dashed lines.

A few application areas focus on the alignment of molecules in polymer liquid crystals. Ordinary polymers have never been able to demonstrate the stiffness necessary to compete against traditional materials like steel. It has been observed that polymers with long straight chains are significantly stronger than their tangled counterparts. Main chain liquid crystal polymers are well suited to ordering processes. For example, the polymer can be oriented in the desired liquid crystal phase and then quenched to create a highly ordered, strong solid. One success story is the development of *Kevlar*, poly(*p*-paraphenylene terephthalamide). This polymer is a synthetic fibre that is five times stronger than steel, weight for weight. Kevlar is very heat resistant and decomposes above 400 °C without melting. It is commonly used in bulletproof vests, in extreme sports equipment, and for composite aircraft construction. It is also used as a replacement for steel cords in car tyres, in fire suits, and as an asbestos replacement.

Kevlar was invented by the DuPont Corporation in the early 1960s, following the work of Stephanie Kwolek. Chemically, Kevlar is an *aramid* liquid-crystalline polymer, which means an aromatic nylon (a nylon containing benzene rings). The chemical structure of Kevlar is shown in Figure 8.27. It derives its strength from intra-molecular hydrogen bonds and aromatic stacking interactions between aromatic groups in neighbouring strands. These interactions are much stronger than the van der Waals interactions found in other synthetic polymers and fibres. Kevlar consists of relatively rigid molecules, which form a planar sheet-like structure similar to silk protein. These properties result in its high mechanical strength and its remarkable heat resistance. Because it is highly unsaturated (Chapter 2, Section 2.4.3), it has a low flammability.

Kevlar molecules have polar groups accessible for hydrogen bonding. Water that enters the interior of the fibre can take the place of bonding between molecules and reduce the material's strength. The groups exposed at the surface lead to good wetting properties. This is important for bonding the fibres to other types of polymer, forming a fibre reinforced plastic. This property also makes the fibres feel more natural and 'sticky' compared to nonpolar polymers like polyethylene. In structural applications, Kevlar fibres can be bonded to one another or to other materials to form a composite.

Problems

8.1 The bright colours found on some insect wings are due to the existence of a thin membrane containing a chiral nematic liquid crystal on their surfaces. Keeping in mind that the light will be reflected by their wings rather than transmitted through them, how do these colours occur?

8.2 Why do long-chain liquid crystal molecules tend to line up in parallel? Why should such molecules align with an electric field?

8.3 In a twisted nematic liquid crystal display, what limits how quickly an electric field can cause the realignment of liquid crystal molecules? When the electric field is turned off, what will determine the time taken for the molecules to return to their twisted configuration?

8.4 Sodium ions diffuse from the glass at the boundaries into a twisted nematic liquid crystal display. Explain how the cell may now lose contrast. Suggest a means to counteract this mode of failure.

8.5 Which of the following molecules is likely to be a calamitic liquid crystal?

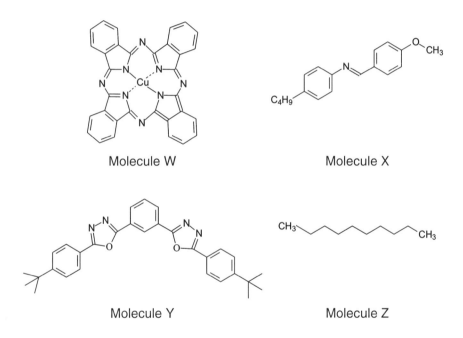

Molecule W   Molecule X

Molecule Y   Molecule Z

References

1 Tredgold, R.H. (1994). *Order in Thin Organic Films*. Cambridge: Cambridge University Press.

2 Zorn, R. and Wu, S.-T. (2005). Liquid crystal displays. In: *Nanoelectronics and Information Technology*, 2nde (ed. R. Waser), 889–909. Weinheim: Wiley-VCH.

3 Raynes, E.P. (1986). The theory of supertwist transitions. *Mol. Cryst. Liq. Cryst. Lett.* 4: 1–8.

4 Clark, N.A. and Lagerwall, S.T. (1980). Submicrosecond bistable electro-optic switching in liquid crystals. *Appl. Phys. Lett.* 36: 899–901.

5 Naumov, A.F. and Love, G. (1999). Control optimization of spherical modal liquid crystal lenses. *Opt. Express* 4: 344–352.

6 Love, G.D. and Maumov, A.F. (2001). Modal liquid crystal lenses. *Liq. Cryst. Today* 10: 1–4.

7 Jiang, H. and Dong, L. (2006). Smart lenses. *Phys. World* 19: 29–31.

Further Reading

Chandrasekhar, S. (1992). *Liquid Crystals*, 2e. Cambridge: Cambridge University Press.

Chen, R.H. (2011). *Liquid Crystal Displays: Fundamental Physics and Technology*. Wiley: Hoboken, NJ.

Choudhury, P.K. (ed.) (2013). *New Developments in Liquid Crystals and Applications*. Hauppauge, NY: Nova Science Publishers.

Collins, P.J. (1990). *Liquid Crystals: Nature's Delicate Phase of Matter*. Bristol: Adam Hilger.

Collins, P.J. and Hird, M. (1997). *Introduction to Liquid Crystals: Chemistry and Physics*. London: Taylor & Francis.

De Gennes, P.G., Prost, J., and Prost, J. (1995). *The Physics of Liquid Crystals*. Oxford: Clarendon Press.

Demus, D., Goodby, J., Gray, G.W. et al. (eds.) (1999). *Physical Properties of Liquid Crystals*. Weinheim: Wiley-VCH.

Kawamoto, H. (2002). The history of liquid-crystal displays. *Proc. IEEE* 90: 460–500.

Khoo, I.-C. and Simoni, F. (eds.) (1991). *Physics of Liquid Crystalline Materials*. Philadelphia: Taylor & Francis.

Lacey, D. (1995). Liquid crystals and devices. In: *An Introduction to Molecular Electronics* (ed. M.C. Petty, M.R. Bryce and D. Bloor), 185–219. London: Edward Arnold.

Lagerwell, S.T. (1999). *Ferroelectric and Antiferroelectric Liquid Crystals*. Wiley-VCH: Weinheim.

Singh, S. (2003). *Liquid Crystals: Fundamentals*. Singapore: World Scientific Publishing.

9

Plastic Electronics

Organic and Molecular Electronics: From Principles to Practice, Second Edition. Michael C. Petty.
© 2019 John Wiley & Sons Ltd. Published 2019 by John Wiley & Sons Ltd.
Companion website: www.wiley.com/go/petty/molecular-electronics2

Sell when you can, you are not for all markets

9.1 Introduction

Since the discovery of semiconducting behaviour in organic materials, there has been a considerable research effort aimed at exploiting this property in electronic and optoelectronic devices. Organic semiconductors can have significant advantages over their inorganic counterparts. For example, thin layers of polymers can easily be made by low-cost methods such as spin-coating or dip-coating. Appropriate processing allows organic thin films to be produced in large areas, and sometimes in the form of freestanding films. High temperature deposition from vapour reactants is generally needed to form thin films of inorganic semiconductors. The synthetic organic chemistry also offers the possibility of designing new materials with different band gaps, important for applications in light-emitting devices and photovoltaic cells. The mobilities of the charge carriers in organic field effect transistors are low when compared to inorganic semiconductors such as Si and GaAs (Chapter 3, Section 3.4.1). Nevertheless, the simple fabrication techniques for organic materials have attracted companies to work on transistor applications, such as data storage and thin film arrays to address displays.

9.2 Organic Diodes

Semiconducting organic compounds may be used in a similar fashion to inorganic semiconductors in metal-semiconductor-metal structures. Perhaps the simplest example is that of a diode, or rectifier, which can be made by sandwiching a semiconductor between different metals. Such devices are examined in the following section. The special case of 'molecular diodes' is explored in Chapter 11, Section 11.5.

9.2.1 Schottky Diode

A *Schottky barrier* is formed when a metal and semiconductor, with appropriate *work functions* (energy needed to remove an electron from a solid), are brought into contact. This process is frequently carried out by the thermal evaporation of a metal onto the surface of the semiconductor in a vacuum (Chapter 7, Section 7.2.2). Consider the case for an n-type semiconductor, where the work function of the metal is greater than the work function of the semiconductor. The potential energy band diagrams before and after contact is made are shown in Figure 9.1. The work function Φ is the energy difference between the *vacuum level* and the Fermi level. The vacuum level defines the energy at which the electron is free from that particular solid, and where the electron has zero kinetic energy.

For the metal, the work function Φ_m is the minimum energy required to remove an electron from the solid. Typical values of Φ_m for very clean surfaces are 4.3 eV for Al and 5.1 eV for Au. However, these figures are modified if the surfaces become contaminated and also differ for the same substance in different morphological forms (e.g. single crystal and polycrystalline). In the metal, there are electrons at the Fermi level E_{Fm}, but in the semiconductor there are usually none located at E_{Fs} (i.e. the semiconductor Fermi level is probably located within the band gap). Nonetheless, the semiconductor work function Φ_s still represents the energy required to remove an electron from the semiconductor. In Figure 9.1 the energy difference between the lowest lying state in the conduction band and the vacuum level is the electron affinity X

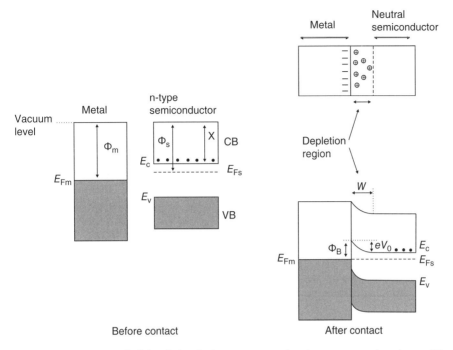

Figure 9.1 Formation of a Schottky barrier between a metal and an n-type semiconductor. CB and VB are the conduction and valence bands, respectively. E_c and E_v are the edges of the conduction and valence bands, respectively. E_{Fs} and E_{Fm} are the Fermi levels of the semiconductor and metal, respectively. Φ_s and Φ_m are the work functions of the semiconductor and metal, respectively. X is the electron affinity of the semiconductor. Φ_B is the Schottky barrier height and V_0 is the contact potential.

(Chapter 3, Section 3.4.1), for the particular case depicted, $\Phi_m > \Phi_s$. Therefore, when the two solids come into contact, the more energetic electrons in the conduction band of the semiconductor can transfer to the metal, into lower empty energy levels (just above E_{Fm}). These transferred electrons accumulate at the surface of the metal and leave behind an electron-depleted region of width W in which there are uncompensated positively-charged donor atoms, i.e. a net positive space-charge. The positive charge within this so-called *depletion region* matches the negative charge on the metal. A *contact potential* V_0 is established between the metal and semiconductor, which prevents further electron diffusion from the semiconductor conduction band to the metal. In equilibrium, the value of V_0 is given by

$$V_0 = \left(\Phi_m - \Phi_s \right) / e \tag{9.1}$$

Associated with the contact potential is an internal electric field directed from the semiconductor to the metal surface. This field is non-uniform and will have its maximum value at the metal–semiconductor interface (where there are a maximum number of field lines directed from positive to negative charge).

The Fermi level throughout the metal and semiconductor must be uniform in equilibrium (a Fermi level difference across the metal-semiconductor junction would result in an electric current flowing in an external circuit). In order for this to occur, the energy bands in the semiconductor must bend, as depicted in Figure 9.1. At the semiconductor edge of the depletion region, the semiconductor is still n-type. However, the carrier concentration in the conduction band n decreases towards the metal–semiconductor interface as $E_c - E_{Fs}$ increases (for intrinsic or undoped material, E_{Fs} is located close to the middle of the band gap, Chapter 3, Section 3.3.5). The potential energy required for an electron to move from the semiconductor bulk to the metal is called the *Schottky barrier height* Φ_B given by

$$\Phi_B = \Phi_m - X = eV_0 + \left(E_c - E_{Fs} \right) \tag{9.2}$$

which is greater than eV_0.

Provided that no external connections are made to the Schottky barrier (open-circuit conditions) no net current will flow through the metal–semiconductor interface. The number of electrons thermally emitted over the potential energy barrier Φ_B from the metal to the semiconductor is equal to the number of electrons thermally emitted over eV_0 from the semiconductor to the metal. However, when the semiconductor side of the junction is connected to the negative terminal of an external DC power source (e.g. a battery) and the metal to the positive terminal, the effect will be to reduce the contact potential from V_0 to $(V_0 - V)$, where V is the magnitude of the external voltage. It is now easier for the electrons to overcome the potential energy barrier into the metal. This situation is called *forward bias* and the resulting forward current is given by.

$$I = I_0 \left[\exp\left(\frac{eV}{k_B T} \right) - 1 \right] \tag{9.3}$$

where I_0 is a constant that is related to the barrier height of the metal–semiconductor junction. When the polarity of the external power source is reversed, the Schottky barrier becomes *reverse biased*. Now, the contact potential is increased by an amount V and it is more difficult for a current to flow. Under these conditions, the current is very low and is primarily due to thermal emission of electrons over the barrier Φ_B from the metal to the semiconductor. Figure 9.2 shows the band structures of the Schottky barrier under (Figure 9.2a) forward and (Figure 9.2b) reverse bias, together with the predicted current versus voltage behaviour of the device (Figure 9.2c). The *I-V* characteristics exhibit rectifying properties, and the device is called a *Schottky diode*.

A Schottky diode can also be made on a p-type semiconductor (many organic semiconductors exhibit p-type conductivity). In this case, the requirement is that the work function of the metal is less than that of the semiconductor, $\Phi_m < \Phi_s$. The band diagram is shown in Figure 9.3. At equilibrium, the alignment of the Fermi levels requires a positive charge on the metal and an equal negative charge distributed in the depletion region of the semiconductor; this is provided by

(a)

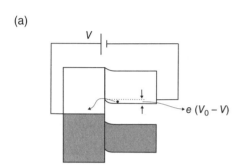

(b)

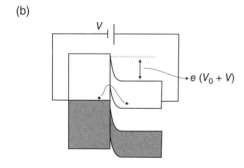

(c)

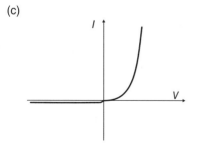

Figure 9.2 Schottky diode. (a) Bands in forward bias. (b) Bands in reverse bias. *V* is the applied voltage and V_0 is the contact potential. (c) Current *I* versus voltage *V* characteristics.

ionized acceptors left uncompensated by holes. For both the p- and n-type Schottky diodes, the forward current is due to injection of *majority carriers* (i.e. electrons in the n-type structure and holes in the case of the p-type device) from the semiconductor into the metal.

Practical Schottky diodes can exhibit different current versus voltage characteristics to those predicted for the ideal structure (e.g. Eq. (9.3) for the forward current). For forward bias voltages greater than $k_B T/e$, the current often takes the form:

$$I = I_0 \exp\left(\frac{eV}{nk_B T}\right) \quad (9.4)$$

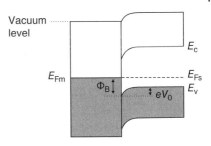

Figure 9.3 Band diagram for a Schottky barrier formed on a p-type semiconductor. E_c and E_v are the edges of the conduction and valence bands, respectively. E_{Fs} and E_{Fm} are the Fermi levels of the semiconductor and metal, respectively; Φ_B is the Schottky barrier height and V_0 is the contact potential.

where *n* is called the *ideality factor* of the diode and usually has a value between 1 and 2 for diodes based on Si. Significantly higher values are often encountered for organic semiconductors. An example of the current versus voltage behaviour of organic diode based on the polymer poly(3-hexylthiophene) is shown in Figure 9.4 [1]. The polymer was spin-coated onto a gold substrate and subsequently doped with $FeCl_3$; a metallic top electrode was established using an indium pressure contact. The ideality factor was 3.8 for measurements taken in as short a time as possible after the doping process. Subsequent changes to the electrical characteristics were attributed to the formation of a thin insulating layer between the top metal contact and the polymer. Although the charge carrier mobilities in organic semiconductors are relatively low, organic diodes can operate to reasonably high frequencies. For example, a pentacene diode has been successfully used to rectifying an ac signal at 50 MHz and it has been argued that GHz operating frequencies are within reach for such devices [2].

The simple analysis of a Schottky barrier assumes that the metal is in intimate contact with a semiconductor surface, with no interfacial layer between. This is not usually the case in practice (as suggested for the poly(3-hexylthiophene) diode described above). A semiconductor

Figure 9.4 Current versus voltage characteristics for an In-polythiophene-Au diode. *Source:* Reprinted from Kuo et al. [1] Copyright (1994), with permission from the Japanese Society of Applied Physics.

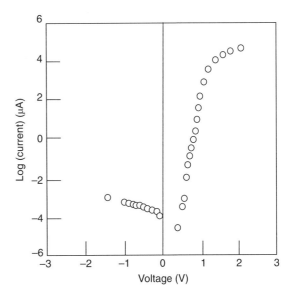

surface will possess defects due to incomplete covalent bonds (Chapter 2, Section 2.5.7) and other effects. These are so-called *surface states*, which can acts as traps for electrons or holes. Furthermore, the contact is seldom an atomically sharp discontinuity between the semiconductor and the metal. There is typically a thin interfacial layer, which is neither semiconductor nor metal. For example, silicon crystals are covered by a thin (1–2 nm) oxide layer, even after etching or cleaving in atmospheric conditions. Because of such effects, it is difficult to fabricate Schottky diodes with barrier heights predicted from work function differences (i.e. Eq. (9.2)) and with ideality factors of unity. In some cases, for example inorganic compound semiconductors such as GaAs, the density of surface states within the band gap is sufficient to produce an effect called *Fermi level pinning*, resulting in barrier heights that are independent of the work function of the metal contact.

An ultrathin insulating layer may deliberately be inserted between a metal and a semiconductor to produce a so-called *metal–insulator–semiconductor* (MIS) structure (see Section 9.3). The effect of the insulator is to control transport and to shape the electric field in the device. The application to electroluminescent (EL) display devices is discussed in Section 9.7.2.

9.2.2 Ohmic Contacts

An *ohmic contact* is a junction between a metal and a semiconductor that does not restrict the current flow. The resistance of the semiconductor outside the contact region rather than the thermal emission rate of carriers essentially limits the current across a potential barrier at the contact. Ohmic contacts exhibit linear *I-V* characteristic in both biasing directions. Often such contacts are more difficult to fabricate than Schottky barriers! Ideal metal–semiconductor contacts are ohmic when the charge induced in the semiconductor in aligning the Fermi levels is provided by majority carriers. This requires the use of low work function metals (e.g. Al, $\Phi_m \approx 4.3$ eV) for n-type semiconductors and high work function metals (e.g. Au, $\Phi_m \approx 5.1$ eV) for p-type semiconductors. An approach that is often used to make ohmic contacts is to dope the semiconductor heavily in the contact region. If a barrier is then formed at the interface, the depletion width is sufficiently small to allow carriers to tunnel through the barrier (Chapter 3, Section 3.5.2).

9.3 Metal–Insulator–Semiconductor Structures

The operation of many switching devices, such as the *metal–insulator–semiconductor transistor* or MISFET, is based on a phenomenon called the *field effect* (NB: the term MOSFET refers to the situation in which the insulator is an oxide, typically silicon dioxide). This can best be understood by first considering the operating principles of a simple MIS device.

9.3.1 Idealized MIS Devices

The band structure for an MIS device, based on a p-type semiconductor, is shown in Figure 9.5. This device differs from the MIS structure described in Section 9.2.1 only in the thickness of the insulating layer. For the MIS device shown, the insulator is assumed to be perfect and relatively thick so that charge transport between the metal and semiconductor does not occur. The structure is essentially that of a parallel plate capacitor in which one plate is a semiconductor. For the idealized case (Figure 9.5), it is assumed that the work function of the metal is equal to that of the semiconductor ($\Phi_m = \Phi_s$). With no external bias, the Fermi levels of the metal and semiconductor align and the various bands are flat throughout the MIS structure – *flat-band condition*.

Figure 9.5 Energy band diagram for the ideal metal–insulator–semiconductor (MIS) structure based on a p-type semiconductor at zero applied bias. In this case, it is assumed that the work function of the semiconductor is equal to the work function of the metal ($\Phi_s = \Phi_m$). E_c and E_v are the edges of the conduction and valence bands, respectively. E_{Fs} and E_{Fm} are the Fermi levels of the semiconductor and metal, respectively; and E_{Fi} represents the position of the Fermi level for intrinsic material.

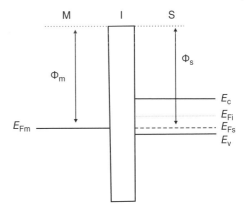

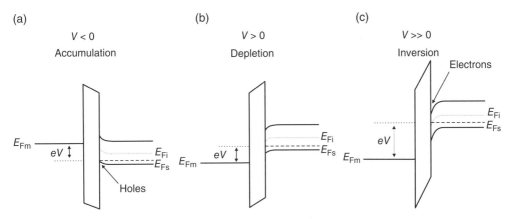

Figure 9.6 Energy band diagrams for a metal–insulator–semiconductor (MIS) structure based on a p-type semiconductor with different voltages applied to the gate electrode. (a) Negative voltage applied to gate, holes accumulate at semiconductor surface. (b) Positive voltage applied to gate, depletion layer forms at semiconductor surface. (c) Large positive voltage applied to gate, layer of negative charge (inversion layer) forms at semiconductor surface. V = applied voltage. E_{Fm} = metal Fermi level; E_{Fs} = semiconductor Fermi level; E_{Fi} = intrinsic Fermi level.

Figure 9.6 shows the effect of applying an external voltage to the MIS structure. If a negative voltage is applied to the metal contact (with respect to the semiconductor), a negative charge will be produced on the metal. Using the capacitor analogy, an equal and opposite charge must therefore be produced at the semiconductor surface. In the case under consideration, i.e. that of a p-type semiconductor, this occurs by hole (majority carrier) *accumulation* at the semiconductor–insulator interface (Figure 9.6a). The detailed explanation for this is as follows. Since the applied negative potential reduces the electrostatic potential of the metal relative to the semiconductor, the electron energies are raised in the metal relative to the semiconductor. As a result, the metal Fermi level E_{Fm} lies above its equilibrium position by an amount eV, where V is the applied voltage. As no current flows through the ideal MIS structure, there can be no variation in the semiconductor Fermi level E_{Fs} within the semiconductor. The conduction and valence bands in the semiconductor are displaced upwards with respect to E_{Fs}. The carrier density in the valence band depends exponentially on the energy difference between the valence band edge E_v and E_{Fs} [3]. Therefore, the band bending causes an accumulation of majority carriers (holes) near the semiconductor surface. This makes the semiconductor surface layer more p-type – leading to a hole accumulation layer.

If a positive voltage is now applied to the metal, a negative charge appears in the semiconductor. In the p-type material, this arises from the depletion of holes from the region near the surface, leaving behind uncompensated ionized acceptors. The situation is called *depletion* and is similar to that for the Schottky diode discussed in Section 9.2.1. In the depleted region, shown in Figure 9.6b, the hole concentration decreases, bending down the bands at the semiconductor–insulator interface.

If the positive voltage on the gate is increased further, the semiconductor bands bend down more strongly. At some point, E_{Fi}, the Fermi level position for the intrinsic material becomes lower than E_{Fs} at the surface. The semiconductor region adjacent to the surface then has the conduction properties of n-type material. This situation is referred to as *inversion* and is illustrated in Figure 9.6c. The inverted layer, separated from the underlying p-type material by a depletion region, is the key to the operation of many MIS transistor devices.

9.3.2 Effect of Real Surfaces

When MIS structures are fabricated from real materials, departures from the idealized case described above will occur. The main effects encountered are due to work function differences between the metal and the semiconductor and to the presence of charges at the insulator–semiconductor interface and within the insulator. If $\Phi_m \neq \Phi_s$, then the bands shown in Figure 9.5 will no longer be flat. An externally applied voltage (the polarity depending on the sign of $(\Phi_m - \Phi_s)$) will be needed to restore the flat-band condition. For example, if $\Phi_m < \Phi_s$ the bands in the semiconductor will bend down near the semiconductor surface. A negative voltage must then be applied to the metal to obtain the flat-band condition. This voltage is called the *flat-band voltage*, V_{fb}, and is given by

$$e V_{fb} = \Phi_m - \Phi_s \tag{9.5}$$

Charges located within the insulator layer may be fixed or mobile. In the latter case, the charges will move under the influence of an electric field; careful processing of the insulator can reduce their concentration. In addition to the charges in the insulator, charges can exist from the presence of surface states (also known as interface states). These generally result from the discontinuity of the bonding within the semiconductor and insulator layers. If the various insulator and interface charges can be considered as an effective positive charge located at the semiconductor–insulator interface and of magnitude Q_i C m^{-2}, then this will induce an equivalent negative charge in the semiconductor. An additional component must therefore be added to the flat-band voltage:

$$e V_{fb} = \Phi_m - \Phi_s - \frac{e Q_i}{C_i} \tag{9.6}$$

where C_i is the capacitance per unit area of the insulating layer. For the case under discussion, the difference in work function and the positive interface charge both tend to bend the bands down at the semiconductor surface; a negative voltage must be applied to the metal relative to the semiconductor to achieve the flat-band condition shown in Figure 9.5.

9.3.3 Organic MIS Structures

The different regimes in an MIS structure may be monitored conveniently by measuring the *differential capacitance* (i.e. $\partial C / \partial V$) of the device. Figure 9.7 shows the capacitance at a measurement frequency of 111 Hz and a temperature of 270 K as a function of bias for an Au- silicon

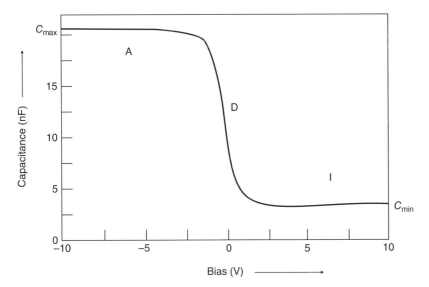

Figure 9.7 Capacitance versus voltage behaviour for a metal–insulator–semiconductor (MIS) structure comprising Si-SiO$_2$-polythiophene-Au measured at 111 Hz and at 270 K. The accumulation A, depletion D, and inversion I regions are indicated. *Source:* Reprinted from Grecu et al. [4] Copyright (2004), with permission from Elsevier.

dioxide-poly(3-hexylthiophene) MIS structure [4]. The accumulation A, depletion D, and inversion I regimes are marked on the figure. The decrease in the capacitance in the depletion region is due to the capacitance of the depletion layer (which decreases as the applied bias moves towards positive voltages and the width of the depletion layer grows) appearing in series with the SiO$_2$ capacitance. When minority carriers (electrons in this case) are generated in the inversion regime, the applied electric field cannot further penetrate the depletion region, which remains a constant thickness. The capacitance therefore remains constant, as shown in Figure 9.7. At very low frequencies, the capacitance in inversion would be expected to rise to the value in accumulation, as the minority carriers respond to the AC measuring voltage [3]. However, this situation is not usually observed for MIS devices based on organic semiconductors. The capacitance versus voltage scan shown in Figure 9.7 showed little dependence on the direction of the voltage sweep [4]. However, MIS structures incorporating organic materials (as either the insulator, semiconductor layer, or both) often show large hysteresis effects. These can be attributed to mobile ions in the organic layer, to trapping of charge at the interface between the semiconductor and insulator and/or to polarization of the insulating layer. The effect of fixed insulator charges and work function differences between the insulator and semiconductor is to displace the measured *C-V* curve along the voltage axis.

9.4 Organic Field Effect Transistors

Field effect transistors are three-terminal structures: a voltage applied to a metallic *gate* affects an electric current flowing between *source* and *drain* electrodes. Figure 9.8 shows how the field effect transistor (FET) device is derived from that of the MIS structure described in the previous section. Application of a voltage to the gate electrode V_g alters the charge density at the semiconductor surface. The mode of operation of organic FETs (OFETs) is invariably in the accumulation region. Hence, the majority carrier current flowing between the source and

(a)   (b)

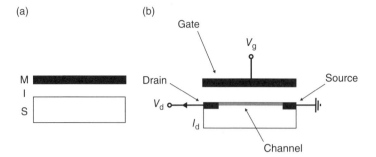

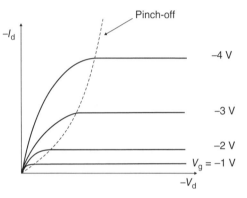

Figure 9.8 Contrast between (a) metal–insulator–semiconductor |(MIS) structure; and (b) metal–insulator–semiconductor transistor (MISFET).

Figure 9.9 Drain current I_d versus drain voltage V_d characteristics for a metal–insulator–semiconductor transistor (MISFET) based on a p-type semiconductor. The curves are shown for different applied gate voltages. The dashed line indicates where the channel of the FET becomes pinched off as a result of increasing V_d.

drain electrodes I_d is changed by the gate voltage. This control feature allows the amplification of small AC signals or the device to be switched from an on state to an off state and back again. These two key operations, amplification and switching, are the basis of many electronic functions.

The drain current versus voltage characteristics for a typical MISFET based on a p-type semiconductor are shown in Figure 9.9; these particular curves are called the transistor *output characteristics*. The electrical behaviour can be reasonably well described by models developed for inorganic semiconductors [3, 5]. When the gate voltage is biased negatively with respect to the grounded source electrode, the accumulation of holes occurs at the semiconductor–insulator interface. For low drain voltage V_d, the channel acts as a resistance; the drain current I_d increases linearly with V_d (the linear regime) and is given approximately by the equation:

$$I_d = \frac{WC_i}{L}\mu\left(V_g - V_t - \frac{V_d}{2}\right)V_d \tag{9.7}$$

where L and W are the length and width, respectively of the conducting channel, V_t is a constant (for a particular device) called the *turn-on* or *threshold voltage*, and μ is the field effect charge carrier mobility. The latter can be calculated in the linear regime from the *transconductance*, g_m:

$$g_m = \left(\frac{\partial I_d}{\partial V_g}\right)_{V_d=\text{constant}} = \frac{WC_i}{L}\mu V_d \tag{9.8}$$

Figure 9.10 Schematic diagrams showing the effect of applying an increasing drain voltage to a metal–insulator–semiconductor transistor (MISFET) (constant gate bias). (a) Linear regime – channel extends between source and drain electrodes. (b) Pinch-off – cross-sectional area of channel becomes zero at the drain electrode. (c) Saturation region – carriers must cross the depletion layer to reach the drain electrode.

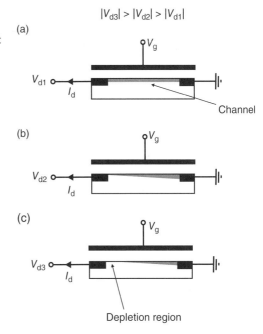

As the magnitude of V_d increases, I_d tends to saturate, as shown in Figure 9.9. This is the result of *pinch-off* of the accumulation layer. The phenomenon is illustrated in Figure 9.10. The combination of the voltages applied to the gate and drain will produce a wedge-shaped accumulation layer. At pinch-off, the effective cross-sectional area of the conduction channel, becomes zero at the drain. The drain current beyond the pinch-off point remains essentially the same and can be modelled using

$$I_d = \frac{WC_i}{2L}\mu\left(V_g - V_t\right)^2 \tag{9.9}$$

In the saturation regime, the carrier mobility can be calculated from the slope of the plot of $|I_d|^{0.5}$ versus V_g. However, the extraction of carrier mobility values for devices that do not exhibit the ideal current versus voltage characteristics in the linear and saturation regimes (i.e. Eqs (9.7) and (9.9)) must be treated with some caution [5]. The transition between the saturated and the linear region appears approximately at $V_d = V_g - V_t$ and is indicated in Figure 9.9 with a dashed line. When the gate electrode for the FET device discussed above is positively biased, the channel is depleted of carriers and little current flows between the source and the drain. The ratio of the current flowing in the accumulation regime to that in depletion is often referred to as the I_{on}-I_{off} ratio of the device.

Looking at the expression for the drain current in the saturation regime given by Eq. (9.9), it appears that I_d becomes zero when V_g is reduced to V_t. In reality, there is still some drain current conduction below threshold, and this is known as the *subthreshold conduction*. In OFETs, this current is related to the conductivity of the organic semiconductor. For state-of-the-art MOSFETs, the slope of the plot of $\log(I_d)$ versus V_g in the subthreshold region is typically 70 mV per decade. This means that a change in the input voltage of 70 mV will change the output I_d by an order of magnitude. Clearly, the smaller the value of this subthreshold slope, the better the transistor as a switch.

For efficient transistor operation, charge must be injected easily from the source electrode into the organic semiconductor and the carrier mobility should be high enough to allow useful quantities of source-drain current to flow. The organic semiconductor and other materials with which it is in contact must also withstand the operating conditions without thermal, electrochemical, or photochemical degradation. There are four main types of organic thin film transistor architecture, depending on the relative positions of the three electrodes to the semiconductor and insulating layers. These are contrasted in Figure 9.11. In the bottom gate structures, depicted in Figure 9.11a and b, the semiconductor layer is formed after the metal gate and gate insulator. This has the advantage of the organic thin film being deposited in the final stage of device processing; it will therefore not be exposed to subsequent thermal or chemical processing steps. However, exposure of the top surface to the environment can lead to device degradation. The top gate structures, shown in Figures 9.11(c) and (d), can reduce or eliminate this problem. It is important to note that the electronic nature of the metal-organic semiconductor junction can depend on whether the semiconductor is deposited onto the metal or vice-versa [6]. Top contact transistors generally exhibit the lowest contact resistances. This probably results from the increased metal–semiconductor contact area. A major contribution to contact resistance in the top contact architectures is the *access resistance*. This is a consequence of the requirement that charge carriers must pass between the source contact to the accumulation layer (channel) and then back to the drain contact to be extracted. To minimize access resistance, the thickness of the organic semiconductor should not be too large.

If the contact resistance R_c becomes larger than the channel resistance $R_{channel}$ (this usually occurs if the channel length of the device is reduced to about 1 μm), then the effective mobility of the transistor μ_{eff} may drop significantly below the intrinsic mobility μ_0 of the organic semiconductor. The relationship between these two parameters is given by [7]:

$$\mu_{eff} \approx \mu_0 \left[1 - \left\{ \frac{\mu_0 C_i W R_c \left(V_g - V_t \right)}{L + \mu_0 C_i W R_c \left(V_g - V_t \right)} \right\}^2 \right] \qquad (9.10)$$

Equation (9.10) accounts for the common observation that measured (effective) mobility of short channel organic transistors is often less than $1\,cm^2\,V^{-1}\,s^{-1}$, even though the semiconductor is known to possess a much larger intrinsic mobility. This is also the reason that the highest operating frequencies of many devices are limited to a few MHz. Judicious attention to device design is need to increase the operating frequency, which may be increased to 10–30 MHz [8].

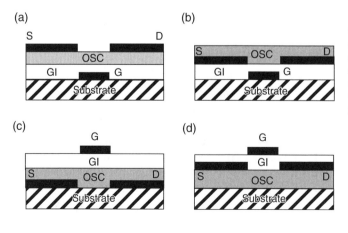

(a)
(b)
(c)
(d)

Figure 9.11 Schematic diagrams of different organic thin film transistor architectures. (a) Bottom gate, top contact; (b) bottom gate, bottom contact; (c) top gate, bottom contact; (d) top gate, top contact. S = source; D = drain; G = gate; OSC = organic semiconductor; GI = gate insulator.

Several OFET architectures have been proposed to reduce the process complexity, in particular the patterning of the semiconductor. One approach is to use a *Corbino* gate design [9]. This transistor configuration uses a nested ring for the source and drain. The outer electrode, the source, acts as a guard ring and is at the same potential as other sources in the vicinity, eliminating leakage through the unpatterned semiconductor layer. Alternative OFET structures use vertically spaced source and drain electrodes, instead of the common in-plane arrangement – the so-called *static induction transistor* [9]. Such devices can operate at relatively high powers and high frequencies, because of the excellent characteristic of the very short distance between the electrodes.

The semiconducting and insulating layers in OFETs can be produced using a variety of different thin film techniques, including spin-coating, thermal evaporation, and Langmuir–Blodgett assembly (Chapter 7). Figure 9.12 illustrates a series of processing steps that can be used to fabricate an OFET using the method of inkjet printing. In the first stage, a substrate (either rigid or flexible) is patterned (Chapter 7, Section 7.4). A conductive compound is inkjet-printed to form the transistor source and drain electrodes. A layer of organic semiconductor is then printed over the channel defined by the source and drain, and a layer of dielectric is deposited by spin-coating. Finally, a conducting gate electrode is inkjet printed.

The operating characteristics of organic transistors have improved markedly since they were first demonstrated in the 1980s [5]. This has been brought about by both improvements in the material synthesis and in the thin film processing techniques, and a realization that organic

Figure 9.12 Processing steps to form an OFET using the technique of inkjet printing. (a) Prepatterned substrate. (b) Organic source and drain electrodes printed. (c) Organic semiconducting layer printed. (d) Gate insulator deposited by spin-coating. (e) Organic gate electrode printed.

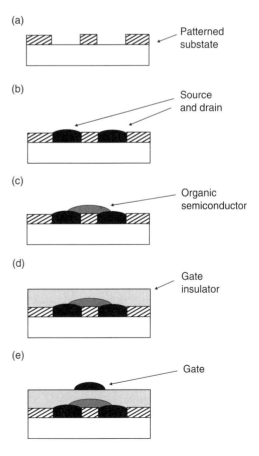

compounds can be degraded by water and/or oxygen in the atmosphere. Consequently, the processing of devices and their subsequent storage usually takes place in an inert ambient. The lifetimes of organic devices can also be considerably extended by some kind of encapsulation. Figure 9.13 shows the output characteristics for an OFET based on thermally evaporated pentacene [10]. This particular structure (channel length = 50 μm and width 500 μm) was fabricated using a cross-linked polymer gate insulator (33 nm thickness) deposited by spin-coating. The electrical characteristics show features that are typically encountered for organic transistors. First, some hysteresis is evident between the forward and reverse voltage scans of some of the curves. This instability reflects dispersive transport and charge trapping in the organic semiconductor or at the interface [5]. The slight curvature of the output curves near the origin ('crowding' of the current versus voltage plots) is indicative of contact resistance. Figure 9.14 shows the dependence of I_d on V_g and $I_d^{0.5}$ on V_g for the same device. The former plot is known as the *transfer characteristic* of the transistor. The latter plot is reasonably linear for high values of V_g and reveals an average mobility figure of approximately $1.1\,\mathrm{cm^2\,V^{-1}\,s^{-1}}$. However, close examination of this curve reveals a slight decrease in its slope as the magnitude of the gate bias increases. A possible explanation is that at lower gate voltages the accumulation

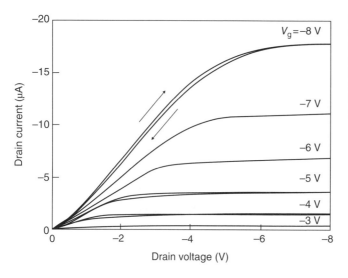

Figure 9.13 Drain current versus drain voltage characteristics of an OFET based on pentacene and using cross-linked PMMA as the gate insulator. Forward and reverse scans are indicated. *Source:* Reprinted from Yun et al. [10] Copyright (2009), with permission from Elsevier.

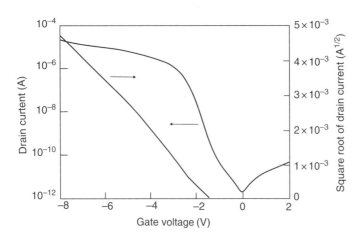

Figure 9.14 Drain current and square root of drain current versus gate voltage characteristics of an OFET based on pentacene and using cross-linked PMMA as the gate insulator; the I_d versus V_d characteristics of this device have been given in Fig. 9.13. Reverse scan is shown. *Source:* Reprinted from Yun et al. [10] Copyright (2009), with permission from Elsevier.

region is not as tightly confined to the interface and extends further into the bulk than at high gate voltages [5]. Alternatively, this phenomenon may also be attributed to contact resistance effects. The pentacene FET has a subthreshold swing of 219 mV per decade and an on-off current ratio of about 10^6.

State-of-the-art OFETs may possess characteristics similar to or greater than those of devices prepared from hydrogenated amorphous silicon, with mobilities around $1\,cm^2\,V^{-1}\,s^{-1}$ and on-off ratios greater than 10^6. Many of the highest performing p-type transistors using small molecules are based on acenes (e.g. pentacene, rubrene) and fused heteroacene materials. A challenge is to make these materials solution processible in order to aid device manufacture. Much progress has been achieved by exploiting the excellent solubility characteristics of the triisopropylsilylethynyl (TIPS) group by substitution on the central pentacene core [5, 11]. Promising polymer semiconductors include donor-acceptor copolymers, such as isoindigo (IID)- and indacenodithiophene (IDT)-based compounds [5, 12]. While these materials generally exhibit lower mobilities than their small molecule counterparts, they can be more readily processed from solution. A further approach uses blends of small molecules and insulating polymers [13]. Carbon nanotube-based OFETs with relatively high mobilities (above $20\,cm^2\,V^{-1}\,s^{-1}$) and operating frequencies (GHz range) can be conveniently fabricated using techniques such as inkjet printing [14]. The exploitation of both carbon nanotubes and graphene in transistor devices is covered separately in Chapter 11, Section 11.10.

Thin film transistors based on organic semiconductors are unlikely to compete with single crystal inorganic semiconductors such as Si or GaAs for fabricating very fast switching devices. However, these may find a role in niche areas, such as components of plastic circuitry for use as display drivers in portable computers and pagers, and as memory elements in transaction cards and identification tags. Such prospects are discussed in the next section.

9.5 Organic Integrated Circuits

A significant step in the microelectronics industry was the ability to manufacture a complete circuit, an *integrated circuit*, on the same substrate. Simple integrated circuits can be fabricated using organic materials. Figure 9.15 shows the characteristics of an early *inverter* (i.e. a circuit that produces logic '1' for logic '0' input and vice-versa) based on two interconnected pentacene

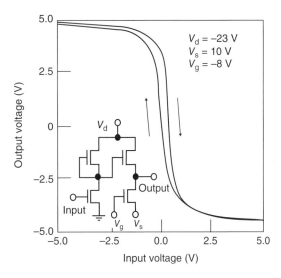

Figure 9.15 Output versus input voltage of an inverter device based on pentacene. The circuit of the structure is shown inset. *Source:* Reprinted with permission from Klauk et al. [15] Copyright (2003) American Institute of Physics.

MISFETs [15]. One of the transistors acts as the load for the switching device. The resulting inverter has sufficiently large gain, but due to the positive switch-on voltage, the input and output levels do not match. Also evident is a hysteresis of a few hundred millivolts in the inverter characteristic, which is possibly due to mobile charges in the gate dielectric. This situation can be improved by the incorporation of a voltage-level-shifting network in the integrated circuit. These simple inverter structures can form the basis of more complex oscillator circuits.

A particularly useful silicon device for digital applications is a combination of n-channel and p-channel MOS transistors on adjacent regions of an integrated circuit. This *complementary MOS* (referred to as CMOS) circuitry dominates the market for applications such as micro-processors (Chapter 1, Section 1.3.1). Complementary circuits are also of interest for less sophisticated applications, especially those that are battery-powered, since they can provide very low static power dissipation and therefore extend battery life. Although a number of p-channel organic MISFETs are available, with performance comparable to amorphous silicon, n-channel devices have not received the same attention [16]. Materials with high electron affinities (Chapter 4, Section 3.4.1) are needed for n-type behaviour. Many such compounds exhibit in instability (oxidation) in air. To some extent, this can be circumvented by the developing materials that favour dense molecular packing. The introduction of electron-withdrawing groups, such as cyano or carbonyl, is a common strategy for lowering the lowest unoccupied molecular orbital (LUMO) energy level. Aromatic di-imides are an example of useful n-type organic semiconductors, exhibiting high electron affinities, high mobilities, and excellent stabilities [16]. Other n-type materials can be based on perylene, tetracyanoquinodimethane (TCNQ), and fullerenes. By combining low-voltage p-channel and n-channel organic thin-film transistors in a complementary arrangement, simple integrated circuits such as digital-to-analogue converters may be achieved [17].

9.5.1 Radiofrequency Identification Tags

Radiofrequency identification (RFID) is a method of remotely storing and retrieving data using devices called RFID tags. An RFID tag is a small object, such as an adhesive sticker, which can be attached to, or incorporated into, a product. The tags contain aerials to enable them to receive and respond to radio-frequency queries from an RFID transceiver. In the roadmap for printed electronics, RFID has an important place due to the enormous available market covered by bar codes. A typical system is depicted in Figure 9.16. These tags can be either *active* or *passive*. Passive RFID tags do not have their own power supply: the minute electrical current induced in the aerial by the incoming radio-frequency scan provides enough power for the tag to send a response. Due to power and cost concerns, the response of a passive RFID tag is necessarily brief: typically just an ID number. Lack of its own power supply makes the device quite small, and commercially available products exist that can be embedded under the skin. Passive tags have practical read ranges that vary from about 10 mm up to about 5 m.

Active RFID tags, on the other hand, must have a power source, and may have longer ranges and larger memories than passive tags, as well as the ability to store additional information sent by the transceiver. The smallest active tags can approach the size of a grain of rice. Identification tags can operate at different frequencies, ranging from about 100 kHz to the microwave region (around 2 GHz).

Figure 9.17 shows (a) the circuit and (b) a photograph of an RFID system based on pentacene thin-film transistors [18]. This comprises of a seven-stage ring-oscillator and a five-stage output buffer. The gate length for all the transistors was nominally 20 μm, and the gate width for each of the ring-oscillator inverters was 60 and 300 μm for the load and driver

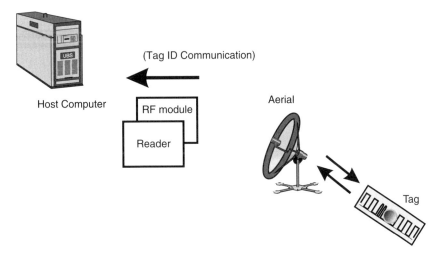

Figure 9.16 Schematic diagram of a radiofrequency identification (RFID) system. The aerial captures the tag ID number, first as analogue radio frequency waves, then it is converted to digital information.

transistors, respectively. An integrated 13.56 MHz RFID tag in a printed complementary organic transistor technology has also been successfully developed [19]. This circuit is relatively complex, with more than 250 organic transistors on the same substrate, and operates at a supply voltage of 24 V.

9.6 Transparent Conducting Films

Transparent conducting films are an important component in a number of electronic devices including liquid-crystal displays, organic light emitting displays, touchscreens, and photovoltaic solar cells. While indium-tin-oxide (ITO) is widely used, this material has many drawbacks, including its tendency to crack under strain, limiting its applications in flexible electronics products. ITO can also decompose in high electric fields and/or at elevated temperatures during device operation. Additionally, indium is scarce and has an associated high cost. Alternatives include other transparent conductive oxides, ultra thin metal films, and metal grids. Relevant to this text, nanowire meshes based on conductive polymers (e.g. poly(3,4-ethylenedioxythiophene poly(styrene sulfonic acid)) (PEDOT:PSS), Chapter 5, Section 5.3), carbon nanotubes, and graphene are also attracting interest.

The minimum standards required for a transparent conducting material to be industrially useful are a sheet resistance (Chapter 3, Section 3.2.2) $R_s < 100\,\Omega$ square^{-1}, together with a transmittance of more than 90% in the visible range. It is also important that the conductive material has a smooth surface (low surface roughness) to prevent short circuits. For touch screen applications, where sheet resistances of less than $500\,\Omega$ square^{-1} are required, these specifications are adequate. However, transparent conducting electrodes in flat panel displays require much lower sheet resistances, down to $R_s \approx 100\,\Omega$ square^{-1}. It is evident from Figure 5.5, Chapter 5, that ITO fulfils these criteria. However, the quality of graphene electrodes is steadily improving, while ITO will become more expensive. Furthermore, the endurance of graphene and its high fracture strain (10 times higher than ITO) give this material a key advantage in bendable and rollable devices.

(a)

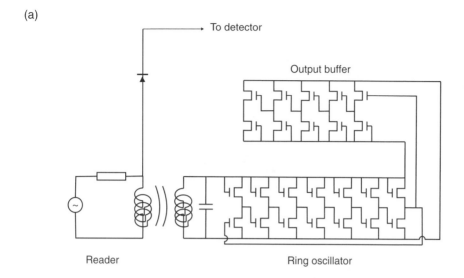

(b)

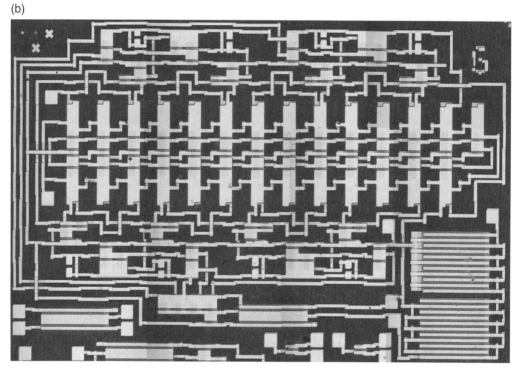

Figure 9.17 (a) Circuit diagram of a ring oscillator, fabricated using pentacene organic FETs (OFETs). (b) Photograph of the organic integrated circuit. *Source:* Reprinted with permission from Baude et al. [18] Copyright (2003) American Institute of Physics.

9.7 Organic Light-emitting Devices

Organic light-emitting devices (OLEDs) represent a significant display technology. These require no backlighting and currently complete with LC displays in the consumer (e.g. TV) market. There is also potential for important medical applications [20, 21].

Figure 9.18 Schematic energy band structure of an organic light-emitting device (OLED). The recombination of electrons and holes results in the emission of light of frequency ν and energy $h\nu$.

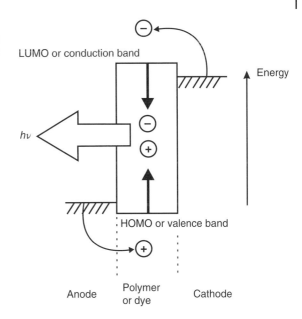

Reports of light emission from organic materials on the application of an electric field (electroluminescence) go back many years. In 1987, scientists at Kodak reported efficient low voltage electroluminescence in an organic thin film incorporating dye molecules [22]. Interest intensified in the 1990s following the report of OLEDs incorporating the conjugated polymer poly(p-phenylenevinylene) (PPV) (Chapter 5, Section 5.3) [23]. The simplest OLED is an EL compound, such as a polymer or dye, sandwiched between metals of high and low work function, as depicted in Figure 9.18. In practice, just like all other electronic devices based on organic compounds, OLEDs are fabricated in an inert atmosphere, e.g. nitrogen, and are encapsulated to prevent the inclusion of water and oxygen, both of which can lead to short operating lifetimes [24].

On application of a voltage, electrons are injected from the low work function cathode into the LUMO level (or conduction π^* band in the case of an organic compound possessing a delocalized electron system) of the organic compound and holes from the high work function anode into the highest occupied molecular orbital (HOMO) level (or valence π band in the case of an organic compound possessing a delocalized electron system). The recombination of these oppositely charged carriers then results in the emission of light. For organic materials, the electrons and holes are usually bound by a few meV to form excitons (Chapter 4, Section 4.4.3). Therefore, their radiative recombination leads to emission energies slightly lower than the HOMO-LUMO separation. Efficient injection of both electrons and holes is a prerequisite to obtain high performance devices. Current displays exploit both low molecular weight organic molecules and polymers as EL materials [25, 26]. The chemical formulae of some of the materials that have been used as the EL compounds in OLED devices are shown in Figure 9.19. The HOMO and LUMO levels and emissive colours are indicated [27, 28]. However, the reader should treat the energy level values as approximate, as these are obtained from different sources, some experimental and some theoretical. The ability to fabricate thin films from such compounds is an important practical issue. For example, the addition of the side chain to the PPV monomer, to form poly[2-methoxy-5-(2-ethylhexyloxy)-1,4-phenylenevinylene] (MEH-PPV), allows the latter compound to be easily spin-coated onto a variety of substrates. The non-polymeric compounds, such as tris-(8-hydroxyquiniline) aluminium (Alq$_3$) (Figure 9.19), are usually formed into thin films by the process of thermal evaporation.

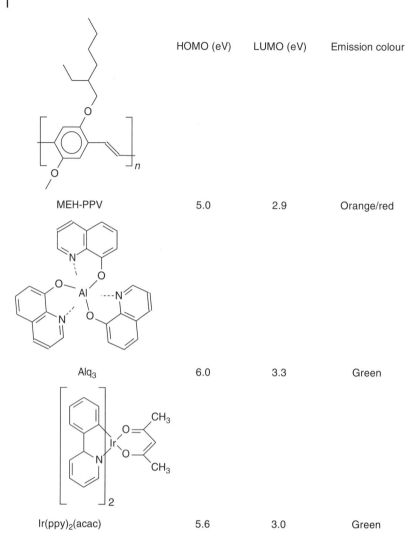

	HOMO (eV)	LUMO (eV)	Emission colour
MEH-PPV	5.0	2.9	Orange/red
Alq₃	6.0	3.3	Green
Ir(ppy)₂(acac)	5.6	3.0	Green

Figure 9.19 Organic compounds used as the emissive layer (EML) in organic light-emitting devices (OLEDs). The HOMO and LUMO energies and emission colours are indicated.

Radiative emission in OLEDs may also be characterized as either fluorescence or phosphorescence, depending on the relaxation path (Chapter 4, Section 4.4.5). Many of the fluorescent dyes that have been incorporated into OLEDs were originally developed as laser dyes, since these materials were designed to possess high photoluminescence quantum efficiencies in dilute solution, along with good stability. However, in general, their solid-state fluorescence is extremely weak due to concentration quenching, and this reduces the EL efficiency. One solution to this problem is to add bulky side groups to the molecule, to prevent aggregation by increasing the steric hindrance. Unfortunately, this often leads to poor charge transport through the material. A further remedy is to dope the emissive dye into an organic matrix. This effectively dilutes the concentration of the emissive dye (the dopant) and thus prevents aggregation. So long as the dopant is red-shifted (i.e. has its emission at a lower energy) compared to the host, excitons formed in the host material will tend to migrate to the dopant prior to relaxation. This results in emission that is predominantly from the dye.

Phosphorescent materials (Chapter 4, Section 4.4.5), such as rare earth complexes, may be exploited to improve the OLED efficiency [25]. This is because, for quantum mechanical reasons, 75% of recombination events are associated with triplet states that, in most cases, do not emit photons when they decay to the ground state. Emission in these materials is from atomic states in the metal ion (due to the strong spin-orbit interaction), so there is little interaction with the organic ligands and the emission bands are very narrow. As a result, both singlet and triplet states can be used to harvest light emission (Chapter 4, Section 4.4.5). A common approach uses a triplet emitting, heavy atom containing, (electro)phosphorescent metal–organic complexes, as typified by bis(2-(2-pyridinyl-N)phenyl-C) (acetylacetonato)iridium(III), Ir(ppy)$_2$(acac), the structure of which is shown in Figure 9.19. These 'dopants' (in this context, the term should not be confused with that used when a material is added to alter the electrical conductivity of an organic semiconductor – Chapter 3, Section 3.4.2) must have high luminescence quantum yields and relatively short lifetimes, of the order 1 µs, to avoid exciton-ion quenching.

Figure 9.20 shows the forward current versus voltage and EL versus voltage characteristics of an OLED based on ITO-PPV-Al [29]. The ITO electrode is positively biased to obtain these characteristics. Negligible current and light emission result with the opposite polarity of applied voltage. The OLED therefore acts as a rectifying device. The steep rise in the forward current between 1 and 2 V can be modelled with the Schottky diode (Eq. (9.4)), with an ideality factor typically in the range 1.6–2.4 for different devices. The deviation from this exponential behaviour above 2 V results from the limiting series resistance associated with the device. The EL from the OLED is observed as the forward voltage is increased beyond about 1.6 V. Generally, OLEDs based on other materials exhibit rectifying electrical characteristics, but the results sometimes suggest other conductivity mechanisms. For example, space-charge limited conduction and Fowler-Nordheim conduction (Chapter 3, Sections 3.5.4 and 3.5.5) are often reported [30].

By judicious choice of compound, EL emission at different wavelengths across the visible spectrum is possible. A convenient way in representing the colour output of an OLED is to use

Figure 9.20 Current density versus voltage and electroluminescence power versus voltage characteristics for an indium tin oxide (ITO)-MEH-poly(*p*-phenylenevinylene) (PPV)-Al organic light-emitting device (OLED). *Source:* Reprinted with permission from Karg et al. [29] Copyright (1997) American Institute of Physics.

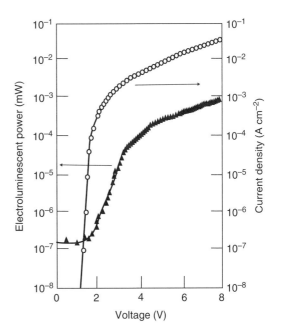

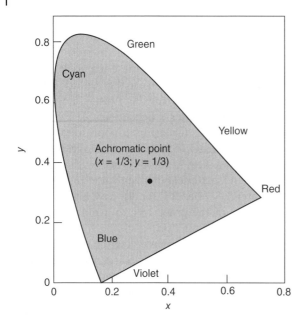

Figure 9.21 Commission International de l'Eclairage (CIE) Chromaticity diagram.

the CIE (Commission International de l'Eclairage) *chromaticity diagram* (Figure 9.21) [31]. Although the human eye has three different types of colour-sensitive cones, it is found that any colour can be expressed in terms of just two colour coordinates x and y. The boundary represents maximum saturation for the spectral colours, and the diagram forms the limit of all perceivable hues. The calculation of the CIE chromaticity coordinates for an OLED requires the multiplication of its spectral power at each wavelength times the weighting factor from each of the three colour matching functions.

A further characteristic of a light source is its *colour-rendering index* (CRI). This is a measure of the ability of the light source to show object colours 'realistically' compared to a familiar reference source, either incandescent light or daylight. The CRI is calculated from the differences in the chromaticities of eight CIE standard colour samples when illuminated by the light source and by a reference; the smaller the average difference in chromaticities, the higher the CRI. A CRI of 100 represents the maximum value. Lower CRI values indicate that some colours may appear unnatural when illuminated by the lamp. Fluorescent lamps have lower CRI values than incandescent lights, which can have a CRI above 95.

9.7.1 Device Efficiency

There is proliferation of units associated with the characterization of light emitting devices and much confusion abounds. A very brief introduction is therefore provided in this section.

First, it is necessary to make a distinction between *photometric* and *radiometric* units [32]. Radiometry is the measurement of optical radiation, which is electromagnetic radiation within the frequency range of 3×10^{11}–3×10^{16} Hz. This corresponds to wavelengths between 0.01 and 1000 μm, and includes the ultraviolet, visible, and infrared regions. Photometry on the other hand is the measurement of light, defined as electromagnetic radiation that is detectable by the human eye. It is therefore restricted to the wavelength range from about 380–780 nm. Photometry is essentially radiometry, except that everything is normalized to the spectral response of the eye, known as the *photo-optic* response. Radiometry uses the familiar unit of the watt as a measure of power (rate of energy emission) or radiant flux. The photometric equivalent of the watt is the *luminous flux*, measured in *lumens* (lm). The *luminous intensity* measured in *candelas* (cd) refers to the flux emitted into a unit solid angle in space or *steradian*. Since there are 4π steradians in a sphere, in the case of a light source emitting equally in all directions (isotropic), the relationship between lumens and candelas is $1\,cd = 4\pi\,lm$. The amount of light leaving a surface is the luminance, measured in $cd\,m^{-2}$; these units are often called nits (from the Latin 'nitere' = to shine). A typical computer screen has a luminance of about $100\,cd\,m^{-2}$, while the luminance of an average clear sky will be around $8000\,cd\,m^{-2}$.

The CIE has adopted, as a standard, an average eye with a predictable response to light at various frequencies. As a result, one light watt consists of 683 lm. This represents the photo-sensitivity of the 'standard' human eye; this response has a peak value of $683\,lm\,W^{-1}$ at a wavelength of 555 nm. According to this function, 700 nm red light is only about 4% as efficient as 555 nm green light (i.e. 1 W of 700 nm red light is only 'worth' $683 \times 0.04 \approx 27\,lm$). Table 9.1 provides a list of the photometric quantities encountered in OLED work.

Quantum efficiencies are often quoted for OLEDs. An important distinction must be made between the definitions of the external and internal efficiencies. For display applications, the commonly accepted definition for the *external quantum efficiency*, η_{ext}, is the ratio of the number of photons emitted by the OLED into the viewing direction to the number of electrons injected. While this external quantum efficiency could also be defined as the ratio of the total number of photons emitted from the device (in all directions) to the number of electrons injected, this definition is not useful for display devices. A large fraction of the light can be waveguided by the organic layer(s)-substrate combination, ultimately emerging out at the edge of the device (Section 9.7.3). Thus, the total amount of light emitted from the device will be significantly higher than the light emitted in the viewing direction; leading to an efficiency based on total light emitted, which can be up to four times larger than η_{ext}. The *internal quantum efficiency* is the ratio of the total number of photons generated within the structure to the number of electrons injected. The internal and external efficiencies therefore differ by the fraction of light coupled out of the structure into the viewing direction.

A convenient measure of the properties of an OLED is the *luminous efficiency*, η_l, in $cd\,A^{-1}$. In many respects, η_l is equivalent to η_{ext}, with the exception that η_l weights all incident photons according to the photoptic response of the eye. In this case

$$\eta_l = \frac{AL}{I} \tag{9.11}$$

where L is the luminance of the OLED (in $cd\,m^{-2}$), A is the device active area (not necessarily equal to the area of light emission), and I is the current.

Table 9.1 Common photometric units.

Property	Description	Units
Energy	Total amount of light emitted from source	Joules (J)
Luminous flux	Rate of energy emitted from source	Lumen (lm)
Luminous intensity	Flux emitted from a point source per unit solid angle	Candela (cd) ≡ Lumen per steradian
Luminance	Flux emitted per unit surface area of extended source per unit solid angle	$cd\,m^{-2}$

Table 9.2 Performance of polymer-based organic light emitting devices (Cambridge Display Technology Ltd. 2016) [33].

	Red		**Green**		**Blue**	
Efficiency (cd A^{-1})	28	24	112	92	10	9.3
Colour (CIE: x, y)	$x = 0.66$; $y = 0.34$	$x = 0.66$; $y = 0.34$	$x = 0.33$; $y = 0.62$	$x = 0.33$; $y = 0.62$	$x = 0.14$; $y = 0.12$	$x = 0.14$; $y = 0.11$
T95 lifetime @ 1000 cd m^{-2} (hrs)	Ongoing	5800	Ongoing	17 000	700	400

Device structure: indium tin oxide (ITO) (45 nm)/Hole Injection Layer (30–65 nm)/Interlayer (20 nm)/Light Emitting Polymer (60–90 nm)/Cathode.
Hole injection, interlayer and emitter are all processed from solution.

Another frequently used display efficiency unit is the *luminous power efficiency* or *luminosity*, η_p, measured in lm W^{-1}. This is the ratio of luminous power emitted in the forward direction (in lm) to the total electrical power required top drive the OLED at a particular voltage. Table 9.2 gives the luminous efficiencies and T95 lifetimes (the time taken for the light output to fall to 95% of its initial value) of some commercial polymer-based OLEDs [33].

9.7.2 Device Architectures

The external quantum efficiencies of the early OLEDs were around 10^{-2}%. This value has now increased to over 20%. A breakthrough in terms of improved OLED performance was achieved when the functions of charge transport and light emission were separated. The overall structure of an OLED can be quite complex, as shown in Figure 9.22. The two electrodes can each be modified to improve the carrier injection (electron and hole injection layers – EIL and HIL); furthermore, carrier-transporting layers (electron transport layer (ETL) and hole transport layer (HTL)) can provide efficient transfer of electrons and holes to the layer that emits the light (emissive layer (EML)). The role of the injection and transport layers can be combined. For example, the carrier transport layers may be doped (Chapter 3, Section 3.4.2) to provide low resistance contact regions. Common p-type dopants are strong electron acceptors such as F$_4$-TCNQ, while alkali metals such as Li or Cs can be effective as n-type dopants. The resulting *p-i-n* devices can exhibit high efficiencies and operating voltages that are close to the thermodynamic limit, i.e. the photon energy divided by the electronic charge [34]. The efficiency gain from the multilayer structure depicted in Figure 9.22 must, of course, be offset against the increased fabrication costs.

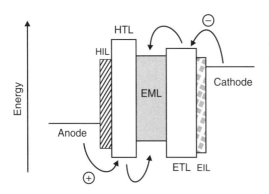

Figure 9.22 Multilayer organic light-emitting device (OLED). HIL = hole injection layer. HTL = hole transport layer. EML = emissive layer. ETL = electron transport layer. EIL = electron injection layer.

Electrodes

The work function of the anode in an OLED must be matched to the HOMO level of the organic semiconductor for effective hole injection. This electrode is typically ITO, since this is transparent and highly conductive. However, the material has a rough surface and is treated with a thin layer of a hole conductor (see below) such as PEDOT:PSS (Chapter 5, Section 5.3), phthalocyanine, or polyaniline, to improve both the contact to the EML and the device stability. Another advantage of these layers is that they smooth out the relatively rough surface of the ITO, thereby preventing any local short-circuiting that would otherwise cause the device to fail. In many OLEDs, the hole current can far exceed the electron current. This results in significant energy wastage, since the excess holes cannot combine with electrons to generate light. Efficiency improvements can be achieved by coating the anode with a thin layer that reduces the hole current (hole blocking layer).

A low work function metal or alloy is used for the cathode in OLEDs (to match with the LUMO level of the organic semiconductor). Unfortunately, metals with very low work functions, such as Li and Ca, are reactive and require careful encapsulation. A thin inorganic insulating layer such as LiF, or an organic monolayer may be inserted between the cathode and the emissive material. The introduction of such an interfacial layer can modify the electronic band structure in a number of ways. For example, the layer may introduce a layer of fixed charge, affecting the height of the barrier to electron injection (i.e. modifying the work function of the metal). An important difference between MIS and MS structures is that, in the former case, the metal Fermi level is no longer 'tied' to the energy band structure of the semiconductor (see earlier discussion in Section 9.3). This means that, as the forward applied voltage is increased, the cathode Fermi level can move with respect to the LUMO level in the organic EML, aligning filled electron states in the metal with vacant states in the organic semiconductor. As long as the interfacial layer is 'transparent' to electrons, the EL will increase. This, of course, imposes some restrictions on the nature of the interfacial layer: it must support a DC voltage, but at the same time be sufficiently thin that the minority electron current can pass through, e.g. by tunnelling.

Hole and Electron Transport Layers

The HTL in an OLED serves two purposes. First, it provides a path for holes to be injected into the emitting layer. Second, it also acts as an electron-blocking layer to confine electrons within the emitting layer. Figure 9.23 shows the chemical formulae of three organic compounds that have been used as hole transport materials and to modify the anode electrode (see above) [25, 26]. Hole transport materials generally possess a shallow LUMO level to prevent electrons entering from the EML, and a HOMO level matching that of the EML material and the work function of the anode. Typical hole-injecting electrodes, such as ITO or PEDOT:PSS, have work functions between 4 and 5 eV and compounds that have HOMO levels matching the values of air stable weak electron donors. One of the most common classes of compound that are used are the arylamines, of which *N-N′*-diphenyl-*N-N′*- bis(3-methylphenyl)-(1-1′-biphenyl)-4-4′-diamine (TPD, shown in Figure 9.23) is one example. However, this particular compound has a glass transition temperature, T_g (Chapter 2, Section 2.6) of only 63 °C, which can result in recrystallization and delamination from the ITO anode.

The roles of the ETL are to transport injected electrons to the emitting layer and to serve as a hole-blocking layer to confine holes in the emitting layer. Some of the electron transport materials that have been used in OLEDs are depicted in Figure 9.24 [27, 35, 36]. Many of these contain oxadiazole groups (i.e. the five-membered rings containing one oxygen and two nitrogens); these are very electron deficient, which allows them to block holes and transport electrons very well.

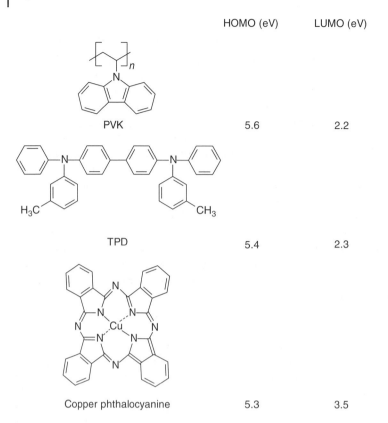

	HOMO (eV)	LUMO (eV)
PVK	5.6	2.2
TPD	5.4	2.3
Copper phthalocyanine	5.3	3.5

Figure 9.23 Hole-transporting molecules used in organic light-emitting devices (OLEDs). The HOMO and LUMO levels are indicated.

Triplet Emission

The external quantum efficiency of devices fabricated using fluorescent materials is generally limited to 25%, because of the 1 : 3 singlet : triplet ratio. To transfer energy efficiently to a phosphorescent dopant, both the singlet- and triplet-excited states of the host material must be higher in energy than the triplet-excited state of the dopant. This wide band gap requirement for the host can make it difficult to optimize simultaneously the charge injection and light emitting properties of the device. The host material must also be able to transport electrons and holes effectively and be energy-level matched to the adjacent transport layers. Figure 9.25 shows an example of an OLED architecture based on the phosphorescent emitter Ir(ppy)$_2$(acac) [27]. In this case, the emitter is doped into a host material TAZ (Figure 9.24). The HTL is HMTPD (a similar molecule to TPD, which shown in Figure 9.23), while the ETL is Alq$_3$ (Figure 9.19). Luminous efficiencies can be high for phosphorescent OLEDs, e.g. > 50 cd A^{-1} [37]. However, high efficiencies are generally only achieved at low current densities. With increasing currents, the efficiency gradually decreases due to a growing influence of different quenching effects, of which triple-triplet annihilation is regarded as being of particular importance (Chapter 4, Section 4.4.5).

A further strategy to improve OLED efficiency, which does not rely on heavy metal complexes, is to exploit *thermally activated delayed fluorescence* (TADF). In this process, the emitting material has singlet and triplet energies sufficiently close in energy so that conversion of triplets to light-emitting singlets is efficient [38]. High external quantum efficiencies are achievable with suitably designed emitters: 30% for green [39] and 20% for blue [40].

HOMO (eV)   LUMO (eV)

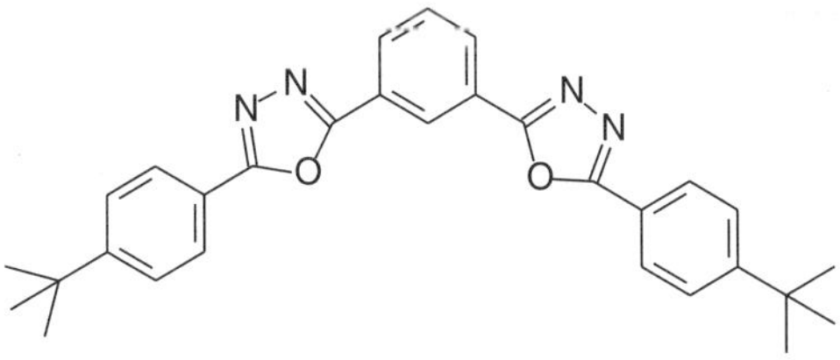

OXD-7   6.5   2.8

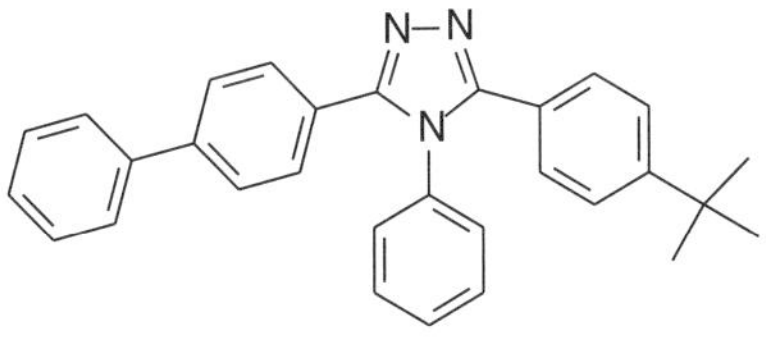

TAZ   6.6   2.6

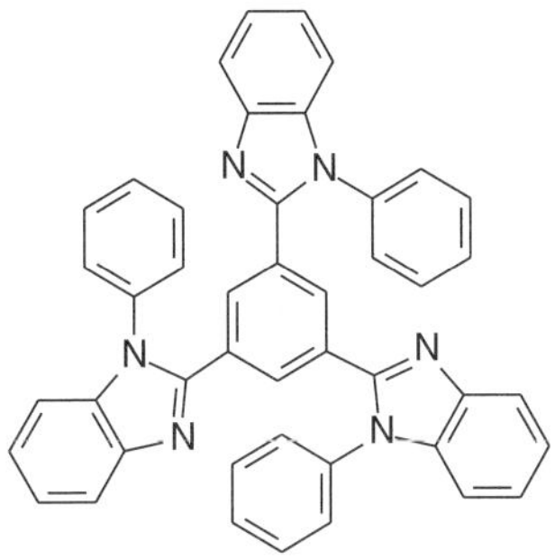

TPBI   6.2   2.7

Figure 9.24 Electron-transporting molecules used in organic light-emitting devices (OLEDs). The HOMO and LUMO levels are indicated.

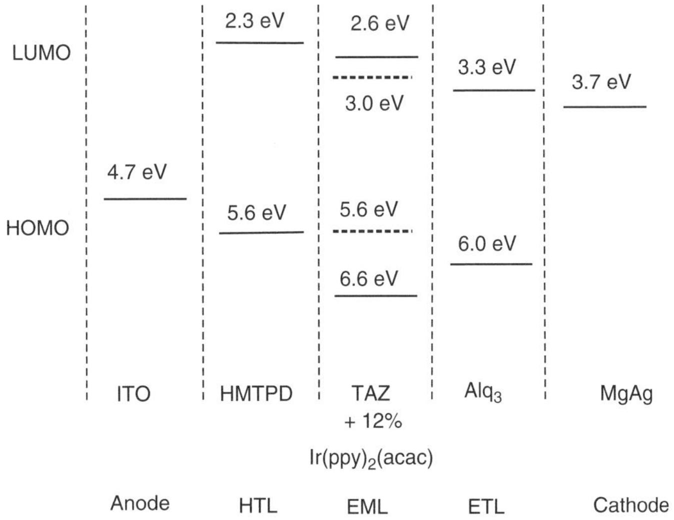

Figure 9.25 HOMO-LUMO energy level diagram for multilayer organic light-emitting device (OLED) based on the phosphorescent molecule Ir(ppy)$_2$(acac). The HOMO and LUMO levels of the emitter are indicated by dashed lines. *Source:* From Yersin and Finkenzeller [27] Copyright (2008) Wiley-VCH; 2008. Reproduced with permission.

Blended Layer and Hybrid Molecular Structures

As an alternative to multilayer OLEDs, such as those based on the structure shown in Figure 9.22, devices incorporating blended single layers of emissive and charge transport materials may be fabricated [28]. These generally possess lower efficiencies than multilayer structures, but they have the considerable advantage of ease of manufacture. Electron transport and hole transport moieties can also be covalently incorporated into the emissive material [41]. Figure 9.26 shows an example of such a molecule – a donor-acceptor dyad [42]. Carbazole is a popular hole-transporting chemical group, while the oxadiazole unit acts as an electron transporter. A simple device architecture based on this molecule (ITO-PEDOT:PSS-carbazole-oxadiazole compound-Ca/Al) exhibited a deep-blue EL, with a high external quantum efficiency.

9.7.3 Increasing the Light Output

Efficiency Losses

The standard OLED architecture consists of two or more layers of organic material placed between two electrodes (Figure 9.22). Although the internal quantum efficiency of the EL material can be high, only a fraction of the light generated finds its way out of the device structure. Figure 9.27 indicates the origin of the main losses [43]. Due to total internal reflections caused by the mismatch in the refractive index between each layer, as well as plasmon coupling at the metal cathode, less than 20% of the emitted light may leave the device. The metal cathode

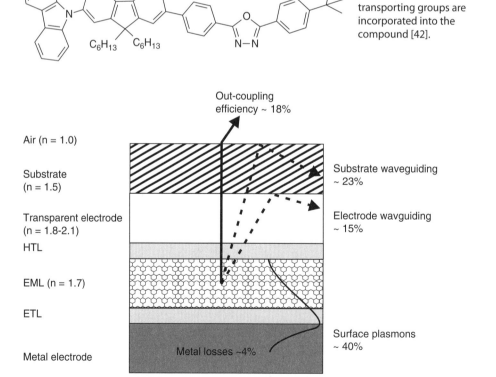

Figure 9.26 Donor-acceptor dyad for emission. Electron and hole transporting groups are incorporated into the compound [42].

Figure 9.27 Main efficiency losses in an organic light-emitting device (OLED). *Source:* Reproduced with permission from Hong and Lee [43] Copyright (2011) Springer.

is a focus of significant loss, as approximately 40% of the total emitted light can couple to surface plasmons (Chapter 4, Section 4.7). Total internal reflection can occur at the substrate-transparent anode and the substrate-air interfaces (Chapter 4, Section 4.5.2) and light can be confined within the OLED layer architecture by waveguided modes (Chapter 4, Section 4.6). Furthermore, a small proportion of the EL (4%) is absorbed at the electrodes.

Extracting light from the surface plasmon modes, thereby increasing the OLED external efficiency, is challenging. Three different strategies have been investigated: index coupling; prism coupling, and grating coupling [44]. In the first method, emission from a green dye can excite an orange dye across a thin silver layer via the plasmon modes. The prism technique is essentially a reverse Kretschmann configuration (Chapter 4, Section 4.7.2), in which dye-emission plasmon modes propagate across a thin silver layer. Rather than exciting another dye molecule, the plasmon mode emits into a silica substrate. This emission is subsequently released into air by a prism matched to the silica substrate. The third and perhaps most useful technique is to scatter the surface plasmon modes by incorporating a periodic structure onto the organic metal interface, such that the subsequent scattering recovers light from some of the surface plasmon modes. Other methods exploit dielectric materials with high refractive index values (e.g. WO_3), in an attempt to suppress the plasmon modes.

Methods to reduce the EL lost through total internal reflection are examined in the following section.

Microlenses and Shaped Substrates

Substrate waveguided light can be relatively easily extracted from the OLED by any means that disrupts the planar nature of the substrate waveguide. This disruption can be achieved by either modification of the substrate-air interface or by incorporating scattering centres into the substrate itself. Early work involved attaching a small lens directly above the OLED. Microlense arrays can also be utilized. Increases in the light output by a factor of approximately two can be achieved using such approaches [44].

A further increase in efficiency can be achieved by shaping the substrate to ensure that the maximum possible fraction of light emitted from the OLED leaves the device in the forward-scattered direction. Patterning of the substrate can enhance the effective emission angle, thereby scattering additional light energy into the viewing direction. One particularly useful structure is a preformed glass mesa, an example of which is shown schematically in Figure 9.28 [45].

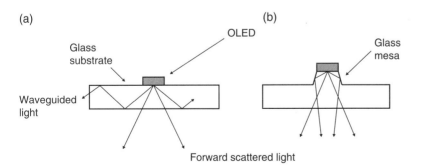

Figure 9.28 Schematic cross-section (not to scale) of an organic light-emitting device (OLED) fabricated on a shaped substrate (mesa) designed to increase the proportion of light emitted by the OLED in the forward direction. (a) Much of the emitted light is waveguided in the glass substrate and lost. (b) Most of the waveguided light is directed into the viewing direction by internal reflection from the walls of the glass mesa. *Source:* From Burrows et al. [45] Copyright (1977) by permission of IEEE.

Compared with the substrate modes, it is significantly more difficult to recover the EL lost through the waveguided modes through the anode-organic layer (e.g. ITO-HTL). Approaches that have been used include: (i) use of a substrate with a refractive index higher than the emitting layer (or reducing the index of the organic layer below that of the substrate), thereby combining the anode-organic waveguided modes with the substrate modes; (ii) disrupting the planar nature of the anode-organic waveguide by introducing scattering centres or internal reflecting surfaces; and (iii) employing photonic band gap structures within the OLED [44]. The second approach has the best combination of potential gain in light enhancement, process simplicity, and cost-effectiveness.

Microcavities

The molecules responsible for light emission exist in a medium bounded on both sides by reflective surfaces and interfaces. The optical medium is usually composed of organic layers with a total thickness of several 100 nanometers. As a consequence, the OLED stack constitutes a *microcavity*.

In Chapter 4, Section 4.5.4, interference effects from the reflection of light from the upper and lower surfaces of a thin film were shown to give rise to a series of maxima and minima in the transmitted (or reflected) electromagnetic (EM) wave. By using a stack of thin films, formed from alternate materials of different refractive indexes, these interference fringe maxima become very sharp. The particular arrangement, in which the dielectric stack is formed between two mirrors, is called a *Fabry-Perot resonator*; the structure is an example of a one-dimensional photonic crystal described in Chapter 4, Section 4.8. The term optical microcavity is also used widely to refer to such a structure with dimensions of the order of light. These have been used successfully with OLEDs to both increase their EL output and to tune their emission wavelength. The schematic structure of such an arrangement is shown in Figure 9.29. The SiO$_2$ and TiO$_2$ have different refractive indices (1.4 and 2.3, respectively) and are alternated to provide a total thickness of one-quarter of a wavelength. The OLED is then fabricated on top of this stack. The EL output of a microcavity OLED based on Alq$_3$ is contrasted with that from a reference device in Figure 9.30 [46]. The form of the EL output for the OLED fabricated on the microcavity is of narrow and intense peaks. The wavelengths of these can be predicted using calculations based on classical optics. By varying the stack, the emission colour of the OLED can be altered, providing a means to produce a range of colours from the same OLED.

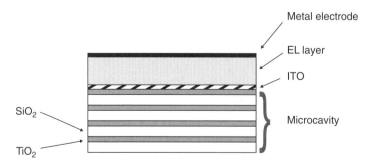

Figure 9.29 Organic light-emitting device (OLED) fabricated on an optical microcavity.

Figure 9.30 Electroluminescent (EL) output for an Alq$_3$ OLED fabricated on a microcavity compared to a reference device. *Source:* Reproduced from Nakayama [46] Copyright (1997), with permission from Taylor & Francis Group LLC.

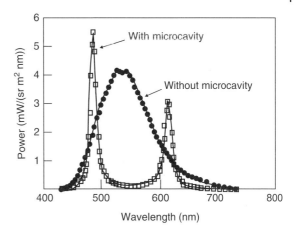

9.7.4 Full-colour Displays

An important application area for OLEDs is in flat panel displays – OLED technology that might eventually replace active-matrix LCDs (Chapter 8, Section 8.4.4). A display consists of a matrix of contacts made to the bottom and top surfaces of each organic light-emitting element, or pixel. The individual pixels may be addressed either actively or passively. In the former case, the display is addressed one line at a time, so that if a display has 2140 lines, then an individual pixel can only be emitting for 1/2140th of the time. High drive currents are needed, leading to heating problems and to expensive driver circuits. However, these issues are to some extent offset by the simplicity of the technique.

Active addressing schemes involve using a device, such as a thin film transistor, attached to each pixel. As a consequence, the pixel can remain emitting for the entire frame rather than for a small fraction of it. Organic EL technology is suited to active matrix addressing, since it is a low voltage technology and OFETs are likely to provide the drive circuitry.

To generate a full-colour image, it is necessary to vary the relative intensities of three closely spaced, independently addressed pixels, each emitting one of the three primary colours of red, green, or blue. Some of the different approaches that have been used are shown in Figure 9.31 [45]. Perhaps the simplest scheme (Figure 9.31a) is the side-by-side-positioned red, green, and blue (R, G, B) subpixels. However, this arrangement requires each of the three closely spaced OLEDs to be sequentially grown and patterned. Alternatively, optical filtering of white OLEDs can produce red, green, and blue emission (Figure 9.31b). This method is not particularly efficient, as a significant amount of light is absorbed in the filters. An alternative approach (Figure 9.31c) exploits a single blue or ultraviolet OLED to pump organic fluorescent wavelength down-converters. Each of these 'filters' consists of a material that efficiently absorbs the blue light and re-emits the energy as either green or red light.

An elegant way of achieving full-colour displays is to stack the red, green, and blue pixels on top of each other, as shown in Figure 9.31d. Since there are no subpixels, this has the advantage of increasing the resolution of the display by a factor of three. However, the method requires semi-transparent electrodes that are compatible with high current densities. One problem encountered with stacking numerous transparent organic layers is the formation of unwanted optical cavities whose resonances alter the emission spectra of OLEDs (see section above on microcavities). Such effects can be eliminated by careful control of layer thickness and composition.

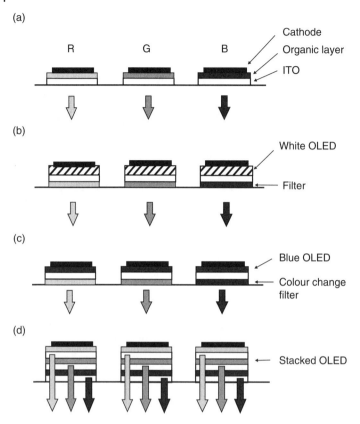

(a)

R   G   B

Cathode
Organic layer
ITO

(b)

White OLED

Filter

(c)

Blue OLED

Colour change
filter

(d)

Stacked OLED

Figure 9.31 Schemes for generating full-colour displays. (a) Separate red, green and blue emitters (R, G, and B) providing pixels side-by-side. (b) The light from white- emitting organic light-emitting devices (OLEDs) is filtered to provide R, G, and B emission. (c) The light from blue-emitting OLEDs is used to generate R, G, and B emission through colour changing filters. (d) Stacked OLEDs emit R, G, and B [45].

9.7.5 OLED Lighting

There are keen commercial and environmental interests in developing white light organic displays for lighting applications. For example, a large-area white light-emitting device will provide a solid-state light source that should compete with conventional lighting technologies [47, 48]. The OLED specifications for lighting applications are somewhat different than those for display applications. The colour, brightness, lifetime, efficiency, and CRI are all critical factors. The CIE co-ordinates required are approximately 0.31 and 0.32, although the interpretation of 'white' does vary throughout the world. For lighting, OLEDs need to deliver significantly higher luminance values than displays. In the case of the latter, values of $200 \, \mathrm{cd \, m^{-2}}$ are probably adequate, whereas a luminance of at least 10 times this figure is required if the OLED is to be used in lighting. This presents a challenge in terms of device operating lifetime. The luminous power efficiencies for incandescent light bulbs are typically less than $20 \, \mathrm{lm \, W^{-1}}$, while the figures for fluorescent tubes are in the $50{-}100 \, \mathrm{lm \, W^{-1}}$ range. Clearly OLED lights need to target the higher efficiencies of fluorescent tubes. To this end, individual white OLEDs with power efficiencies of more than $100 \, \mathrm{lm \, W^{-1}}$ and $15 \, \mathrm{cm} \times 15 \, \mathrm{cm}$ lighting panels with a power efficiency of more than $60 \, \mathrm{lm \, W^{-1}}$ have already been demonstrated [37]. Finally, for a source to be human eye-friendly, it should possess a CRI of more than 80.

Different methods of making an intrinsically white-emitting OLED by blending emissive species, either in single or multiple layers, have been studied. Alternatively, a blue OLED can be used with one or more down-conversion layers. Many white OLEDs exploit phosphorescent emitters as the fraction of excitons that can be radiative can approach 100%. There is evidence, however, that fluorescent emitters can produce more than the 25% emitting excitons predicted theoretically [46]. The triplet-triplet annihilation process can convert some of the nonradiative triplet excitons into radiative singlet excitons.

All-phosphorescent white devices can be fabricated using multiple dopants in the same emitting layer. Alternatively, a blue fluorescent emitter can be combined with phosphorescent emitters of other colours, creating a hybrid white OLED. In this case, the management of singlet and triplet excitons is crucial – to channel most of the triplet energy to the phosphorescent molecules, retaining the singlet energy on the blue fluorescent dopant [49].

One of the desirable features of OLED lighting is the ability for users to adjust its luminance (dimming). This can be achieved either by adjusting the amount of current used to drive the device or by changing the duty cycle using a pulse-width modulated driver. Unfortunately, as OLED devices are fabricated using multiple emitters, and the proportion of photons from the individual emitters can change as the current level changes, the output colour will vary with dimming.

9.7.6 Light-emitting Electrochemical Cells

In a light-emitting electrochemical cell (LEC), the light emitting layer consists of a blend of an EL organic compound (polymer, small molecule, or an ionic transition metal complex) with an electrolyte [50]. Under applied bias, the mobile anions and cations of the electrolyte accumulate at the electrodes, forming electric double layers. The effect is to reduce the charge injection barriers at the electrodes. One of the advantages of LECs over OLEDs is that they have relatively low operating voltages, which can approach E_g/e, where E_g is the band gap of the organic EL material (as for the p-i-n doped structures discussed in Section 9.7.2). LECs also provide much better EL efficiencies, due to superior charge injection balance from the two electrodes. Unlike OLEDs, LECs do not require the use of specific work-function-selected electrodes, but rather are able to make use of more stable and inert metallic materials.

The most commonly used electrolyte groups in LECs are alkali metal salts dissolved in ether-based ion transporters and ionic liquids. However, polymerizable electrolytes and mixed ion and electron conductors have been used to address specific problems, such as a slow turn-on time and phase separation.

9.7.7 Organic Light-emitting Transistors

Organic light-emitting transistors (OLETs) combine the functionalities of OFETs and OLEDs [51, 52]. In the case of the OFET, the injection of either electrons or holes from the source and drain contacts is sufficient for successful device operation. In contrast, the simultaneous injection of electrons and holes, followed by their subsequent recombination, are required in an OLET. Figure 9.32a depicts a typical OLET architecture. Electrons and holes are injected from the source and drain contacts and move towards each other under suitable bias conditions. In practice, a good match is needed between the work functions of the electrodes and the LUMO, HOMO levels of the organic semiconductor.

The materials requirements for OLETs also differ from those in OLEDs. For the latter devices, the carrier mobilities in the light-emitting layer are not generally important parameters, since the 'channel' length of an OLED is typically a few tens of nanometres. However, the electrons and holes must travel significantly further distances in the transistor structure and good carrier mobilities are needed. Organic semiconductors that have been used successfully in OLET architectures include fluorine derivatives, PPV-based polymers, acenes, oligothiophenes, furan-incorporated oligomers, spirobifluorenes, and various phosphorescent compounds [51].

(a)

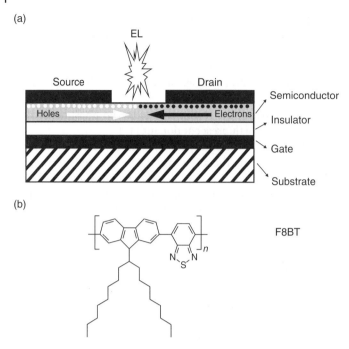

Figure 9.32 (a) Architecture of an organic light-emitting transistor (OLET). (b) Chemical structure of the green-emitting polymer, poly(9,9-di-*n*-octylfluorene-*alt*-benzothiadiazole) (F8BT) used in OLETs.

(b)

F8BT

Both unipolar and ambipolar device operation have been demonstrated. The majority of OLETs operate in the unipolar (p-channel) regime, in which the hole density extends all the way across the transistor channel, such that the hole accumulation layer functions as the anode for the OLED with electron injection by tunnelling from the drain electrode. Early work on OLETs exploited acenes such as tetracene to achieve light emission. For unipolar OLETs, the light generation is restricted to a region close to the minority carrier injection electrode; this can reduce the light emission due to metal contact-induced exciton quenching. Improvements in the carrier injection may be achieved by using asymmetric work function source and drain contacts, for example Al for electrons and Au for holes.

The first ambipolar OLETs were fabricated using a bulk heterojunction (BHJ) as the active component, by mixing two materials with complementary properties. In such devices, the position of the exciton recombination zone can be controlled by using appropriate (gate) biasing conditions. Efficient photoluminescent organic compounds, which also possess balanced electron and hole mobilities, have now been developed. An example is the green-emitting polymer, poly(9,9-di-*n*-octylfluorene-*alt*-benzothiadiazole) (F8BT), the chemical structure of which is depicted in Figure 9.32b. Bilayer OLETs, which have separate layers for the charge transport and EL, have also been reported. Other modifications to improve the operation of the basic OLET architecture, shown in Figure 9.32a, include the use of a split-gate architecture, non-planar electrodes, and a vertical device configuration [51].

As high current densities can be obtained and the light emission can be chosen to occur far away from absorbing metal electrodes, OLETs are a versatile and attractive architecture, not only for low-loss light signal transmission in optoelectronic integrated circuits but also for the realization of electrically pumped organic lasers. OLETs may also constitute a key element for the development of next-generation organic active matrix display technology (Section 9.7.4), since these devices combine electrical switching with light emission.

9.7.8 Electronic Paper

Perhaps the ultimate goal for portable displays in terms of readability, ease-of-access and use, ruggedness, and ultra-low power consumption is the electronic emulation of paper – *e-paper* [53, 54]. Paper is a reflective 'device' with no energy consumption. Numerous e-paper technologies have now been developed that surpass conventional LCDs in sunlight viewability and in low-power consumption. The two main approaches that have been successfully developed are *electrophoretic* and *electrowetting*.

In the simplest implementation of an electrophoretic display, titanium dioxide particles approximately 1 μm in diameter are dispersed in a hydrocarbon oil between two conductive plates. A dye is added to the oil, together with other compounds that cause the particles to become charged in an electric field. With a voltage applied, the particles drift towards the oppositely-charged electrode. When the particles are located at the viewing side of the display, this appears white because of the light scattered from the high refractive index TiO_2. If particles are located at the rear of the display, this appears dark because the incident light is absorbed by the dye. Examples of commercial electrophoretic e-paper devices include the high resolution active matrix displays used in the Amazon Kindle and the Sony Reader.

An electrowetting display (as produced, for example, by Liquavista) exploits the controlling of the shape of a confined water-oil interface with an electric field. With no voltage applied, the (coloured) oil forms a flat film between the water and a hydrophobic insulating coating on the electrode, resulting in a coloured pixel. If a voltage is applied between the electrode and the water, the interfacial tension between the water and the coating changes. This results in a partly transparent pixel. Displays based on electrowetting can respond sufficiently fast to display video information.

Rollable e-paper is a very appealing electronic product. This requires a bending radius of 5–10 mm, which is easily achievable by a graphene electrode. In addition, graphene's uniform absorption across the visible spectrum is appropriate for colour e-papers. However, the contact resistance between the graphene electrode and the metal line of the driving circuitry is a problem.

9.8 Organic Photovoltaic Devices

Concerns over global climate change, local air pollution, and resource depletion are making photovoltaics (PVs) an increasingly attractive method of energy supply. The current technology is based on single crystal silicon solar cells. These have developed since the 1940s and offer conversion efficiencies of 15–20% for off-the-shelf commercial devices (although higher figures are reported in the laboratory). However, the technology is more expensive than conventional power generation. Options for reducing photovoltaic generating costs include working on incremental advances to improve silicon efficiencies, but also developing fundamentally different devices based on alternative materials. The field of organic photovoltaics (OPVs) dates back to 1959, when it was discovered that anthracene could be used to make a solar cell [55]. Photovoltaics based on organic compounds, such as polymers or dyes, offer the possibility of large-scale manufacture at low temperature coupled with low cost [56–59]. Until the end of the twentieth century, little progress had been made with OPV devices, and energy conversion efficiencies of up to only about 1% were achieved. The availability of new conductive organic materials and different OPV designs has significantly improved on this figure, with a figure of 13.2% reported in 2016 [60].

9.8.1 Photovoltaic Principles

The structure of a simple organic solar cell is depicted in Figure 9.33. In essence, the device is very similar to the OLED described in the previous section. If the incoming photons have energy greater than the band gap of the polymer (or greater than the HOMO-LUMO separation in the case of organic molecular materials) then these will be absorbed creating electrons and holes. In an inorganic photovoltaic cell, these electrons and holes would be generated within, or close to, a depletion region in the semiconductor and they would be free to migrate to opposite electrodes, where they can do useful work in an external electrical load. However, in the organic material, the electrons and holes are bound together in excitons (Chapter 4, Section 4.4.3). An immediate problem in OPV cells is to split these excitons. This can be conveniently accomplished at an interface, the simplest interface being the junction between the electrodes and the organic material. Under open-circuit conditions, holes are collected at the high work function electrode (e.g. ITO) and electrons at the low work function electrode (e.g. Al).

The idealized current versus voltage behaviour of the OPV cell is shown in Figure 9.34. In the dark (Figure 9.34a), the device behaves as a diode. Under illumination, the I-V characteristic is

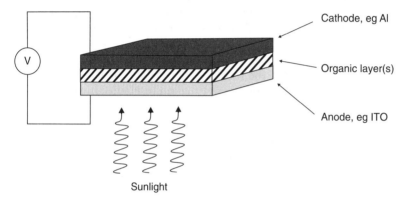

Figure 9.33 Schematic diagram showing the structure of a photovoltaic cell based on a semiconductive organic compound.

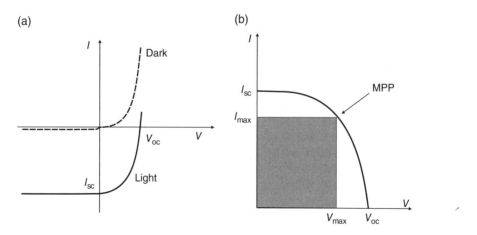

Figure 9.34 (a) Current I versus voltage V characteristics for a photovoltaic structure in the dark and in the light. V_{oc} = open-circuit voltage. I_{sc} = short-circuit current. (b) Expanded region of I-V curve showing the maximum voltage V_{max} and the maximum current I_{max} available from the solar cell for optimized power output. The shaded area represents the maximum power available from the cell. MPP = maximum power point.

displaced down the current axis as a photocurrent is generated. Two important device parameters, shown in Figure 9.34a, are V_{oc}, the open-circuit voltage, and I_{sc}, the short-circuit current. The former represents the voltage appearing across the terminals of the solar cell when no current is drawn from the device; this is the maximum attainable voltage that the OPV device can provide. When the terminals are short-circuited, I_{sc} represents the maximum current that the solar cell can provide. The *I-V* characteristic under illumination is shown in more detail in Figure 9.34b. The actual operating point on a given solar-cell *I-V* characteristic is determined by the load resistance that is connected between its terminals. To maximize the output power, it is desirable to choose the value of load resistance such that the operating point is at the maximum power point, MPP in Figure 9.34b. In this case, the current and voltage outputs of the photovoltaic cell are I_{max} and V_{max}, respectively. The energy conversion efficiency η is given by

$$\eta = \frac{I_{max}V_{max}}{P_i A} = \frac{I_{sc}V_{oc}F}{P_i A} = \frac{J_{sc}V_{oc}F}{P_i} \tag{9.12}$$

where P_i is the incident solar power density, J_{sc} is the short-circuit current density, and A is the area of the solar cell. The ratio of $(I_{max}V_{max})$ to $(I_{sc}V_{oc})$ is defined as the fill-factor, F, of the solar cell. It is clearly desirable to produce devices with F values as close to unity as possible. Figure 9.35 shows the equivalent circuit of a simple OPV structure. Imperfections in the device leading to current leakage are represented by a shunt resistance, R_{shunt}, and parasitic series resistance effects are represented by R_{series}. The current generated by the incoming EM radiation is taken into account by the incorporation of a constant current generator of value I_l. Note that a forward dark current I_d will flow through the rectifying junction. The load resistance, voltage, and current are given by R_{load}, V_{load}, and I_{load} in Figure 9.35. Using simple circuit theory, it can be shown that [61]

$$I_{load}\left(1 + \frac{R_{series}}{R_{shunt}}\right) = I_l - I_d - \frac{V_{load}}{R_{shunt}} \tag{9.13}$$

This equation can be solved numerically to ascertain the effects of R_{series} and R_{shunt} on the cell output characteristics. However, it is evident that, in general, it is important for R_{series} to be small and for R_{shunt} to be large.

9.8.2 Bulk Heterojunctions

Although OPV devices can be based on a single organic semiconductor sandwiched between high and low work function metals (in which case the device resembles the Schottky diode

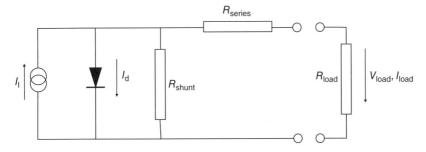

Figure 9.35 Equivalent circuit of a photovoltaic solar cell. I_l is the current generated by the solar radiation. I_d = forward biased current through rectifying barrier. R_{series}, R_{shunt} and R_{load} represent the series, shunt, and load resistors. V_{load} and I_{load} are the voltage and current, respectively, associated with the load resistor.

described in Section 9.2.1), a more common configuration for the active organic layer is a bilayer blend of an electron donor material and an electron acceptor material. The difference between the LUMO energy levels between the materials (rather than the work function differences of the electrodes) then creates the driving force, whereby dissociation by rapid electron transfer from the donor to the acceptor can occur. Following exciton dissociation, the free carriers created at the interface are ideally transported to their respective electrodes under the influence of the internal electrical field. Two important loss mechanisms can occur to limit this. The first, termed *geminate recombination*, is that the electron–hole pair created at the interface recombines by the transfer of the electron back from the LUMO level of the acceptor to the HOMO level of the donor. The other possibility, termed *bimolecular recombination* (or *non-geminate recombination*) occurs when the dissociated free carriers recombine before reaching the electrodes; this process may be trap-assisted, as in Shockley-Read-Hall recombination, Chapter 3, Section 3.3.5. For soluble organic materials such as polymers, the exciton diffusion length is relatively small, of the order of 10 nm or so. If the exciton does not encounter a donor-acceptor interface during its lifetime, it can undergo non-radiative decay in a loss mechanism for the cell. For organic devices, film thicknesses of 100–200 nm are required to absorb all of the incident light (thicker layers will provide an unacceptable series resistance). In the case of a simple bilayer architecture, illustrated in Figure 9.36a, this would result in the majority of the excitons being created away from the donor-acceptor interface and they would not be harvested. The answer to this problem is to mix the donor and acceptor species in a so-called *bulk heterojunction* (BHJ). Figure 9.36b illustrates such a theoretical blend of donor and acceptor materials. A more practical way to achieve the desired material morphology is to exploit solution processing, in which rapid drying leads to a random network of donor-acceptor domains (Figure 9.36c). However, continuous percolation pathways are required through each phase to transport the electrons and holes to their respective electrodes. This charge transport will be strongly influenced by the local order and crystallinity within each phase, with crystalline morphologies generally leading to the highest charge carrier mobilities.

One method of controlling the nanoscale morphology of a BHJ device is to add a small fraction of a higher boiling point co-solvent to the blend solution before spinning. This can allow the BHJ morphology to continue to evolve in a solvent-rich environment. Post deposition methods can also be used to enhance the phase separation between the donor and acceptor species. These include thermal annealing and solvent annealing. In the latter technique, the organic material is exposed to solvent vapours that have the ability to dissolve selectively in one of the components of the thin film, mobilizing it sufficiently to change the morphology. With appropriate control of the nanoscale morphology of the blend, all excitons can be created within a diffusion length of a donor-acceptor interface, and hence be harvested.

Figure 9.36 shows two materials that have been used in BHJ OPV cells – poly(2-methoxy-5-(3′,7′-dimethyloctyloxy)-1,4-phenylenevinylene) (MDMO-PPV) copolymer as the donor blended with [6,6]-phenyl-C_{61} butyric acid methyl ester (PCBM) as the acceptor. However, numerous other compounds have been studied. Polymer donors include MEH-PPV, polythiophenes, fluorene-containing polymers, carbazole-based polymers, and low-band gap compounds such as polymers containing benzothiadiazole moieties. Acceptor molecules are generally based on fullerenes, such as C_{60} and C_{70} [62]. The material requirements for the BHJ structures include the ability for the donor and acceptor species to form the required nano-morphology, and the resulting domains to be sufficiently conductive to allow the charges to be transported efficiently to the electrodes. The optical absorption of the photoactive layer must also be tuned to match the incoming photon flux from the solar radiation. The peak in the solar spectrum corresponds to a wavelength of approximately 600 nm (depending on location on the Earth's surface) and it has a useful range of about 300–2000 nm. For photovoltaic devices, the most

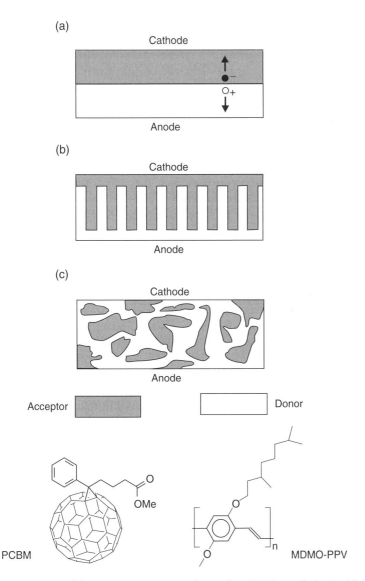

Figure 9.36 Bulk heterojunction organic photovoltaic (OPV) morphologies. (a) Bilayer structure. (b) Theoretical ordered blend of donor and acceptor domains. (c) Random domain network. The chemical structures of [6,6]-phenyl-C_{61} butyric acid methyl ester (PCBM) (acceptor) and poly(2-methoxy-5-(3′,7′-dimethyloctyloxy)-1,4-phenylenevinylene) (MDMO-PPV) (donor) are shown at the bottom.

effective conversion of this radiation to electrical energy is achieved using semiconductors with a band gap of around 1000 nm (i.e. $E_g \approx 1.2\,\text{eV}$).

For BHJ type OPV devices, the upper limit to V_{oc} is determined by the energy difference between the LUMO level of the donor and the HOMO level of the acceptor. Exciton binding and other factors further reduce it. High band gap materials will possess higher V_{oc} values, but will harvest fewer incoming photons than low band gap materials. Reported open-circuit voltages for simple BHJ devices are generally between 0.5 and 1 V. However, significantly higher figures (> 1.5 V) are achieved using tandem cells [59]. These architectures generally use a wide band gap material as the first charge separation layer and a narrow band gap material as the second charge separation layer with a thinner front cell than back cell so that the photocurrents are balanced. If the overall cell thickness is too high, the effect is to reduce both J_{sc} and F. The short-circuit current density values achieved with OPV architectures are usually less than $15\,mA\,cm^{-2}$. There have been some exceptions, usually achieved using low band gap materials [59]. The fill-factor of an OPV cell is a complex function of the electrical resistance elements in the device (Figure 9.35), which in part depend on the sheet resistance of the electrodes. The F values of OPV cells vary greatly, but are mostly less than 0.6 [59]. It has been suggested that the energy conversion efficiency (in %) of simple OPV cells is limited to 0.55 × short-circuit current density (in units of $mA\,cm^{-2}$), a rule obeyed by many inorganic photovoltaic cells; however, for tandem cells, the maximum efficiency could be doubled to ≈ $1.1 \times J_{sc}$ [59].

One significant development in organic solar cells has been the shift from the 'normal' geometry of the OPV stack, where electrons exit from the top electrode (usually Al, as shown in Figure 9.33) and holes at the bottom (usually ITO) to an inverted configuration where electrons exit at the bottom and holes at the top [63]. Electron and hole extraction layers are needed either side of the active organic layer in order to direct the carriers to the appropriate electrodes. The change in architecture allows the use of other metals (i.e. Au or Ag) to be used as the top electrode (now the anode), which are much more resistant to oxygen and water. The result is usually a far more stable OPV device that, with proper encapsulation, may last for many years. A further variation on the OPV structure shown in Figure 9.33 is to use top illumination. Here, the top electrode must be semi-transparent and generally takes the form of an ultra-thin metal film/grid or a highly conductive organic material such as carbon nanotubes or graphene.

While marked progress has been achieved with OPVs over the last decade, significant problems remain. The stability of the devices is considered a major problem. A further challenge is the transition for the square millimetre devices prepared in the laboratory to large-scale technical production. The devices will also need to compete with other low-cost and large area PV technologies, such as polycrystalline silicon, amorphous silicon and perovskite-based cells [64].

The OPV configuration has also been used successfully to produce organic photodiodes for imaging technologies. Such structures offer cheaper processing methods and devices that are light, flexible, and compatible with large (or small) areas [65, 66]. The photodiodes can be integrated onto metal–semiconductor–semiconductor compatible electrodes [67]. A further intriguing device, up converting near-infrared radiation to visible light, can be realized by the combination of a BHJ architecture and a phosphorescent OLED [68].

9.8.3 Dye-sensitized Solar Cell

In 1991, Grätzel and co-workers manufactured an artificial photosystem containing a light-harvesting antenna, using a dye-sensitized titanium dioxide thin film with a redox electrolyte [69, 70]. This dye-sensitized photovoltaic device operates in the following manner, depicted schematically in Figure 9.37. Upon photon absorption, a dye electron is elevated to an excited state. The dye then injects the excited electron directly into the conduction band of a semiconductor, which subsequently creates a current in an external circuit. The regeneration of the oxidized dye takes place at the counter electrode by means of an electron-donor species (typically a liquid electrolyte based on the redox couple iodide-triiodide). The early cells used a thin film of nanoparticles of titanium dioxide as the semiconductor, offering a high interfacial surface area leading to rapid and efficient charge transfer.

Two factors combined to give a high, 7%, conversion efficiency for Grätzel's cell. First, the ruthenium based dye and the titanium oxide film were able to absorb 46% of the incoming solar

Figure 9.37 Carrier transport in dye-sensitized solar cell.

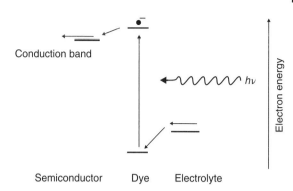

flux. By using a light-harvesting antenna, Grätzel was also able to separate light absorption and charge transport. This provided the cell with a more superior fill factor than a conventional silicon solar cell, since exciton recombination losses were minimized as there was no exciton travelling through the semiconductor defects. In 2015, conversion efficiencies of up to 13% were achieved for organic solvent-based liquid electrolyte cells [71]. However, a major problem is the contamination of the cell by moisture/water. Even the most robust encapsulation arrangements have been unable to completely eliminate this. Several alternatives to organic solvents have been proposed, for example solid-state conductors, plastic crystals, and gels. An interesting development is the replacement of the traditionally used organic solvents with those based on water. Photoanode modifications, the introduction of additives and surfactants, the selection of specifically conceived redox couples, the preparation of suitable cathodes, and the stabilization of electrolytes, have progressively led to the fabrication of 100% aqueous cells, with efficiencies of up to about 5% [71].

9.8.4 Luminescent Concentrator

The luminescent concentrator has been under study since the 1970s [72–74]. Such a structure can eliminate the visual impact and minimize the amount of expensive semiconductor material required in photovoltaic solar cells. The concept is based on incorporating, in a transparent matrix, materials that are both absorbing and luminescent. The refractive index of the transparent matrix is chosen to be larger than the surrounding medium (air). As the luminescent materials emit light in all directions, part of the emitted light will be internally reflected at the matrix-air interface, as shown schematically in Figure 9.38. By designing geometries for which the length and width of the transparent matrix are larger than its thickness (as for a plate configuration), geometric concentration is accomplished. The emitted light is totally internally reflected and will be concentrated within the matrix where it can be harvested at the edges. There, it can be transformed into electrical energy by a photovoltaic cell. Although increases in photon flux by a factor of one hundred, compared with exposing the same photocell to direct sunlight, have been predicted, only modest overall system efficiencies ($\approx$ 5%) have been achieved [73]. (It is important to note that the efficiency figures are somewhat deceptive, as they are entirely dependent on the nature of the attached photovoltaic cell.) This is partly due to the fact that for the organic luminescent materials used, the luminescent wavelengths are only slightly higher than the absorption wavelengths. Under these conditions, a significant part of the emitted light is lost by re-absorption. However, the design of newer materials, based on quantum dots and photonic crystals, is offering renewed interest in this area.

In 2014, researchers at Michigan State University designed and fabricated the first visibly-transparent luminescent concentrator devices, which selectively harvest near-infrared photons

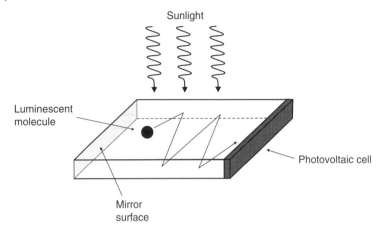

Figure 9.38 Schematic diagram of a luminescent flat plate concentrator. Direct and diffuse sunlight are absorbed by luminescent molecules. The emission is confined to the thin plate by total internal reflection. The radiation is collected by a photovoltaic cell at one end of the plate.

based on fluorescent organic salts (cyanine derivatives) [75]. These devices exhibit a clear visible transparency similar to that of glass. It is suggested that efficiencies of over 10% are possible due to the large fraction of photon flux in the near-infrared spectrum.

9.9 Other Application Areas

9.9.1 Conductive Coatings

Conductive coatings have applications as diverse as electromagnetic shielding and corrosion control. Traditionally, the simplest systems have consisted of fine metallic particles dispersed in an insulating polymer matrix. However, as discussed in the earlier parts of this chapter, the development of intrinsically conductive polymers, which can be processed by methods such as spin-coating and solution casting, has opened up further opportunities. Carbon nanotubes and graphene can also provide highly conductive, albeit expensive, thin films [76].

Static electrical charge can accumulate on the surfaces of insulating plastic (and other) surfaces by build up on insulating surfaces by the *triboelectric effect*. These lead to the attraction of dust, the difficulty in handling very thin films and, in some circumstances, to the generation of electric shocks and sparks. Static charging can be reduced significantly or eliminated by the addition of *antistatic agents*. These compounds work either by making the surface of the material itself slightly conductive, or by absorbing moisture from the air. In the former case, a conductive polymer such as PEDOT:PSS (Chapter 5, Section 5.3) can be conveniently used. Surfactants, possessing both hydrophilic and hydrophobic groups (Chapter 8, Section 8.2.2), can be exploited to promote the uptake of water on the surface of an insulator. The hydrophobic moiety interacts with the surface of the material, while the hydrophilic part of the molecule interacts with the air moisture and binds the water molecules. Common antistatic agents are based on long-chain aliphatic amines and amides and quaternary ammonium salts. Antistatic agents are also added to some aircraft jet fuels, to impart electrical conductivity and to avoid the build-up of static charges, which could lead to sparking igniting the fuel vapours. The development of textiles that can dissipate static electricity, and can also screen the microwave radiation from mobile telephones, has led to the incorporation of conductive polymers into fibres.

9.9.2 Batteries, Supercapacitors, and Fuel Cells

Lithium-ion batteries (sometimes abbreviated to Li-Ion) are a type of rechargeable battery commonly used in consumer electronics [77]. These are currently one of the most popular types of battery, with one of the best energy-to-weight ratios, no memory effect, and a slow loss of charge when not in use. However, one drawback of the Li-ion battery is that its life span is dependent upon ageing from time of manufacturing (shelf life), regardless of whether it is charged, and not just on the number of charge-discharge cycles.

Lithium-ion polymer batteries, or more commonly *lithium polymer batteries* (abbreviated Li-Poly or LiPo), are rechargeable and have technologically evolved from lithium-ion batteries. Ultimately, the lithium salt electrolyte is not held in an organic solvent like in the proven lithium-ion design, but in a solid polymer composite such as polyacrylonitrile. There are many advantages of this design over the classic lithium-ion design, including the fact that the solid polymer electrolyte is not flammable (unlike the organic solvent that the Li-Ion cell uses). Since no metal battery cell casing is needed, the battery can be lighter and can be specifically shaped to fit the device it will power. Because of the denser packaging without intercell spacing between cylindrical cells and the lack of metal casing, the energy density of Li-Poly batteries is over 20% higher than that of a classical Li-Ion battery and approximately three times better than nickel-cadmium (NiCd) nickel metal hydride (NiMH) batteries. Commercialized technologies use an ion-conducting polymer instead of the traditional combination of a microporous separator and a liquid electrolyte. This promises not only better safety, as polymer electrolyte does not burn as easily, but also the possibility of making battery cells very thin, as they do not require pressure applied to 'sandwich' the cathode and anode together. The polymer electrolyte seals both electrodes like glue. Other organic batteries that have been developed use one metallic electrode, for example lithium, and one polymer electrode, such as polypyrrole or polyaniline. Graphene may also be used. However, further research is needed before lithium polymer devices are competitive with other battery technologies.

Supercapacitors (or *ultracapacitors*) are devices capable of managing high power rates compared to batteries [78, 79]. Although supercapacitors provide 10^2–10^3 times higher power in the same volume, they are not able to store the same amount of charge as batteries do. This makes supercapacitors suitable for those applications in which power bursts are required. Supercapacitors do not use the conventional solid dielectric of ordinary capacitors. Instead, they exploit electrostatic double-layer capacitance or electrochemical pseudo-capacitance, or a combination of both. Electrostatic double-layer capacitors use carbon electrodes or derivatives with much higher electrostatic double-layer capacitance than electrochemical pseudo-capacitance, achieving separation of charge in a *Helmholtz double layer* at the interface between the surface of a conductive electrode and an electrolyte. The separation of charge is typically 0.3–0.8 nm, much smaller than in a conventional capacitor. Electrochemical capacitors consist of two electrodes separated by an ion-permeable membrane (separator), and an electrolyte ionically connecting both electrodes. When an applied voltage polarizes the electrodes, ions in the electrolyte form electric double layers of opposite polarity to the electrode's polarity. For example, positively polarized electrodes will have a layer of negative ions at the electrode-electrolyte interface, along with a charge-balancing layer of positive ions adsorbing onto the negative layer. The opposite is true for the negatively polarized electrode.

Electrochemical pseudo-capacitors use metal oxide or conducting polymer electrodes with a high amount of electrochemical pseudo-capacitance. The capacitance is achieved by Faradaic electron charge-transfer with redox reactions, intercalation, or electrosorption. Hybrid capacitors, such as the lithium-ion capacitor, use electrodes with differing characteristics: one exhibiting mostly electrostatic capacitance and the other mostly electrochemical capacitance.

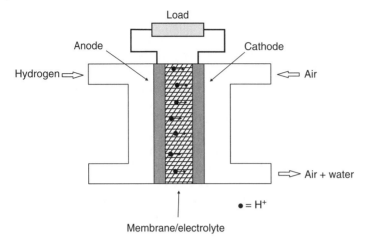

Figure 9.39 Schematic diagram of a hydrogen fuel cell.

A variety of conductive polymers, including polypyrrole, polyaniline, PEDOT, polythiophene, and PPV, have been studied for use as electrode materials in supercapacitors. However, the relatively low conductivities of these materials has limited their usefulness. Other organic materials under investigation include carbon nanotubes and graphene [80].

A *fuel cell* is an electrochemical energy conversion device similar to a battery, but designed for continuous replenishment of the reactants consumed. This device produces electricity from an external supply of fuel and oxygen as opposed to the limited internal energy storage capacity of a battery. Additionally, the electrodes within a battery react and change as a battery is charged, or discharged, whereas a fuel cell's electrodes are catalytic and relatively stable. Perhaps the most important device is the hydrogen fuel cell, shown schematically in Figure 9.39. The anode and cathode reactions are as follows:

$$H_2 \rightarrow 2H + 2e^-$$
$$O_2 + 4H^+ + 4e^- \rightarrow 2H_2O$$

(9.14)

On the anode side, hydrogen diffuses to the anode catalyst, where it dissociates into protons and electrons. The protons are conducted through the membrane to the cathode, but the electrons are forced to travel in an external circuit (supplying power), because the membrane is electrically insulating. On the cathode catalyst, oxygen molecules react with the electrons (which have travelled through the external circuit) and protons to form water. In this example, the only waste product is water vapour and/or liquid water. In addition to pure hydrogen, there are hydrogen-carrying fuels for fuel cells, including diesel, methanol, and chemical hydrides, and the waste product with these types of fuel is carbon dioxide.

Fuel cells with solid polymer electrolyte membranes separating the anodic (fuel) compartment from the cathodic (oxidant) compartment were first introduced in the Gemini space programme in the early 1960s. Those cells were expensive and had short lifetimes, because of the poor stability of the membranes (based on sulphonated polystyrene–divinylbenzene). More successful polyelectrolytes subsequently developed are based on perfluorinated polymers, the first of which was Nafion®, with the structure depicted in Figure 9.40a [81]. Nafion's unique ionic properties are a result of incorporating perfluorovinyl ether groups terminated with sulphonate groups onto a tetrafluoroethylene backbone. There is a range of similar polyelectrolytes with the same generic structure, differing only in the values of x, y, m, and n and either with or

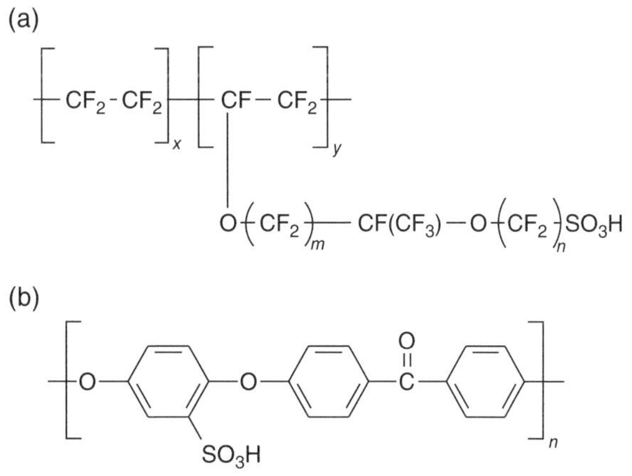

Figure 9.40 Ionic conductors: (a) Nafion® and (b) PEEK – poly(etheretherketone).

(a)

(b)

without —(CF(CF$_3$))— moieties in the pendant. Such material readily absorbs water to achieve conductivities as high as $1\,S\,m^{-1}$. Other, lower-cost proton-conducting electrolytes using aromatic polymers, which exhibit high thermal and chemical stability, have been developed. One example, poly(etheretherketone) (PEEK), is shown in Figure 9.40b; this compound may be sulphonated to produce a polyelectrolyte. PEEK is a thermoplastic with extraordinary mechanical properties. The Young's Modulus is 3.6 GPa and its tensile strength is 170 MPa. PEEK is partially crystalline, and has a glass transition temperature of 143 °C, a melting temperature of 334 °C, and is highly resistant to thermal degradation. The material is also resistant to both organic and aqueous environments, and is used in bearings, piston parts, pumps, compressor plate valves, and cable insulation applications.

Problems

9.1 The eye has sensitivity to light of about 10^{-17} watt. What is the minimum number of photons per second (flux) needed for the eye to see something?

9.2 Consider a silicon MOSFET with a gate oxide thickness of 5 nm. If the transistor's gate has both width and length of 100 nm, estimate the number of electrons in the channel for an applied gate bias of 1 V. Hence, calculate the change in gate voltage for the addition of 10 electrons to the channel. Comment on your answer.

9.3 A multilayer OLED is based on the architecture Al-Alq$_3$-NPB-MTDATA-ITO (you will need to research some of these compounds). What is the function of each layer? The thicknesses of the Alq$_3$ and 4,4′,4″-tris[phenyl(m-tolyl)amino]triphenylamine (MTDATA) are both fixed at 70 nm, while that of the NPB layer is varied. To achieve a fixed current density of 80 mA cm^{-2} the voltage applied to the OLED (drive voltage) increases with the thickness of the NPB layer, as shown in the table below.

NPB layer thickness [nm]	Driving voltage [V]
10	12.0
20	12.8
40	14.4
60	16.0

Assuming that the voltage drop in the MTDATA layer is negligible, calculate the electric field in the NPB layer. Estimate the electric field in the Alq$_3$ layer.

9.4 A BHJ OPV cell uses P3HT as the donor and PCBM as the acceptor components. The LUMO levels for these compounds are −3.2 eV (P3HT) and −3.7 eV (PCBM), whereas the HOMO levels are −5.0 eV (P3HT) and −6.0 eV (PCBM). Draw a potential energy band diagram for the interface between the donor and acceptor. What wavelengths of incoming solar radiation will these materials absorb? Estimate the maximum open-circuit voltage for the heterojunction device. How might the HOMO and LUMO levels of the donor and acceptor compounds be modified to improve the device efficiency?

9.5 An OFET uses an evaporated p-type organic semiconductor (hole mobility = 0.207 cm^{2}V^{-1}s^{-1}) as the semiconductor and a solution processed polymer (dielectric constant = 2.5) as the gate insulator. The channel length L = 20 μm, channel width W = 200 μm, and the gate insulator thickness is 250 nm. If the threshold voltage V_t = −1.0 V, calculate the drain current in nA (to 3 significant figures) for an applied gate voltage V_g = −2.0 V.

References

1 Kuo, C.S., Wakin, F.G., Sengupta, S.K., and Tripathy, S.K. (1994). Schottky and metal-insulator-semiconductor diodes using poly(3-hexylthiophene). *Jpn. J. Appl. Phys.* 33: 2629–2632.

2 Steudel, S., Myny, K., Arkhipov, V. et al. (2005). 50 MHz rectifier based on an organic diode. *Nat. Mater.* 4: 597–600.

3 Sze, S.M. (1969). *Physics of Semiconductor Devices*. New York: Wiley.

4 Grecu, S., Bronner, M., Opitz, A., and Brütting, W. (2004). Characterization of polymeric metal-insulator-semiconductor diodes. *Synth. Met.* 146: 359–363.

5 Sirringhaus, H. (2014). 25th anniversary article: organic field effect transistors: the path beyond amorphous silicon. *Adv. Mater.* 26: 1319–1335.

6 Watkins, N.J., Yan, L., and Gao, Y. (2002). Electronic structure symmetry of interfaces betweenpentacene and metals. *Appl. Phys. Lett.* 80: 4384–4386.

7 Benor, A. and Knipp, D. (2008). Contact effects in organic thin film transistors with printed electrodes. *Org. Electron.* 9: 209–219.

8 Ante, F., Kälblein, D., Zaki, T. et al. (2012). Contact resistance and megahertz operation of aggressively scaled organic transistors. *Small* 8: 73–79.

9 Kymissis, I. (2009). *Organic Field Effect Transistors: Theory, Fabrication and Characterization*. New York: Springer.

10 Yun, Y., Pearson, C., Cadd, D.H. et al. (2009). A cross-linked poly(methyl metacrylate) gate dielectric by ion-beam irradiation for organic thin-film transistors. *Org. Electron.* 10: 1596–1600.

11 Yi, H.T., Payne, M.M., Anthony, J.E., and Podzorov, V. (2012). Ultra-flexible solution-processed organic field-effect transistors. *Nat. Commun.* 3: 1259.

12 Zhang, W., Smith, J., Watkins, S.E. et al. (2010). Indacenodithiophene semiconducting polymers for high-performance, air-stable transistors. *J. Am. Chem. Soc.* 132: 11437–11439.

13 Niazi, M.R., Li, R., Li, E.Q. et al. (2015). Solution-printed organic semiconductor blends exhibiting transport properties on par with single crystals. *Nat. Commun.* 6: 8598.

14 Chen, K., Gao, W., Emaminejad, S. et al. (2016). Printed carbon nanotube electronics and sensor systems. *Adv. Mater.* 28: 4397–4414.

15 Klauk, H., Halik, M., Zschieschang, U. et al. (2003). Pentacene organic transistors and ring oscillators on glass and on flexible polymeric substrates. *Appl. Phys. Lett.* 82: 4175–4177.

16 Zhao, Y., Guo, Y., and Liu, Y. (2013). 25th anniversary article: recent advances in n-type and ambipolar organic field effect transistors. *Adv. Mater.* 25: 5372–5391.

17 Xiong, W., Guo, Y., Zschieschang, U. et al. (2015). A 3-V, 6-bit C-2C digital-to-analog converter using complementary organic thin-film transistors on glass. *IEEE J. Solid-State Circuits* 45: 1380–1388.

18 Baude, P.F., Ender, D.A., Haase, M.A. et al. (2003). Pentacene-based radio-frequency identification circuitry. *Appl. Phys. Lett.* 82: 3964–3966.

19 Fiore, V., Battiato, P., Abdinia, S. et al. (2015). An integrated 13.56-MHz RFID tag in a printed organic complementary TFT technology on flexible substrate. *IEEE Trans. Circuits Syst. Regul. Pap.* 62: 1668–1677.

20 Elze, T., Taylor, C., and Bex, P.J. (2013). An evaluation of organic light emitting diode monitors for medical applications: great timing, but luminance artifacts. *Med. Phys.* 40: 92701.

21 Noctura Sleep Mask. Available from URL. https://noctura.com. Accessed 29 September 2017.

22 Tsang, C.W. and VanSlyke, S.A. (1987). Organic electroluminescent diodes. *Appl. Phys. Lett.* 51: 913–915.

23 Burroughes, J.H., Bradley, D.D.C., Brown, A.R. et al. (1990). Light-emitting diodes based on conjugated polymers. *Nature* 347: 539–541.

24 Moro, L., Boesch, D., and Zeng, X. (2015). OLED encapsulation. In: *OLED Fundamentals: Materials, Devices, and Processing of Organic Light-Emitting Diodes* (ed. D.J. Gaspar and E. Polikarpov), 25–65. Boca Raton, FL: CRC Press.

25 Gaspar, D.J. and Polikarpov, E. (eds.) (2015). *OLED Fundamentals: Materials, Devices, and Processing of Organic Light-Emitting Diodes.* Boca Raton, FL: CRC Press.

26 Li ZR (ed.) (2015). *Organic Light-Emitting Materials and Devices*, 2e. Boca Raton, FL: CRC Press.

27 Yersin, H. and Finkenzeller, W.J. (2008). Triplet emitters for organic light-emitting diodes: basic properties. In: *Highly Efficient OLEDs with Phosphorescent Materials* (ed. H. Yersin), 1–98. Weinheim: Wiley-VCH.

28 Oyston, S., Wang, C., Hughes, G. et al. (2005). New 2,5-diaryl-1,3,4-oxadiazole-fluorene hybrids as electron transporting materials for blended-layer organic light emitting diodes. *J. Mater. Chem.* 15: 194–203.

29 Karg, S., Meier, M., and Riess, W. (1997). Light-emitting diodes based on poly-p-phenylene-vinylene: I charge-carrier injection and transport. *J. Appl. Phys.* 82: 1951–1966.

30 Gong, X. and Wang, S. (2008). Polymer light-emitting diodes: devices and materials. In: *Introduction to Organic Electronic and Optoelectronic Materials and Devices* (ed. S.-S. Sun and L.A. Dalton), 373–400. Boca Raton, FL: CRC Press.

31 Coaton, J.R. and Marsden, A.M. (eds.) (1997). *Lamps and Lighting*, 4the. London: Arnold.

32 Forrest, S.R., Bradley, D.D.C., and Thompson, M.E. (2003). Measuring the efficiency of organic light-emitting devices. *Adv. Mater.* 15: 1043–1048.

33 Cambridge Display Technology Ltd. Available from URL. www.cdtltd.co.uk/technology-scope/oled-displays-lighting. Accessed 21 September 2017.

34 Loeser, F., Tietze, M., Lüssem, B., and Blochwitz-Nimoth, J. (2015). Conductivity doping. In: *OLED Fundamentals: Materials, Devices, and Processing of Organic Light-Emitting Diodes* (ed. D.J. Gaspar and E. Polikarpov), 189–233. Boca Raton, FL: CRC Press.

35 Sun, S.-S. (2008). *Introduction to Organic Electronic and Optoelectronic Materials and Devices* (ed. L.A. Dalton). Boca Raton, FL: CRC Press.

36 O'Brien, D., Bleyer, A., Lidzey, D.G. et al. (1997). Efficient multilayer electroluminescence devices with poly(*m*-phenylenevinylene-co-2,5-dioctyloxy-*p*-phenylenevinylene) as the emissive layer. *J. Appl. Phys.* 82: 2662–2670.

37 Universal Display Corporation. Available from URL http://www.universaldisplay.com. Accessed 21 September 2017.

38 Uoyama, H., Goushi, K., Shizu, K. et al. (2012). Highly efficient organic light-emitting diodes from delayed fluorescence. *Nature* 492: 234–240.

39 Sun, J.W., Lee, J.-H., Moon, C.-K. et al. (2014). A fluorescent organic light-emitting diode with 30% external quantum efficiency. *Adv. Mater.* 26: 5684–5688.

40 Hirata, S., Sakai, Y., Masui, K. et al. (2015). Highly efficient blue electroluminescence based on thermally activated delayed fluorescence. *Nat. Mater.* 14: 330–336.

41 Kamtekar, K.T., Wang, C., Bettington, S. et al. (2006). New electroluminescent bipolar compounds for balanced charge-transport and tuneable colour in organic light emitting diodes: triphenylamine-oxadiazole-fluorene triad molecules. *J. Mater. Chem.* 16: 3823–3835.

42 Linton, K.E., Fisher, A.L., Pearson, C. et al. (2012). Colour tuning of blue electroluminescence using bipolar carbazole-oxadiazole molecules in single-active-layer organic light emitting devices (OLEDs). *J. Mater. Chem.* 22: 11816–11825.

43 Hong, K. and Lee, J.-L. (2011). Recent developments in light extraction technologies of organic light emitting diodes. *Electron. Mater. Lett* 7: 77–91.

44 Lu, M.-H.M. (2015). Microcavity effects and light extraction enhancement. In: *OLED Fundamentals: Materials, Devices, and Processing of Organic Light-Emitting Diodes* (ed. D.J. Gaspar and E. Polikarpov), 299–337. Boca Raton, FL: CRC Press.

45 Burrows, P.E., Gu, G., Bulović, V. et al. (1997). Achieving full-colour organic light-emitting devices for lightweight, flat-panel displays. *IEEE Trans. Electron. Dev.* 44: 1188–1202.

46 Nakayama, T. (1997). Organic luminescent devices with a microcavity structure. In: *Organic Electroluminescent Materials and Devices* (ed. S. Miyata and H.S. Nalwa), 359–389. Amsterdam: Gordon and Breach.

47 Sasabe, H. and Kido, J. (2013). Development of high performance OLEDs for general lighting. *J. Mater. Chem. C* 1: 1699–1707.

48 Tyan, Y.-S. (2015). Design considerations for OLED lighting. In: *OLED Fundamentals: Materials, Devices, and Processing of Organic Light-Emitting Diodes* (ed. D.J. Gaspar and E. Polikarpov), 399–436. Boca Raton, FL: CRC Press.

49 Sun, Y., Giebink, N.C., Kanno, H. et al. (2006). Management of singlet and triplet excitons for efficient white organic light-emitting devices. *Nature* 440: 908–912.

50 Mindemark, J. and Edman, L. (2016). Illuminating the electrolyte in light-emitting electrochemical cells. *J. Mater. Chem. C* 4: 420–432.

51 Zhang, C., Chen, P., and Hu, W. (2016). Organic light-emitting transistors: materials, device configurations, and operations. *Small* 12: 1252–1294.

52 Muccini, M. and Toffanin, S. (2016). *Organic Light-Emitting Transistors: Towards the Next Generation Display Technology*. Hoboken, NJ: Wiley.

53 Jung, S. and Theis, D. (2005). Electronic paper. In: *Nanoelectronics and Information Technology*, 2nde (ed. R. Waser), 953–965. Weinheim: Wiley-VCH.

54 Heikenfeld, J., Drzaic, P., Yeo, J.-S., and Kock, T. (2011). A critical review of the present and future prospects for electronic paper. *J. Soc. Inf. Disp.* 19: 129–156.

55 Kallmann, H. and Pope, M. (1959). Photovoltaic effect in organic crystals. *J. Chem. Phys.* 30: 585–586.

56 Brabec, C. and Dyakonov, V. (2008). *Scherf U (Editors.) Organic Photovoltaics. Materials, Device Physics and Manufacturing Technologies*. Wiley-VCH: Weinheim.

57 Leo, K. (ed.) (2017). *Elementary Processes in Organic Photovoltaics 2016 (Advances in Polymer Science)*. Switzerland: Springer International.

58 Moulé, A.J., Neher, D., and Turner, S.T. (2014). P3HT-based solar cells: structural properties and photovoltaic performance. In: *P3HT Revisited – From Molecular Scale to Solar Cell Devices. Adv. Polym. Sci*, vol. 265 (ed. S. Ludwigs), 181–232. Berlin: Springer.

59 Jørgensen, M., Carlé, J.E., Søndergaard, R.R. et al. (2013). The state of organic solar cells – a meta analysis. *Sol. Energy Mater. Sol. Cells* 119: 84–93.

60 Heliatek. Available from URL. http://www.heliatek.com/en Accessed 21 September 2017.

61 Pulfrey, D.L. (1978). *Photovoltaic Power Generation*. New York: Van Nostrand.

62 Thompson, B.C. and Fréchet, J.M.J. (2008). Polymer-fullerene composite solar cells. *Angew. Chem. Int. Ed.* 47: 58–77.

63 Wang, K., Liu, C., Meng, T. et al. (2016). Inverted organic photovoltaic cells. *Chem. Soc. Rev.* 45: 2937–2975.

64 Morgera AF, Lughi V. (2015). Frontiers of photovoltaic technology: a review. IEEE Clean Electrical Power (ICCEP) Conference; Taormina: 115–121.

65 Jansen-van Vuuren, R.D., Armin, A., Pandey, A.K. et al. (2016). Organic photodiodes: the future of full color detection and image sensing. *Adv. Mater.* 28: 4766–4802.

66 Jahnel, M., Thomschke, M., Ullbrich, S. et al. (2016). On/off-ratio dependence of bulk hetero junction photodiodes and its impact on electro-optical properties. *Microelectron. Eng.* 152: 20–25.

67 Jahnel, M., Thomschke, M., Fehse, K. et al. (2015). Integration of near infrared and visible organic photodiodes on a complementary metal-oxide-semiconductor compatible backplane. *Thin Solid Films* 592: 94–98.

68 Lv, W., Zhong, J., Peng, Y. et al. (2016). Organic near-infrared upconversion devices: design principles and operation mechanisms. *Org. Electron.* 31: 258–265.

69 O'Regan, B. and Grätzel, M. (1991). A low-cost, high-efficiency solar cell based on dye sensitized colloidal TiO_2 films. *Nature* 353: 737–740.

70 Ye, M., Wen, X., Wang, M. et al. (2015). Recent advances in dye-sensitized solar cells: from photoanodes, sensitzers and electrolytes to counter electrodes. *Mater. Today* 18: 155–162.

71 Bella, F., Gerbaldi, C., Barolo, C., and Grätzel, M. (2015). Aqueous dye-sensitized solar cells. *Chem. Soc. Rev.* 44: 3349–3862.

72 Weber, W.H. and Lambe, J. (1976). Luminescent greenhouse collector for solar radiation. *Appl. Opt.* 15: 2299–2300.

73 Debije, M.G. and Verbunt, P.P.C. (2012). Thirty years of luminescent solar concentrator research: solar energy for the built environment. *Adv. Energy Mater.* 2: 12–35.

74 Khamooshi, M., Salati, H., Egelioglu, F. et al. (2014). A review of solar photovoltaic concentrators. *Int. J. Photoenergy* 2014: 958521.

75 Zhao, Y., Meek, G.A., Levine, B.G., and Lunt, R.R. (2014). Near-infrared harvesting transparent luminescent solar concentrators. *Adv. Opt. Mater.* 2: 606–611.

76 Asthana, A., Maitra, T., Büchel, R. et al. (2014). Multifunctional superhydrophobic polymer/carbon nanocomposites: graphene, carbon nanotubes, or carbon black? *Appl. Mater. Interfaces* 6: 8859–8867.

77 Goodenough, J.B. and Park, K.-S. (2013). The li-ion rechargeable battery: a perspective. *J. Am. Chem. Soc.* 135: 1167–1176.

78 Zhong, C., Deng, Y., Hu, W. et al. (2015). A review of electrolyte materials and compositions for electrochemical supercapacitors. *Chem. Soc. Rev.* 44: 7484–7539.

79 González, A., Goikolea, E., Barrena, J.A., and Mysyk, R. (2016). Review on supercapacitors: technologies and materials. *Renew. Sust. Energ. Rev.* 58: 1189–1206.

80 Warner, J.H., Schäffel, F., Bachmatiul, A., and Rümmeli, M.H. (2013). *Graphene: Fundamentals and Emergent Applications*, 409–425. Waltham: Elsevier.

81 Kraytsberg A, Ein-Eli Y. (2014). Review of advanced materials for proton exchange membrane fuel cells. *Energy Fuel* 28: 7303–7330.

Further Reading

Bao, Z. and Locklin, J. (2007). *Organic Field-Effect Transistors*. Boca Raton, FL.: CRC Press.

Bredas, J.-L. and Marder, S.R. (eds.) (2016). *The WSPC Reference on Organic Electronics: Organic Semiconductors. Volume 2: Fundamental Aspects of Materials and Applications*. Hackensack, NJ: World Scientific.

Borsenberger, P.M. and Weiss, D.S. (1998). *Organic Photoreceptors for Xerography*. New York: Marcel Dekker.

Chang, Y.L. (2015). *Efficient Organic Light Emitting-Diodes (OLEDs)*. Boca Raton, FL: CRC Press.

Eisberg, R. and Resnick, R. (1985). *Quantum Physics*, 2e. New York: Wiley.

Klauk, H. (ed.) (2006). *Organic Electronics: Materials, Manufacturing, and Applications*. Weinhein: Wiley-VCH.

Ostroverkhova, O. (ed.) (2013). *Handbook of Organic Materials for Optical and Optoelectronic Devices*. Cambridge: Woodhead.

Schwoerer, M. and Wolf, H.C. (2006). *Organic Molecular Solids*. Berlin: Wiley-VCH.

Sze, S.M. and Kwok, K.N. (2007). *Physics of Semiconductor Devices*, 3e. Hoboken, NJ: Wiley.

Wallace, G.G., Spinks, G.M., Kane-Maguire, L.A.P., and Teasdale, P.R. (2002). *Conductive Electroactive Polymers: Intelligent Materials Systems*. Boca Raton, FL: CRC Press.

Waser R (ed.) (2005). *Nanoelectronics and Information Technology*, 2nd Ed. Weinhein: Wiley-VCH.

10

Chemical Sensors and Physical Actuators

Above the sense of sense

10.1   Introduction

This chapter will focus on chemical sensors and physical actuators that exploit organic materials. Some examples of devices (e.g. chemical sensors) already exist as commercial products, while others (e.g. wearable electronics) are fast developing technologies that will certainly make an impact in the market place over the next decade.

The development of effective devices for the identification and quantification of chemical and biochemical substances for process control and environmental monitoring is a continuing need. Many sensors do not possess the specifications to conform to existing or forthcoming legislation; some systems are too bulky or expensive for use in the field. Inorganic materials such as the oxides of tin and zinc have traditionally been favoured as the sensing element in gas and vapour sensors. One disadvantage of devices based on metallic oxides is that they usually have to be operated at elevated temperatures, limiting some applications. As an alternative,

Organic and Molecular Electronics: From Principles to Practice, Second Edition. Michael C. Petty.
© 2019 John Wiley & Sons Ltd. Published 2019 by John Wiley & Sons Ltd.
Companion website: www.wiley.com/go/petty/molecular-electronics2

there has been considerable interest in trying to exploit the properties of organic materials. Many such substances, in particular phthalocyanine derivatives, are known to exhibit a high sensitivity to gases. Lessons can also be taken from the biological world; for example, a carbon monoxide detector can be designed to simulate the reaction between CO and haemoglobin. A significant advantage of organic compounds is that their sensitivity and selectivity can be tailored to a particular application by modifications to their chemical structure. Moreover, thin film technologies such as self-assembly or molecular imprinting enables ultra-thin layers of organic materials to be engineered at the molecular level. There is also significant interest in exploiting nanoparticles, particularly metallic nanoparticles, in chemical and biological sensors. Such materials provide a high surface-to-volume ratio and can be synthesized with a particular size, shape, and surface functionality.

There are many physical principles upon which sensing systems might be based: changes in electrical resistance (chemiresistors), refractive index (fibre optic sensors), and mass (quartz microbalance) have all been exploited in chemical sensing [1]. The main challenges in the development of new sensors are in the production of cheap, reproducible, and reliable devices with adequate sensitivities and selectivities.

In this chapter, a brief survey of the different sensor technologies and their applications is given. The concept of a sensor or actuator will first be described. Then, some of the most important parameters that define a sensor are introduced. Sensing systems will be classified through the means of transduction and on the basis of the measured quantity.

10.2 Sensing Systems

One of the first issues encountered in describing a sensing system is to provide a satisfactory definition of the term sensor. A *sensor* (the term *transducer* is also used) is a device that is able to convert a variation of any quantity, property, or physical state, which is called the input signal or *measurand*, into a 'useful' output signal (i.e. something able to be read and interpreted). A sensor, therefore, is an element sensitive to the input and able to communicate with a measuring or control system. Sensors generally have their output in an electrical form, while the term *actuator* is used to describe sensors with a non-electrical output. Natural sensors, like those found in living organisms, usually respond to electrochemical signals; their physical nature is based on ion transport. In artificial sensors, information is usually transmitted through the transport of electrons or photons.

It has been proposed by Middlehoek that a sensor or actuator can be classified according to the energy domain of its primary input–output [1]. Six classes of input and output signals can be distinguished: mechanical, thermal, electrical, magnetic, radiation, (bio)chemical.

Sensors are also one of two types: *active* or *passive*. Active sensors directly generate an electrical signal in response to an external stimulus. The input stimulus energy is converted into output energy without the need for an additional power source. Examples are thermocouples, pyroelectric radiation detectors, and piezoelectric pressure sensors. Passive sensors require an external energy source for their operation. The signal is then modified by the sensor to produce a useful output. Passive sensors are sometimes called *parametric* because their properties change in response to an external stimulus, and these changes are subsequently converted into an output signal. For instance, a *thermistor* is a temperature sensitive resistor. This device does not directly generate any electrical energy, but by passing an electric current through it (excitation signal), the resistance can be measured by detecting variations in current and/or voltage across the thermistor. These variations (measured in ohms) are directly related to the temperature change.

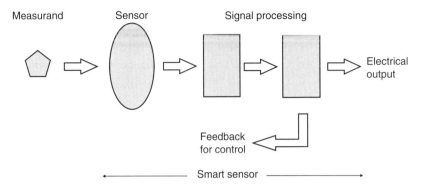

Figure 10.1 Schematic diagram of an electronic sensor system.

Figure 10.1 shows a schematic diagram of a sensing system. Modern devices can be highly miniaturized and the sensing element can be integrated with circuitry for signal processing (e.g. amplification, filtering, Bluetooth communication). A feedback path for control purposes may also be provided. Such a system may be referred to as an *integrated* or *smart sensor*.

A smart sensor can address issues related to the degradation of the device, to thermal drift, and to non-linearity in an automatic manner. Self-diagnosis of the working state and self-calibration are very attractive features in such a measuring system, making it able to adapt to different environments.

10.3   Definitions

A correct definition of the parameters of a sensing system is fundamental for a better understanding of the sensor behaviour and for establishing uniform standards. It is then possible to have a consistent comparison between the performances of different sensing systems. The descriptions also need to be attributed to the correct part of the sensing system. For example, some parameters only characterize the active element(s), while others define the performance of the entire sensing system.

An ideal sensor should have characteristics such as a high *selectivity* and *sensitivity*, complete *reversibility*, long-term *reliability* and *stability*, short *response* and *recovery times*, and a good *signal-to-noise ratio*. Other factors might include multiplexing capabilities, low cost, a minimal complexity for use, and portability for in-situ applications. Compatibility with microcontrollers, flexibility in the assembly of different platforms and the ability to integrate several sensors within a single package, are also highly desirable features.

Reliability, sensitivity, and selectivity are characteristics that depend directly on the material used as sensor. Features such as response time and noise are more complex, since several components of the sensing system play a role. In particular, it is important to note that the main source of noise in a sensing system is, very often, the system itself, with all its electronic components making a contribution. Shielding of the multiple noise sources within the system is, therefore, essential.

The response of a sensor is calculated from the value of the output before the measurement starts, R_0 and the value measured in the presence of a measurand, R_{meas}, according to the following equation:

$$\left| \frac{\Delta R}{R_0} \right| = \left| \frac{R_{meas} - R_0}{R_0} \right| \tag{10.1}$$

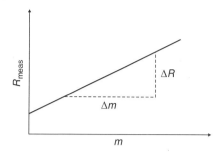

Figure 10.2 Output versus input for a sensor operating in a linear regime. R_{meas} is the output or response of the sensor while m is the input or measurand.

The sensitivity of a sensor is defined as the derivative of the parameter R_{meas} with respect to the measurand, m (the input). If there is a linear variation between the input and the sensor output, as shown in Figure 10.2, then the sensitivity, S, is given by

$$S = \frac{\Delta R}{\Delta m} \tag{10.2}$$

When the response of the sensing system is not linear, then the sensitivity can be expressed as

$$S = S(m) = \frac{dR_{meas}}{dm} \tag{10.3}$$

As shown in Figure 10.3, the selectivity S_{m_0} for a practical value of the measurand m_0 is

$$S = \frac{dR_{meas}}{dm}\bigg|_{m=m_0} = S_{m_0} \tag{10.4}$$

The response time of a sensor can be defined in a number of ways. For example, this may define the time interval for the response signal R_{meas} to change from 10 to 90% of its final saturation value, R_f, after applying the stimulation as a step-function, as shown in Figure 10.4. If the system is reversible, then it is also possible to define a recovery time, as the time for the response signal to change from 90 to 10% of R_f. In other cases, the response and recovery times describe the time until the sensor output signal has reached $1-1/e$ (i.e. approximately 63%) of its final value. The reciprocal of the response time is related to the *bandwidth* of the sensor; the quicker the response, the larger the bandwidth.

The signal-to-noise (S/N) ratio of a measuring system is defined as the ratio of the strength of the signal plus noise carrying information to the unwanted interference (noise). It is measured using the logarithmic decibel (dB) scale and it is equal to 20 times the base-10 logarithm of the amplitude ratio (i.e. voltage of the signal divided by the voltage of the noise, $\frac{V_s}{V_n}$), or 10 times the logarithm of the power ratio. If $V_s = V_n$, then the S/N ratio is equal to zero. In this situation, the signal is difficult to measure directly and sophisticated signal processing techniques must be used to recover the signal.

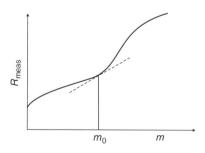

Figure 10.3 Output versus input for a sensor operating in a nonlinear regime. R_{meas} is the output or response of the sensor while m is the input or measurand.

For biochemical sensors, selectivity is very important. This is a measurement of the ability of the sensor to respond to only one input in the presence of interferences. In most biological systems, specificity is achieved by shape recognition, which involves a comparison with a reference. High selectivity means that the contribution from the primary species dominates, and that the interference from other species is minimal. However, an absolutely selective sensor does not really exist and there is always some interference present.

Sensor *resolution* is the minimum change in the measurand that a sensing device can resolve. In other words, it is the size of the incremental steps in the response obtained for a linear increase of the input signal. The resolution is not necessarily constant over the whole measuring range. A further important parameter of a sensing system is its operative or *dynamic range*, i.e. the range of values of the measurand, which can be detected by the sensor.

A number of different sensor classification systems can be found in the literature. One method is to consider all its properties, such as what it measures (measurand or stimulus), its specifications, the conversion mechanism, the material(s) it is fabricated from, and its field of application. Categorization schemes based on the physical or chemical means of transduction have been used, but these can be difficult to follow. Sensors are often grouped according to what they measure. It is then possible to look in detail at the different options possible and to consider the characteristics and features of each sensing system. According to this approach, there are three sensor families or groups: chemical sensors; physical sensors; and biological sensors. The latter will be covered in some detail in Chapter 12, Section 12.8. Here, chemical and physical sensors will be described; the focus, of course, will be on the exploitation of organic materials as the sensing elements.

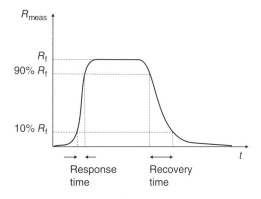

Figure 10.4 Response and recovery times of a sensor following a step-input stimulus.

10.4 Chemical Sensors

Chemical sensors are devices sensitive to stimuli produced by various chemical compounds or elements [2]. The most important property of these sensors is selectivity. Changes in the chemical environment are transformed into sensor properties such as electrical conduction, thermal conductivity, or refractive index. Chemical sensors can be subdivided into gas and liquid sensors. The operation of all these devices depends on the interaction of a chemically sensitive material with the *analyte* (i.e. the measurand). This reaction produces a physical change in the sensor, which is converted into a measurable electronic signal. The interaction between the analyte and the sensor material can be reversible or irreversible. In the former case, the analyte molecules dissociate from the sensor material when the external concentration is removed and, overall, they undergo no net change. Examples of this type include the adsorption of gases onto polymer layers and the interactions of gases with conductive polymers. Here, the reaction between the sensor material and the analyte is determined by intermolecular forces that are the results of one or more of the chemical bonding arrangements described in Chapter 2, Section 2.3. The most selective reactions tend to be those such as the 'lock-key' mechanism that operates in a biological sensor (Chapter 12, Section 12.8).

For irreversible interactions, the analyte undergoes a chemical reaction at the sensor surface catalysed by the sensor material. The analyte is consumed in the sensing process, although the numbers of molecules reacting represent a small proportion of the total number within the sample. The term 'irreversible' in this context can be a little confusing. Removal of the analyte concentration will still reverse the sensing process, but the associated time constant may be longer than for the 'reversible' sensors. The sensitivity and selectivity of this type of irreversible sensor are determined by the choice of the catalytic surface. A good example of this type of sensor is the amperometric electrochemical sensor described in the following section.

10.4.1 Electrochemical Cells

Electrochemical sensors are probably the most versatile chemical sensors. Some of these measure voltage (*potentiometric*); others monitor electric current (*amperometric*). Such detectors are used where a chemical reaction takes place or when the charge transport is modulated by the interaction of the chemical species under observation. Electrochemical gas sensors ionize gas molecules at a three-phase boundary layer (atmosphere, electrode of a catalytically active material, electrolyte). One of the ions, e.g. O^{2-}, H^+, Cl^-, which is involved in the reaction on the surface, can be preferentially conducted in the electrolyte.

In the simplest form of amperometric sensor (sometimes called a micro fuel cell, Chapter 9, Section 9.9.2), two electrodes, a sensing and a counter electrode, are separated by a thin layer of electrolyte. Gas diffusing to the sensing electrode reacts at the surface of the electrode, either by oxidation (removal of electrons) or reduction (addition of electrons). This reaction causes the potential of the electrode to increase or decrease with respect to the counter electrode. With a resistor connected across the electrodes, a current is generated that can be detected and used to determine the concentration of gas present.

One of the conditions required for amperometric sensors to work accurately is that the potential of the counter electrode should remain constant. In practice, the surface reactions at each electrode cause these to polarize (become charged). This process may be small initially, but it increases with the level of reactant gas and effectively limits the concentration range to which the sensor can respond. The effect can be counteracted by the introduction of a reference electrode of stable potential. This is placed within the electrolyte in close proximity to the sensing electrode. As the electrode must maintain a constant potential for correct operation, it is important that no current is drawn from it. A *potentiostatic* feedback circuit is therefore used to measure the potential difference between sensing and reference electrodes. Certain amperometric sensors, e.g. oxygen sensors, do not require a reference electrode, as the reaction involves a chemical change in the counter electrode.

In potentiometric cells, the potential difference generated between the electrodes by different partial pressures on either side of the cell provides the output. An example is the *Gas-FET*, a chemically sensitive field effect transistor (FET). These devices use a FET (Chapter 9, Section 9.4) with a gate metallization exposed to the surrounding atmosphere. In the case of a Pd-gate device, as depicted in Figure 10.5a, hydrogen gas becomes adsorbed on the metal surface and dissociates into atoms, which subsequently diffuse to the oxide-metal interface where a dipole layer is created. This effective change in the work function of the gate electrode results in a shift in the threshold voltage, with a consequent change in the output characteristics of the transistor, as shown in Figure 10.5b.

For solution work, *ion-selective electrodes* are extensively used. These devices consist of membranes that respond to one particular ionic species in the presence of others. A well-known example is the pH electrode. The output is in the form of a voltage V that is related to the concentration of an ion by the *Nernst equation*:

$$V = V_0 + 2.303 \frac{RT}{nF} \log\left(\frac{a_{\text{red}}}{a_{\text{ox}}}\right) \qquad (10.5)$$

where V_0 is a constant to account for all the other potentials in the system, R is the gas constant, n is the number of electrons transferred, and F is Faraday's constant (Chapter 7,

Figure 10.5 (a) Pd-gate gas-sensitive field effect transistor (FET). (b) The effect of exposure to hydrogen gas on the drain current I_d versus gate voltage V_g characteristic.

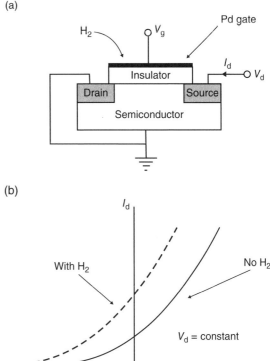

Section 7.2.4: F = electronic charge × Avogadro's number = $9.65 \times 10^7 \, C\,kmol^{-1}$) and a_{red} and a_{ox} are the *activities* of the reduced and oxidized species, respectively. The activity is related to the ion concentration c by the relationship.

$$a = \gamma c \tag{10.6}$$

where γ is the *activity coefficient*. This parameter is essentially a correction factor, introduced to take into account non-ideal behaviour. In dilute solutions, γ approaches unity and the activity can then be identified with concentration.

For a solid electrolyte in equilibrium with ions in solution, a plot of the measured potential versus the logarithm of the ion concentration will give a straight line, with a slope of about 58 mV (= $2.303RT/F$) per decade at room temperature for a reaction involving the transfer of one electron, e.g. a pH sensor.

The complete measurement system consists of an ion-selective electrode, an internal reference electrode and an external reference electrode. Commercial ion-selective electrode systems often combine the two electrodes into one unit. The need for a reference electrode has been outlined above, for the case of an amperometric electrochemical gas sensor. There are three main requirements that such a reference potential must satisfy. It must be stable, reversible and reproducible [2]. The condition for reversibility (good ohmic contact) is satisfied by selecting an electrochemical reaction that is very fast, i.e. has a very high exchange-current density (> $1\,mA\,cm^{-2}$). An example of such a reaction is that utilized by the silver/silver chloride electrode, which comprises silver metal coated with a few microns of silver chloride. The Ag/AgCl reference electrode must be stored in a separate compartment with defined and constant

(a)

(b)

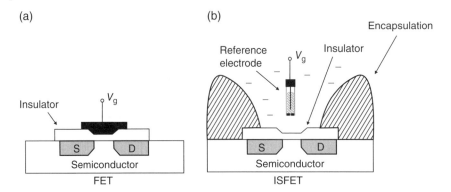

Figure 10.6 Comparison of the structure of (a) field effect transistor (FET) and (b) ion-sensitive field effect transistor (ISFET) devices.

activity of the chloride ion. However, the electrical circuit must be completed; a liquid junction generally realizes this contact. The junction is an open channel, but the outflow of the internal reference solution is very low, typically $2\,nl\,h^{-1}$.

Ion-sensitive field effect transistors, or ISFETs, have been developed since the 1970s [2, 3]. In essence, the ISFET is a MOSFET in which the gate connection is separated from the silicon substrate in the form of a reference electrode. This is inserted in an aqueous solution that is in contact with the gate insulator, as shown in Figure 10.6, which compares the architecture of (a) an FET to that of (b) an ISFET. A change in the interfacial potential at the liquid/gate insulator interface will produce a change in the threshold voltage of the transistor. Gate insulators such as SiO_2, Si_3N_4, Al_2O_3, and Ta_2O_5 are responsive to hydrogen ions. The mechanism of operation is associated with the formation of a thin hydrated surface layer (4–5 nm). In the case of quartz, it is known that surface silanol groups dissociate:

$$\equiv SiOH \Leftrightarrow SiO + H^+ \tag{10.7}$$

Sensitivity to ions other than H^+ can be achieved by the deposition of suitable gate materials. For example, polymeric membranes containing the ionophore valinomycin (Chapter 12, Section 12.6.1) can provide a response to potassium ions. Layers of immobilized enzymes (Chapter 12, Section 12.2.3) are often used to convert a substance, for which no sensor is available, into a substance for which a chemical sensor exists. A well-known example is the urea sensor that makes use of immobilized urease. This converts urea into ammonia ions, carboxyl ions, and hydroxyl ions, and provides the option of monitoring the pH as an indirect method for detecting urea. Such devices are called enzymeFETs or ENFETs. ImmunoFETs or IMFETs make use of highly specific immunological reactions (Chapter 12, Section 12.2.2) between antigens and antibodies.

Although most pH ISFETs have been based on the semiconductor silicon combined with an inorganic insulator, e.g. silicon nitride or tantalum oxide, to provide hydrogen ion selectivity, there are reports of device architectures exploiting both organic semiconductors and insulators [e.g. 4]. Furthermore, other organic layers (e.g. ionophores such as valinomycin) may extend the selectivity of organic field effect transistor (OFET)-based sensors [5]. Such devices offer additional benefits over their inorganic counterparts, such as mechanical flexibility, and, possibly, combination with textiles and biological structures.

One advantage of ISFET technology is that it offers integration with the signal processing required for a complete instrumentation system. Circuitry can readily be incorporated with the chemically sensitive transistor devices to provide improved signal-to-noise ratio, a reduction in drift and some compensation for temperature changes. However, one problem is the need for an external reference electrode. The highly miniaturized solid-state chemical sensors that ISFET technology produces are invariably used with conventional liquid-filled reference electrodes. ISFET products are generally aimed at markets where the vulnerability of glass membrane electrodes presents measurement difficulties, such as the food industry.

10.4.2 Resistive Gas Sensors

A *chemiresistor* sensor exploits the resistance change of a thin layer of a gas-sensitive material. Although the most common type of resistive gas sensor uses semiconducting metal oxides, such as tin oxide, as the sensing element (so-called *Taguchi* sensors), devices can be based on organic materials. The latter respond to a broad range of gases and vapours and may operate at room temperature (chemiresistors based on inorganic semiconductors usually have to be heated). Furthermore, a wide range of compounds can be synthesized, which offers the prospect of tailoring devices to respond to particular analytes. A chemiresistor can be fabricated by depositing a thin film of the active (gas-sensitive) material onto an interdigitated microelectrode structure, such as that shown in Figure 10.7. This structure allows a large surface area of the sample to be exposed to the analyte gas or vapour.

In the case of metal oxide type devices, a model that has been derived involves the gas affecting the space charge region (depletion region, Chapter 9, Section 9.2) adjacent to the surface of the material or at grain boundaries within the material. Figure 10.8 illustrates the process that can occur on exposure to oxygen. The gas will first become chemisorbed (Figure 10.8a) (Chapter 2, Section 2.5.7). The molecules act as surface acceptors, removing electrons from the valence band of the semiconductor and giving rise to an increase in the hole concentration at the surface. The conduction band bends upwards in the energy diagram of Figure 10.8b to provide a positive fixed charge (due to fixed donor atoms) that compensate for the negative charge trapped at the surface.

The above ideas are probably applicable to some organic chemiresistors based on molecular materials. However, charge-transfer reactions (Chapter 3, Section 3.4) can occur in certain organic compounds, such as the phthalocyanines or porphyrins [6, 7]. For the majority of phthalocyanines, which are p-type semiconductors, an increase in conductivity is observed when exposed to electron acceptor gases (i.e. oxidizing gases) such as oxygen or NO_2, while a decrease is seen on exposure to electron donor gases (reducing gases) such as ammonia. The conductivity changes are often accompanied by the appearance of charge-transfer bands in the visible

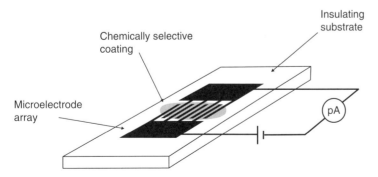

Figure 10.7 Schematic diagram of a chemiresistor device, fabricated by coating a thin film of gas or vapour-sensitive semiconductive material onto a planar microelectrode array.

(a)   (b)

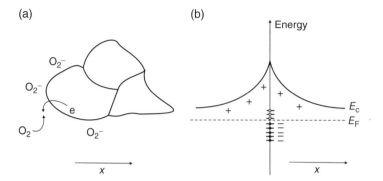

Figure 10.8 (a) Grain structure in a polycrystalline material showing the chemisorption of oxygen at the surface of the grains. (b) The negatively charged oxygen molecules produce an upward curvature of the electronic energy bands at the grain boundary, providing a depletion region.

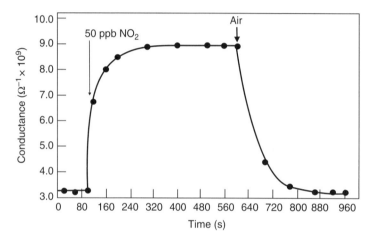

Figure 10.9 Response and recovery of a chemiresistor coated with a vacuum-evaporated lead phthalocyanine films and exposed to 50 ppb (parts per billion) of nitrogen dioxide in air at 150 °C. *Source:* Reprinted from Bott and Jones [8]. Copyright (1984), with permission from Elsevier.

spectra. A typical example for a chemiresistor response is illustrated in Figure 10.9 for exposure of an evaporated lead phthalocyanine film to 50 ppb (parts per billion) of nitrogen dioxide [8]. The charge-transfer reaction also depends on the nature of the central metal ion in the phthalocyanine molecule and is dependent on the morphology of the thin film. The latter must accommodate both the charge-transfer interaction and charge carrier transport. It is believed that the former can occur more readily if the faces of the phthalocyanine or porphyrin ring rather than an edge are available for electron transfer through complex formation. On the other hand, a long-range stacking of cofacially oriented phthalocyanine rings facilitates charge transport.

The extreme thinness of Langmuir–Blodgett (LB) layers (Chapter 7, Section 7.3.1) provides a high surface to volume ratio for the sensor and, therefore, a potential for high sensitivity. Asymmetrically substituted materials may show enhanced response and recovery times in comparison with LB films of unsubstituted compounds. Such results indicate the importance of the peripheral groups (required to provide the solubility for LB processing) in the phthalocyanine molecules.

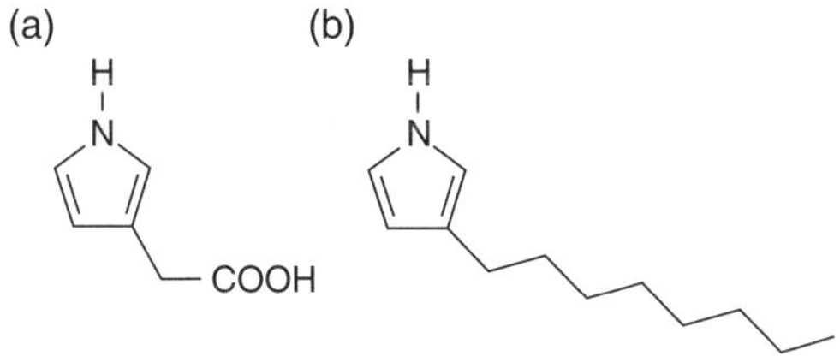

Figure 10.10 Examples of two substituted pyrrole monomers used to make conductive polymers [11]. (a) The incorporation of the carboxylic acid group gives a hydrophilic polymer. (b) The incorporation of the alkyl chain substituent produces a hydrophobic polymer.

Many conductive polymers have been used as gas or vapour sensors [9]. The most common materials exploited are polypyrrole, polyaniline, and polythiophene. Exposure to gases leads to oxidation or reduction of the polymer chains, resulting in a variation of conductivity. The selectivity of such chemiresistors may depend on the method of depositing the organic layer, reflecting a change in the chemical composition and/or morphology of the thin film [10]. Substitution of the basic polymer can also lead to modifications to the sensing response. For example, Figure 10.10 shows two examples of substituted pyrrole monomers [11]. While the attachment of a carboxylic acid group (Figure 10.10a) leads to a more hydrophilic polymer (likely to interact with similar molecular species), the substitution with the hydrocarbon chain (Figure 10.10b) leads to a more hydrophobic material. One of the attractions of using conductive polymers is their ease of preparation. Techniques such as solution casting, spin-coating, and electrochemical deposition are relatively straightforward methods to fabricate sensing devices. Molecular imprinting offers a route to create specific and selective cavities in a three-dimensional polymeric network, which are complementary to the size and shape of the target species [12].

Inkjet printing (Chapter 7, Section 7.2.5) is a further method that is attracting attention as the polymer is only deposited where it is needed. The morphology of such films can be important for very thin chemiresistors. Figure 10.11 shows electrical data for a single inkjet-printed layer of the conductive polymer poly(3,4-ethylenedioxythiophene poly(styrene sulfonic acid)) (PEDOT-PSS) (Chapter 5, Section 5.3) [13]. On exposure to alcohol vapour, the electrical conductivity effectively falls to zero. This irreversible decrease in conductivity is the result of a vapour-induced change in structure of the organic film. The morphology of the as-deposited inkjet printed layer will depend on the contact angle between the PEDOT-PSS and the underlying substrate. This is influenced by the chemical (degree of polarity) and physical (degree of roughness) nature of the substrate. Very thin inkjet-printed PEDOT-PSS layers will be in the form of a series of small islands, as depicted in Figure 10.12. If such a film is exposed to the vapour of an alcohol, the vapour can re-dissolve the organic film in the regions where it is very thin, i.e. between the islands of organic film. Therefore, the electrical connection between the PEDOT-PSS islands is lost and the current passing through the organic film falls effectively to zero. When the vapour is turned off, the polar PEDOT-PSS molecules are preferentially adsorbed onto the existing islands of the polymer rather than onto the substrate. The morphology of the inkjet printed film, shown in the atomic force microscopy (AFM) images in Figure 10.12, has thereby been permanently altered. This type of sensing response has been called a 'chemical fuse'.

The gas sensing mechanism(s) for conductive polymers is complex. Five possible sites of interaction between the vapour and the polymer have been identified [15]. First, the vapour molecules could affect charge-transfer between the polymer and the electrode contact. Second, the vapour could lead to oxidation, or reduction, of the polymer chains influencing the number of charge carriers. Although this mechanism could be important for reactive gases, such as H_2S or ammonia, it is less likely for most organic vapours, such as simple alcohols, which are not

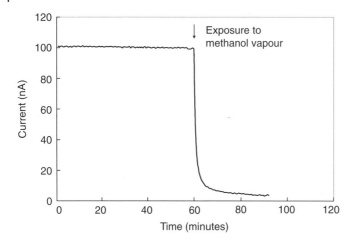

Figure 10.11 The current response of thin PEDOT-PSS layer to the exposure of 5000 ppm (parts per million) methanol. *Source:* Reprinted with permission from Mabrook et al. [13]. Copyright (2005) American Institute of Physics.

strong oxidizing or reducing agents. In the third case, the vapour could interact with the mobile charge carriers on the polymer chains, thereby changing their mobility. Fourth, the vapour molecules could interact with the counterions within the film. Finally, the vapour could influence the interchain hopping of the charge carriers.

The low-dimensionalities of carbon nanotubes and graphene make these materials promising candidates for sensor applications [16, 17]. Such compounds can also be mixed with polymers to improve their dispersion and provide better quality thin films. The interaction of gas molecules with graphene modifies its electronic properties by doping. High sensitivity may be achieved, with the prospect of detecting individual gas molecule biding events. However, the problems of other sensors of this type remain, such as achieving adequate selectivity and avoiding sensor contamination.

The response of organic and inorganic resistive gas detectors is often modelled by the *Langmuir isotherm*, developed by Irving Langmuir in 1916. For molecules in contact with a solid surface at a fixed temperature, this describes the partitioning between the gas phase and adsorbed species as a function of applied pressure. The isotherm is based on three assumptions: that adsorption cannot proceed beyond monolayer coverage; that all sites are equivalent and that the surface is uniform; and, finally, that the ability of a molecule to absorb at a given site is independent of the occupation of neighbouring sites. The model is depicted in Figure 10.13.

The resulting isotherm relates the fraction of surface sites occupied θ ($0 < \theta < 1$) to the gas pressure P:

$$\theta = \frac{bP}{1 + bP} \tag{10.8}$$

where b is the equilibrium constant. Note that $\theta \rightarrow bP$ at low pressures, while $\theta \rightarrow 1$ at high pressures. This is illustrated by the graph shown in Figure 10.14a. Assuming that the conductivity is

(a)

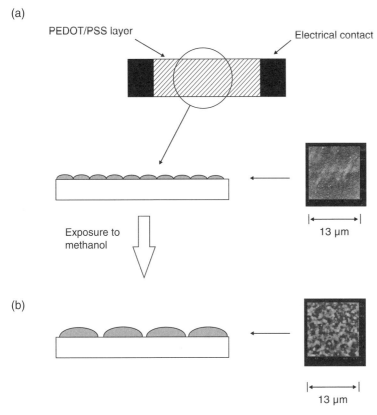

(b)

Figure 10.12 Schematic diagrams of the inkjet-printed PEDOT-PSS 'chemical fuse'. (a) Fresh PEDOT-PSS thin film with an island distribution providing an electrically conductive path through the film. (b) Proposed reorganization of the PEDOT-PSS material where almost all the conductive paths in the film surface are broken due to the presence of alcohol. The atomic force micrographs on the right reveal the film morphology before and after exposure to 5000 ppm of methanol for 30 minutes. *Source:* From *IEEE Sens. J.*, **6**, Mabrook et al. [14]. Copyright (2006). Reproduced by permission of IEEE.

Figure 10.13 Schematic diagram that is the basis for the Langmuir isotherm. Molecules in the gas phase can occupy fixed sites on the surface of a material.

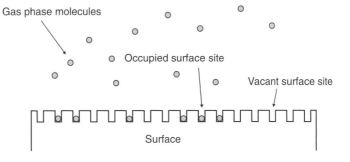

related directly to the number of molecules of the active gas adsorbed on the sample surface, then under a constant voltage, the current as a function of the partial gas pressure can be described by the Eq. (10.8). The change in resistance of a polypyrrole chemiresistor to increasing concentrations of methanol in air is shown in Figure 10.14b [15]. The shape of the response

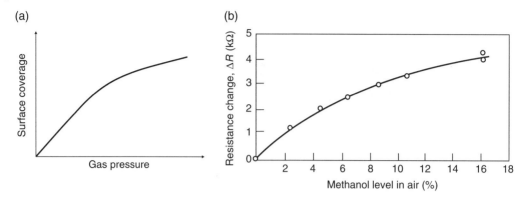

Figure 10.14 (a) Theoretical form of the Langmuir isotherm. (b) Typical isotherm for a conductive polymer gas sensor showing its response to methanol vapour. *Source:* Reprinted from Bartlett and Ling-Chung [15]. Copyright (1989), with permission from Elsevier.

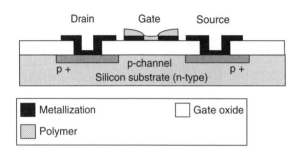

Figure 10.15 Schematic diagram of a charge-flow transistor [18]. Part of the gate metallization in the silicon MOSFET is replaced by a semiconductive organic material.

curve is similar to that of the Langmuir isotherm, Figure 10.14a, showing a linear response at low concentrations of vapour and a saturation of the response at high concentrations.

One problem associated with many chemiresistor devices described above is that the current outputs are low (typically picoamperes), requiring elaborate detection electronics and careful shielding and guarding of components. This difficulty may be overcome by incorporating the organic sensing layer into a silicon FET. A schematic diagram of one type of structure is shown in Figure 10.15 [18, 19]. Note that configuration is quite different from that depicted for the Gas-FET shown in Figure 10.5, which is a potentiometric device based on the change in work function of a gate metal on exposure to a gas. The FET device shown in Figure 10.15 has a 'hole' in its gate metal that is filled with a conductive polymer. The principle of operation of this *charge-flow* transistor is as follows [18]. When a voltage is applied to the gate of the FET, the capacitor that is formed between the gate and the silicon substrate charges in two stages. First, the metallic part of the gate charges very rapidly to the applied voltage. Charge then gradually flows through the conductive polymer (which will have a lower conductivity than the gate metal) until the organic part of the capacitor is uniformly charged to the applied voltage. The time required for this charging process to be complete will depend on the sheet resistance (the resistivity divided by the film thickness – Chapter 3, Section 3.2.2) of the polymer. Therefore, exposure to gas will affect the turn-on time of the FET. Figure 10.16 shows data for such an FET, incorporating a film of a spin-coated phthalocyanine derivative in the gate (gate 'hole' = 35 μm) [19]. Increasing the concentration of NO_2 increases the conductivity of the phthalocyanine and reduces the turn-on time of the transistor. Because of the amplification provided by the transistor configuration, the current flowing in this FET device is significantly higher than that for the

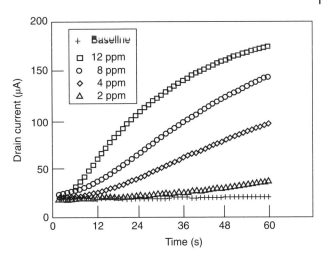

Figure 10.16 Drain current I_d response for a charge flow transistor when a voltage step of $-5\,V$ is applied to the gate electrode. Data are shown for the device exposed to different concentrations of nitrogen dioxide. Width of hole in the gate metallization = 35 μm. A phthalocyanine derivative is spin coated to fill the gate hole. *Source:* Reprinted from Barker et al. [19]. Copyright (1996), with permission from Elsevier.

chemiresistor structure shown in Figure 10.7. Of course, it is also possible to use a chemically-sensitive organic film as the semiconductive layer in a diode or a transistor. However, in these cases the device must be fabricated so that the gas can interact readily with the organic material.

10.4.3 Dielectric Sensors

For chemical sensors, the issue of reversibility is closely related to that of chemical selectivity. Weak interactions between the vapour and sensor coating will produce sensors with good reversibility. However, such sensors will not have sufficient sensitivity and selectivity to be widely useful. Very strong interactions may improve the sensitivity and selectivity, but can result in sensors that are irreversible or only slowly reversible. Hence, if reversibility is important, a compromise must be reached between the selectivity and the reversibility. This balance can be achieved by sensor coatings that interact with vapours via solubility interactions [20]. When a polymer film is exposed to a vapour, as shown in Figure 10.17, the equilibrium distribution of solute molecules between the gas phase and the polymeric phase is measured by the *partition coefficient*, K_p, given by

$$K_p = \frac{c_p}{c_v} \tag{10.9}$$

where c_p is the concentration of solute in the polymer and c_v is the concentration of solute in the vapour phase. Values of $\log K_p$ can be as high as 6–7 for some vapour/polymer combinations [20].

Figure 10.17 Partition of vapour molecules between the gas phase and a polymer film. c_v and c_p are the concentrations of solute in the vapour phase and polymer, respectively.

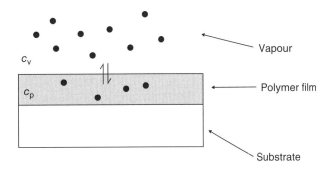

These numbers demonstrate that the concentration of vapour molecules in the sensing material can be one to ten million times more concentrated than in the vapour phase. Absorption is therefore very effective at collecting and concentrating analyte molecules on the sensor's surface. Partition coefficients are strongly temperature dependent. Absorption normally decreases with increasing temperature. As a result, sensors will become less sensitive and less selective as the temperature increases.

Using the principle outlined above, non-conductive polymers (i.e. those not containing π-bonds) can be used as dielectric or capacitance-based sensors. This class of polymer is more wide-ranging than that of conductive polymers, offering more scope for chemical modification to tailor the gas/polymer interaction. The detection principle is based on the dissolution of the gas or vapour into the polymer film with a consequent change of the permittivity. To monitor such changes, the sensor takes the form of a capacitor structure.

Amorphous elastomeric polymers represent an important category of non-conductive sensor coatings. In the elastomeric phase (i.e. above the glass transition temperature – Chapter 2, Section 2.6) constant thermal motion of the polymer chains allows rapid vapour diffusion. Adsorption and desorption of a vapour leaves the material in the same state. The softness of elastomeric materials has an additional advantage for piezoelectric sensors such as surface acoustic wave (SAW) devices (Section 10.4.4) that are sensitive to changes in material stiffness. An example of such materials is the family of polysiloxanes characterized by a repeat unit ($-RR'-Si-O-$), where R and R' are generic functional groups. By modifying the side chains, nonpolar, polar, and polarizable polymers can be obtained. Two examples are shown in Figure 10.18: (a) polycyanopropylmethylsiloxane and (b) a phenyl-methyl-diphenylsiloxane copolymer. The former compound is highly polar (the nitrile or $-CN$ bond is associated with a large dipole moment) and is expected to interact well with polar vapours (e.g. acetonitrile – CH_3CN; or 1,1,1-trifluoroethane – CF_3CH_3), while the copolymer structure depicted in Figure 10.18b has a large polarizability (i.e. a large ease of dipole formation in an electric field) and should interact well with vapours that are similarly polarizable (e.g. benzene).

Figure 10.19 shows the frequency behaviour at room temperature, over the range $10^{-1}–10^{3}$ kHz, of the capacitance of an LB film of a co-ordination polymer formed by the reaction of the bifunctional amphiphilic ligand 5, 5′ methylenebis (*N*-hexadecylsalicylideneamine) and copper ions in an interfacial reaction at the water surface [21]. The measurements were taken in both nitrogen gas and ethanol vapour (3.3%). The inset shows the transient behaviour measured at a

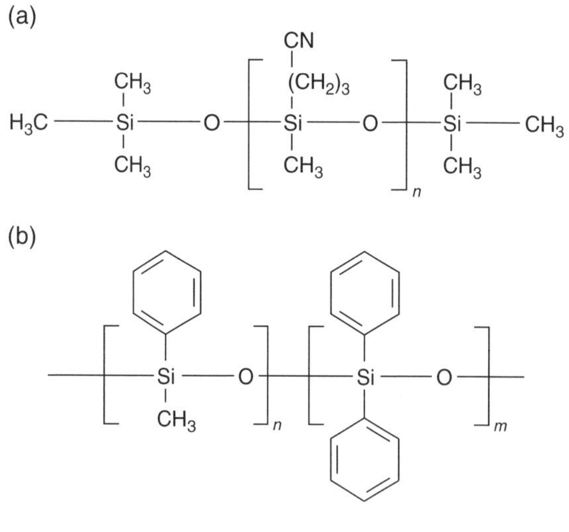

(a)

(b)

Figure 10.18 Two polysiloxanes derivatives. (a) polycyanopropylmethylsiloxane. (b) A phenylmethyldiphenylsiloxane copolymer.

Figure 10.19 Capacitance versus frequency for a co-ordination polymer LB film in nitrogen and exposed to ethanol vapour (3.3%). The inset shows the transient behaviour measured at a fixed frequency (1 kHz) when the ethanol vapour was turned on and off (indicated by the arrows). *Source:* Reprinted from Casalini et al. [21]. Copyright (1999), with permission from Elsevier.

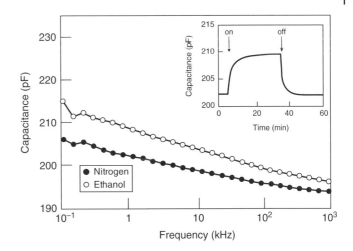

fixed frequency (1 kHz) when the ethanol vapour was turned on and off. The conductance of the dielectric sensor is often measured at the same time as its capacitance. Measurements at different frequencies form the basis of the technique of impedance (or admittance) spectroscopy, a powerful method for the investigation of the dynamic electrical properties of dielectrics (Chapter 3, Section 3.6.2). Discrimination between different vapours may be achieved by monitoring the complex admittance at several frequencies.

The percentage changes in the capacitance and conductance for the co-ordination polymer described above, on exposure to a number of different vapours, and normalized to the vapour concentration, are depicted in Figure 10.20. Such changes may result from a variety of processes, including:

1) a change in the permittivity due to bulk dissolution of the vapour into the LB film;
2) a change in the polymer permittivity induced by the interaction between the polymer and the vapour molecules; and
3) a change in the film thickness (i.e. swelling).

As the dielectric relaxation for the organic solvent is expected to occur at a very high frequency (beyond the maximum used in the experiment), mechanism (1) should produce changes in capacitance and conductance that are constant over the frequency range used. This should also be the case for mechanism (3). From Figure 10.20, it is evident that the capacitance increases when the dielectric sensor is exposed to ethanol and acetonitrile, with a greater fractional increase for acetonitrile. This is consistent with the strongly polar nature of the latter solvent (dipole moment for acetonitrile = 3.92 Debye compared to 1.69 Debye for ethanol). A decrease of capacitance is observed with the non-polar benzene, almost certainly associated with swelling of the film. The frequency dependence of the admittance is not the same for the three vapours, suggesting that mechanism (2) may be significant. However, from a practical viewpoint, it is clear that monitoring the admittance at different frequencies provides a means of discrimination between the vapours studied. This idea is explored in more detail in Section 10.6, dealing with electronic noses.

10.4.4 Acoustic Devices

Measuring mass can lead to highly sensitive chemical sensing devices. Many such *gravimetric* sensors operate at ultrasonic frequencies and register the shift in the resonant frequency $\Delta\nu$ of

(a)

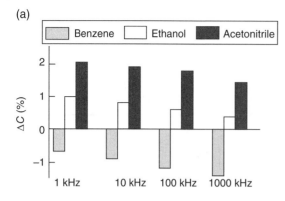

(b)

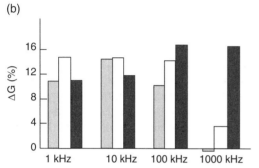

Figure 10.20 Average changes in (a) the capacitance ΔC and (b) the conductance ΔG for a co-ordination polymer LB film, measured at four frequencies for three different vapours. *Source:* Reprinted from Casalini et al. [21]. Copyright (1999), with permission from Elsevier.

a piezoelectric crystal oscillating a frequency ν when an additional mass is deposited on its surface, as shown schematically in Figure 10.21. The relationship between the frequency shift and additional mass Δm is given by the *Sauerbrey equation*, one form of which may be written as

$$\Delta \nu = -\left(B\Delta m \right)\nu^2 \tag{10.10}$$

where B is a calibration constant. (This method is also used as the basis of film thickness measurements for thermal evaporation, Chapter 7, Section 7.2.2.) The negative sign in Eq. (10.10) indicates the resonant frequency decreases as the mass of additional material increases.

This class of sensors is extremely sensitive; for a typical quartz crystal operating at a frequency of 10 MHz, an additional mass of the order of a nanogram can produce a shift in frequency of 1 Hz. To improve the selectivity, a chemically active layer (e.g. a polymer film) can be used to coat the resonant crystal. However, this can complicate the analysis as the vapour will affect the viscoelastic properties of the polymer as well as the mass loading.

Another class of gravimetric sensors is represented by SAW detectors. Their working principle is based on the propagation of mechanical waves along a solid surface that is in contact with a medium of lower density (e.g. air). Figure 10.22 shows a schematic diagram of an SAW sensor. The SAW is propagated from one set of interdigitated electrodes on a quartz substrate by the application of a high frequency (RF or radio frequency signal). The energy of the SAW is confined to a region close to the substrate surface, which is just a few wavelengths thick. A similar electrode arrangement is used to detect the signal. Interaction between a gas or vapour and a selective coating, which is positioned between the two sets of interdigitated electrodes, results in a perturbation of the acoustic wave. This can be detected at the receiving electrodes by a change in the phase or amplitude of the signal. Surface acoustic wave devices typically operate in the frequency range 30 MHz–1 GHz, i.e.

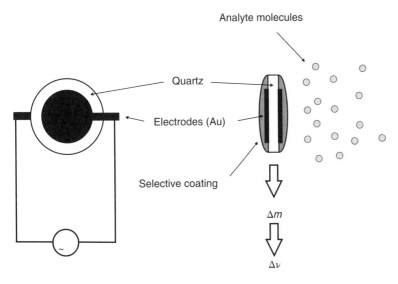

Figure 10.21 Bulk acoustic wave sensor or quartz microbalance. The absorption of the analyte molecules produce a change in the mass Δm and thereby a change in the resonant frequency $\Delta \nu$ of the quartz crystal.

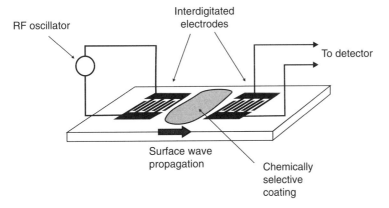

Figure 10.22 Surface acoustic wave (SAW) device for chemical sensing.

much higher than bulk oscillators. As the SAW devices also have a similar $(\Delta\nu)/(\nu)^2 \propto$ mass relationship to Eq. (10.10), a significant increase in sensitivity might be expected. Problems with drift and noise mean that large increases in sensitivity are not always achieved. However, other advantages such as the ability of fabricate more than one SAW device on the same substrate and the fact that the devices can be made smaller than bulk devices have led to much interest in this form of chemical sensor.

There are different types of surface wave that can be exploited in SAW devices. The most widely known is the *Rayleigh wave*, which consists of wave components both parallel (longitudinal) and perpendicular (transverse) to the direction of propagation. As the SAW wave propagates along the substrate, surface particles move in an elliptical path (an analogy may be made with the ripples on a water surface caused by a physical disturbance). A different type of SAW is the *Love wave*. Such waves are purely transverse and are only associated with shear stresses and are generally exploited for liquid sensing.

Piezoelectric materials other than quartz can be used in gravimetric sensing. For example, the piezoelectric polymer poly(vinylidene difluoride) (PVDF) has been introduced in Chapter 2, Section 2.6.3. Such materials can possess low *acoustic impedances* in comparison to inorganic compounds, providing an excellent mechanical match to water and biological systems.

10.4.5 Optical Sensors

Optical chemical sensors translate information from the chemical to the optical domain, and then finally to an electric signal. An optical probe is used to interrogate the chemically active layer; the principle is based on the interaction of electromagnetic waves with a material (Chapter 4, Section 4.4). The optical properties of the sensors (e.g. refractive index, absorption, scattering, and fluorescence) are modified by the chemical reaction with the measurand. In general, the first requirement for this class of sensor is the generation of an optical excitation beam. This could be used either for chemically modifying the active sensors (e.g. photoisomerization) and/or the analyte, or it could be used simply as a probe. For sensors exploiting phenomena such as chemiluminescence or bioluminescence, optical excitation is not required, because it is the chemical reaction itself that generates an electromagnetic signal. For other optical sensing systems, it is necessary to convert the chemical reaction and the molecular selectivity to a second optical signal, through, for example, indicators immobilized on a support. Alternatively, secondary reaction products like oxygen, carbon monoxide, and ammonia might be detected using a bioreactive layer. Finally, waveguides or optical fibres transmit and propagate the optical response to light sensors, which convert it to an electric signal.

Simple optical gas or vapour sensors exploit changes in the optical absorption spectra generated by the interaction of the sensing layer with the analytes (e.g. CO_2, O_2). For example, Figure 10.23 shows the time evolution of the UV–visible spectrum of an LB film of a porphyrin on exposure to 4.4 ppm NO_2 [22]. The optical absorption spectra of porphyrins include an intense absorption band around 425 nm, known as the *Soret band*. A number of bands of lower intensity and longer wavelength, referred to as *Q bands*, are also observed. The relative position and intensity of these Q bands gives information on the position and type of substituent groups present. Figure 10.23 shows that immediately following exposure to the NO_2 gas, there is a rapid decrease in the intensity of the Soret band and the appearance of new bands at around 480 and 700 nm. The stationary point at around 450 nm, indicating zero absorbance change at

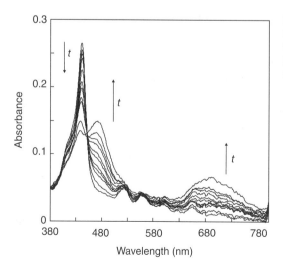

Figure 10.23 The time evolution of the UV–visible spectrum of an LB film of a porphyrin derivative following exposure to 4.4 ppm NO_2 gas. *Source:* Reprinted from Richardson et al. [22]. Copyright (2005), with permission from Elsevier.

this wavelength throughout the exposure cycle, is called an *isosbestic* point. This indicates a two state reaction process:

$$P_A + NO_2 \rightarrow P_B \tag{10.11}$$

The addition of NO_2 causes a change in state for some of the porphyrin (P) molecules from state A to state B, and the intensities of the Soret and 480 nm bands indicate the relative proportions of A and B, respectively.

The rapid decrease in the intensity of the Soret band on exposure to NO_2 shown in Figure 10.23 is followed by a slower decay in the absorbance before full saturation occurs. This is believed to be a result of the difference between surface adsorption and bulk diffusion of the toxic species. Heating the sample in nitrogen can reverse the changes in the optical spectra shown in Figure 10.23.

The manifestation of a sensor based on optical changes such as those described above can take a number of forms. For example, the use of optical fibres results in the development of compact, miniaturized, and low cost sensors for laboratory or in-situ applications. One of the main advantages of this category of chemical sensor is that they do not need to have an electric current passing through the sensing material and consequently the issue of power dissipation is absent.

Every gas, liquid, or solid, with a covalent bound and strong dipolar characteristic, interacts with infrared radiation at a specific frequency (Chapter 6, Section 6.6). On the basis of this principle, infrared absorption sensors have been developed. Whenever a target molecule interacts with the chemically active layer of the sensor, new absorption bands are generated and the original spectrum is modified. Comparison of a series of spectra allows the identification of the analyte and the determination of its concentration.

Fluorescence based sensors are able to detect very low concentrations of chemicals with good selectivity. A polymeric matrix may host the receptors or the matrix itself may be a receptor for the analyte. When the sensor probe is exposed to the analyte, the polarity of the sensing receptor changes and, with it, its emission spectrum. Using an optical fibre coated at one end with the sensing material, the same propagation channel (i.e. the fibre) is used first to interrogate the sensor (exciting signal or input) and then to collect the response (output). A multiwavelength spectral detector, positioned at the end of the fibre, then compares the two optical signals. Optical fibre biosensors are described in Chapter 12, Section 12.8.2.

Specificity may be introduced into chemical sensors for the detection of metal ions in liquids by using metal-binding macrocycles. An important class of these *ionophores* is the crown ether, a macrocyclic polyether containing ($-O-CH_2-CH_2-$)$_n$ repeat units. The self-assembly of one layer of such material can be exploited, using electrochemical principles, for metal ion detection (Chapter 11, Section 11.4). Optical sensors can use *chromoionophores* [23]. These are more sophisticated versions of simple crown ether rings and are constructed from two functionally different chemical groups: an ionophore, recognizing specific ions, and a chromophore, transducing the chemical information produced by the ionophores-ion interaction into an optical signal. Figure 10.24 shows the chemical structure for an amphiphilic chromoionophore based on a benzothiazolium styryl dye containing a 1,10-dithia-18-crown-6 ether group. The presence of the hydrocarbon chain allows this compound to form an insoluble layer at the air/water interface. The optical absorbance of LB films can be affected by complexation with metal ions such as Ag^+. The changes in absorbance at 440 nm of a two bilayer LB film on exposure to different silver concentrations are shown in Figure 10.25 [24]. The response to metal cations was found to depend on whether the LB films were used as-deposited, or kept in acidic water for a few hours prior to complexation study. This treatment was thought to influence the formation of aggregates in the LB film. Complexation with Ag^+ cations then results in the break up of the aggregates. The reaction is fully reversible if the complexed film is re-treated in acidic water for 20 hours.

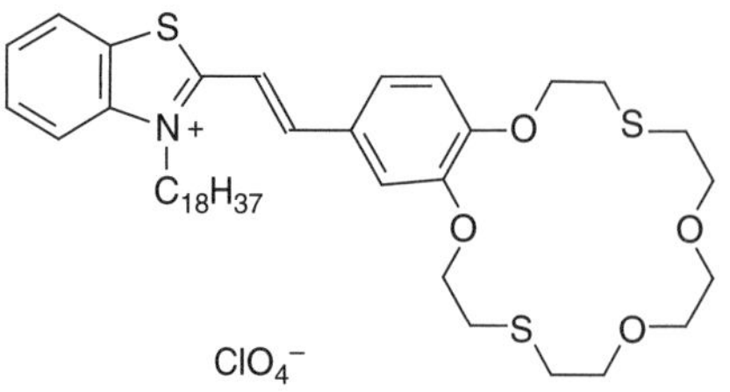

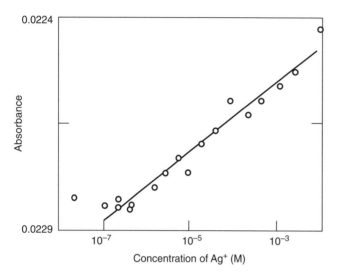

Figure 10.24 An amphiphilic chromoionophore based on a benzothiazolium styryl dye containing a 1,10-dithia-18-crown-6 ether group [23].

Figure 10.25 Change in absorbance of an LB film of the chromoionophore depicted in Figure 10.24 (two bilayers) at 440 nm as a function of AgClO$_4$ concentration in acidic water. The LB film was deposited from an acidic subphase and held in acidic water for 20 hours. *Source:* Lednev and Petty [24]. Copyright (1994) American Chemical Society.

Optical sensors exploiting evanescent waves (Chapter 4, Section 4.7.1) utilize a light beam passing through a waveguide that propagates as an evanescent wave in the medium surrounding the waveguide. For example, evanescent waves are formed when sinusoidal waves travelling in a medium of refractive index n_1 are (internally) reflected from an interface with a second medium of refractive index n_2 at an angle greater than a specific value (critical angle). Internal reflection means that no refracted wave is generated in the second medium, but the incident wave is fully reflected within the first medium. Although no net energy is transferred from one medium to the other, an optical disturbance occurs in the second medium that takes the form of an evanescent wave. The intensity of such radiation decays exponentially with the distance from the interface at which they are formed. If in the second medium (the sample), a chemical species is able to absorb electromagnetic radiation at the wavelength of the probe, then there will be a reduction in the intensity of the reflected light passing through the waveguide. A detector at the end of the optical support will register such a variation. Reduction in the intensity profile can therefore indicate the presence of the analyte and provide information about its concentration.

Particular types of evanescent wave sensors exploit surface plasmon resonance (Chapter 4, Section 4.7.2). This has been the subject of theoretical and experimental research for several decades [25]. The electromagnetic fields of a surface plasmon wave are distributed in a highly asymmetric fashion and the vast majority of the field is concentrated in the dielectric. A surface plasmon propagating along the surface of silver is less attenuated and exhibits higher localization of the electromagnetic field in the dielectric than a surface plasmon supported by gold. However, gold is more suitable for sensing applications in liquids, because it is relatively stable (e.g. does not oxidize). As the excitation of surface plasmons by an optical wave results in resonant transfer of energy into the surface plasmons, there will be a resonant absorption of the energy of the optical wave. Because of the strong concentration of the electromagnetic field in the dielectric (an order of magnitude higher than that in typical evanescent field sensors using dielectric waveguides), the propagation constant of the surface plasmon, and consequently the resonance condition, is very sensitive to variations in the optical properties of the dielectric adjacent to the metal layer supporting the wave. Changes in the optical parameters of the trans-ducing medium can be detected by monitoring the interaction between the surface plasmon and the optical wave.

As noted in Chapter 4, Section 4.7.2, there are several methods available to couple optical energy into surface plasmons. Two of the most common techniques are attenuated total reflec-tion (ATR) in prism or semi-cylinder couplers and diffraction at the surface of a grating. It is also possible to excite surface plasmon resonance (SPR) by an optic fibre, thereby providing a high degree of miniaturization. In the ATR Kretschmann configuration, a simple SPR system for chemical sensing can operate at a fixed angle. This is chosen to be close to the resonant condition, at a point where the reflectivity varies rapidly with angle. The reflected light inten-sity is then monitored as a chemical species interacts with a coating deposited onto the metal. Figure 10.26a shows the reflectivity response to copper ions for a fixed-angle SPR sensing arrangement [26]. The sensing film consists of a polyelectrolyte bilayer (polyethyleneimine/ poly(ethylene-*co*-maleic acid) (PEI/PMAE)) deposited by the electrostatic layer-by-layer

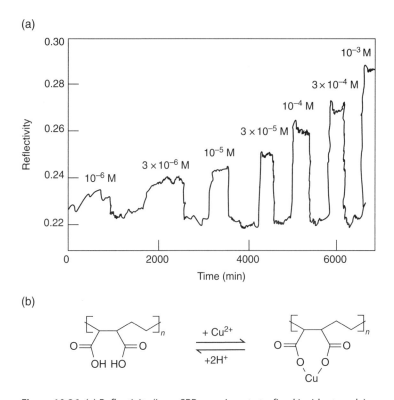

(a)

(b)

Figure 10.26 (a) Reflectivity (in an SPR experiment at a fixed incident angle) versus time for a polyelectrolyte bilayer (PEI/PMAE) exposed to increasing concentrations of copper acetate. The sample was immersed in hydrochloric acid and purged with pure water between the different copper ion concentrations. *Source:* Reprinted from Pearson et al. [26], Copyright (2001), with permission from IOP Publishing Limited. (b) Reaction of copper ions and maleic acid to form a chelate.

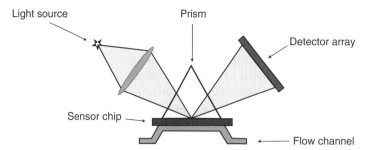

Figure 10.27 Schematic diagram of commercial SPR system with no moving parts. A wedge-shaped light source and a diode array are used to measure the intensity of the reflected light.

technique (Chapter 7, Section 7.3.3). On exposure to copper, it is thought that a chelate is formed with polyethylene-co-maleic acid, one of the components of the organic film. The reaction, depicted in Figure 10.26b, could be fully reversed by exposing the complexed film to hydrochloric acid for 10 minutes.

In scanning angle SPR, the wavelength is fixed and the incident angle is varied. It is thereby possible to have a qualitative picture for the resonance curve following interactions of the vapour with the sensing material. Alternatively, the incident angle can be fixed while the wavelength is varied. Only one wavelength will excite the surface plasmons, resulting in a minimum in intensity of the reflected light. Hence a plot of reflectivity versus wavelength will be similar to that of reflectivity versus incident angle.

SPR systems have now been successfully commercialized and are extensively applied to biochemical analysis. Most of these use prism coupling for the momentum enhancement. The Kretschmann configuration is commonly used in preference to the Otto arrangement, as the analyte can easily come into contact with the sensing layer. One commercial SPR system is equipped with a continuous flow arrangement in which four channels are coupled in series [27]. One of the channels can be used to provide an in-line reference. An automatic sample needle delivers buffer solution and analyte to the surface of the sensing 'chip'. The continuous flow ensures that no changes in the analyte concentration occur during the measurement. A wedge-shaped light source and a diode array are used to measure the relative intensities of the reflected light, as shown in Figure 10.27. In this way, the optical interface has no moving parts.

With an increased interest in optical sensors for use in remote and portable control instrumentation, there is much work on the development of low-cost SPR systems. Work is also focused on multichannel SPR sensing, in which different sensing layer are exposed simultaneously to the same analyte. Figure 10.28 shows the definition of several channels on an SPR chip. The different regions can be achieved using various patterning methodologies, including layer-by-layer self-assembly [28].

10.5 Biological Olfaction

The human *olfactory system* has many receptors cells (sensors) that are individually non-specific; signals from these are fed to the brain via a network of primary and secondary neurons for processing. It is generally believed that the selectivity of the olfactory system is a result of a high degree of parallel processing in the neural architecture. In humans, the process of olfaction begins with the inhalation of molecules and their transport to the olfactory receptor cells, located within a specialized membrane, of approximate dimensions 2 by 5 cm, called the

Figure 10.28 Schematic diagram of an arrangement for a multi-channel SPR sensing chip.

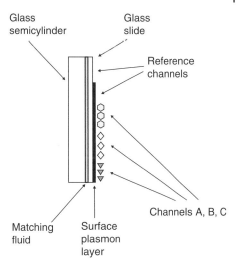

Glass semicylinder

Glass slide

Reference channels

Channels A, B, C

Matching fluid

Surface plasmon layer

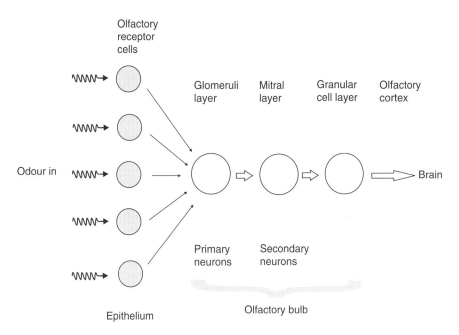

Olfactory receptor cells

Glomeruli layer

Mitral layer

Granular cell layer

Olfactory cortex

Odour in

Brain

Primary neurons

Secondary neurons

Epithelium

Olfactory bulb

Figure 10.29 Schematic representation of the human olfactory system.

epithelium, in the nose. The subsequent interactions generate electrical signals, which propagate down the axon (Chapter 12, Section 12.7) of the olfactory receptor cells to the olfactory bulb, as depicted in Figure 10.29. Further layers, referred to as glomeruli, process the signals. Here, the nature of the odorous stimulus is encoded as a specific combination of activated glomeruli, which takes the form of a two-dimensional map. Particular odorants are associated with specific topographic patterns of activity. Although mammals possess in the order of 1000 different receptor proteins, it is established that they can detect more than this number of different odorants. This suggests that any given odorant is able to interact with more than one receptor protein and, conversely, that any single receptor is able to interact with more than one odorant.

It is worth noting that olfactory receptor cells are replaced every 30–60 days, so at any one time the population of olfactory receptor cells contributing to a particular glomerus will possess a range of ages. This will provide the system with a mechanism to overcome the problem of a loss of sensitivity of the receptor cells with time. Although the sensitivity of any given receptor is likely to decrease with time, the summed response from a population of cells of all ages will be more stable [11].

The map of glomerular activity, as detected by the *mitral* cells, then feeds into the next level of information processing, *the granular cell* level, before passing via the *olfactory cortex* to the brain. Olfactory information travels not only to the *limbic system*, primitive brain structures that govern emotions, behaviour, and memory storage, but also to the brain's *cortex*, or outer layer, where conscious thought occurs. In addition, it combines with taste information in the brain to create the sensation of flavour.

10.6 Electronic Noses

While many chemical-sensing devices can show adequate sensitivities, the selectivity can be poor. For example, a semiconductive polymer (chemiresistor) may show a similar change in electrical resistance to a range of oxidizing (reducing) gases. To circumvent this difficulty, one approach is to use an array of sensing elements, rather than a single device. This is the method favoured by nature, as described above. The *electronic nose* is an attempt to imitate the human olfactory system [11]. Emulation of *gustation*, the sense of taste, would similarly lead to an *electronic tongue* [29]. Individual sensors can be based on polymer films. Each element is treated in a slightly different way during deposition, so that it responds uniquely on exposure to a particular gas or vapour. The pattern of resistance changes in the sensor array can then be used to fingerprint the vapour. Alternatively, data for a single sensor, but measured under different conditions, such as the conductivity of a chemiresistor at different frequencies (Figure 10.20), may be used to provide a suitable fingerprint.

Figure 10.30 shows a schematic diagram of an electronic nose. If more than two linear sensors are used to characterize a sample, a *multivariate* evaluation is usually applied to visualize

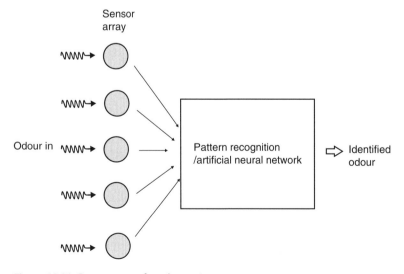

Figure 10.30 Components of an electronic nose.

and evaluate the data [11]. There are many such techniques in use. In electronic noses, pattern recognition approaches are generally used. Unsupervized pattern recognition algorithms try to cluster the results, which are usually visualized in a two-dimensional plot. A common technique that is applied is *principal component analysis*. This involves the use of a mathematical procedure to transform a number of (possibly) correlated variables into a (smaller) number of uncorrelated variables called principal components. Another approach to multivariate analysis uses an *artificial neural network*, ANN [30]. This is a highly parallel information processing system that has been inspired from our understanding of the biological nervous system. The network consists of a lattice of information processing elements (neurons – Chapter 12, Section 12.7) that are connected together in a certain way. The strengths of these connections are called *synaptic weights* and are determined either during a training (or learning) phase for supervized neural networks, or by an algorithm for unsupervized neural networks.

10.7 Physical Sensors and Actuators

Physical sensors are selective and sensitive to physical phenomenon, such as pressure, force, acceleration, specific heat, current, electric charge, magnetic field, electrical resistance, and light. These form a major part of the sensor industry, with an increasing number of applications. Micro-Electro-Mechanical Systems (MEMS or the 'nano' equivalent NEMS) is the integration of mechanical elements, sensors, actuators, and electronics on a common silicon substrate through microfabrication technology. This section will be confined to a description of pressure sensors and polymer actuators as far as such structures are being developed with organic materials [31].

10.7.1 Touch Sensors

Touch, or tactile, sensors detect not only the presence of a touch, but also its position. A common touch sensor, exploited in handheld computing devices, is the touchscreen. This can operate on either resistive or capacitive principles. Figure 10.31 shows the principle of the former arrangement. The device consists of a two resistive layers separated by an insulating (usually

Figure 10.31 Principles of a resistive touch sensor based on a conductive polymer.

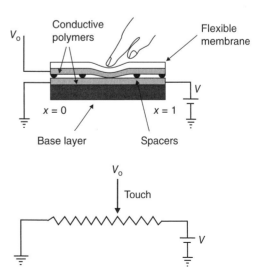

air) spacer layer. Some systems use conductive indium-tin-oxide as the resistive layer, although many touchscreens are now being developed based on electrically conductive polymers or graphene. If a flexible non-conductive polymer base plate is used, the touchscreen assembly becomes fully flexible. The touchscreen is essentially a variable resistor, where V is the applied voltage and V_o is the output voltage – typically fed, via an analogue to digital converter, to a computer. The touch position x is related to the measurement voltage by $x = V_o/V$. Even if the electrical contact is made between the flexible membrane and the base plate at several locations, there will only be one measured voltage and hence only one intermediate touch location measured.

Capacitive touchscreens are another example of tactile sensors. A capacitor is formed when a user's finger touches a dielectric coating deposited onto a resistive base (e.g. a conductive polymer). One electrode is this resistive layer, while the other electrode is the user's finger. The capacitance is typically detected by observing a shunt to ground of current at frequencies in the kHz range. Position information is obtained by monitoring the magnitude of this current.

Tactile sensors are very important in the development of automated industrial processing equipment. For example, a robot hand must be able to detect and subsequently grip a work piece with sufficient force to hold it firmly, but not to use excessive force that may cause damage. The distributed sensing in a human hands allow the manipulation of an egg with enough force that it does not slip, but without so much force that it breaks. One of the goals in robotics is to develop an array of sensors that provides the systems with a similar source of tactile feedback, i.e. 'smart skin'. The majority of tactile sensors that have been developed for robots have been based on the resistance changes associated with compression of a conductive elastomer (e.g. a carbon loaded rubber) or foam. There are also examples of sensors that exploit conductive polymers. When a force is applied to the sensor element, the resistance changes as a result of the *piezoresistive* effect (which is different from the piezoelectric effect observed in insulating polymers such as PVDF, Chapter 2, Section 2.6.3).

10.7.2 Polymer Actuators

Polymers that can be activated to change shape or size have been available for many years [31, 32]. The activation mechanisms can include chemical, thermal, pneumatic, optical, and magnetic processes. The first documented experiment with electroactive polymeric materials is attributed to Roentgen where, in 1880, he observed a length change in a rubber-band (with one fixed end and a mass attached to the free end) that he subjected to an electric field. Following the observation of a substantial piezoelectric activity in PVDF, other polymer systems have been investigated, and a number of effective materials have been developed.

Generally, electroactive polymer actuators can be divided into two major groups based on their activation mechanism: ionic (involving mobility or diffusion of ions) and electronic (driven by electric field). The former category is often referred to as 'wet', because the actuators use aqueous component materials. The electronic polymers (electrostrictive, electrostatic, piezoelectric, and ferroelectric) are driven by electric fields and can be made to hold the induced displacement under activation of a DC voltage, allowing them to be considered for robotic applications. These materials have a greater mechanical energy density and they can be operated in air with no major constraints, but they require a high activation field ($> 100 \, \text{V} \, \mu\text{m}^{-1}$) close to the breakdown level. Ionic materials (gels, polymer-metal composites, conductive polymers, and carbon nanotubes) are driven by diffusion of ions and these require an electrolyte for the actuation mechanism. Their major advantage is their operation at drive voltages as low as 1–2 V. However, there is a need to maintain their wetness and, apart from conductive polymers and carbon nanotubes, it is difficult to

Figure 10.32 Composition of a polyelectrolyte gel.

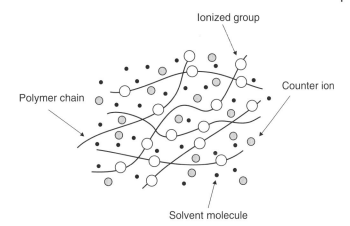

sustain DC-induced displacements. The induced displacement of both the electronic and ionic polymer can be geometrically designed to bend, stretch, or contract.

Polymeric gels are two-phase systems composed of a solid phase or elastic matrix permeated by a fluid, generally water, and a number of different types have been developed and studied. An example is shown in Figure 10.32. These systems exhibit plastic contraction with changes in temperature, pH, magnetic, or electrical field, and have a vast number of applications; for example soft actuators in the biomedical field or for controlled drug release. Their application in smart lenses has already been noted in Chapter 8, Section 8.7. Gel actuators are characterized by large strains (50% or more) and lower forces than conducting polymers. Many soft, wet, ionic, and reactive artificial dense gels have been developed in the past two decades, based on organic compounds ranging from conductive polymers to graphene. The material composition (reactive polymer, ions, and water) mimics, in its simplest form, the composition of living cells. Most of these materials are biocompatible with different biological cells and tissues.

The poor mechanical properties often found in extension restrict the application of electropolymerized films and gels. Therefore, many of the devices described in the literature are *bimorphs*, with a stronger supporting layer. An example is depicted in Figure 10.33, in which a conductive polymer is deposited onto a gold layer. Such microactuators have been used in quite sophisticated applications. Figure 10.34 shows the use of a microrobot to move and position a small object in the laboratory [33]. The microrobot arm is made of polypyrrole microactuators and consists of an 'elbow', a 'wrist', and a 'hand' with two to four 'fingers'. Each joint consists of two separately controllable micromuscles. Using this arrangement, a 100 μm glass bead could be lifted and moved over a system of polyurethane tracks. This microactuator might be considered as the 'big brother' version of the molecular actuators described in Chapter 11, Section 11.11.

Figure 10.33 Conductive polymer/Au bimorph actuator. (a) No applied voltage. (b) Voltage applied (e.g. polymer is oxidized).

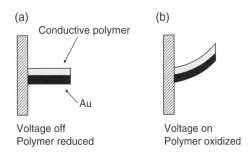

(a)

(b)

(c)

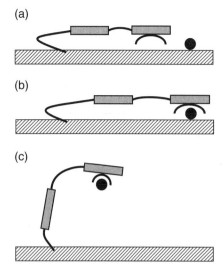

Figure 10.34 (a) A microrobot arm made of polypyrrole microactuators. The microrobot can be used to (b) locate and (c) lift a 100 μm glass bead [33].

The electronic (or so-called dry) actuators usually respond to electric fields (as opposed to charge or mass transport). The resulting strain is approximately proportional to the square of the electric field. If the response is determined by the field-induced reorientation of the crystalline or semi-crystalline structure, then the polymer is said to be *electrostrictive*. However, if the response is dominated by the interaction of the electrostatic charges on the electrodes (often called the *Maxwell stress*), then the polymer is called a *dielectric elastomer*. It should be noted that the phenomenon of electrostriction is different to that of piezoelectricity, introduced in Chapter 2, Section 2.6.3, which is exploited by certain acoustic sensors (Section 10.4.5). The latter only occurs in certain types of crystal structure (those without a centre of symmetry) and results in a strain that is proportional to the applied electric field. Unlike piezoelectricity, there is also no inverse electrostrictive effect, i.e. a deformation does not produce an electric field.

10.7.3   Lab-on-a-Chip

A promising tool for analysing proteins and protein complexes in the biology laboratory of the future is a microfluidic device commonly called a *lab-on-a-chip* [34–36]. These 'laboratories' are fabricated using photolithographic processes developed in the microelectronics industry to create circuits of miniature chambers and channels in a quartz, silica, or glass chip. They direct the flow of liquid chemical reagents just as semiconductors direct the flow of electrons. These reagents can be diluted, mixed, reacted with other reagents, or separated – all on a single chip.

The flow of a fluid through a *microfluidic channel* can be characterized by the *Reynolds number*, *Re*, defined as

$$Re = \frac{Lv\rho}{\mu} \tag{10.12}$$

where L is the most relevant length scale, μ is the viscosity, ρ is the fluid density, and v is the average velocity of the flow. For many microchannels, L is equal to $4A/P$ where A is the cross-sectional area of the channel and P is the wetted perimeter of the channel. Due to the small dimensions of microchannels, the Re is usually much less than 100, often less than 1.0. In this Re regime, the flow is completely laminar and no turbulence occurs. The transition to turbulent flow generally occurs at Re in the region of 2000. Laminar flow provides a means by which molecules can be transported in a relatively predictable manner through microchannels. However, even at $Re < 100$, it is possible to have momentum-based phenomena such as flow separation.

There are two common methods by which fluid actuation through microchannels can be achieved. In pressure driven flow, the fluid is pumped through the device via positive displacement pumps, such as syringe pumps. One of the basic laws of fluid mechanics for pressure driven laminar flow, the so-called no-slip boundary condition, states that the fluid velocity at the walls must be zero. This produces a parabolic velocity profile within the channel. Pressure-driven motion, termed *Poiseuille flow*, is well understood, but designing and fabricating reliable mechanical pumps in the traditional materials of silicon and glass has been difficult. These devices require multiple levels of fabrication, and are easily damaged by particles of dust and contaminants in the fluid.

Another common technique for pumping fluids is that of electro-osmotic pumping. If the walls of a microchannel have an electric charge (most surfaces do), an electric double layer of counter ions will form at the walls. When an electric field is applied across the channel, the ions in the double layer move towards the electrode of opposite polarity. This creates motion of the fluid near the walls and transfers via viscous forces into convective motion of the bulk fluid. If the channel is open at the electrodes, as is most often the case, the velocity profile is uniform across the entire width of the channel. However, if the electric field is applied across a closed channel (or a backpressure exists that just counters that produced by the pump), a recirculation pattern forms in which fluid along the centre of the channel moves in a direction opposite to that at the walls. In closed channels, the velocity along the centreline of the channel is 50% of the velocity at the walls. Electrically driven flow has a number of drawbacks: sensitivity to impurities that adsorb on the wall of the channel, ohmic generation of heat in the fluid, and the need for high voltages (on the order of kilovolts).

Microfluidic devices can be fabricated from a variety of materials. Based on fabrication techniques developed in the microelectronics industry, silicon has been used extensively to create such devices. Polymers offer a lot of advantages for microsensors, microactuators, and microfluidics. These are relatively low-cost materials, fabrication techniques are simple, and there is no need for special clean-room and/or high temperature processes. Polymers can be deposited on various types of substrates, and there is a wide choice of molecular structures. This enables films to be produced with various physical and chemical properties, including sensing and actuation behaviour. A popular material for fabrication of microfluidic device is the silicone polymer, poly(dimethylsiloxane) (PDMS) (Section 10.4.4). Microfluidic devices can be fabricated from PDMS by pouring the liquid over a mould (usually silicon or photoresist) and curing to cross-link the polymer. The result is an optically clear, relatively flexible material that can be stacked onto other cured polymer slabs to form complex three-dimensional geometries.

Common fluids used in microfluidic devices include whole blood samples, bacterial cell suspensions, protein or antibody solutions, and various buffers. The use of microfluidic devices for biomedical research and clinically important technologies has a number of significant advantages. First, because the volume of fluids within these channels is very small, usually several nanolitres, the amount of reagents and analytes used is quite small. This is especially significant for expensive reagents. The fabrication techniques used to construct microfluidic devices are relatively cheap and are very amenable both to highly elaborate, multiplexed devices and also to mass production. In a manner similar to that for microelectronics, microfluidic technologies enable the fabrication of highly integrated devices, useful for performing several different functions on the same substrate chip. One of the long-term goals in the field of microfluidics is to create integrated, portable clinical diagnostic devices for home and bedside use, thereby eliminating time consuming laboratory analysis procedures.

Methods of manipulating particles, such as living cells, on the chip surface include *electrophoresis* and *dielectrophoresis* [37, 38]. Electrophoretic forces arise from the interaction of a particle's charge in an electric field, while dielectrophoresis originates from the polarizability (Chapter 4, Section 4.3.2) and occurs with uncharged particles. Figure 10.35 shows a schematic diagram illustrating the principle of dielectrophoresis. A particle is suspended within an asymmetric electrode arrangement. Under the influence of the field, the particle

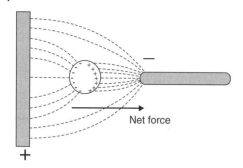

Figure 10.35 Origin of dielectrophoresis. The application of a non-uniform electric field **E** to an uncharged particle produces a polarization. The interaction of the induced dipoles and the external field causes a force to act on the particles.

will become polarized. The interaction between the induced charges and the local electric field will then result in a force on the particle. Because of the inhomogeneous nature of the electric field in Figure 10.35, this force is greater on the side facing towards the point electrode than that on the side facing the plane electrode, and there is net motion towards the point electrode. This effect is called positive dielectrophoresis. However, if the particle is less polarizable than the surrounding medium, the induced dipole will align in the opposite direction to the field and the particle will be repelled from the high-field regions – negative dielectrophoresis. The force that is created in dielectrophoresis is dependent on the induced dipole and is unaffected by the direction of the electric field, responding only to the field gradient. Since the alignment field is irrelevant, AC fields can also generate the force. This has the advantage of eliminating any electrophoretic force a net particle charge. The time-averaged dielectrophoretic force F_{DEP} acting on a spherical body of radius r is given by [38]:

$$F_{\text{DEP}} = 2\pi r^3 \varepsilon_m' \, \text{Re}\left[K(\omega)\right] \nabla E_{\text{rms}}^2 \tag{10.13}$$

where ε_m' is the real part of the complex permittivity (i.e. the dielectric constant – Chapter 3, Section 3.6.1) of the surrounding medium, E_{rms} is the root-mean-squared value of the electric field, ∇ is the mathematical *Del vector operator* $\left(\nabla = \dfrac{\partial}{\partial x} + \dfrac{\partial}{\partial y} + \dfrac{\partial}{\partial z}\right)$ and $\text{Re}[K(\omega)]$ is the real part of the complex permittivity factor given by

$$K(\omega) = \frac{\varepsilon_p - \varepsilon_m}{\varepsilon_p + 2\varepsilon_m} \tag{10.14}$$

where ε_p, ε_m are the complex permittivities of the particle and medium, respectively.

By careful construction of the electrode geometry that is used to generate the field, it is possible to create electric field morphologies so that regions of increasing field strengths bound potential energy minima. In such electrode arrangements, particles experiencing positive dielectrophoresis are attracted to the regions of maximum field (e.g. the electrode edges), while particles experiencing negative dielectrophoresis are trapped in isolated field minima.

Uncharged particles can also rotate under the influence of applied fields via *electro-orientation* and *electro-rotation effects* [38]. The former results from non-spherical particles, whilst the latter exploits a rotating electric field. *Travelling-wave dielectrophoresis* is a linear analogue of electro-rotation. The electrodes are arranged in a track and the phase of the electric field is advanced along the track. This produces an electric field wave that travels along the track. Particles can then be moved along these tracks.

10.8 Wearable Electronics

An increasing number of electronic products are being carried around by people, e.g. mobile phones, laptops, personal hi-fis, Personal Digital Assistants (PDAs), with more such gadgets under development. It makes sense, therefore, to integrate such products into our clothes. Smart or intelligent clothing is a combination of electronics and clothing textiles [39, 40]. The first step is to integrate existing products into clothing simply by sewing these in. For example, orientation, communications, and geographic positioning electronics can all be incorporated into outdoor clothing. The functional architecture is implemented using *GPS* (Global Positioning System) for navigation and *GSM* (Global System for Mobile communications) for communication. Power supplies and the user interface can be incorporated in the form of a supporting vest.

New fibre and textile materials and miniaturized electronic components make it possible to create truly usable smart clothes. These intelligent clothes are worn like ordinary clothing, providing help in various situations according to the designed application. Significant challenges remain in this field. Fabric sensors must be washable, non-toxic, and resistant to surface shear. Mechanical and electrical connections must also be washable and resistant to not only water but also to bending, torsion, and other factors that a wearable item may be subjected to.

Problems

10.1 (a) What is the importance of microfluidics in chemical sensing? Give at least three important reasons for downscaling to the micro-domain.
   (b) When downscaling from the macro-scale to the micro-domain, several physical phenomena begin to dominate. Name at least two of these and explain their influence on fluidic flow.

10.2 A parallel plate capacitive vapour sensor is fabricated by sandwiching a $1\,\mu m$ thick film of a polysiloxane (dielectric constant = 2.5) between two metal electrodes, each of dimensions $1 \times 1\,mm$. The sensor is exposed to chloroform vapour. Calculate the maximum change in capacitance of the sensor. Comment critically on the sensor design.

10.3 (a) Section 10.3 discusses the *sensitivity* of a sensing device. What is the *cross-sensitivity* of a sensor?
   (b) Distinguish between the *operational lifetime* and the *shelf lifetime* of a chemical sensor.

10.4 A chemiresistor, as described in Section 10.4.2, is designed to change its resistance in the presence of a gas or vapour. However, the device will also respond to changes in the ambient temperature. How can this unwanted response be removed?

10.5 A quartz crystal microbalance vibrating at $1\,MHz$ has a mass sensitivity equal to $2.0 \times 10^8\,cm^2\,g^{-1}\,s^{-1}$. Calculate the additional mass per cm^2 in units of ng that will produce a frequency shift of $1\,Hz$. If the crystal oscillating frequency increases to $3\,MHz$, what frequency shift in Hz will be produced by the same additional mass?

References

1 Middelhoek, S. and Noorlag, D.W. (1981–82). Three-dimensional representation of input and output transducers. *Sensors and Actuators* 2: 29–41.

2 Janata, J. (2009). *Principles of Chemical Sensors*, 2e. New York: Springer.

3 Jimenez-Jorquera, C., Orozco, J., and Baldi, A. (2010). ISFET based microsensors for environmental monitoring. *Sensors* 10: 61–83.

4 Ritjareonwattu, S., Yun, Y., Pearson, C., and Petty, M.C. (2010). Enhanced sensitivity of an organic field-effect transistor pH sensor using a fatty acid Langmuir-Blodgett film. *Organic Electronics* 11: 1792–1795.

5 Ritjareonwattu, S., Yun, Y., Pearson, C., and Petty, M.C. (2012). An ion sensitive organic field-effect transistor incorporating the ionophore valinomycin. *IEEE Sensors Journal* 12: 1181–1186.

6 Snow, A.W. and Barger, W.R. (1989). Phthalocyanine films in chemical sensors. In: *Phthalocyanines, Properties and Applications* (ed. C.C. Leznoff and A.B.P. Lever), 341–391. New York: Wiley-VCH.

7 Ray, A.K. (2006). Organic materials for chemical sensing. In: *Handbook of Electronic and Photonic Materials* (ed. S. Kasap and P. Capper), 1241–1266. Würzburg: Springer.

8 Bott, B. and Jones, T.A. (1984). A highly sensitive NO_2 sensor based on electrical conductivity changes in phthalocyanine films. *Sensors and Actuators* 5: 43–53.

9 Yoon, H. (2013). Current trends in sensors based on conducting polymer nanomaterials. *Nanomaterials* 3: 524–549.

10 Agbor, N.E., Petty, M.C., and Monkman, A.P. (1995). Polyaniline thin films for gas sensing. *Sensors and Actuators B* 28: 173–179.

11 Gardner, J.W. and Bartlett, P.N. (1999). *Electronic Noses: Principles and Applications*. Oxford: Oxford University Press.

12 Uzun, L. and Turner, A.P.F. (2016). Molecularly-imprinted polymer sensors: realizing their potential. *Biosensors and Bioelectronics* 76: 131–144.

13 Mabrook, M.F., Pearson, C., and Petty, M.C. (2005). An inkjet-printed chemical fuse. *Applied Physics Letters* 86: 013507.

14 Mabrook, M.F., Pearson, C., and Petty, M.C. (2006). 'Inkjet-printed chemical sensors for the detection of organic vapours. *IEEE Sensors Journal* 6: 1435–1444.

15 Bartlett, P.N. and Ling-Chung, S.K. (1989). Conducting polymer gas sensors. Part II: Response of polypyrrole to methanol vapour. *Sensors and Actuators* 19: 141–150.

16 Warner, J.H., Schäffel, F., Bachmatiul, A., and Rümmeli, M.H. (2013). *Graphene: Fundamentals and Emergent Applications*. Waltham: Elsevier.

17 Salavagione, H.J., Diez-Pascual, A.M., Lázaro, E. et al. (2014). Chemical sensors on polymer composites with carbon nanotubes and graphene: the role of the polymer. *Journal of Materials Chemistry A* 2: 14289.

18 Senturia, S.D., Sechen, C.M., and Wishneusky, J.A. (1977). The charge-flow transistor: a new MOS device. *Applied Physics Letters* 30: 106–108.

19 Barker, P.S., Petty, M.C., Monkman, A.P. et al. (1996). A hybrid phthalocyanine/silicon field effect transistor sensor for NO_2. *Thin Solid Films* 284–285: 94–97.

20 Grate, J.W. and Abraham, M.H. (1991). Solubility interactions and the design of chemically selective sorbent coatings for chemical sensors and arrays. *Sensors and Actuators B* 3: 85–111.

21 Casalini, R., Wilde, J.N., Nagel, J. et al. (1999). Organic vapour sensing using thin films of a co-ordination polymer: comparison of electrical and optical techniques. *Sensors and Actuators* 57: 28–34.

22 Richardson, T.H., Dooling, C.M., Jones, L.T., and Brook, R.A. (2005). Development and optimization of porphyrin gas sensing LB films. *Advances in Colloid and Interface Science* 116: 81–96.

23 Lednev, I.K. and Petty, M.C. (1996). Langmuir monolayers and Langmuir-Blodgett multilayers containing macrocyclic ionophores. *Advanced Materials* 8: 615–629.

24 Lednev, I.K. and Petty, M.C. (1994). Aggregate formation in Langmuir-Blodgett films of an amphiphilic benzothiazolium styryl chromoionophore. *Langmuir* 10: 4185–4189.

25 Homola, J. (2006). *Surface Plasmon Resonance Based Sensors.* Berlin: Springer.

26 Pearson, C., Nagel, J., and Petty, M.C. (2001). Metal ion sensing using ultrathin organic films prepared by the layer-by-layer adsorption technique. *Journal of Physics D: Applied Physics* 34: 285–291.

27 Biacore. Available from URL: https://proteins.gelifesciences.com. Accessed 6 October 2017.

28 Palumbo M, Petty MC. (2006). Polyelectrolytes-based SPR sensors. In: CA Grimes, EC Dickey, MV Pishko (eds). *Encyclopaedia of Sensors.* Vol. 8. California: American Scientific Publishers; pp. 33–52.

29 Méndez, M.L.R. (2016). *Electronic Noses and Tongues in Food Science.* London: Academic Press.

30 Barker, P.S., Chen, J.R., Agbor, N.E. et al. (1994). Vapour recognition using organic films and artificial neural networks. *Sensors and Actuators B* 17: 143–147.

31 Ionov, L. (2015). Polymeric actuators. *Langmuir* 31: 5015–5024.

32 Otero, T.F., Martinez, J.G., and Arias-Pardilla, J. (2012). Biomimetic electrochemistry from conducting polymers. A review. Artificial muscles, smart membranes, smart drug delivery and computer/neuron interfaces. *Electrochimica Acta* 84: 112–128.

33 Jager, E.W.H., Smela, E., and Inganäs, O. (2000). Microfabricating conjugated polymer actuators. *Science* 290: 1540–1545.

34 Whitesides, G.M. (2006). The origins and future of microfluidics. *Nature* 442: 368–373.

35 Chin, C.D., Linder, V., and Sia, S.K. (2012). Commercialization of microfluidic point-of-care diagnostic devices. *Lab on a Chip* 12: 2118–2134.

36 Bocquet, L. and Tabeling, P. (2014). Physics and technological aspects of nanofluidics. *Lab on a Chip* 14: 3143–3158.

37 Voldman, J. (2006). Electrical forces for microscale cell manipulation. *Annual Review of Biomedical Engineering* 8: 425–454.

38 Hughes, M.P. (2000). AC electrokinetics: applications for nanotechnology. *Nanotechnology* 11: 124–132.

39 Castano, L.M. and Flatau, A.B. (2014). Smart fabric sensors and e-textile technologies: a review. *Smart Materials and Structures* 23: 053001.

40 Stoppa, M. and Chiolerio, A. (2014). Wearable electronics and smart textiles: a critical review. *Sensors* 14: 11957–11992.

Further Reading

Fraden, J. (2010). *Handbook of Modern Sensors*, 4e. New York: Springer.

Nabook, A. (2005). *Organic and Inorganic Nanostructures.* Boston: Artech House.

Saha, K., Agasti, S.S., Kim, C. et al. (2012). Gold nanoparticles in chemical and biological sensing. *Chemical Reviews* 112: 2739–2779.

Spinks, G.M., Whitten, P.G., Wallace, G.G., and Truong, V.-T. (2008). An introduction to conducting polymer actuators. In: *Introduction to Organic Electronic and Optoelectronic Materials and Devices* (ed. S.-S. Sun and L.R. Dalton), 733–763. Boca Raton. FL: CRC Press.

Waiser, R. (ed.) (2005). *Nanoelectronics and Information Technology.* Weinheim: Wiley-VCH.

Wright, R.H. (1982). *The Sense of Smell.* Boca Raton, FL: CRC Press.

11

Molecular and Nanoscale Electronics

Organic and Molecular Electronics: From Principles to Practice, Second Edition. Michael C. Petty.
© 2019 John Wiley & Sons Ltd. Published 2019 by John Wiley & Sons Ltd.
Companion website: www.wiley.com/go/petty/molecular-electronics2

O brave new world!

11.1 Introduction

Molecular electronics recognizes the spectacular size reduction in the individual processing elements in integrated circuits over recent years (Chapter 1). The 'bottom-up' approach to nanotechnology offers many intriguing prospects for manipulating materials at the nanometre scale, thereby providing opportunities to build up architectures with predetermined and unique physical and/or chemical properties. The literature is packed with examples of these activities, far too many to condense into a single chapter of a book. A selection of the ideas is presented here to provide an illustration of what might be possible as progress is made in the twenty-first century.

11.2 Nanosystems

The development of molecular electronics, as with all other areas of scientific endeavour, is constrained by the laws of physics; in most cases, these are the laws of classical or Newtonian physics. However, on the nanoscale, other principles become more important, even crucial. Some of the ideas of quantum mechanics, which are responsible, for example, for the chemical bonding that holds solids together, have been introduced in Chapters 2 and 3. Quantum mechanical tunnelling (Chapter 3, Section 3.5.2) becomes a significant electrical conduction process at dimensions of less than 5 nm and therefore will have a key role as molecular scale electronics develops. In the following sections, some other concepts are introduced that might be significant for electronic and/or optoelectronic devices operating at nanometre dimensions.

11.2.1 Scaling Laws

The magnitudes of physical quantities characterizing nanoscale systems differ considerably from those familiar from the macroscale world. Some of these quantities can be estimated by applying scaling laws to the values for macroscale configurations [1, 2]. For example, the strength of a structure and the force it exerts can be assumed, in the first instance, to scale with its cross-sectional area. Nanoscale devices accordingly exert only small forces: a stress of $10^{10}\,\mathrm{N\,m^{-2}}$ equates to $10^{-8}\,\mathrm{N\,nm^{-2}}$ or $10\,\mathrm{nN\,nm^{-2}}$.

Of particular relevance to this chapter is the scaling of classical (macroscopic) electromagnetic systems. Here, it is convenient to assume that electrostatic field strengths (hence electrostatic stresses) are independent of scale. The onset of strong field-emission currents (Chapter 3, Section 3.5.5) from conductors limits the electrostatic field strength permissible at the electrodes of nanoscale systems; values of $10^{9}\,\mathrm{V\,m^{-1}}$ can readily be tolerated. At this field strength, one nanometre corresponds to a potential difference of 1 V.

If all the dimensions of a material are reduced by a constant K, then the effects on various important electrical parameters can be calculated. For example, electrical resistance R will scale with K, while the capacitance C (for a parallel plate capacitor) will scale as K^{-1} (assuming of course that the resistivity and permittivity of the material remain unchanged as its dimensions are reduced). An important consequence is that the *time constant* of a resistor capacitor combination, i.e. the RC product, will remain unchanged. This is also the case for the current density J (J = current/area, and both the current and area decrease as K^{-2} with scaling). Current densities in aluminium interconnections in microelectronic circuitry are limited to 10^{10} A m^{-2} or less by electromigration, which is a diffusive process that redistributes metal atoms under the influence of electrical forces and eventually interrupts circuit continuity. This current density equates to 10 nA nm^{-2}.

The above represent general scaling laws. Similar arguments can be applied to particular electronic devices, such as the field effect transistor, as described in Chapter 1, Section 1.3.1 [3]. In some models, as the device dimensions are reduced by K, the doping density is *increased* by K in order to keep the resistance constant. In this case, the RC time constant varies as K^{-1}, thereby providing faster switching times. A disadvantage, however, is that the current density now varies as K. The assumption of a constant electric field also provides problems as this leads to a reduction in operating voltages, which means that compatibility with standard voltage levels used in other parts of the electronic system will be lost.

The diffusion of matter (i.e. the net transport of particles from a region of higher concentration to a region of lower concentration) is a macroscopic manifestation of Brownian motion (Chapter 2, Section 2.7). Brownian motion can be described by a *random walk*, in which a particle moves a fixed distance, hits something and then sets out in a totally different (random) direction. The consequence is that the distance travelled is proportional to the square root of the number of steps (in contrast to a normal walk in which, for travel in a straight line, the distance travelled is proportional to the number of steps). This has important consequences on the nanoscale. The diffusion coefficient of oxygen in water is 18×10^{-10} m^2 s^{-1}. The time taken for a molecule to diffuse a given distance is approximately the (distance)^2 divided by the diffusion coefficient. Hence, an oxygen molecule will move 10 nm in about 50 ns, but will take around 90 minutes to move 1 cm. Bigger molecules diffuse somewhat more slowly than smaller ones. The diffusion coefficient for a sugar molecule (sucrose) in water is about 5×10^{-10} m^2 s^{-1}, while for a macromolecule like a protein, the diffusion coefficient is likely to be less than 10^{-10} m^2 s^{-1}.

11.2.2 Interatomic Forces

Solids are held together by strong ionic or covalent forces in the macroworld (Chapter 2, Section 2.3). Other bonds, such the van der Waals bond and the hydrogen bond, are much weaker, but these can become important as nano-dimensions are approached. The classical van der Waals bond is a very short-range interaction with the energy associated with it varying as r^{-6}, where r is the distance between the interacting molecules. However, if a single molecule or atom is interacting with a large body, such as a plane or a large sphere, the dependence of the interaction on the spacing from the point to the plane or sphere will become r^{-3}. Other possibilities are shown in Figure 11.1 [1, 4].

(a) Atom-surface

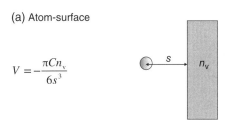

$$V = -\frac{\pi C n_v}{6 s^3}$$

(b) Sphere-surface

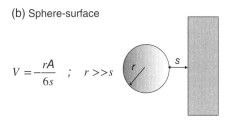

$$V = -\frac{rA}{6s} \quad ; \quad r >> s$$

(c) Two parallel chains

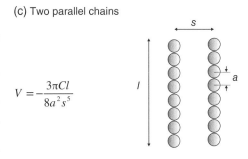

$$V = -\frac{3\pi C l}{8 a^2 s^5}$$

Figure 11.1 Potential energy of the van der Waals attractions V for three different extended geometries: (a) atom-surface; (b) sphere-surface; and (c) two parallel chains. n_v is the number of atoms per unit volume, A is the Hamaker constant, a tabulated material parameter, and C is the van der Waals constant [1, 4].

A further force, the *Casimir* force, operates between metallic surfaces, forcing them together. This is associated with the modes of oscillation of the electromagnetic field in an enclosed region. The force is vanishingly small except for extremely closely spaced surfaces, those that are relevant to molecular scale devices. For example, for two metal surfaces spaced by 10 nm, there is an attractive force of about 100 kPa (one atmosphere) arising from the Casimir effect.

11.3 Engineering Materials at the Molecular Level

In this section, some ideas are explored for introducing functionality into materials by molecular engineering. This endeavour might be considered as a sort of half-way house between organic electronics and molecular electronics. For example, molecular scale processing (Langmuir–Blodgett (LB) or Layer-by-Layer (LbL) deposition) may be exploited to confer a particular functionality into a macroscopic sample (e.g. an organic film of 100 nm dimensions).

11.3.1 Polar Materials

A simple molecular architecture – an organic 'superlattice' – is that of the alternate-layer molecular film. This can be realized in the laboratory by LB deposition (Chapter 7, Section 7.3.1) or using self-assembly exploiting electrostatic forces (LbL films, Chapter 7, Section 7.3.3). The polar heads of the LB molecules are associated with an electric dipole moment (Chapter 2, Section 2.3.5). In an alternating Y-type *ABABABA*…film, the dipoles of the two different molecules *A* and *B* will not cancel out and the multilayer will exhibit an overall polarization in the direction normal to the substrate plane. This is illustrated in Figure 11.2 for an LB film comprising an alternating arrangement of monolayers of a long chain fatty acid and a long chain fatty amine. The figure indicates that the deposition can result in a proton transfer from the acid to the amine head group, thereby creating a large dipole moment. The polarization

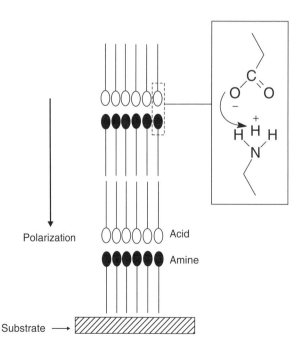

Figure 11.2 Schematic diagram of an organic superlattice formed by the alternate-layer Langmuir–Blodgett deposition of a long-chain acid and a long-chain amine. The orientations of the polar head groups are shown on the right.

Figure 11.3 Surface potential of acid/amine alternate layer LB arrays as a function of the number of layers. Data are shown for two complementary orientations of the polar axis. *Source:* Reprinted from Christie et al. [5]. Copyright (1986) American Institute of Physics.

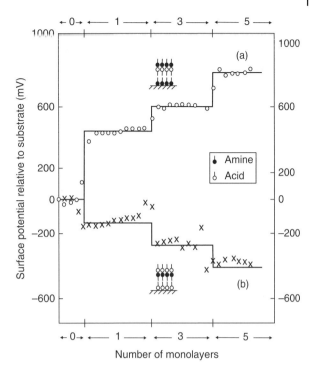

within the multilayer film can be measured using a *Kelvin probe*; data are shown in Figure 11.3 for acid/amine assemblies in the form of surface potential, i.e. the effective potential difference across the film, as a function of the film thickness (number of monolayers) [5]. The only difference between the two sets of data is the order in which the two materials were deposited. The surface potential values for data set (a) were obtained for a structure in which the layer of fatty amine was deposited first on the substrate (aluminium-coated glass); for the data set (b), the acid was transferred first. Thus, the surface potential appears to reflect the dipole orientations in the two multilayer assemblies.

If the temperature of an acentric thin film structure, such as that shown in Figure 11.2, is varied, then the polarization across the film will change. This is the basis of the pyroelectric effect, which is only observed in materials that do not possess a centre of symmetry (Chapter 5, Section 5.6). If a metallic contact is provided on top of the alternate-layer film, the magnitude of the current I flowing through the film (i.e. perpendicular to the substrate plane) is given by

$$I = pA \frac{dT}{dt} \tag{11.1}$$

where p is the pyroelectric coefficient of the film, A is the sample area, and dT/dt is the rate of temperature change. This equation can be derived easily from the definition of p given by Eq. (5.14), Chapter 5, Section 5.6.1. Note that a current will only be produced while the temperature is varied. Figure 11.4 shows the pyroelectric current obtained by heating and cooling an LB acid/amine assembly [6]. A positive current is measured on heating, while if the multilayer film is cooled, the direction of current is reversed. The pyroelectric coefficients of LB films are of the order $1-10 \, \mu C \, m^{-2} \, K^{-1}$. These are considerably less than seen in inorganic single crystals and ceramics and slightly less than observed for organic materials produced in other forms. For example, poly(vinylidene difluoride) (PVDF) (Chapter 5, Section 5.6.2) possesses a

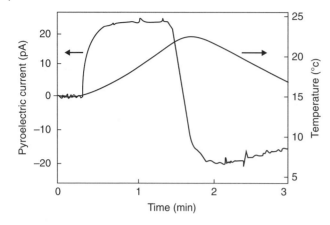

Figure 11.4 Time variation of temperature and pyroelectric current for a 99-layer acid/amine LB film. The arrows show which axis should be consulted for each curve. *Source:* From Jones et al. [6]. Copyright (1988). Reproduced with permission of IEEE.

pyroelectric coefficient of 40 μC m^{-2} K^{-1}. However, the LB film structure has the advantage that it is relatively easy to fabricate – PVDF must be poled in order (Section 2.6.3) to produce a non-centrosymmetric structure. The LB film also possesses a relatively small dielectric constant, ε_r: an important figure of merit for pyroelectric materials is p/ε_r [6]. Asymmetric LB structures, such as those depicted in Figure 11.2, are also expected to be piezoelectric, i.e. they will become polarized if they are subjected to mechanical stress.

11.3.2 Nonlinear Optical Materials

The alternating molecular architecture described in the previous section may also form the basis of thin organic layers possessing second-order nonlinear optical properties (Chapter 4, Section 4.3.2). Such films will exhibit phenomena such as second-harmonic generation and the linear electro-optic (Pockels) effect. Conjugated materials possess π-electrons that are loosely bound to the molecules and can contribute considerably to the molecular polarizabilities. As a consequence, some organic compounds exhibit large nonlinear susceptibility coefficients, often considerably larger than those of conventional inorganic dielectrics. However, the measurable nonlinear optical effects are generally quite small. The second-order coefficients of organic films can be obtained by exciting the film with a source of intense radiation (e.g. a laser) and monitoring the amount of second-harmonic radiation that is produced. An example of the result of such an experiment has already been given in Chapter 4, Section 4.5.4. The SPR technique, also described in Chapter 4 (Section 4.7), can be adapted for the measurement of the electro-optic coefficient of an LB film. An electric field, E, is applied to the organic film as the angle of incident of the p-polarized light is changed. A small change in permittivity, due to the Pockels effect, is produced. The second-order nonlinear susceptibility is then given by [7]

$$\chi^{(2)}\left(-\omega; \omega, 0\right) = \frac{\Delta\varepsilon' + j\Delta\varepsilon''}{2E} \tag{11.2}$$

The frequency argument in parenthesis and to the right of the semicolon represent the applied fields (an optical frequency ω and a dc field in this case), whereas the field resulting from the interaction (which may be regarded as emitted, hence the negative sign) is given to the left of the semicolon.

SPR and Pockels data are shown in Figure 11.5 for monolayers of a hemicyanine and a nitrostilbene dye (both mixed with a fatty acid) [8]. In contrast to second-harmonic measurements, the Pockels experiment gives both the magnitude and the sign of the molecular hyperpolarizability. Comparison of Figures 11.5a and b reveal that the β coefficient for the hemicyanine layer is in

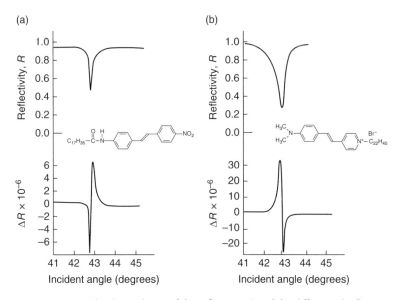

Figure 11.5 Angular dependence of the reflectivity R and the differential reflectivity ΔR for monolayer films of a fatty acid mixed with (a) a nitrostilbene compound and (b) a hemicyanine compound; the molecular structures are depicted. The experimental data and theoretical fits are indistinguishable on the scales shown. *Source:* Reprinted from Petty and Cresswell [8]. Copyright (1991) World Scientific.

the opposite sense to that of the nitrostilbene layer, i.e. the differential reflectivity curves for the two dyes are inverted.

The second-order nonlinear optical effects noted for monolayers may be retained in thicker films if the molecules are arranged in a non-centrosymmetric manner. This can be achieved by alternating an 'active' compound with a 'passive (e.g. a fatty acid) spacer layer, alternating two active materials that have been chosen so that their β coefficients are additive, or exploiting compounds that deposit as X-type or Z-type films, Chapter 7, Section 7.3.1. If the molecular alignment is preserved as the LB film is built up, then a quadratic relationship between the second-harmonic intensity and the film thickness should result. Figure 11.6 shows an example

Figure 11.6 Variation of the square root of the second-harmonic intensity, $\sqrt{I^2\omega}$, with the number of deposited layers for Z-type LB films of a cationic hemicyanine dye. *Source:* After Ashwell et al. [9]. Copyright (2000). Reproduced by permission of The Royal Society of Chemistry.

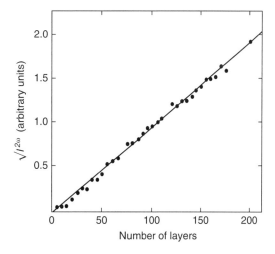

of this, obtained for Z-type LB films of a cationic hemicyanine dye [9]. The quadratic dependence on the second-harmonic intensity is preserved in excess of 200 Z-type layers. These films possess high nonlinear susceptibilities and have the additional advantage that they are transparent at the fundamental wavelength (1.06 μm in this case).

11.3.3 Photonic Crystals

Self-assembly approaches can also be used to build photonic crystal architectures (Chapter 4, Section 4.8) [10]. Simple structures can be fabricated using arrays of SiO$_2$ spheres. These synthetic *opals* (the gemstone opal is made up of tiny spherical particles of quartz) can be assembled from colloidal suspensions or by using the LB technique. In this colloidal self-assembly, spheres of equal size in suspension are used. These spheres have a diameter that is comparable to the wavelength of light. When they are in an appropriate solvent they are charge-stabilized; they have a repulsive interaction at short range that prevents them from sticking together. Because these particles are so small and their density is not too different from the solvent, they settle very slowly under the influence of gravity and their movement is mostly Brownian motion (Chapter 2, Section 2.7). When the colloidal suspension settles down, the spheres naturally arrange themselves in a face-centred cubic (FCC) structure (Chapter 2, Section 2.5.4). This immediately provides a three-dimensional dielectric that has a length scale comparable to the wavelength of light. The colloidal structure can then be impregnated with a high refractive index material. Once the original SiO$_2$ template has been removed, an *inverse opal* structure is obtained in which a regular array of air spheres is embedded in a substance that has a high refractive index.

Figure 11.7 shows an example of a synthetic opal created by the LB technique [11]. In this case, the silicon dioxide particles were first made hydrophobic using a surface treatment before being spread on the surface of a water subphase. The initial layer was formed by the Langmuir-Schaefer method (Chapter 7, Section 7.3.1), in which the glass substrate was slowly moved upward in a horizontal position. Subsequent layers were built up by the normal

Figure 11.7 Scanning electron microscope image of a synthetic opal formed by the Langmuir–Blodgett technique. The particles have a diameter of approximately 300 nm. *Source:* After Bardosova and Tredgold RH [10]. Copyright (2002). Reproduced by permission of The Royal Society of Chemistry.

(vertical) LB method. A highly ordered, close-packed arrangement of the individual opal particles is evident in Figure 11.7.

11.4 Molecular Device Architectures

As noted in Chapter 1, the evolution of silicon microelectronics, the major enabling technology of the information revolution, is threatened not only by technical issues that may restrict the further miniaturization of integrated circuits, but also from the increased fabrication cost that accompanies shrinking the feature size below 100 nm. To overcome these limitations, a world-wide research effort is underway, which is focused on electronic device structures that go beyond conventional device architectures. The aim is to find materials and concepts that lead to devices that are scalable for at least several generations below 50 nm and fast (ns and less). The technology and materials must also be compatible with present-day and future genera-tions of complementary metal-oxide-semiconductor (CMOS).

Many scenarios for future electronic devices envisage organic molecules as the 'switches' in computational architectures. Different approaches have been used to exploit the electrical properties of individual molecules or assemblies of molecules. An example of the latter is the cross-bar architecture depicted in Figure 11.8, in which a layer of molecules (deposited by, say, chemical self-assembly) is sandwiched between metallic stripes. One immediate problem is that of depositing the top electrode without damaging the organic layer [12, 13]. However, there are many other issues to be faced by the experimentalist using this deceptively simple device structure. For example, the electrodes may possess surface oxide layers that have similar thicknesses and to the organic layer (e.g. aluminium oxide on aluminium). The electrical resistances of such surface layers can be high, leading to large contact resistances. If dissimilar electrodes are used to provide the top and bottom contacts (e.g. gold for the bottom contact, and evaporated aluminium for the top), then the differences in work functions of the metals will lead to some asymmetry in the current versus voltage characteristics of the devices. Even if the same electrode material is used for both the top and bottom electrodes, it cannot be assumed that work function differences will be eliminated. The work function of a metal film will depend on the method of preparation (Chapter 9, Section 9.2.1). For example, the work function of gold deposited onto glass may differ from that of the same metal deposited onto an organic material [14].

The electrical conductivity of many organic compounds is also affected by the presence of moisture and/or oxygen in the atmosphere. Certain chemical sensors (Chapter 10) exploit this fact. Stringent precautions are taken to encapsulate plastic electronics devices, such as field effect transistors and electroluminescent displays, to increase their lifetimes. Similar approaches

Figure 11.8 Schematic diagram of cross-bar structure for electrical conductivity studies.

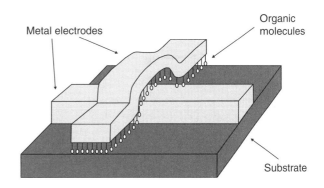

Metal electrodes

Organic molecules

Substrate

are essential when exploring the electrical behaviour of assemblies of organic molecules. The samples must be stored and measured in dry, and in some cases dark, environments.

A further problem, common to all areas of microelectronics, is that of structural defects in the organic film. While atomic force microscopy can reveal a very high degree of molecular order over distances of many nanometres (e.g. Chapter 6, Section 6.8), the order may not be preserved as the scale is increased. Defects, such as pinholes and grain boundaries, may act as nucleation points for the growth of metallic filaments.

Many reports of 'unusual' electrical behaviour in organic thin films have appeared in the literature over the last 50 years. Some of these are almost certainly the result of the poor quality of the organic layer and/or the lack of attention to the issues discussed above. Before any electrical data on assemblies of organic molecules are interpreted, a fundamental question that should always be asked is this: Are these electrical characteristics a property of the organic molecules, the metallic electrode, or of the presence of any interfacial (e.g. oxide) layer or a combination of these?

Other methods of measuring the electrical properties of molecular structures use the tip of a scanning probe microscope to provide the top contact. This circumvents some of the difficulties outlined above. Figure 11.9 shows a technique in which a monolayer on a gold surface is used in conjunction with gold nanoparticles [15]. A self-assembled monolayer of octanethiol on a gold surface is first obtained. Molecules of 1,8-octanedithiol are then inserted into the octanethiol monolayer using a replacement reaction, whereby one of the two thiol groups becomes chemically bound to the gold surface. The octanethiol monolayer acts as a molecular insulator, isolating the dithiol molecules from one another. Incubating the monolayer with a suspension of gold nanoparticles then derivatizes the thiol groups at the top of the film. A gold-coated conducting atomic force microscope (AFM) probe is finally used to locate and contact individual particles bonded to the monolayer. Measurements on over 4000 nanoparticles produced only 5 distinct families of curves. Figure 11.10 shows representative curves from each family. The curves correspond to multiples of a fundamental curve, which is ascribed to a situation in which a single dithiol molecule links the gold nanoparticle to the underlying gold substrate. In the low voltage region (between ±0.1 V), the single molecule has a resistance of $900 \pm 50 \, M\Omega$.

Alternative measurement techniques use molecular assemblies in contact with liquid electrodes (e.g. Hg or liquid electrolytes) and take advantage of electrochemical principles. The combination of the self-assembly process with molecular recognition offers a powerful route to the development of nanoscale systems that may have technological applications as sensors, devices, and switches. For example, the complexation of a neutral or ionic guest at one site in a molecule may induce a change in the redox properties of the system. Figure 11.11 shows a schematic arrangement of a self-assembled layer that includes a derivative of the charge-transfer molecule tetrathiafulvalene (TTF) (Chapter 3, Section 3.4.1). The incorporation of the metal-binding macrocycle is to enable the molecule to function as a metal-cation

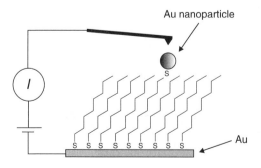

Au nanoparticle

Figure 11.9 Representation of an experimental arrangement for the measurement of the conductivity of a single molecule. *Source:* After Cui et al. [15].

Figure 11.10 Current versus AFM tip bias, $I(V)$, curves measured with the equipment shown in Figure 11.9. The five curves shown are representative of distinct families, $N\,I(V)$ that are integer multiples of a fundamental curve $I(V)$ (N = 1, 2, 3, 4, and 5). *Source:* From Cui et al. [15]. Reprinted with permission from The American Association for the Advancement of Science.

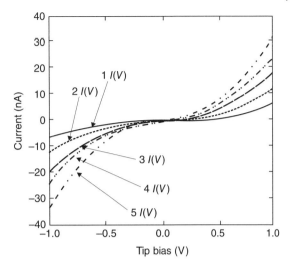

Figure 11.11 A self-assembled film containing a cation-sensitive, redox-active molecule [16].

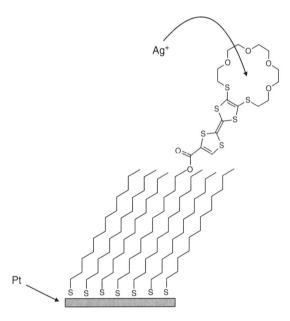

sensor. Monolayers assembled onto platinum have been shown to exhibit electrochemical recognition to Ag^+ ions [16].

One important property of any switching element is that of the switching speed. This is determined by the RC time constant of the device. The limiting factors, restricting the switching speed bandwidth of electronic components, are the carrier mobilities in the materials and the parasitic capacitances (requiring constant charging and discharging) associated with the device structure. In inorganic semiconductor devices, these parasitics are not small, as a result of the high dielectric constants and very thin junction depletion regions. However, silicon transistors can still switch fast because of the relatively high charge carrier mobilities, leading to high current densities (exceeding $100\ \mathrm{MA\ m^{-2}}$). For organic semiconductors, time constants are not expected to be high, as the charge carrier mobilities in organic materials are relatively small (Chapter 3, Section 3.2.3).

11.4.1 Break Junctions

A popular approach to measuring the electrical behaviour of small number of molecules, or even single molecules, is to use a *break junction* [13, 17]. This normally consists of two metal wires separated by a very thin gap, on the order of the inter-atomic spacing (less than a nanometre). The gap can be obtained by physically pulling a single wire apart or through chemical etching or electromigration. As the wire breaks, the separation between the electrodes can be indirectly controlled by monitoring the electrical resistance of the junction. Figure 11.12 shows one example of the technology. A length of a metallic wire is fixed on a flexible substrate, called a bending beam. Making a notch near the middle of the wire reduces its cross-section. The bending substrate is normally fixed at both ends by counter supports. A vertical movement of the push rod, which can be precisely controlled by a piezoelectric actuator or motor, can exert a force on the bending beam. As this is bent, the wire starts to elongate, resulting in the reduction of the cross-section at the notch and finally producing complete fracture of the metal wire. After breaking the metal wire, two clean facing nanoelectrodes are generated. The bending or relaxing of the substrate controls the distance between these. After integrating the molecules into the gap, these may bridge the two electrodes and the electronic properties of the molecules can be determined. Both two-terminal and three-terminal break junctions have been fabricated [17]. The break junction can be incorporated into a high vacuum system and integrated with measurement techniques, such as Raman spectroscopy, inelastic tunnelling spectroscopy, or noise spectroscopy, to obtain fingerprint information of the molecules [17].

A significant disadvantage of mechanically controllable break junctions is the uncontrollable nature of the breaking process. For instance, the local shape of the electrodes and the atomic configurations of the electrodes are unknown. Break junctions cannot be scaled in a manner similar to crossbars to address a specific cross-section, which limits their potential applications. Many challenges need to be overcome before single-molecule devices can be used as commercial products. Nevertheless, as a fundamental research technique, break junctions provide an important platform for investigations of single molecule behaviour.

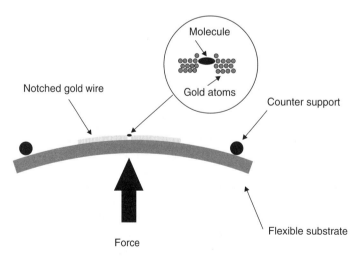

Figure 11.12 A mechanically controllable break junction for molecular electronics.

11.5   Molecular Rectification

The concept of molecular rectification, i.e. asymmetrical current versus voltage behaviour within a molecule, was predicted in 1974 by Aviram and Ratner [18]. Since this time, other models (e.g. the Kornilovitch–Bratkovsky–Williams model and the Datta–Paulsson model) have been developed to explain the rectification of individual molecules [13]. The original proposal was that an asymmetric organic molecule containing a donor and an acceptor group separated by a short σ-bonded bridge, allowing quantum mechanical tunnelling, should exhibit diode characteristics. An example of such a molecule, together with the energy band structure, is given in Figure 11.13. The ground highest occupied molecular orbital (HOMO) and excited lowest unoccupied molecular orbital (LUMO) states of the molecules are E_{D1}, E_{D2} (for the donor) and E_{A1}, E_{A2} (for the acceptor), respectively. The presence of the σ-bridge is essential to the device operation. If the electronic systems of the donor and acceptor are allowed to interact strongly with one another, a single donor level will exist on the timescale of the experiment. The donor and acceptor sites should effectively be insulated from one another in order for the device to function.

If the molecule is sandwiched between two metallic electrodes, the passage of electrons through the device can be considered as a three-step process: metal cathode to acceptor, acceptor to donor, and finally from the donor to the anode. As the applied voltage is increased (the acceptor side of the molecule is biased negatively with respect to the donor side), electrons will be transferred from the cathode to the acceptor molecule (to energy state E_{A2}) and from the donor (from E_{D1}) to the anode. Electrons will move from the acceptor molecule LUMO level (E_{A2}) to the donor HOMO level (E_{D1}) by the process of quantum mechanical tunnelling (Chapter 3, Section 3.5.2) if the σ bridge is sufficiently short. If the direction of the applied bias is reversed, conduction can only take place at much higher voltages, i.e. the HOMO level of the donor would have to be higher than the LUMO level of the acceptor. This leads to predicted diode characteristics.

There have been many attempts to demonstrate this effect in the laboratory, particularly in organic thin films [13, 19]. Asymmetric current versus voltage behaviour has certainly been recorded for many metal-insulator-metal structures, although these results are often open to several interpretations as a result of the experimental issues highlighted in the previous section. For example, the asymmetry in the electrical characteristics may originate from work function differences or from Schottky barrier (Chapter 9, Section 9.2.1) formation between the

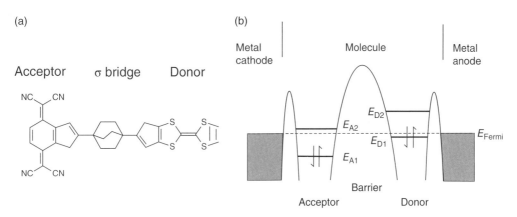

Figure 11.13 (a) An example of a molecular rectifier molecule. (b) Energy band structure of molecular rectifier.
Source: Reprinted from Aviram and Ratner [18], Copyright (1974), with permission from Elsevier.

organic material and one of the electrodes. It is unlikely that the two electrodes forming a break junction will be perfectly symmetric, as the two electrodes will have different shapes or surface features. A key test is to relate the observed rectification to a property of the molecule under test. Figure 11.14 shows the chemical structure of two similar self-assembling monolayers consisting of a bulky 4-methoxynaphthyl donor that is linked via a —CH=CH— π-bridge to a bulky quinolinium acceptor [20]. The steric hindrance provided by the donor and acceptor of the molecule depicted in Figure 11.14a enforces the required non-planarity, which, in turn provides an effective electron-tunnelling barrier between the electroactive ends of the molecule. Self-assembled monolayers of this molecule exhibit rectification ratios of about 30 at ±1 V, as shown in Figure 11.15. Significantly, the electrical asymmetry is suppressed in self-assembled monolayers of a less bulky analogue (Figure 11.14b).

It has been noted that mechanisms other than asymmetric couplings between the molecule and the electrodes may be the origin for the observed rectification in molecular junctions [19].

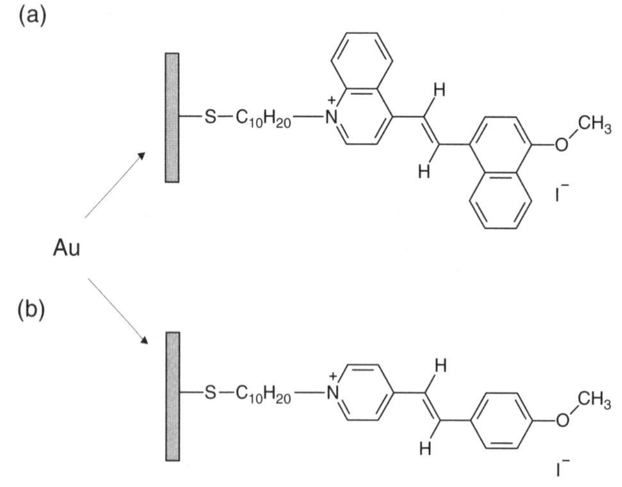

(a)

(b)

Au

Figure 11.14 Chemical structures of self-assembled monolayers of (a) rectifying and (b) non-rectifying molecules. The counterion in each case is iodide. *Source: From Ashwell et al. [20]. Copyright (2005). Reproduced by permission of The Royal Society of Chemistry.*

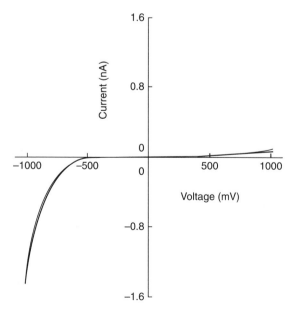

Figure 11.15 Current versus voltage characteristics of the iodide salt of the molecule depicted in Figure 11.14a. The polarity relates to the substrate electrode and the higher forward bias current corresponds to electron flow from the gold-coated substrate to the contacting tip. *Source: From Ashwell et al. [20]. Copyright (2005). Reproduced by permission of The Royal Society of Chemistry.*

For example, a molecular bias drop may be responsible for the experimental data. First, some of the applied bias drops across the molecule, indicating that the electrode-molecule-electrode junction has a (permanent or induced) dipole. In effect, the molecular channel energies change with the bias: a positive (negative) bias might bring a channel closer to resonance, thereby increasing the current, whereas a negative (positive) bias would push the channel away from resonance and decrease the current. Asymmetric electrode couplings may indirectly lead to rectification, even though they are not directly responsible for it. A bias drop across the molecule can be attributed to a (permanent or induced) dipole in the molecular junction. It is probable that the different linker groups used to produce asymmetric couplings may also lead to different induced dipoles with an applied bias. The change in induced dipole from one system to the next might then change the rectification ratio of the junction, giving the illusion that asymmetric couplings lead to rectification.

The forward current in a molecular rectifier is a result of quantum mechanical tunnelling through the σ-bonded barrier, and the response of this current will be essentially instantaneous. However, rectifiers must not only conduct when they become forward biased, but also cease to conduct when they are reverse biased. The device current will include a component flowing through the parasitic capacitance that is inherently present. This will not be a problem at low switching frequencies, but will limit the maximum frequency for which the device has useful characteristics. An estimate of the limiting frequency of the device proposed by Aviram and Ratner – about 1.5 kHz – can be obtained from the *RC* time constant of the device and by using some of the original parameters [21]. The key factor is the current density through the device. Even if the parasitic capacitance is large, it can be charged and discharged faster if the device current is increased. Further increases lead to further reductions in switching time, up to as point where transit delays due to charge carrier movement start to dominate.

11.6 Electronic Switching and Memory Phenomena

Memories represent by far the largest part of electronic systems; for example, memories in silicon chips can occupy more that 70% of the chip area. Memory devices are inherently *volatile* or *non-volatile* [22]. In the former case, the information is lost if the power supply is removed. In silicon technology, examples of volatile memory are the *static random access memory* (SRAM) and the *dynamic random access memory* (DRAM). Data stored in the cells of volatile memory must periodically be refreshed. In contrast, *flash memory* (the term 'flash' refers to the erase process, in which blocks of memory are erased in one operation) is non-volatile, storing information (in the ideal case indefinitely) if the power is turned off. In addition, flash memory offers fast read access times (although not as fast as volatile DRAM memory) and better shock resistance than hard disk memories.

It is expected that the majority of non-volatile memories in the early part of the twenty-first century will be based on flash technologies. Moore's Law (Chapter 1, Section 1.3.2) will continue to drive transistor-based memory technology scaling but technology complexity will increase. Scaling will eventually become difficult due to the high electric fields required for the programming and erase operations and the stringent requirements for long-term charge storage. These demands are imposing fundamental scaling limitations on the memory cell operating voltages and on the physical thickness of the tunnelling dielectric. Overcoming such restrictions will require innovations in cell structure and device materials. A scenario can be envisaged where logic is implemented in CMOS (or in polymer FETs, in the case of plastic electronics) and the memory is added to the system using a different technology.

11.6.1 Resistive Bistable Devices

Organic memory devices are generally formed by interposing thin organic layers between two electrodes [22]. The cross-bar arrangement described in Section 11.4 is popular, as it is highly scalable. This architecture permits the closest packing of bit-cells, with each occupying an area of $4F^2$, where F is the minimum feature size (the line-width and spacing of the electrodes). A variation uses the tip of a scanning probe microscope instead of a (thermally evaporated) metallic top electrode, leading to ultrahigh data storage density. Write-once/read-many-times organic memory can be fabricated by burning polymer 'fuses'. For example, a thin conductive polymer layer, or fuse, of poly(3,4-ethylenedioxythiphene) (PEDOT – Chapter 5, Section 5.3) is sandwiched between two electrodes [23]. The as-deposited device shows high conductance due to the polymer, but when a burning-voltage pulse is applied (the write process), the PEDOT fuses will 'blow' and cause the device to be in an open-circuit condition. The write process can be as short as microseconds, depending on the thickness of the PEDOT layer and the amplitude of the voltage pulse.

An intriguing AFM-based data storage concept (called the 'Millipede') that has potentially ultrahigh density was developed by IBM in 2000 [24]. Indentations, 30–40 nm, are made in a thin (50 nm) polymethylmethacrylate (PMMA) layer, resulting in a data storage density of 2500–3000 Gbits cm^{-2}. The very large-scale integration of micro-nanomechanical devices on a single chip leads to an impressive two-dimensional arrangement of AFM cantilevers (100 mm^{-2}) with integrated read and write functionality. However, this type of memory is largely of academic interest. Because of competing data storage technologies, no commercial product has been developed.

Obtaining reliable organic bistable memory devices poses an even greater challenge. Reports of switching and memory effects in thin films go back to the 1960s [25]. The film thicknesses are generally less than 1 µm and the phenomena are observed in different types of material (inorganic compounds, such as silicon dioxide, chalcogenides, and metal oxides, and organic compounds, such as polymers and charge-transfer complexes). A device 'formation' process is often required before switching effects are observed. At least six different types of switching characteristics have been identified [26]. Furthermore, the thin films have been formed using a variety of techniques (e.g. spin-coating, thermal evaporation). Many recipes for switching are based on the incorporation of metallic nanoparticles within an organic film [22, 25, 26]. It is now clear that the intentional addition of such nanoparticles is not necessary to observe memory effects, although these may improve the switching performance. The only experimental parameter that appears to be common to all the structures studied is the presence of metallic electrodes (e.g. Al) below and on top of the thin film. Unfortunately, comparison of data produced by different groups is often difficult as experiments are conducted in various ways, e.g. different voltage ranges or different measurement protocols.

Figure 11.16a shows the chemical structure of an organic compound – 7-{4-[5-4-*tert*-butyl-phenyl-1,3,4-oxadiazol-2-yl]phenyl}-9,9-dihexyl-*N*,*N*-diphenyl-fluoren-2-amine – that has been used in an organic resistive memory [27]. The device is formed by sandwiching a thin film (~50 nm in thickness and formed by spin-coating) of this compound between aluminium electrodes. The current versus voltage behaviour, shown in Figure 11.16b, is reasonably symmetric to the polarity of the applied voltage and exhibits both *negative differential resistance* (NDR) and memory effects. NDR does not imply a negative value for the resistance; it is simply the part of the *I*-*V* characteristic where the current decreases as the voltage increases. For the data shown, the On state is obtained by applying a voltage V_{max}, close to the current maximum (just before the NDR region) and reducing it to zero (write). In contrast, switching from the high conductivity On state to the Off state is accomplished by selecting a voltage V_{min},

Figure 11.16 Electrical behaviour of a bistable metal–organic thin film-metal device. (a) The organic material is the polyfluorene derivative shown and the metal electrodes are both Al. (b) Current versus voltage characteristics. The 'On' and 'Off' states are set at voltages close to V_{max} and V_{min}, respectively. The transition from the Off to the On state occurs at the threshold voltage V_{th}. *Source:* Reprinted from Dimitrakis et al. [27], Copyright (2008) American Institute of Physics.

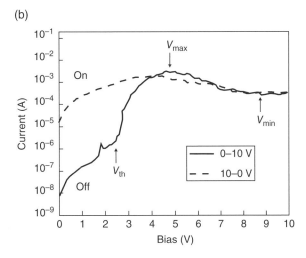

corresponding to the current minimum in the NDR region and reducing this rapidly to zero (erase). The state of the device (On or Off) is determined by measuring the current at a low value of applied voltage, say 1 V (read).

There is no consensus in how resistive thin film memories, such as that described above, operate [22, 25–28]. The explanations generally fall into two distinct categories: (i) charge injection and storage in the thin film; and (ii) metallic filament formation. The direct observation of metallic filaments by electron microscopy and energy dispersive X-ray spectroscopy suggests that the latter hypothesis may be the more likely [29, 30]. Three-dimensional cross-bar polymer resistive devices have been demonstrated [31]. A number of companies and research organizations (e.g. Adesto, Crossbar, HP) are pursing resistive random access memory (RRAM or ReRAM) or *memristor* technologies, albeit using inorganic materials as the switching medium, for commercial development [32–34].

Memory devices can also exploit the movement of parts of molecules or even of individual atoms. Organic bistable rotaxane molecules have been used in a number of these studies; an example is shown in Figure 11.17 [35]. This molecule is amphiphilic and the ring component can move between the polar (right-hand side of the molecule) and nonpolar (central sulphur-rich group) regions of the main part of the molecule. The molecules can be assembled on an electrode using the LB approach and a top electrode then deposited to form a cross-bar structure.

A high density memory can be made by using the electrostatic attraction between suspended, crossed carbon nanotubes [36] and a company, Nantero, has been established to exploit this technology [37]; this is discussed in Section 11.10. A further device, based on silver sulphide (Ag_2S), works by controlling the formation and annihilation of an atomic bridge at the crossing point between two electrodes [38]. Figure 11.18 shows a schematic diagram of one such

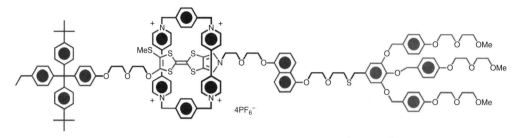

Figure 11.17 Bistable rotaxane structure. The 'ring' component can move between the hydrophobic sulphur-rich region shown centrally to the hydrophilic 'stopper' region on the far right of the molecule.

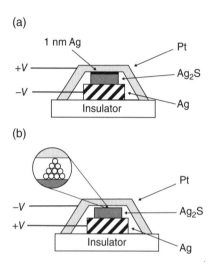

Figure 11.18 A rewritable memory bit based on the properties of a silver sulphide mixed ionic conductor. (a) A one-nanometre-thick silver layer deposited on top of the Ag_2S layer is incorporated into the sulphide layer when the Pt electrode is positively biased. (b) A bridge of Ag atoms is formed locally when the Pt electrode is negatively biased. *Source:* Reprinted by permission from Macmillan Publishers Ltd.: van Ruitenbeek [38], Copyright (2005).

memory cell. A 1 nm thick silver layer deposited on top of the Ag_2S layer disappears into the sulphide layer when a current flows from the platinum electrode to the silver electrode (Figure 11.18a). This results in loss of contact between the two electrodes and initializes the device. A bridge of silver atoms is locally formed by applying a voltage of the opposite sign, re-establishing contact between the silver sulphide and the platinum (Figure 11.18b). The conductance through the device can be as small as one quantum unit of conductance, suggesting that the silver bridge can touch the platinum electrode with just one atom. Although this structure is based on inorganic compounds, the principle of operation may have wider applicability.

If molecular components are to be used as functional elements in place of semiconductor-based devices, they must compete with inorganic semiconductors under the extreme conditions required for processing and operating a practical device. There is some concern as to whether organic materials possess adequate stability to meet the extreme performance conditions required for any type of practical device, e.g. high-temperature processing steps during manufacture ($\sim$400 °C); relatively high temperature operating conditions (up to 140 °C); and very large numbers of operational cycles over a lifetime ($\sim$10^{12}). However, as discussed in Chapter 5, there are organic compounds that exhibit good chemical and thermal stability. In this respect, results exploiting the redox behaviour of self-assembled porphyrin molecules are encouraging [39]. Here, information is stored in the discrete oxidation states of the molecules. (See also the discussion in Section 11.8.3 on chemical switching.)

11.6.2 Flash Memories

Charge storage is the physical basis of the flash memory. This device is similar in structure to a MOS field effect transistor (MOSFET), except that it has two gate electrodes, one on top of the other. The top electrode forms the control gate, below which a 'floating gate' is capacitatively

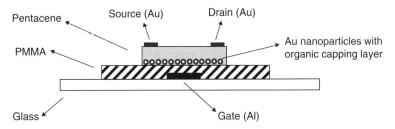

Figure 11.19 Schematic diagram showing the structure of a thin film memory transistor based on gold nanoparticles.

coupled to the control gate and the underlying silicon. The memory cell operation involves putting charge on the floating gate or removing it, corresponding to two logic levels. Nanoflash devices utilize single or multiple nanoparticles as the charge storage elements. These are usually embedded in the gate oxide of a field effect transistor and located in close proximity to the transistor channel. Figure 11.19 depicts a simple version of this device. A very thin (< 3 nm) insulating capping layer surrounding the gold nanoparticles allows charge (electrons or holes) to move between the pentacene semiconductor and the nanoparticles by the process of quantum mechanical tunnelling.

Figure 11.20 shows the output and transfer characteristics of the organic memory transistor depicted in Figure 11.19 [40]. The data are shown for devices both with and without nanoparticles. The hysteresis evident in the nanoparticle-containing transistors can be attributed to the charging and discharging of the nanoparticles. When a negative gate bias is applied, holes are injected from the pentacene layer into the nanoparticle layer, charging up the latter and programming the memory device. In contrast, when a negative gate bias is applied, holes are injected from the nanoparticle layer through the pentacene layer, resulting in an erase process. The injection/ejection of charges to/from the nanoparticle layer will lead to a shift in the threshold voltage. Arrays of such nano floating gate memories have been fabricated. Other notable achievements include the development of solution processing to build up devices on flexible supports [41, 42]. For commercial development, the challenge is to fabricate robust architectures with sufficient charge retention times to compete with inorganic memory devices.

11.6.3 Ferroelectric RAMs

Longer retention times in capacitive DRAM-memory cells can be achieved if the dielectric layer has ferroelectric properties [43, 44]. As noted in Chapter 5, Section 5.6.2, a ferroelectric material is one that has an in-built polarization; furthermore, this can be reversed by an applied electric field. An important organic material that exhibits such ferroelectric behaviour is PVDF; the chemical structure of this polymer and its different crystalline phases has been discussed in Chapter 2, Section 2.6.3. A one-bit memory cell is in the form of a parallel plate capacitor, with the ferroelectric material as the dielectric. The application of a strong electric field (poling) results in an overall dipole moment. Writing a binary logic state '1' or a '0' to the ferroelectric memory element requires the application of a high voltage across the ferroelectric capacitor. Depending on the polarity of the voltage, this polarizes the capacitor in one particular state. To read the cell, a further voltage is applied. If the polarity of this voltage matches the polarization direction in the ferroelectric material, nothing happens. However, in the case that the applied voltage polarity opposes the dipole direction, the ferroelectric will be re-polarized and charge will flow to the sensing circuit. The reading

(a)

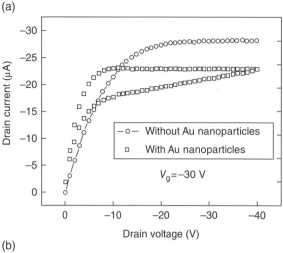

Figure 11.20 (a) Output characteristics of memory transistor, shown in Figure 11.19, with and without gold nanoparticles. (b) Transfer characteristics of the memory transistor with and without gold nanoparticles. *Source:* Reprinted from Mabrook et al. [40], Copyright (2009) American Institute of Physics.

(b)

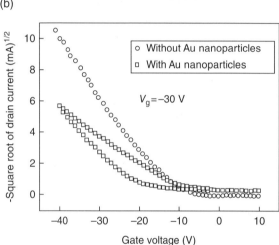

process – destructive readout – results in the loss of data, which must be rewritten. Complex addressing circuitry is therefore needed.

A significant problem in the reliability of ferroelectric memories is that of fatigue. This is the tendency of the polarization to decrease as a result of repeated switching. A further issue relates to scaling; a reduction in the capacitor size will reduce the average current during a switching event. Combining a ferroelectric polymer with a transistor may circumvent some of these problems. The different polarization states of the ferroelectric gate insulator lead to a change in the drain-to-source current of the transistor. Reading the memory sate is therefore a non-destructive process, so the device lifetime is determined only by the number of times that the memory is written.

In 2011, Thin Film Electronics together with Palo Alto Research Center (PARC), a Xerox company, announced a working prototype of a printed non-volatile ferroelectric memory addressed with complementary organic circuits, the organic equivalent of CMOS circuitry [45]. The commercial products are targeted at smart consumables. Examples are labels holding product information such as serial numbers, expiration dates, and geographic codes. Each label can store up to 6.8×10^{10} distinct data combinations with 10 year retention.

11.6.4 Spintronics

Spintronics (an acronym for SPIN TRansport electrONICS) exploits the spin of an electron to process and store digital information [46–48]; the subject is also called *magnetoelectronics*. The simplest method of generating a spin polarized current is to inject the current through a ferromagnetic material. Much of the interest in the area started with the company IBM exploiting magnetically-induced resistance or magnetoresistance in the early 1990s. The most successful spintronic device to date is the *spin valve*. This uses a layered structure of thin films of magnetic materials (common inorganic materials are Fe/Cu and Co/Cu), which changes electrical resistance depending on applied magnetic field direction. When the two magnetization vectors (Chapter 5, Section 5.7.1) of the ferromagnetic layers are aligned, then a relatively high current will flow, whereas if the magnetization vectors are antiparallel, then the resistance of the system is much higher. The magnitude of the change ((antiparallel resistance − parallel resistance)/parallel resistance) is called the giant magnetoresistance (GMR) ratio. The 2007 Nobel Prize in Physics was awarded to Albert Fert and Peter Grünberg for the discovery of GMR. Devices have been demonstrated with GMR ratios as high as 200%, with typical values greater than 10%. This is a vast improvement over the anisotropic magneto-resistance effect in single layer materials, which is usually less than 3%.

Figure 11.21 shows a schematic diagram for a non-volatile memory device based on a magnetic tunnel junction. This has been developed by IBM and other companies and exploits a spin-dependent tunnelling phenomenon. The devices have two ferromagnetic layers separated by a thin insulating barrier. The first layer polarizes the spins of current-carrying electrons, which cross the barrier to the second layer by quantum mechanical tunnelling when both layers are magnetically aligned. When the magnetism of the second

Figure 11.21 Schematic diagram of a magnetic tunnel junction. (a) The fixed ferromagnetic layer polarizes the spins of current-carrying electrons, which cross the barrier to the second layer by quantum mechanical tunnelling when the magnetism of both layers is aligned. (b) When the direction of magnetism in the second layer is reversed, the tunnelling is reduced.

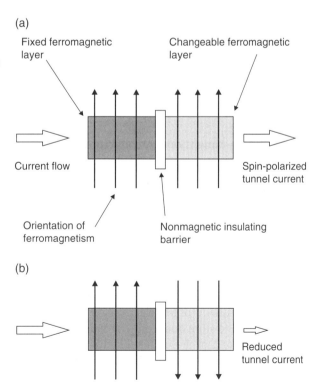

layer is reversed, the tunnelling is reduced. The magnitude of the current in the memory cell can be used to indicate a '0' or a '1'.

A *magnetic random access memory*, or MRAM, product prototype based on spintronics principles was announced in 2003 and the first ever MRAM commercial products were launched in 2006 [47]. The early devices were based on 180 nm process technology. A number of developments suggest that MRAM can be scaled to 60 nm and below. The most notable is the discovery of the *spin-momentum-transfer effect* [47]. This exploits the net angular momentum that is carried by a spin-polarized current and the transfer of this momentum to the magnetization of the second layer. It offers the potential of significantly lower switching currents and, therefore, a much lower energy for writing. The spin-momentum-transfer effect becomes important when the minimum dimension of the memory cell is less than 100 nm and becomes more efficient as the cell size is reduced (the opposite to what occurs with the use of conventional magnetic field switching).

There are also many proposals for a transistor that works by switching electrons between spin states. In one option, a gate controls whether most of the electrons have the same spin state. It is also possible to exploit magnetic domain wall motion (Chapter 5, Section 5.7.1) to form computational logic elements. For this case, information is stored through the presence or absence of a domain wall in a linear array of domain walls in a magnetic thin film loop confined to a channel in a silicon chip. Such a memory might eventually provide a solid-state replacement for the magnetic hard drive used in computers.

Although most of the materials developed for spintronic devices are inorganic, a case has been made for exploiting organic compounds [46, 49]. This is based on the weak spin-orbit and hyperfine interactions in organic molecules, which leads to the possibility of preserving spin-coherence over times and distances much longer than in conventional metals or semi-conductors. In 2004, spin-valves using a tris-(8-hydroxyquiniline) aluminium (Alq_3) (Chapter 9, Section 9.6) layer sandwiched between ferromagnetic electrodes, were shown to possess a GMR value of about 40% at 11 K [50]. Since this time, other groups have reported similar results using Alq_3 and other organic semiconductor materials [49]. However, the observed magnetoresistance is often very variable, in both magnitude and sign, even for the same device structure. This indicates key differences in the ferromagnetic electrode-organic semiconductor interface properties, which can be induced by the intrinsic material properties, interface interactions, or even differences in the fabrication conditions.

11.6.5 Three-dimensional Architectures

Molecular scale electronics may also offer increased device densities by fabricating three-dimensional architectures. The cross-bar structure described in Section 11.4 offers a simple means of realizing this. For example, if the 'active' material can be deposited in thin film form at low temperatures, such devices could be built up in three-dimensions, since the additional processing to form a new memory layer will not affect the underlying layer(s). The principle of a three-dimensional memory based on LB films has also been suggested [51]. The device requires a molecule with a central conjugated region of high electron affinity (for an n-type material, the electron affinity is the energy difference between the bottom of the conduction band and the vacuum level – Chapter 3, Section 3.4.1) surrounded by aliphatic substituents of low electron affinity. A multilayer structure, as shown in Figure 11.22, might be used to store one N-bit word, the presence or absence of charge on the nth layer representing a 0 or 1 of the nth bit. The LB film could be assembled on the gate of an FET and, on application of an electric field; transport of bits across the layers may be detected as induced charge on the gate.

Figure 11.22 (a) Molecular memory holding *N* bits (top) compared to a conventional silicon memory holding 1 bit (bottom). (b) Electron energy versus perpendicular distance for molecular memory with no applied electric field. *Source:* Reprinted from Burrows et al. [51], with permission from Elsevier.

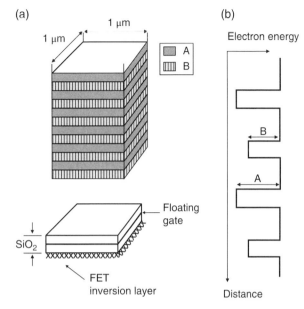

11.7 Single-electron Devices

Single-electron devices are those that can control the motion of even a single electron and consist of quantum dots or nanoparticles associated with tunnel junctions [52]. However, the terminology can be a little confusing [53]. The name suggests that the devices work with only electron, which is not quite correct. Even a single atom contains a number of electrons and a metal- or semiconductor-based 'single-electron' device will have a huge number of electrons that have relevance to its operation. The principle of *Coulomb blockade* enables electrons to be localized on an isolated island or transported to one so that a precise number are transferred. In the single-electron transistor (SET) based on this effect, a capacitatively-coupled gate electrode is used to control this transfer. The simplest device is the arrangement depicted in Figure 11.23. Electrons can tunnel between a reservoir and an island. The gate voltage controls this electron motion. Moving an electron from the reservoir to the island is accomplished by applying a voltage to the gate V_g, where

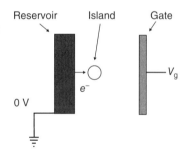

Figure 11.23 Principle of Coulomb blockade. Electronic charge is isolated on an 'island'.

$$V_g = \frac{e}{C} \tag{11.3}$$

where e is the electronic charge and C is the capacitance of the island due to the gate electrode. The charging energy E is given by the expression for the energy associated with a capacitor, i.e.

$$E = \frac{e^2}{2C} \tag{11.4}$$

The capacitance of a sphere scales with its radius, r:

$$C = 4\pi\varepsilon_r\varepsilon_0 r \tag{11.5}$$

(Students of electrostatics might think it unusual that the capacitance of an isolated sphere has a finite value of capacitance; however, deriving the capacitance between two concentric spheres and then allowing the radius of the outer sphere to become infinitely large can easily demonstrate the above formula.)

For a 1 μm island (sphere) in free space, $C \sim 5 \times 10^{-17}$ F and $E \sim 2 \times 10^{-22}$ J or about 1.4 meV. At 300 K, the thermal energy $k_B T$ is about 25 meV, an order of magnitude larger. However, if the island is very small (< 10 nm) or the temperature is very low, the charging energy becomes greater than the thermal energy and single electrons may be isolated on the island. Strong Coulomb repulsion – Coulomb blockade – will block the transfer of a second electron.

A second requirement necessary for the observation of Coulomb blockade is that the quantum fluctuations in the electron number n on the island are sufficiently small that the charge is localized on the island. This translates to a lower limit for the resistance of the tunnelling barrier, R_t.

If the typical time to charge or discharge the island is Δt:

$$\Delta t = R_t C \tag{11.6}$$

The Heisenberg Uncertainty principle (Chapter 3, Section 3.3.1):

$$\Delta E \Delta t \approx \frac{e^2}{C} R_t C \geq h \tag{11.7}$$

implies that R_t should be much larger than the resistance quantum (von Klitzing constant – Chapter 3, Section 3.3.8):

$$R_t \geq \frac{h}{e^2} \tag{11.8}$$

in order for the energy uncertainty to be much smaller than the charging energy.

By depositing small metal particles onto an insulating self-assembled monolayer on a metal substrate, the tunnelling of single electrons through a metal-molecule-metal junction can be studied [54]. Using current versus voltage measurements on Pd clusters on a decanethiol monolayer on Au (111) at 77 K, Coulomb blockade is observed; the experimental data are shown in Figure 11.24. The tunnelling of an electron to the metal particle is blocked until the applied voltage reaches a value equal or higher than the charging energy of the particle. As a consequence, the current will increase in a step-wise manner, and the current versus voltage curve will display the shape of a staircase, the so-called *Coulomb staircase* depicted in Figure 11.24a. Measurements taken with uncoated 'reference' monolayer exhibit no such features. From the derivative of the *I-V* data, *dI/dV* (Figure 11.24b), the fractional charge on the cluster can be determined. As seen in the curves, the experiment and the theory are in good agreement. Coulomb blockade has been suggested as a mechanism that might explain the nonlinear *I-V* behaviour of conductive polymers [55].

Metrologists have pursued electron-counting standards for capacitance and current using single-electron devices [56]. Although a single-electron box can control the number of electrons in the island, it does not have the properties of a switching device. *Single-electron transistors* are therefore three-terminal switching devices, which can transfer single electrons from source

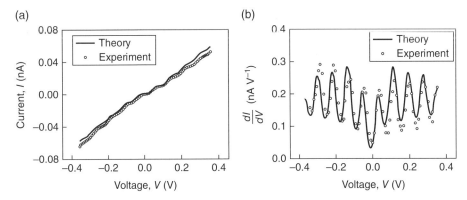

Figure 11.24 (a) Current versus voltage characteristic showing Coulomb staircase. (b) Derivative dI/dV for the data presented in (a). In each case, the experimental points are shown as open circles, while the full curves represent the theoretical fits. The experimental I versus V curve is taken on top of a Pd cluster at 80 K. *Source:* Reprinted from Oncel et al. [54], Copyright (2005) American Institute of Physics.

to drain; a schematic diagram of such a device is shown in Figure 11.25. To make a memory cell, the SET is coupled to a MOSFET and used to transfer precise electron numbers to its gate or to remove them. The MOSFET current is a function of the presence or absence of these electrons on the gate and has two levels that define memory states. It is possible to observe this memory effect with just one extra electron present on the memory node, but in practical cases 10 to 100 electrons may be used. The development of arrays of SETs for computing applications is a challenge. Success will require exceptional performance, uniformity, and stability from the devices [56].

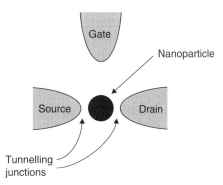

Figure 11.25 Schematic diagram of a single-electron transistor (SET).

11.8 Optical and Chemical Switches

The three most important stimuli that can be used to switch a chemical compound are electrical energy (electrons or holes), light energy (photons), and chemical energy (in the form of protons, metal ions, specific molecules, etc.). Some examples from the first category have been provided in Section 11.6. Any molecular-level system that can be reversibly switched between two different states by use of an external stimulus can be exploited for storing information. In the case of a photochemical input, the most common switching processes are related to photo-isomerization or photo-induced redox reactions. If electrochemistry is exploited, then the induced processes are, of course, redox reactions.

Switching processes can take place under thermodynamic or kinetic control. In the former, the molecule responding to the stimulus is in thermodynamic equilibrium with its surroundings, so that when the stimulus is removed the molecule reverts to its initial state. An example is a fluorescence sensor, the emission intensity of which is modulated by the presence of a particular substrate. Kinetic control, which means that the two states are separated by some kinetic barrier, is usually found in systems responding to photonic stimulation. The kinetic control can operate over very different timescales, from picoseconds (for some electron excited states) to years (for some electrochromic systems). For systems under thermodynamic control, it is not possible to address a single molecule because of the rapid equilibration between the states.

The following sections contain some examples of optical and chemical switches in molecular materials.

11.8.1 Fluorescence Switching

Sensors based on the control of fluorescence have been described in Chapter 10, Section 10.4.6. Figure 11.26 illustrates the operating principle [57]. A potentially fluorescent unit, e.g. an anthracene moiety, is linked to an electron-donating group, e.g. an amine. Photo-induced electron-transfer from the HOMO orbital of the donor will quench the fluorescence (Figure 11.26a). However, when the HOMO level of the donor is bound to an appropriate molecule or ion (in the case of an amine, by protonation, for example) fluorescence can be observed because the HOMO level of the donor is reduced in energy and electron transfer can no longer occur (Figure 11.26b).

11.8.2 Photochromic Systems

Photochromic materials reversibly change their absorption spectra, i.e. their colour, in response to incident light [58]. In photochromic systems, the interconverting species are isomers, because the photoreaction simply causes rearrangement of the electronic and nuclear structure. A general photochromic reaction is depicted in Figure 11.27. Light of energy $h\nu_1$ causes switching of a stable isomer A to a higher energy isomer B. After photochemical conversion (a process that can be performed in a few femtoseconds), a spontaneous back reaction is expected to occur. This can be fast or slow, depending on the system. The reverse reaction can also occur by illuminating with a different frequency of light, ν_2, or by means of a thermal process, providing energy ΔE. Photochromic materials may also display *thermochromism*

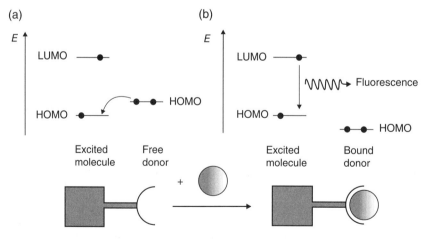

Figure 11.26 Representation of fluorescence quenching. (a) An electron from the HOMO level of a donor group quenches the fluorescence from an excited molecule. (b) The energy of the donor group is lowered as a result of a chemical interaction. The group can no longer donate electrons to the HOMO level of the excited molecule and fluorescence can be observed. *Source:* Reprinted from Balzani et al. [57]. Copyright (2008), with permission from Wiley-VCH.

Figure 11.27 Left – a photochromic reaction where A represents the initial unswitched state of the photochromic material and B the switched material. The relative energy levels of the states A and B are depicted on the right.

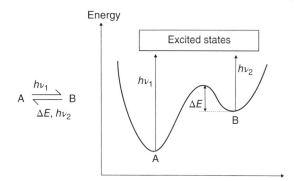

and *electrochromism*. In the former case, colour change is induced by heat that causes reorganization of the electronic structure (e.g. in spiropyrans) or dissociation (organo-metallic compounds). In electrochromism, the colour is modified or generated by an electric field; this is discussed in more detail further on in this section.

The above describes ideal photochromic behaviour. In practice, photochromic systems all exhibit 'fatigue' to a greater or lesser extent. The time required to switch between colourless and coloured species gradually increases and the maximum colouration is reduced after repeated cycling between A and B. Fatigue is caused by unwanted side reactions under the influence of light and heat. These effects become noticeable after any number of switching cycles, but some organic systems can survive more than 10^4 cycles without damage.

A large number of photochromic materials exist and these, broadly, fall into six separate classes, depending on their switching mechanism. These are hydrogen tautomerism, dissociation, dimerization, *cis-trans* isomerization, cyclization, and charge-transfer. Examples of two systems are shown in Figure 11.28. Figure 11.28a illustrates *cis-trans* isomerization (Chapter 2, Section 2.4.2) in the azobenzene family. This process is also responsible for light-induced development processes, in higher plants and for vision in animals. The photochromic behaviour of bacteriorhodopsin, a natural light-harvesting protein, and the possible applications in memory devices are described in some detail the Chapter 12, Section 12.10.1.

Azobenzene derivatives in the form of LB films have been widely studied. In these, UV illumination of the *trans* isomer at the wavelength of the $\pi \rightarrow \pi^*$ transition near 330 nm causes this band to diminish and a less intense $n\pi^*$ band of the *cis* isomer near 440 nm to grow. Heat, UV irradiation at 250 nm and irradiation at wavelengths greater than 400 nm (i.e. radiation of the $n\pi^*$ band) recovers the *trans*-isomer. One problem is that photochromic *cis-trans* isomerization

Figure 11.28 Two classes of photochromic compounds: (a) azobenzenes; (b) spiropyrans.

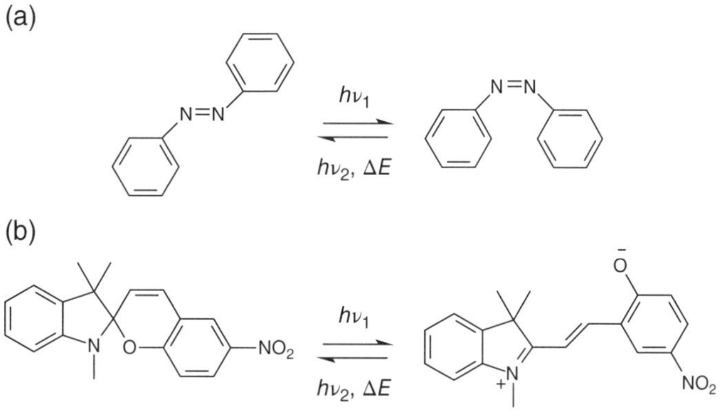

(a)

(b)

reactions that occur in solution are often suppressed in the semi-rigid environment of a thin solid film. Some novel solutions have been proposed to circumvent this difficulty. Most of these rely on methods to increase the area-per-molecule available for switching, for example by incorporating the azobenzene molecules into the cylindrical cavity provided by amphiphilic β-cylocodextrin molecules.

Figure 11.28b shows a particularly important class of photochromic compounds, the spiropyrans. The photochromism results from the presence of a conjugated ring, which is photochemically opened and closed – the process of cyclization. The majority of spiropyran compounds are positively photochromic; under UV irradiation changing from a pale yellow to a coloured form, as shown in Figure 11.29 for a spiropyran compound [59]. As LB films, spiropyrans are of interest because of their narrow H- and J-aggregate absorption bands (Chapter 4, Section 4.4.2), which cover a range of wavelengths and offer some potential for the realization of memory devices.

In certain photochromic compounds, a reversible photochemical transformation can lead to a change in their chirality (Chapter 2, Section 2.4.2); these can form the basis of *chiroptical* switches [57]. In a chiral photochromic system, the left-handed and right-handed forms of a chiral compound represent two distinct states in a binary logic element.

The term electrochromic is applied to compounds that can be interconverted by reversibly redox (reduction–oxidation processes) between two different absorption spectra. Whereas photochromic systems usually involve two forms of a molecule, in electrochromic systems, several successive switching processes can occur. A number of different electrochromic

(a)

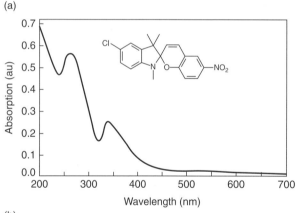

(b)

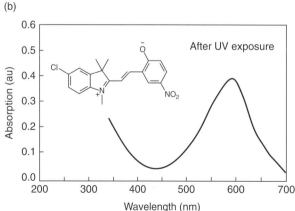

Figure 11.29 Absorption spectra and molecular structures of (a) the closed and (b) the open forms of a spiropyran compound. *Source:* From Srinivasan [59]. Copyright (2000) John Wiley & Sons Limited. Reproduced with permission.

technologies have been developed. First, there are solution-based systems, which rely on organic electrochromic species dissolved in the electrolyte compartment of an electrochemical cell comprising at least one transparent electrode. The most commonly used compounds are the viologens, salts of 4,4'-bipyridines. These are synthetically tuneable, which allows for different colours, and have intrinsically high extinction coefficients, yielding excellent colouration intensities. The switching speed depends on the diffusion of these and other redox active species in the electrolyte to the electrodes and is typically in the order of seconds. Because the redox active species are dissolved in an electrolyte, these mobile molecules will diffuse to both electrodes once an appropriate electrical potential has been applied to the circuit. When the potential has been removed, the charged species mix, transfer their charges, and the colour dissipates from the system. Therefore, there is no open circuit memory in these devices and power must be applied continuously to maintain colouration.

A different approach relies on the intercalation, or insertion and bonding, of ions into the crystal lattice of materials. This technology is based on electrochromic cells in which at least one of the electrodes is a thin but compact layer of a metal oxide, typically tungsten oxide WO_3, which exhibits electrochromic properties. Colouration is achieved when ions, typically H^+ or Li^+, and electrons are injected electrochemically into a WO_3 layer by means of an applied voltage. An advantage of this intercalation electrochromic technology compared to solution-based technologies is that it offers a memory capability in devices. The condition is that the complementary redox species are bound to the counter electrode surface or a charge storage layer is available to support the charging of the device (without dissipation upon removal of the applied potential). The intercalation process is slow due to the diffusion of the ions in the electrolyte and into the thin film. In addition, the metal oxides used do not form very highly coloured complexes, so the contrast ratio is intrinsically lower than for solution based organic systems.

An important application is the electrochromic window. This can block the glare of the sun or provide instant privacy with the flip of a switch. Electrochromic windows are part of a new generation of technologies called switchable glazing, or 'smart' windows, which change the light transmittance, transparency, or shading of windows in response to an environmental signal. Nanomaterials can be used to provide devices with improved colouration, faster switching times, longer lifetimes, and potentially reduced costs [60].

The spectroscopic technique of *hole burning* has been studied as a method to store data on the molecular level. In this technique, either a photochemical or a photophysical process leads to a change in the electronic structure, and hence in the optical absorption spectrum of a molecule in a polymer matrix. Since the molecule is just one member of an ensemble which gives rise to an inhomogeneously broadened absorption spectrum, the change reduces the absorption at a specific wavelength within the band, i.e. 'burns' a hole in the absorption band, as shown in Figure 11.30.

11.8.3 Chemical Control

Chemical stimuli can be exploited in a variety of ways to achieve switching. For example, electronic communication between metal centres across a bridging molecular wire can be controlled by protonation/deprotonation of an acid or basic site on the bridge. Supramolecular species comprising donor and acceptor units connected by means of noncovalent forces can be disassembled and re-assembled by modulating the reactions that keep the two components together [57].

There have been a number of suggestions that utilize the properties of the hydrogen bond. This is nearly unique in its range of energies (~0.02–1.3 eV per atom) and is ubiquitous in

(a)

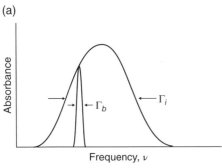

(b)

Figure 11.30 (a) Spectral profile of an inhomogeneously broadened absorption band of width Γ_i is composed of numerous homogeneous bands of width Γ_b but different centre frequencies. (b) Hole-burning causes the absorption intensity to be removed from bands at frequencies ν_1 and ν_2 and transferred to product bands P_1 and P_2.

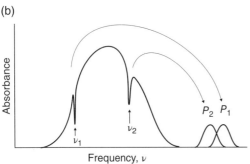

Figure 11.31 Tautomeric forms of a hemiquinone derivative proposed as a molecular switch [61].

biological systems. Figure 11.31 shows a proposed switch based on hydrogen bonding in an hemiquinone molecule [61]. The two isomers of the molecule depicted differ only in the placement of the protons are called *tautomers*. The hydrogen atoms that are involved in the tautomerism are suspended in a perfectly symmetrical double well potential and are not completely localized in one particular well, or in one part of the molecule, but move to the left and right, giving rise to an oscillation of the structure shown in Figure 11.31. In the proposed operation of the device, an applied electric field perturbs the double well potential, which becomes asymmetric and therefore allows the two states of the molecule to be distinguished. The protons move between the potential wells by tunnelling.

Many organic molecules, such the fullerene C_{60}, exhibit multiple reduction-oxidation states, offering the possibility of 'multibit' information storage [62]. An example of this is shown Chapter 5, Section 5.5.2, Figure 5.12. One bit of data can be associated with each redox state.

11.9 Nanomagnetics

Magnetic behaviour depends on the dimensionality of the system and low-dimensional materials can give rise to phenomena not observed in isotropic solids. In this respect, there have been a number of studies on the magnetic behaviour of ultra-thin film systems, particularly LB films [63]. The initial experiments, in the late 1970s, focused on a study of manganese stearate; the motivation was to investigate theoretical predictions of magnetic order in a

Figure 11.32 Hysteresis loop at 3 K in the plot of magnetization versus field for an LB film based on an amphiphilic lipid and $Cu_3[Fe(CN)_6]_2$. *Source:* Reprinted from Mingotaud et al. [63]. Copyright (2001), with permission from Wiley-VCH.

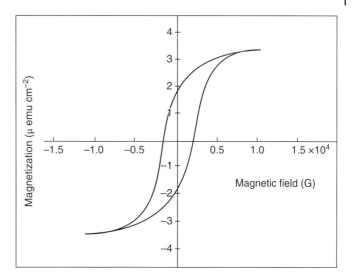

two-dimensional system. However, such a magnetic monolayer might be considered the ultimate magnetic memory storage device.

Other compounds that have been studied in LB and self-assembled films include metallic phosphonates, derivatives of phosphonic acid, and charge-transfer compounds such as TTF and bis(ethylenedithio)tetrathiafulvalene (BEDT-TTF) (Chapter 5, Section 5.4) and bimetallic compounds with the general formula $Ak[B(CN)_6].nH_2O$, where A and B can be divalent or trivalent transition metals. The CN ligands are good mediators of magnetic exchange, and examples of these mixed valence bimetallic compounds show ferrimagnetic ordering or ferromagnetic ordering (Chapter 5, Section 5.7.1) at critical temperatures, which range from liquid helium temperatures to above room temperature. LB films can be formed by using a positively charge lipid spread on a subphase containing a dilute colloidal suspension of the inorganic compound. Figure 11.32 shows a hysteresis loop at 3 K for such an LB film based on $Cu_3\{Fe(CN)_6\}_2$; the remanent magnetization is approximately 2.3 μemu cm^{-2} and the coercive field is 2100 G [63].

Interest is also focused on 'high-spin' molecules. Cluster of such molecules exhibit large magnetic hysteresis and may provide the means for achieving bistability and information storage at the molecular level (see Section 11.14). Furthermore, these nanomagnets provide interesting architectures for observing quantum magnetization and electron tunnelling through a potential barrier from one molecular state to another. The most thoroughly studied single molecule magnets are the mixed valence manganese clusters based on the $Mn_{12}O_{12}$ core. These are formed by an internal tetrahedron of four Mn^{IV} (i.e. valency of four) ions ($S = 3/2$) surrounded by eight Mn^{III} (valency of three) ions ($S = 2$). Exchange interactions within the cluster results in a ground state with a large spin ($S = 10$), which encounters a thermal barrier for reversal in the direction of magnetization along the uniaxial magnetic axis. In the crystalline state, these neutral clusters, such as $Mn_{12}O_{12}$ (acetate)$_{18}$, exhibit a stable hysteresis loop with a coercive field as large as 1.5 T at 2 K [63]. Such neutral clusters can be incorporated into LB films by mixing them with a fatty acid matrix. Plots of magnetization versus applied magnetic field for LB films exhibit a hysteresis loop at 2 K. Other approaches to the formation of self-organized magnetic materials have included the LbL electrostatic deposition and methods based on classical colloidal chemistry.

11.10 Nanotube and Graphene Electronics

Carbon nanotubes and graphene are unique materials (Chapter 5, Section 5.5). A vast number of papers is continuing to appear on the properties and applications of these compounds [64, 65].

Both experiments and theory have shown that single-walled-carbon-nanotubes can be either metals or semiconductors. The remarkable electrical properties of single wall (carbon) nanotubes (SWNTs) originate from the unusual electronic structure of the two-dimensional material graphene. When wrapped to form a SWNT, the momentum of the electrons moving around the circumference of the tube is quantized. The result is either a one-dimensional metal or a semiconductor, depending on how the allowed momentum sates (i.e. wavevector or k states) compare with the preferred directions for conduction.

The physics and technology of electron transport in nanodevices are fully explained by the seminal work of the late Rolf Landauer, an eminent IBM scientist, who derived the following formula for the conductance G of a one-dimensional ballistic conductor (Chapter 3, Section 3.2.1):

$$G = \frac{2e^2}{h} MT \tag{11.9}$$

where M is the number of modes or channels and T is the transmission coefficient, or the average probability that an electron injected at one end of the conductor will reach (be transmitted to) the other end. The factor of 2 in Eq. (11.9) takes into account the spin degeneracy of the electrons. Note the appearance again of the term h/e^2, the resistance quantum. In the case of a SWNT, taking into account sublattice degeneracy of graphene and assuming that the contacts are perfect ($T = 1$) gives [66].

$$G = \frac{4e^2}{h} = 155 \ \mu S \tag{11.10}$$

This corresponds to a resistance of about 6.5 kΩ. Experiments reveal that the resistance of SWNTs varies considerably, from approximately 6 kΩ to many megohms, the differences mostly due to contact resistances between the electrodes and the nanotubes. When contact resistances are eliminated, the measurements reveal a resistance per length of 4 kΩ μm^{-1}, a mean free path of 2 μm, and a room temperature resistivity of approximately 10^{-6} Ω cm. The conductivity of metallic nanotubes can therefore be equal to, or even exceed, the conductivity of metals like copper at room temperature.

Semiconducting nanotubes can possess very high charge carrier mobilities, e.g. hole mobilities of 10^3–2×10^4 cm^2 V^{-1} s^{-1} for tubes grown by chemical vapour deposition [67]. These values should be contrasted to the figures for inorganic and organic semiconductors given in Chapter 3, Section 3.4.1, Table 3.1. The conductivity of SWNTs is mainly p-type, but n-type conduction is also observed. Chemical doping can be achieved in a number of ways. For example, doping with alkali metals, which donate electrons to the nanotubes, can be used to achieve n-type material.

Figure 11.33a shows a schematic diagram of a field effect transistor based on a SWNT; an AFM image of a nanotube in such a device is shown in Figure 11.33b. The drain current versus drain voltage characteristics for a SWNT FET are shown in Figure 11.34. In this particular case, a thin layer of a high dielectric material, ZrO_2, was used as the gate dielectric [66, 67]. Standard FET behaviour is evident (Chapter 9, Section 9.4), with the drain current initially increasing in a linear fashion with the drain voltage, and then saturating. The nanotube exhibits excellent

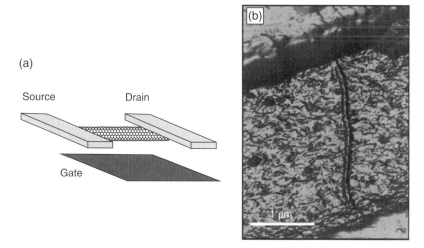

Figure 11.33 (a) Schematic diagram of a single-walled carbon nanotube transistor. (b) Atomic force microscope image of the device showing a nanotube connecting the source and drain electrodes. *Source: From McEuen and Park [66]. Reproduced by permission of the MRS Bulletin.*

Figure 11.34 Drain current I_d versus drain voltage V_d for different values of gate bias for a SWNT field effect transistor with a ZrO_2 gate insulator. *Source: Reprinted by permission from Macmillan Publishers Ltd: Javey et al. [67].*

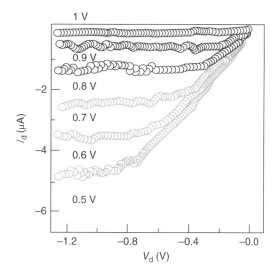

electrical properties, with a maximum transconductance $g_m = 12\,\mu A\,V^{-1}$ at $V_g = 0.4\,V$. The performance of such devices is superior, in some respects, to state-of-the-art silicon MOSFETs. The remarkable behaviour can be attributed, in part, to the lack of surface states (Chapter 2, Section 2.5.7) in SWNT devices. Other work has shown that SWNT FETs with sub-10 nm channel lengths can be fabricated [68]. Such devices outperform silicon transistors with more than four times the diameter-normalized current density ($2.41\,mA\,\mu m^{-1}$) at a low operating voltage of 0.5 V.

Under certain conditions (e.g. with a thin gate insulator), *ambipolar* FETs can be fabricated [69]. For these devices, the conduction is by electrons when a positive bias is applied to the gate, but by holes when a negative voltage is applied. It is therefore possible to inject electrons and holes simultaneously from opposite ends of the SWNT channel. These injected carriers are confined to the nanotube and, when they meet, they recombine with the emission of radiation; the

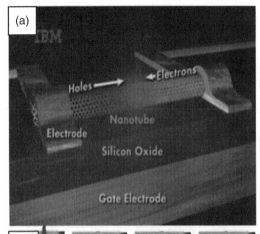

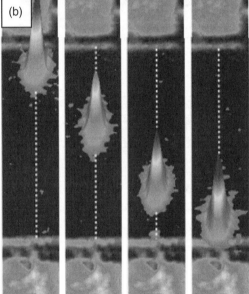

Figure 11.35 (a) Electrons and holes can be injected from opposite ends of a carbon nanotube to create a single molecule, electrically controlled light source. (b) The light emission can be moved between the two metal electrodes by varying the gate voltage. *Source:* Reprinted from Avouris and Appenzeller [69]. Copyright (2004) American Institute of Physics.

experiment is depicted in Figure 11.35 [69]. The diameter of the nanotube defines the wavelength of the emitted light, typically in the infrared region, and the position of the emitting spot along the length of the nanotube can be varied by changing the gate bias. This three-terminal device is essentially a single-molecule, electrically-controlled light source.

Despite the promising performance of individual SWNT devices, the control of nanotube diameter, chirality, density and placement remains inadequate for microelectronics production, particularly where large area coverage is required. Transistors incorporating tens of thousands of individual nanotubes are more practical. FETs with charge carrier mobilities of 10s of $cm^2\,V^{-1}\,s^{-1}$ can be produced, some on flexible supports [65]. Such devices are attractive for driving organic light-emitting device (OLED) displays as they possess a higher mobility than amorphous silicon ($\sim 1\,cm^2\,V^{-1}\,s^{-1}$) and can be fabricated by low temperature, non-vacuum methods.

It has already been noted that (Section 11.6.1) memory devices might be made from nanotubes. One scheme uses a crossbar arrangement of rows and columns of nanotubes separated by supporting blocks [36, 37]. The application of an appropriate voltage between the desired column and row bends the top tube into contact with the bottom nanotube. Van der Waals forces maintain the contact, even after the voltage has been removed. Separation of the SWNTs is then achieved by the application of a voltage pulse of the same potential. Assuming a minimum cell size to be a square of 5 nm length, a packing density of 10^{12} elements cm^{-2} can be achieved. The inherent switching time is estimated to be in the 100 GHz range. The memory is as fast as and denser than DRAM and has essentially zero power consumption in stand-by mode.

A further electronic application of carbon nanotubes is as microelectronics interconnects, in which their high current-carrying capacity and resistance to electromigration offer advantages over copper (Chapter 1, Section 1.3.3). However, if carbon nanotubes are to find their way into commercial electronic devices, their potential toxicity needs to be addressed [70]. The biocompatibility of the nanotubes is strongly influenced by their geometry and surface chemistry. It may, therefore, be possible to engineer the materials so that they are biocompatible and degradable [71].

Graphene is under intense investigations as a material for high frequency transistor applications [64]. However, the exploitation of graphene is difficult, as the material does not possess a band gap. The graphene energy band structure leads to both low on/off current ratios and the inability to switch the device to an off state of minimal drain current. Further challenges include the deposition of graphene over large areas with controlled grain size, thickness, and crystallographic orientation and the development of a suitable gate insulator with a high-quality interface with graphene. A number of research strategies are being targeted in order to open up the band gap such as using graphene nanoribbon and chemical modification. However, these approaches appear limited and may lead to degradation in the carrier mobility. The problem of low on/off ratio may be resolved using new transistor architectures, which exploit the modulation of the graphene work function, allowing control over vertical (rather than in-plane) carrier transport [64]. Despite all these difficulties, graphene transistors operating with high cut-off frequencies (> 400 GHz) have been reported and transistor fabrication procedures have been developed that allow a graphene integrated circuit to perform practical wireless communication functions, receiving and restoring digital text transmitted on a 4.3 GHz carrier signal [72].

11.11 Molecular Actuation

Mechanical movement is essential, both in nature and in the technological world. Enzymes such as myosin and kinesin are natural linear motors that convert chemical energy into mechanical work. Rotary movements are also observed in nature. These 'devices' are explored in more detail in Chapter 12, Section 12.11. However, systems exhibiting simple movements can be developed using the materials and fabrication techniques described in the earlier chapters of this book.

11.11.1 Dynamically Controllable Surfaces

An electrically-driven system for the dynamical control of interfacial properties (e.g. wettability) is illustrated in Figure 11.36 [73]. This exploits a transition between a straight and bent conformation of a self-assembled surfactant molecule on the application of a voltage. Self-assembly generally leads to a dense packing arrangement of the aliphatic chains, leaving little room for conformation transitions (Chapter 7, Section 7.3.2). To create a less densely packed monolayer,

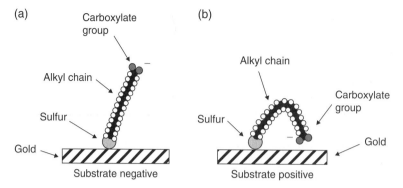

Figure 11.36 Idealized representation of the transition between (a) a straight (hydrophilic) and (b) bent (hydrophobic) molecular conformations of a long chain thiol molecule self-assembled onto a gold surface. *Source:* After Lahann et al. [73].

a long-chain thiol molecule with a very bulky terminal group (at the opposite end from the sulphur termination) is used. Subsequent cleavage of the space-filling end groups leaves a relatively low-density of the self-assembled monolayer. Upon application of a positive potential to the gold substrate, the negatively charged carboxylate groups experience an attractive force towards the substrate, causing the hydrophobic chains to undergo conformational changes. Reversible transitions can be confirmed at the macroscopic level using contact angle measurements. Other switchable surfaces have also been developed, including those based on biological molecules [74]. Potential applications for such smart surfaces may be as diverse as chemical or biochemical sensors, substrates to study important biochemical processes, and functional units for microfluidic devices (e.g. valves).

11.11.2 Rotaxanes

Rotaxane molecules have already been encountered earlier in this chapter, in Section 11.6.1. A rotaxane is a large molecule that comprises of a macrocyclic and a dumbbell-shaped component [35, 75]. The latter is encircled by the macrocycle, which is prevented from disengagement by two bulky end groups. The two components of the rotaxane cannot therefore separate from one another, even though they are not linked by covalent chemical bonds. However, the two components may move relative to one another. Simple motions are depicted in Figure 11.37 [57]. When two recognition sites on the dumbbell differ in their chemical properties, a rotaxane can exist in two different conformations and under an appropriate stimulus may switch between these states. This is shown schematically in Figure 11.37. The model is one of a molecular shuttle, with the 'train' moving between 'stations'. Examples of two recognition sites are provided in Figure 11.15, where the two sites have polar and nonpolar properties. In this compound, the

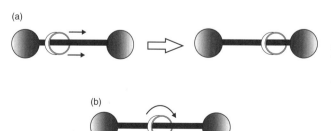

Figure 11.37 Schematic diagram showing movements that can occur in a rotaxane molecule. *Source:* Reprinted from Balzani et al. [57]. Copyright (2008), with permission from Wiley-VCH.

transition from one state to the other is achieved by the application of an electric field. In other rotaxanes, the linear movement can be controlled by chemical or photochemical stimuli.

Rotation of the 'wheel' component of the rotaxane (Figure 11.37b) is usually a spontaneous process. However, this can be artificially controlled. For example, chemical and electrochemical reactions can be used to make the ring component of a rotaxane oscillate between two positions [57]. Producing motors that continually spin in the same direction has been an important goal in the development of molecular machines. In 2011, a molecule with four functional units that undergo continuous and defined conformational changes upon sequential electronic and vibration excitation was developed [76]. This 'nanocar' could move forward over a surface. The field of molecular machinery resulted in the award of the 2016 Nobel Prize for Chemistry to Jean-Pierre Sauvage, Sir J. Fraser Stoddart, and Bernard L. Feringa.

11.11.3 Optical Tweezers

Optical trapping and manipulation of micrometre-sized objects was first reported in 1970 [77]. Since then, one important line of research has been the development of optical tweezers [78]. Such devices are simple and elegant examples of molecular 'machines'. Figure 11.38 shows an example of molecular tweezers based on the *cis-trans* isomerization of an azobenzene derivative [79]. In the *trans* configuration, the compound shown has weak co-ordinating capacity for large cations. Excitation with light causes *trans* to *cis* isomerization, leading to a molecular configuration suitable for enclosing large metal ions between the two crown ethers, with a strong increase in the co-ordination capacity. Somewhat related to tweezer-like movements are the conformational changes caused by electrostatic attractive forces generated by photo-induced electron transfer in dyads (bivalent molecules) with semiflexible bridges.

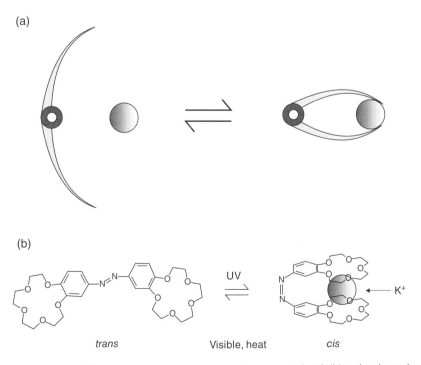

Figure 11.38 (a) Tweezers on the macroscopic scale contrasted with (b) molecular scale tweezers based on the *cis-trans* isomerization of an azobenzene derivative. *Source:* Adapted from Shinkai et al. [79].

In these cases, photo-induced electron transfer can result in the formation of charges of opposite sign. As a consequence of the resulting electrostatic attraction, conformational distortion occurs and the two ends of the molecule approach each other – a molecular 'harpoon' [57].

11.12 Molecular Logic Circuits

In most of the experiments on molecular-scale electronics, discrete devices are studied, i.e. the inorganic materials associated with silicon microelectronic devices are replaced by organic counterparts, albeit at the molecular scale. Such individual devices need to be connected together to form the logic circuitry that is the foundation of digital computers. In the latter, information is encoded in electrical signals. Threshold values and logic conventions are established for each signal. In a positive logic convention, a '0' is used to represent a signal that is below the threshold value and a '1' indicates that a signal is above the threshold. The logic circuits of silicon microprocessor systems process such binary data through a sequence of logic gates. Although it is not necessarily true that the components of a molecular computer will have to operate in an analogous manner to a silicon-based computer, much effort is being directed to the design, synthesis, and characterization of molecular systems that mimic conventional logical operations [57, 80].

The three basic types of logic gate are the NOT, AND, and OR gates. Other gates, which are a combination of these basic three gates, also exist: examples are the NAND (AND gate plus a NOT gate) and the NOR (OR plus NOT). Each has been designed to perform according to a set of rules delineated in a so-called *truth table*, which is a list of outputs that the gate should give in response to the complete range of input combinations. The NOT gate simply inverts the signal at its input, i.e. a logic 0 input results in a logic 1 output, and vice versa. In chemical systems, this function is common. For example, a luminescence output can be quenched by a chemical input [81].

A molecular AND gate is shown in Figure 11.39 [82]. This logic function has two inputs, A and B, and one output. Logic 1 is only obtained on the output when the input signals are both

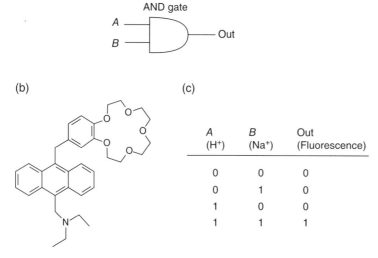

(a)

AND gate

A ——
B —— Out

(b)

(c)

A (H$^+$)	B (Na$^+$)	Out (Fluorescence)
0	0	0
0	1	0
1	0	0
1	1	1

Figure 11.39 A chemically-based AND logic gate. (a) Circuit symbol. (b) Anthracene derivative. (c) Truth table for AND function. *Source:* Reprinted from Balzani et al. [57]. Copyright (2008), with permission from Wiley-VCH.

at 1, i.e. the output is 1 if, and only if, input A *and* input B are at 1. The molecular AND gate is based on an anthracene derivative. In methanol, and in the presence of H^+ and Na^+, the fluorescence quantum yield of this compound is high (output state 1 in the truth table). However, the three output states 0 have a low fluorescence output. The photo-induced electron-transfer process involves the amine moiety in the first two states of the truth table and the crown ether in the third. The crown ether alone cannot quench the anthracene fluorescence, but when the amine is protonated, the process becomes thermodynamically allowed and does occur.

Many other examples of chemically based logic operations can be found in the literature [57]. In some of the early examples, it was proposed to exploit conformational changes in molecules. The *soliton switch*, introduced by Carter [83, 84], is perhaps the best-known example of this. A soliton in a conjugated polymer, such as polyacetylene, is a defect that separates chain segments with different bond alternation (Chapter 3, Section 3.4.3). These segments may be considered as logical states and a passing soliton 'switches' the chain from one state to another. Figure 11.40 shows how solitons in polymer networks might be switched and steered – soliton valving. The figure illustrates the change of state of a three-state network following the passage of a soliton. The passage of the soliton from A to B (or from B to A) moves the double bond at the branch carbon from the A chain to the B chain. In the upper right of the Figure 11.41, a soliton moving from B to C moves the double bond to the C chain.

Other approaches to molecular logic exploit the electrical properties of organic molecules and propose to construct circuitry by chemically linking the various components (e.g. molecular rectifier molecules, Section 11.5). These and other ideas have attracted their fair share of

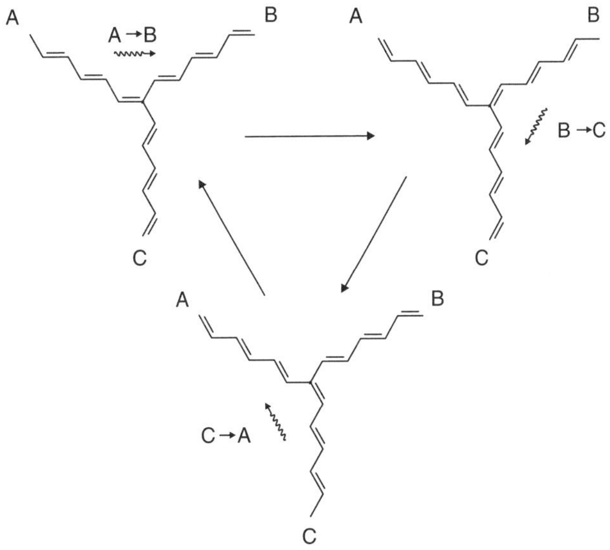

Figure 11.40 Soliton valving. The propagation of a soliton from chain A to chain B corresponds to a clockwise rotation by 120° of the upper left-hand configuration [61].

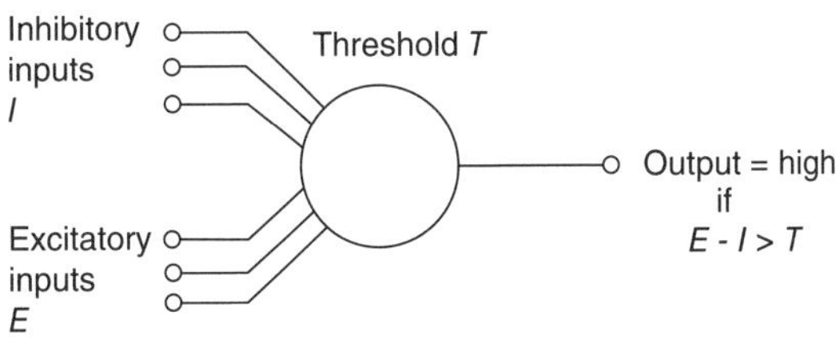

Figure 11.41 Representation of neural logic.

criticism for being impractical to realize. The problems of interconnecting individual devices, be these molecular diodes, switches, or simple logic circuits, are formidable. There is also the issue of how to read-in and read-out data at the molecular level. For example, in the case of the soliton switching, in order to 'read' the state of a chain segment, two adjacent carbon atoms have to be marked and the bond between them has to be inspected.

There have also been many speculations about self-assembling, self-repairing, fault-tolerant molecular systems; none have yet been well specified or fabricated. The steps needed have been outlined [85]. First, a high-level logic function associated with a molecule must be identified. A means is then required to place such molecules, with precision, to make an interface with the molecules and to verify both the position and communication. It is then essential to be able to interconnect and/or isolate the molecules without destroying their functionality. Finally, an efficient assembly scheme for very large numbers of such molecules must be developed.

11.13   Computing Architectures

Most computers use the stored-program concept designed by John Von Neumann. The *von Neumann architecture* is a model for a computing machine that uses a single storage structure to hold both the set of instructions on how to perform the computation and the data required or generated by the computation. Such machines are also known as stored-program computers. The separation of storage from the processing unit is implicit in this model. In it, programs and data are stored in a slow-to-access storage medium (such as a hard disk) and work on these is undertaken in a fast-access, volatile storage medium, such as random access memory (RAM).

The brain (nature's molecular computer) utilizes parallel processing, instead of the serial approach in von Neumann systems. This means that the brain can send a signal to hundreds of thousands of other neurons (Chapter 12, Section 12.7) in less than 20 milliseconds, even though it takes a million times longer to send a signal than a computer switch. The brain works mainly by non-linear computation using the rate of pulse production by a neuron or nerve cell as the information signal being sent to another cell. There are about 10^{11} neurons in the human brain, and each is connected to 10^3–10^4 others. This gives a crude 'bit-count' of 10^{11} to 10 [15]. An equivalent artificial 'brain' might therefore be built from $10^5 \times 8$ Gbit chips, with a power dissipation of many MW!

Neural logic may be either analogue or digital. In the latter form, the neuron is designed to respond to the sum of its N inputs (which may be inhibitory or excitory). Figure 11.41 shows a schematic diagram of such a gate. Provided that the sum exceeds a given threshold T the neuron will output logic 1, otherwise it produces logic 0. By combining neurons into totally connected networks – neural networks – it is possible to construct adaptive learning systems, control systems, and pattern recognition systems. In Chapter 10, Section 10.6, artificial neural networks were used in an electronic nose, emulating the human olfactory system. Some elementary properties of neurons can be mimicked by simple chemical systems, although no computational system exploiting these has been built. Neural networks are very similar to *cellular automata*. These are regular arrays of *finite-state machines* [86] and provide a useful architectural target from the standpoint of fabrication and logical issues.

In a cellular automaton, each finite-state machine is connected to a finite number of its neighbours using the same interconnection net. Von Neumann first investigated such systems as models for completely discrete physical dynamical systems such as the brain. Figure 11.42 shows an example of a two-dimensional cellular automaton layout. It consists of a regular spatial array of identical cells, and each characterized by a finite discrete-valued

Figure 11.42 A cellular automaton. An individual cell, shown on the right, is a finite-state machine.

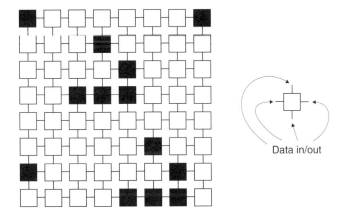

Data in/out

state function. At successive discrete intervals of time, each cell makes a transition to a new state, which depends on the previous state values of the connected neighbours.

The best-known example of a cellular automaton is the *Game of Life*, devised by the mathematician John Conway. The game is played on a field of cells, as in Figure 11.42, each of which has eight neighbours (adjacent cells). In the standard Game of Life, a cell is 'born' if it has exactly three neighbours, stays alive (survival) if it has two or three alive neighbours, and dies otherwise. Ever since its publication, it has attracted much interest because of the surprising ways the patterns can evolve. The cellular automaton is an example of emergence and self-organization.

A major issue associated with the manufacture of any molecular electronic device or system will almost certainly be their relatively poor fabrication yield resulting from problems such as the influence of background charge, difficulties to make reliable contacts to these devices, and lithographic inaccuracies when it comes at the nanometre level. Although it is expected that solutions to the above problems can be found through research, it seems intuitively difficult to find such solutions at the device level without an increase of the fabrication cost. A reasonable consequence of the above is that the search for an alternative to the expensive but very robust MOS technology in the form of a much less expensive but less robust technology should be accompanied by fault-tolerant schemes, permitting the functionality of the chips even if some of the devices are not even operational. A good example of such an approach is the 'Teramac' computer that was designed and fabricated by HP in USA [87]. This system was based on a high redundancy of wires that connect 'unreliable' switching devices, so that even if some of the paths contain non-operational devices there is always one functional path. Once built, the system is checked by software and the operational paths identified. Consequently, the problem of expensive fabrication costs to make perfect devices is transferred to the time-consuming procedure of identifying the operational paths by software. The above scheme was mainly designed with wires and switching devices that permit a reconfiguration of the system when a non-identified non-functional path is found.

For molecular electronics to be competitive with silicon, the technology will clearly have to offer something other than speed and bit density. Three-dimensional memory/logic is often mooted as a potential advantage for molecular systems and that certainly could lead to higher bit densities than for two-dimensional architectures. But it is much more likely that the functional properties of molecular materials will prove more significant. This is already obvious with liquid crystal and electroluminescent organic displays. One might imagine a future molecular electronic technology that integrates electronic logic/functionality with sensing and/or display capability in a foldable or, in medical applications, implantable packaging.

11.14 Quantum Computing

The unit of classical information is the bit, which takes one of the two possible values, 0 or 1. Any amount of classical information can be expressed as a sequence of bits. A classical computer executes a series of simple operations (gates), each of which acts on a single pair of bits. By executing many gates in succession, the computer can evaluate more complex mathematical operations on a set of input bits. Quantum information can also be reduced to elementary units, called *quantum bits* or *qubits*. A qubit is a two-level quantum system, such as the spin on an electron (exploited in the emerging technology of spintronics described in Section 11.6.4). The idea behind *quantum computing* is to use an array of such quantum 'particles' to perform mathematical operations [88, 89].

A simple quantum computer might use an array of quantum dots. Such a system is very similar to the cellular automata described in the previous section. To transfer the principles of cellular automata from the software level to real computing systems, two steps must be taken. First, the quantum dots, e.g. nanoparticles, must be arranged into regular two-dimensional arrays. Connections must then be established between the quantum dots by means of some physical interaction, for example, optical, magnetic, or electrical. Finally, the matrix of quantum dots has to be connected to input and output devices.

A simple example of arrays of quantum dots interacting electrostatically might be used as the basis of a quantum computer [90]. In the array of nanoparticles, which are firmly placed on the substrate and separated by small gaps comparable with the electron tunnelling distance ranging from 2 to 3 nm, the electron charge can propagate along the chain following the relay mechanism, as shown in Figure 11.43a. The driving force of such electron transfer is the electrostatic interaction between two negatively charged neighbouring particles. As a result of this electrostatic repulsion, the electron will be transferred to the next particle. However, the electron relay transfer in such a simple chain of nanoparticles is not reliable. Much better stability can be achieved using a cell containing five particles, depicted in Figure 11.43b. This cell has two stable configurations, which can be assigned the logical states '1' and '0'. As shown in Figure 11.43c, the situation of two neighbouring cells having different configurations '1' and '0' (or '0' and '1') is not energetically favourable (and therefore not stable) and should be transformed to the more energetically situations of '1' and '1' (or '0' and '0'). Therefore, information can be transported along the row of five-dot cells. This computing scheme is still binary and suitable to sequential computing, but it does not require the wiring of every cell.

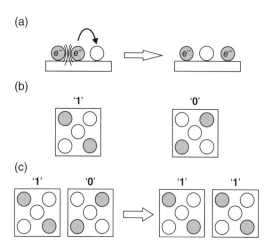

Figure 11.43 (a) Electrostatic interaction in a linear array of quantum dots. (b) Two stable logic states in a five-dot cell. (c) Switching process in the five-dot cell.

11.15   Evolvable Electronics

Living systems are able to achieve incredible feats of computation with remarkable speed and efficiency. Many of these tasks have not yet been adequately solved using algorithms running on the most powerful computers. Natural evolution is a bottom-up design process. Living systems assemble themselves from molecules and are extremely energetically efficient when compared to man-made computational devices. The technological drive to produce ever-smaller devices (Moore's law, Chapter 1, Section 1.3.2) is leading to the construction of machines at the molecular level. However, the basic computational paradigm is still von Neumann.

In contrast, *evolution-in-materio* aims to emulate nature and use computer-controlled evolution to create information processors [91]. Disordered material is trained via external stimuli, usually electric fields, to perform a specific computational task. Most research to date has used the training phase to optimize the electrical connections to the material, as for example in *field-programmable gate arrays* (FPGAs). Functions such as tone discriminators and maze solving can be emulated within liquid crystal displays and the operation of logic gates and the solution to a variety of computational problems solved using carbon nanotube/polymer composites [92, 93]. There is also interest in exploiting memristors in unconventional computing applications [94]. All these materials/devices may be considered as 'static'. The application of electric field does not significantly perturb their morphology and the process of evolution is essentially that of optimization of the magnitudes and distribution of the applied voltages.

Dynamic *evolution-in-materio* processors allow the morphology of the material to be altered during the training process. An example is that of a thin film composite in which single-walled carbon nanotubes are suspended in a liquid crystal. The nanotubes act as a conductive network, while the liquid crystal provides a host medium to allow the conductive network to reorganize when voltages are applied [95]. Figure 11.44 depicts an example of the experimental set-up. The diagram shows the 16 tracks and electrical contact pads (diameter of 50 μm and pitch of 100 μm) of a micro-electrode array [95]. Two of the contact pads are used as inputs, two as outputs, while the remainder are configurational inputs (for training). The magnified region in Figure 11.44 shows four electrodes with a material identified by the black bars. Before training, the individual carbon nanotube bundles are randomly distributed and connections between the electrodes are sparse. During the training process the material undergoes a physical

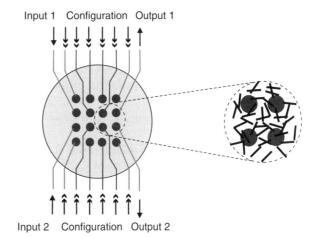

Figure 11.44 Evolution-in-materio of single-walled carbon nanotube/liquid crystal composites. Schematic diagram of experimental arrangement with inputs, outputs, and configuration signals to a micro-electrode array. An expanded area around four electrodes is shown on the right. The black bars correspond to bundles of carbon nanotubes. *Source:* Reprinted from Massey et al. [93].

reorganization as new electrical pathways are formed to perform the desired computation – in this example, a classification task. Other approaches to dynamic *evolution-on-materio* processors exploit processes in biological systems, such as slime moulds [96]. Such computational systems are unlikely to compete with silicon-based processors for speed, but may provide a complementary technology suited to solve particular tasks.

Problems

11.1 A single electron transistor works by transferring electrons from the channel to metallic particles of radius 0.4 μm. Can this device be operated satisfactorily at room temperature? Explain your reasoning.

11.2 According to the scaling laws, would 1000 engines occupying the same volume as one engine be more or less powerful?

11.3 A floating-gate nonvolatile memory transistor, which uses metallic nanoparticles as the charge storage medium, has a control gate to floating gate capacitance of 2.5 fF. Approximately, how many electrons are needed to be injected and stored on the floating gate to shift the measured threshold voltage by 0.5 V?

11.4 The cut-off frequency f_c of a forward biased diode is given by

$$f_c = \frac{1}{2\pi RC}$$

where R, C are the resistance and capacitance, respectively. Using data provided in the original paper by Aviram and Ratner (*Chem. Phys. Lett.* 1974; **29**:277–283) and reasonable values for other quantities, calculate the cut-off frequency for a molecular diode.

11.5 Using a cipher inspired by this book, a message is encoded:
BCHPQGOE.
Decode the message.

References

1 Drexler, E.K. (1992). *Nanosystems*. New York: Wiley.
2 Rogers, B., Adams, J., and Pennathur, S. (2015). *Nanotechnology: Understanding Small Systems*. Boca Raton, FL: CRC Press.
3 Kasap, S.O. (2017). *Principles of Electrical Engineering Materials and Devices*, 4e. Boston: McGraw-Hill.
4 Reitman, E.A. (2001). *Molecular Engineering of Nanosystems*. New York: Springer.
5 Christie, P., Roberts, G.G., and Petty, M.C. (1986). Spontaneous polarization in organic superlattices. *Appl. Phys. Lett.* 48: 1101–1103.
6 Jones, C.A., Petty, M.C., and Roberts, G.G. (1988). Langmuir-Blodgett films: a new class of pyroelectric materials. *IEEE Trans. Sonics Ultrason.* 35: 736–740.
7 Petty, M.C. (1996). *Langmuir-Blodgett Films*. Cambridge: Cambridge University Press.

8 Petty, M.C. and Cresswell, J. (1991). Deposition and characterization of multilayer films for nonlinear optics. In: *Materials for Photonic Devices* (ed. D. Andrea, A. Lapiccirella, G. Marletta and A. Viticoli), 259–269. Singapore: World Scientific.

9 Ashwell, G.J., Ranjan, R., Whittam, A.J., and Gandolfo, D.S. (2000). Second-harmonic generation from alternate-layer and Z-type Langmuir-Blodgett films: optimization of the transparency/efficiency trade-off. *J. Mater. Chem.* 10: 63–68.

10 Bardosova M, Wagner T (eds). (2015). *Nanomaterials and Nanoarchitectures*. Proceedings of NATO Advanced Study Institute on Nanomaterials and Nanoarchitectures, Cork, Ireland, 2013. Dordrecht: Springer.

11 Bardosova, M. and Tredgold, R.H. (2002). Ordered layers of monodispersive colloids. *J. Mater. Chem.* 12: 2835–2842.

12 Haick, H. and Cahen, D. (2008). Contacting organic molecules by soft methods: towards molecule-based electronic devices. *Acc. Chem. Res.* 41: 359–366.

13 Xiang, D., Wang, X., Jia, C. et al. (2016). Molecule-scale electronics: from concept to function. *Chem. Rev.* 116: 4318–4440.

14 Watkins, N.J., Yan, L., and Gao, Y. (2002). Electronic structure symmetry of interfaces between pentacene and metals. *Appl. Phys. Lett.* 80: 4384–4386.

15 Cui XD, Primak A, Zarate X, Tomfohr J, Sankey OF, et al. (2001). Reproducible measurement of single-molecule conductivity. *Science* 294: 571–574.

16 Moore, A.J., Goldenberg, L.M., Bryce, M.R. et al. (1998). Cation recognition by self-assembled layers of novel crown-annelated tetrathiafulvalenes. *Adv. Mater.* 10: 395–398.

17 Xiang, D., Jeong, H., Lee, T., and Mayer, D. (2013). Mechanically controllable break junctions for molecular electronics. *Adv. Mater.* 25: 4845–4867.

18 Aviram, A. and Ratner, M.A. (1974). Molecular rectifiers. *Chem. Phys. Lett.* 29: 277–283.

19 Zhang, G. and Ratner, M.A. (2015). Is molecular rectification caused by asymmetric electrode couplings or by a molecular bias drop? *J. Phys. Chem. C* 119: 6254–6260.

20 Ashwell, G.J., Mohib, A., and Miller, J.R. (2005). Induced rectification from self-assembled monolayers of sterically hindered π-bridged chromophores. *J. Mater. Chem.* 15: 1160–1166.

21 Peterson, I.R. (1992). Langmuir-Blodgett films: a route to molecular electronics. In: *Nanostructures Based on Molecular Materials* (ed. W. Göpel and C. Ziegler), 195–208. Weinheim: Wiley-VCH.

22 Petty, M.C. (2013). Organic electronic memory devices. In: *Handbook of Organic Materials for Optical and Optoelectronic Devices* (ed. O. Ostroverkhova), 618–653. Cambridge: Woodhead.

23 Möller, S., Perlov, C., Jackson, W. et al. (2003). A polymer/semiconductor write-once read-many-times memory. *Nature* 426: 166–169.

24 Vettiger, P., Despont, M., Drechsler, U. et al. (2000). The 'millipede' – more than one thousand tips for future AFM data storage. *IBM J. Res. Dev.* 44: 323–340.

25 Zhu, L., Zhou, J., Guo, Z., and Sun, Z. (2015). An overview of materials issues in resistive random access memory. *J. Materiomics* 1: 285–295.

26 Scott, J.C. and Bozano, L.D. (2007). Nonvolatile memory elements based on organic materials. *Adv. Mater.* 19: 1492–1463.

27 Dimitrakis, P., Normand, P., Tsoukalas, D. et al. (2008). Electrical behaviour of memory devices based on fluorine-containing organic thin films. *J. Appl. Phys.* 104: 004510.

28 Cho, B., Song, S., Yongsung, J. et al. (2011). Organic resistive memory devices: performance enhancement, integration and advanced architectures. *Adv. Funct. Mater.* 21: 2806–2829.

29 Cho, B., Yung, J.-M., Song, S. et al. (2011). Direct observation of Ag filamentary paths in organic resistive memory devices. *Adv. Funct. Mater.* 21: 3976–3981.

30 Pearson, C., Bowen, L., Lee, M.-W. et al. (2013). Focused ion beam and field-emission microscopy of metallic filaments in memory devices based on thin films of an ambipolar organic compound consisting of oxadiazole, carbazole, and fluorine units. *Appl. Phys. Lett.* 102: 213301.

31 Song, S., Cho, B., Kim, T.-W. et al. (2010). Three-dimensional integration of organic resistive memory devices. *Adv. Mater.* 22: 5048–5052.

32 Crossbar Inc. Available from URL: http://www.crossbar-inc.com. Accessed 20 October 2017.

33 Adesto Technologies. Available from URL: http://www.adestotech.com. Accessed 13 August 2018.

34 HP Labs. Available from URL: http://www.hpl.hp.com. Accessed 20 October 2017.

35 Yang, W., Li, Y., Liu, H. et al. (2012). Design and assembly of rotaxane-based molecular switches and machines. *Small* 4: 504–516.

36 Rueckes, T., Kim, K., Joselevich, E. et al. (2000). Carbon nanotube-based nonvolatile random access memory for molecular computing. *Science* 289: 94–97.

37 Nantero. Available from URL. http://nantero.com Accessed 20 October 2017.

38 van Ruitenbeek JM. (2005). Silver nanoswitch. *Nature* 433: 21–22.

39 Liu, Z., Yasseri, A.A., Lindsey, J.S., and Bocian, D.F. (2003). Molecular memories that survive silicon device processing and real-world operation. *Science* 302: 1543–1545.

40 Mabrook, M.F., Yun, Y., Pearson, C. et al. (2009). A pentacene organic thin film memory transistor. *Appl. Phys. Lett.* 94: 173–302.

41 Leong, W.L., Mathews, N., Tan, B. et al. (2011). Towards printable organic thin film transistor based flash memory devices. *J. Mater. Chem.* 21: 5203–5214.

42 Tho, L.V., Baeg, K.-J., and Noh, Y.-Y. (2016). Organic nano-floating gate transistor memory with metal nanoparticles. *Nano Converge* 3: 10.

43 Heremans, P., Gelinck, G.H., Müller, R. et al. (2011). Polymer and organic non-volatile memory devices. *Chem. Mater.* 23: 341–358.

44 Mai, M., Ke, S., Lin, P., and Zeng, X. (2015). Ferroelectric polymer films for organic electronics. *J. Nanomater.* 812538.

45 Thin Film Electronics ASA. Available from URL http://thinfilm.no/products-memory Accessed 20 October 2017.

46 Rocha, A.R., García-Suárez, V.M., Bailey, S.W. et al. (2005). Towards molecular spintronics. *Nat. Mater.* 4: 335–339.

47 Wolf, S.A., Lu, J.W., Stan, M.R. et al. (2010). The promise of nanomagnetics and spintronics for future logic and universal memory. *Proc. IEEE* 98: 2155–2168.

48 Hoffmann, A. and Bader, S.D. (2015). Opportunities at the frontiers of spintronics. *Phys. Rev. App.* 4: 047001.

49 Zhan, Y. and Fahlman, M. (2012). The study of organic semiconductor/ferromagnetic interfaces in organic spintronics: a short review of recent progress. *J. Polym. Sci. B Polym. Phys.* 50: 1453–1462.

50 Xiong, Z.H., Wu, D., Vardeny, Z.V., and Shi, J. (2004). Giant magnetoresistance in organic spin-valves. *Nature* 427: 821–824.

51 Burrows, P.E., Donovan, K.J., and Wilson, E.G. (1989). Electron motion perpendicular to Langmuir-Blodgett multilayers of conjugated macrocyclic compounds: the organic quantum well. *Thin Solid Films* 179: 129–136.

52 Uchida, K. (2005). Single-electron devices for logic applications. In: *Nanoelectronics and Information Technology*, 2e (ed. R. Waser), 423–441. Weinheim: Wiley-VCH.

53 Yano, K., Ishii, T., Sano, T. et al. (1999). Single-electron memory for giga-to-tera bit storage. *Proc. IEEE* 87: 633–650.

54 Oncel, N., Hallbäck, A.-S., Zandvliet, H.J.W. et al. (2005). Coulomb blockade of small Pd clusters. *J. Chem. Phys.* 123: 004703.

55 Akai-Kasaya, M., Okuaki, Y., Nagano, S. et al. (2015). Coulomb blockade in a two-dimensional conductive polymer monolayer. *Phys. Rev. Lett.* 115: 196801.

56 Stewart, M.D. Jr. and Zimmerman, N.M. (2016). Stability of single electron devices: charge offset drift. *Appl. Sci.* 6: 187.

57 Balzani, V., Credi, A., and Venturi, M. (2008). *Molecular Devices and Machines: Concepts and Perspectives for the Nanoworld*, 2e. Weinheim: Wiley-VCH.

58 Tian, H. and Zhang, J. (eds.) (2016). *Photochromic Materials: Preparation, Properties and Applications*. Wenheim: Wiley-VCH.

59 Srinivasan, M.P. (2000). Organic photochromism: spiro compounds as functional molecules. In: *Functional Organic Polymeric Materials* (ed. T.H. Richardson), 273–293. Chichester: Wiley.

60 Runnerstrom, E.L., Llordés, A., Lounis, S.D., and Milliron, D.J. (2014). Nanostructured electrochromic smart windows: traditional materials and NIR-selective plasmonic nanocrystals. *Chem. Commun.* 50: 10555–10572.

61 Aviram A, Seiden PE, Ratner MA. (1981). Theoretical and experimental studies of hemiquinones and comments on their suitability for molecular storage elements. In: FL Carter (ed.). *Proceedings of the Molecular Electronic Devices Workshop*. Naval Research Laboratory Memorandum Report 4662. Washington, DC: Naval Research Laboratories; pp. 3–16.

62 Echegoyen, L. and Echegoyen, L.E. (1998). Electrochemistry of fullerenes and their derivatives. *Acc. Chem. Res.* 31: 593–601.

63 Mingotaud, C., Delhaus, P., Meisel, M.W., and Talham, D.R. (2001). Magnetic Langmuir-Blodgett films. In: *Magnetism: Molecules to Materials. Molecule-Based Materials*, vol. 2 (eds. J.S. Miller and M. Drillon), 457–484. Weinheim: Wiley-VCH.

64 Novoselov, K.S., Fal'ko, V.I., Colombo, L. et al. (2012). A roadmap for graphene. *Nature* 490: 192–200.

65 De Volder MFL, Tawfick SH, Baughman RH, Hart AJ. (2013). Carbon nanotubes: present and future commercial applications. *Science* 339: 535–539.

66 McEuen, P.L. and Park, J.Y. (2004). Electron transport in single-walled carbon nanotubes. *MRS Bull.* 29: 272–275.

67 Javey, A., Kim, H., Brink, M. et al. (2002). High-*K* dielectrics for advanced carbon-nanotube transistors and logic gates. *Nat. Mater.* 1: 241–246.

68 Franklin, A.D., Luisier, M., Han, S.-J. et al. (2012). Sub-10 nm carbon nanotube transitor. *Nano Lett.* 12: 758–762.

69 Avouris, P. and Appenzeller, J. (2004). Electronics and optoelectronics with carbon nanotubes. *Ind. Phys.* 10: 18–21.

70 Madani, S.Y., Mandel, A., and Seifalian, A.M. (2013). A concise review of carbon nanotube's toxicity. *Nano Rev.* 4: 21521.

71 Bianco, A., Kostarelos, K., and Prato, M. (2011). Making carbon nanotubes biocompatible and biodegradable. *Chem. Commun.* 47: 10182–10188.

72 Han, S.-J., Garcia, A.V., Oida, S. et al. (2014). Graphene radio frequency receiver integrated circuit. *Nat. Commun.* 5: 3086.

73 Lahann, J., Mitragotri, S., Tran, T.N. et al. (2003). A reversibly switching surface. *Science* 299: 371–374.

74 Rant, U. (2012). Sensing with electro-switchable biosurfaces. *Bioanal. Rev.* 4: 97–114.

75 Neri, P., Sessier, J.L., and Wang, M.-X. (eds.) (2016). *Callixarenes and beyond*. Cham, Switzerland: Springer.

76 Kudernac, T., Ruangsupapichat, N., Parschau, M. et al. (2011). Electrically driven directional motion of a four-wheeled molecule on a metal surface. *Nature* 479: 208–211.

77 Marangò, O.M., Jones, P.H., Gucciardi, P.G. et al. (2013). Optical trapping and manipulation of nanostructures. *Nat. Nanotechnol.* 8: 807–819.

78 Pacoret, C. and Régnier, S. (2013). A review of haptic optical tweezers for an interactive microworld exploration. *Rev. Sci. Inst.* 84: 081301.

79 Shinkai, S., Nakaji, T., Ogawa, T. et al. (1981). Photoresponsive crown ethers. 2. Photocontrol of ion extraction and ion transport by a bis(crown ether) with a butterfly-like motion. *J. Amer. Chem. Soc.* 103: 111–115.

80 Lorente, N. and Joachim, C. (eds.) (2013). *Architecture and Design of Molecule Logic Gates and Atom Circuits*. Heidelberg: Springer.

81 De Silva AP, De Silva SA, Dissanayake AS, Sandanayake KRAS. (1989). Compartmental fluorescent pH indicators with nearly complete predictability of indicator parameters – molecular engineering of pH sensors. *J. Chem. Soc. Chem. Commun.* 1054–1056.

82 De Silva AP, Gunaratne HQN, McCoy CP. (1997). Molecular photoionic AND logic gates with bright fluorescence and 'off-on' digital action. *J. Am. Chem. Soc.* 119: 7891–7892.

83 Carter FL (ed.) (1981). *Proceedings of the Molecular Electronic Devices Workshop*. Naval Research Laboratory Memorandum Report 4662. Washington, DC: Naval Research Laboratories.

84 Roth, S. (1995). *One-Dimensional Metals*. Weinheim: Wiley-VCH.

85 Barker, J.R. (1995). Molecular electronic logic and architectures. In: *An Introduction to Molecular Electronics* (ed. M.C. Petty, M.R. Bryce and D. Bloor), 345–376. London: Edward Arnold.

86 Amos, M. (2006). *Genesis Machines*. London: Altantic Books.

87 Heath, J.R., Kuekes, P.J., Snider, G.S., and Williams, R.S. (1998). A defect-tolerant computer architecture: opportunities for nanotechnology. *Science* 280: 1716–1721.

88 Ladd, T.D., Jelezko, F., Laflamme, R. et al. (2010). Quantum computers. *Nature* 464: 45–53.

89 Debnath, S., Linke, N.M., Figgatt, C. et al. (2016). Demonstration of a small programmable quantum computer with atomic qubits. *Nature* 536: 63–70.

90 Montemerlo MS, Love JC, Opiteck GJ, Goldhaber-Gordon DJ, Ellenbogen JC. (1996). *Technologies and Designs for Electronic Nanocomputers*. MITRE Technical Report No. 96W0000044. McLean, VA: MITRE.

91 Miller, J., Harding, S., and Tufte, G. (2014). Evolution-in-materio: evolving computation in materials. *Evol. Intel.* 7: 49–67.

92 Massey, M.K., Kotsialos, A., Qaiser, F. et al. (2015). Computing with carbon nanotubes: optimization of threshold logic gates using disordered nanotube/polymer composites. *J. Appl. Phys.* 117: 134903.

93 Mohid, M., Miller, J.F., Harding, S.L. et al. (2016). Evolution-in-materio: solving computational problems using carbon nanotube-polymer composites. *Soft. Comput.* 20: 3007–3022.

94 Vourkas, I. and Sirakoulis, G.C. (2016). *Memristor-Based Nanoelectronic Computing Circuits and Architectures*. Cham, Switzerland: Springer.

95 Massey, M.K., Kotsialos, A., Volpati, D. et al. (2016). Evolution of electronic circuits using carbon nanotube composites. *Sci. Rep.* 6: 32197.

96 Adamatzky, A. (ed.) (2016). *Advances in Physarum Machines: Sensing and Computing with Slime Mould*. Cham, Switzerland: Springer.

Further Reading

Bar-Cohen, Y. (ed.) (2006). *Biomimetics: Biologically Inspired Technologies*. Boca Raton, FL: Taylor & Francis.

Brand, O., Fedder, G.K., Hierold, C. et al. (eds.) (2008). *Carbon Nanotube Devices*. Weinheim: Wiley-VCH.

Jones, R.A.L. (2004). *Soft Machines*. Oxford: Oxford University Press.

Metzger, R.M. (2018). Quo vadis, unimolecular electronics? *Nanoscale*. 10: 10316–10332.

Nicolini, C. (ed.) (1996). *Molecular Manufacturing*. New York: Plenum Press.

Ostroverkhova, O. (ed.) (2019). *Handbook of Organic Materials for Optical and Optoelectronic Devices*, 2e. Cambridge: Woodhead.

Schwoerer, M. and Wolf, H.C. (2006). *Organic Molecular Solids*. Berlin: Wiley-VCH.

Warner, J.H., Schäffel, F., Bachmatiuk, A., and Rümmeli, M.H. (2013). *Graphene: Fundamentals and Emergent Applications*. Waltham, MA: Elsevier.

Waser, R. (ed.) (2005). *Nanoelectronics and Information Technology*, 2e. Weinheim: Wiley-VCH.

Wolf, E.L. (2004). *Nanophysics and Nanotechnology*. Weinheim: Wiley-VCH.

12

Bioelectronics

Organic and Molecular Electronics: From Principles to Practice, Second Edition. Michael C. Petty.
© 2019 John Wiley & Sons Ltd. Published 2019 by John Wiley & Sons Ltd.
Companion website: www.wiley.com/go/petty/molecular-electronics2

Such stuff as dreams are made on

12.1   Introduction

Challenging and longer-term goals of molecular electronics are to emulate some of the sophisticated processes occurring in nature and to produce operational bioelectronic devices. In this chapter, the biological world is reviewed and its relationship with molecular electronics is explored. A brief overview of the most important biomolecules is first provided. Examples of some biological processes that might be exploited in molecular electronic devices, particularly in the areas of sensors, optical memories, and energy conversion are described later in the chapter.

12.2   Biological Building Blocks

The *cell* (Section 12.4) is the fundamental structural and functional unit of all living organisms. Many different molecules are found here. The detailed structure and conformation of each compound determines in which chemical reactions it can participate, and therefore its role in the life of the cell. Important classes of biomolecules include nucleic acids, proteins, carbohydrates, and lipids. Other compounds perform functions such as transporting energy from one part of the cell to another, or utilizing the sun's energy to drive chemical reactions. All these molecules, and the cell itself, are in a state of constant change. A cell cannot remain healthy unless it is continually forming and breaking down proteins, carbohydrates, and lipids, repairing damaged nucleic acids, and using and storing energy. Such energy-linked reactions are collectively known as *metabolism*.

12.2.1   Amino Acids and Peptides

Amino acids are a class of organic compounds that contain both the amino (NH_2) and carboxyl (COOH) chemical groups. The primary building blocks of all proteins, regardless of their species of origin, are the group of 20 different amino acids listed in Table 12.1. The amino and carboxyl groups are both attached to a single carbon atom, called the α-alpha carbon. The α-amino group is free or unsubstituted in all the amino acids except one, proline. A further variable group, R is attached to the α-carbon; it is in their R groups that the molecules of the 20 amino acids differ from one another. The simplest of the acids, glycine, contains an R group that comprises a single hydrogen atom. Since the amino and carboxylic acid groups are basic and acidic, respectively (Chapter 5, Section 5.2.2), amino acids are *zwitterions*, with both negative and positive charges. The structures of the neutral (undissociated) and charged forms of glycine are shown in Figure 12.1. Zwitterions are highly polar substances for which intermolecular electrostatic attraction leads to strong crystal lattices. With the exception of glycine, all of the amino acids are chiral molecules (Chapter 2, Section 2.4.2).

Two amino acid molecules may be joined to yield a *dipeptide* through a *peptide bond*, formed by the removal of a water molecule from the carboxyl group of one amino acid and the α-amino group of the other by the action of strong condensing agents. Figure 12.2 shows the peptide formed from two molecules of glycine, glycylglycine. Like glycine, this exists as a zwitterion.

Higher peptides are also possible; a tripeptide contains three amino acids, a tetrapeptide four, and so on. Peptides are named from the sequence of their constituent amino acids, beginning from the amino-terminated end. When many amino acids are joined in a long chain, this

Table 12.1 The 20 amino acids common in proteins.

Amino acid	Three-letter symbol
Alanine	Ala
Arginine	Arg
Asparagine	Asn
Aspartic acid	Asp
Cysteine	Cys
Glutamine	Gln
Glutamic acid	Glu
Glycine	Gly
Histidine	His
Isoleucine	Ile
Leucine	Leu
Lysine	Lys
Methionine	Met
Phenylalanine	Phe
Proline	Pro
Serine	Ser
Threonine	Thr
Tryptophan	Trp
Tyrosine	Tyr
Valine	Val

is called a *polypeptide*. Such compounds contain only one free α-amino acid group and one free α-carboxyl group at their ends. An example is given in Figure 12.3, the pentapeptide serylglycltyrosinylalanylleucine.

In addition to the amino acids that form proteins, many other amino acids have been found in nature, including some that have the carboxyl and amino groups attached to separate carbon atoms. These unusually structured amino acids are most often found in fungi and higher plants.

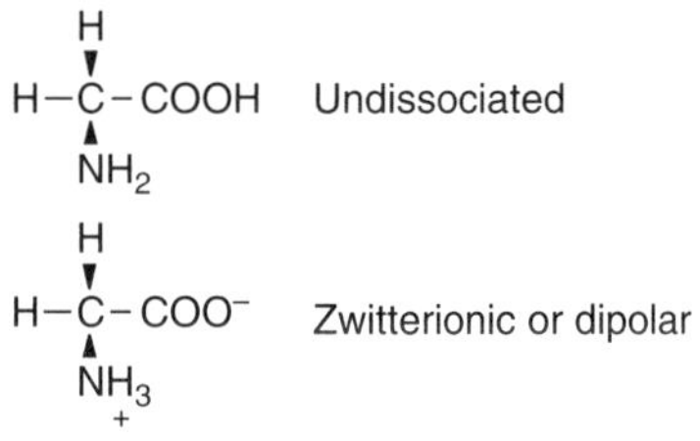

12.2.2 Proteins

A *protein* is a complex, high-molecular-mass, organic compound consisting of amino acids joined by peptide bonds. These molecules are the most abundant species in most cells and generally constitute 50% of their dry weight. Proteins are versatile cell components, some are *enzymes*, some serve as structural components, and

$$
\begin{array}{c}
H \\
| \\
H-C-COOH \quad \text{Undissociated} \\
| \\
NH_2
\end{array}
$$

$$
\begin{array}{c}
H \\
| \\
H-C-COO^- \quad \text{Zwitterionic or dipolar} \\
| \\
NH_3 \\
^+
\end{array}
$$

Figure 12.1 Undissociated and zwitterionic forms of the amino acid glycine.

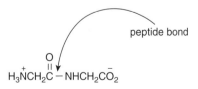

$$
\overset{+}{H_3}NCH_2\overset{\overset{O}{\|}}{C}-NHCH_2CO_2^-
$$

peptide bond

Figure 12.2 Glycylglycine – a dipeptide. The figure shows the peptide bond formed between the two molecules of glycine.

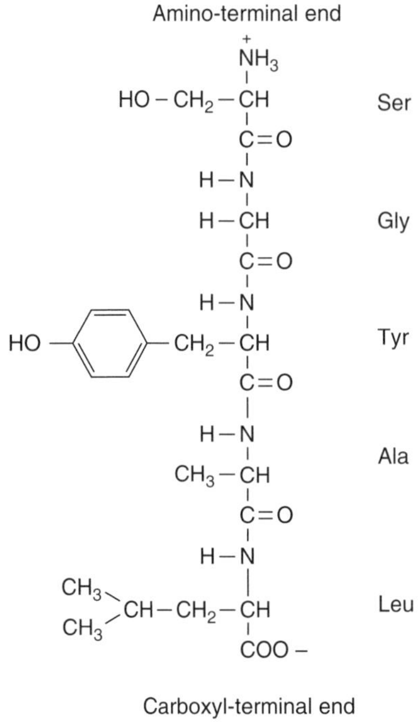

Amino-terminal end

Figure 12.3 Structure of the polypeptide serylglycyltyrosinylalanylleucine. Peptides are named beginning with the NH_2-terminal amino acid.

some have *hormonal* activity. Table 12.2 provides examples of the different types of proteins, classified according to their biological function.

A protein may be a single polypeptide chain, or it may consist of several such chains held together by weak molecular bonds. The R groups of the amino acid subunits determine the final shape of the protein and its chemical properties. Using only 20 different amino acids, a cell constructs thousands of different proteins, each of which has a highly specialized function.

Proteins serve two important biological functions. First, they can act as a structural material. The structural proteins tend to be fibrous in nature, i.e. the polypeptide chains are lined up more or less parallel to each other and are joined to one another by hydrogen bonds. These are physically tough and are normally insoluble in water. Depending on the actual three-dimensional arrangement of the individual protein molecule and its interaction with other similar molecules, a variety of structural forms may result. Typical examples of fibrous proteins are *α-keratin*, the major component of hair, feathers, nails and skin, and *collagen*, the major component of tendons.

The other important property of proteins is their role as biological regulators. Here, the proteins are responsible for controlling the speed of biochemical

Table 12.2 Classification of proteins according to their biological function.

Class	Examples
Enzymes	Ribonuclease, Trypsin
Storage proteins	Ovalbumin (eggs), Casein (milk)
Transport proteins	Haemoglobin, Serum albumin
Contractile proteins	Actin, Myosin
Protective proteins of blood	Antibodies, Fibrinogen
Toxins	Botulinus toxin, Snake venoms
Hormones	Insulin, Adreneocorticotrophin
Structure proteins	Keratins, Collagen

Figure 12.4 Schematic diagram showing the Y-shaped structure of an antibody.

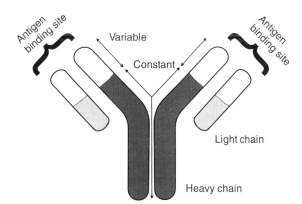

reactions and the transport of various materials throughout the organisms. The catalytic proteins (enzymes) and transport proteins tend to be globular in nature. The polypeptide chain is folded around itself in such a way to give the entire molecule a rounded shape. Globular proteins are soluble in aqueous systems and diffuse readily.

Antibodies (also referred to as immunoglobulins) are produced by white blood cells. These special proteins respond to a specific *antigen* (bacteria, virus, or toxin). Each antibody has a section in its chemical makeup that is sensitive to a particular antigen and binds to it in some way. An antibody consists of four polypeptides – two heavy chains and two light chains joined to form a Y-shaped molecule, as depicted in Figure 12.4. The amino acid sequences in the tips of the Y vary significantly among different antibodies. This variable region, composed of 110–130 amino acids, gives the antibody its specificity for binding an antigen. The variable region includes the ends of the light and heavy chains.

The sequence of amino acids in the covalent backbone of a protein is called its primary structure. The secondary structure refers to the specific geometric arrangement of the polypeptide chain along one axis. Two common arrangements are the *α-helix*, shown in Figure 12.5, and the *β-conformation*, shown in Figure 12.6, in which the polypeptide

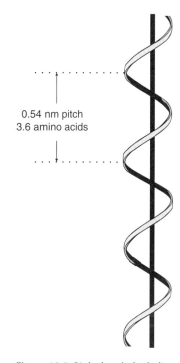

Figure 12.5 Right-handed α-helix.

chains are in an extended zig-zag configuration called a *pleated sheet*. The α-helix is right-handed, with a pitch of 0.54 nm or 3.6 amino acid units. Both the helix and pleated sheet are very stable structures, held together by hydrogen bonding. The specific configurations of the polypeptide chains are stable because of particular amino acid sequences. For example, an α-helix tends to form spontaneously only in the case of polypeptide chains in which consecutive R groups are relatively small and uncharged, as in α-keratin.

A tertiary structure for proteins also exists. This term is used to refer to the three-dimensional structure of globular proteins, in which the polypeptide chain is tightly folded and packed into a compact spherical form. The molecule tends to orient itself so that the nonpolar side chains lie inside the bulk of the structure, where they attract each other by van der Waals forces. The polar side chains are usually found on the surface of the molecule; consequently, these can

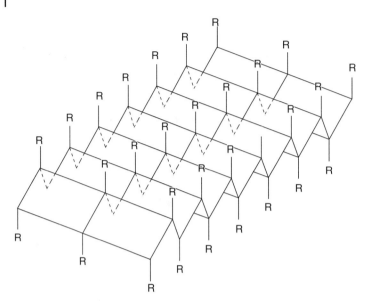

Figure 12.6 Schematic representation of three parallel chains in β structure, showing the pleated sheet arrangement. All the R groups project above or below the plane of the figure.

hydrogen bond to the solvent molecules and confer the necessary water solubility. The process by which a linear protein chain forms its secondary and tertiary structure is called *protein folding.*

12.2.3 Enzymes

The enzymes make up the largest and most highly specialized class of proteins. They act as catalysts for thousands of chemical reactions. These reactions would otherwise occur at extremely low rates. Enzymes have traditionally been named according to the substance that they act on, called the *substrate*, or according to the nature of the reaction catalysed. Thus, urease catalyses the hydrolysis of urea and arginase catalyses the hydrolysis of arginine. However, in many cases, enzymes have been given names that are not so informative, such as pepsin or trypsin. Enzymes are grouped into six major classes, indicated in Table 12.3.

Enzymes have molecular weights ranging from about 12 000 to over 1 million. Some enzymes consist only of one or more polypeptide chains, but others contain an additional component needed for activity, called a *cofactor*. This may be a metal such as Mg, Mn, Zn, or Fe, or it may be a complex organic molecule, usually called a *coenzyme*. Enzymes use sophisticated molecular recognition and work via a 'lock-key' mechanism by which only a specific substrate will have the correct shape to fit within the enzyme's catalytic site. Therefore, most enzymes only serve a single well-defined catalytic function.

12.2.4 Carbohydrates

Carbohydrates are the basic fuel molecules of the cell. These molecules contain carbon, hydrogen, and oxygen in approximately equal amounts. Green plants and some bacteria use *photosynthesis* to make simple carbohydrates (sugars) from carbon dioxide, water, and sunlight (Section 12.10.2). Animals, however, obtain their carbohydrates from foods. Once a cell

Table 12.3 Classification of enzymes.

Classification of enzymes
Oxido-reductases (electron-transfer reactions)
Transferases (transfer of functional groups)
Hydrolases (hydrolysis reactions)
Lyases (addition to double bonds)
Isomerases (isomerization reactions)
Ligases (formation of bonds with ATP cleavage)

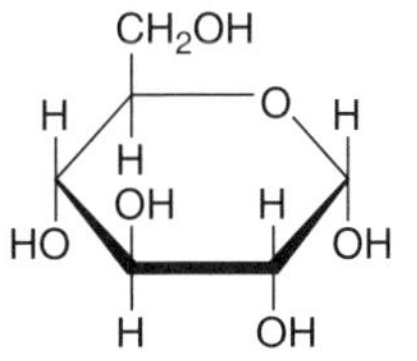

Figure 12.7 Structure of α-D-glucose.

possesses carbohydrates, it may break these molecules down to yield chemical energy or use them as raw material to produce other biomolecules.

The term carbohydrate is used loosely to characterize an entire group of natural products that are related to simple sugars. There are three major classes of carbohydrates: *monosaccharides*, *oligosaccharides*, and *polysaccharides*. Glucose is an example of a monosaccharide and is the most important fuel molecule for most organisms. This is a ring structure, as shown in Figure 12.7. The particular form of the molecule depicted is that used extensively by sugar chemists: the sugar ring is written as a planar hexagon with the oxygen in the upper right vertex and additional chemical groups are indicated by straight lines through each vertex, either above or below the plane. The carbohydrates occur in optically active form (generally true for natural products) and only one enantiomer (Chapter 2, Section 2.4.2) is found in nature. The usual form of glucose found in nature is dextrorotatory–D-glucose.

Sucrose, or cane sugar, is a disaccharide. The most abundant carbohydrates in nature are the plant polysaccharides cellulose and starch, which are polymers of glucose. Cellulose is the constituent of plant walls that provides stiffness and strength. It is important that plants do not loose this strength, thus they do not possess enzymes capable of digesting cellulose. In contrast, starch forms a staple food of many plants and animals; all plants and many animals possess enzymes that allow them to digest starch. Starch contains two types of polysaccharides: α-amylose and amylopectin. The former consists of long unbranched chains of D-glucose units, while amylopectin is highly branched. Despite the widespread differences between cellulose and starch, these molecules have exactly the same chemical formula. The difference is in the way the rings are joined. The cellulose molecules are in such a geometrical arrangement that they produce stiff molecules, packed tightly together and are held in place by strong hydrogen bonds. In contrast, the orientation of the joints between the sugar rings in the starch molecules leads to a more open helical structure and fewer internal bonds. While amylose and amylopectin molecules do pack together, they do so in a looser and weaker fashion and are easily separated, to be broken down by the specific enzymes that all plants and most animals possess for the purpose.

12.2.5 Lipids

Lipids are fatty substances that play a variety of roles in the cell. Some are held in storage for use as high-energy fuel; others serve as essential components of the cell membrane. Examples are shown in Figure 12.8. A simple type of lipid is a fatty acid (Figure 12.8a), which is used as the basis of Langmuir–Blodgett (LB) film deposition (Chapter 7, Section 7.3.1). This consists of a long hydrocarbon chain terminated in a carboxylic acid group. The chain may be saturated or it may contain one or more double bonds; a few fatty acids contain triple bonds. Nearly all fatty acids in nature have an even number of carbon atoms and have chains that are between 14 and 22 carbon atoms long.

The simplest and most abundant lipids, which contain fatty acids as building blocks, are the neutral lipids, also called *fats*, *triglycerides*, or *triacylglycerols* (Figure 12.8b). These are esters of the alcohol glycerol, with three fatty acid molecules, and bear no net electrical charge. Triacylglycerols are the major component of fats in plant and animal cells. These compounds are excellent forms of energy because of the high number of reduced CH groups available for oxidation-dependent energy generation processes.

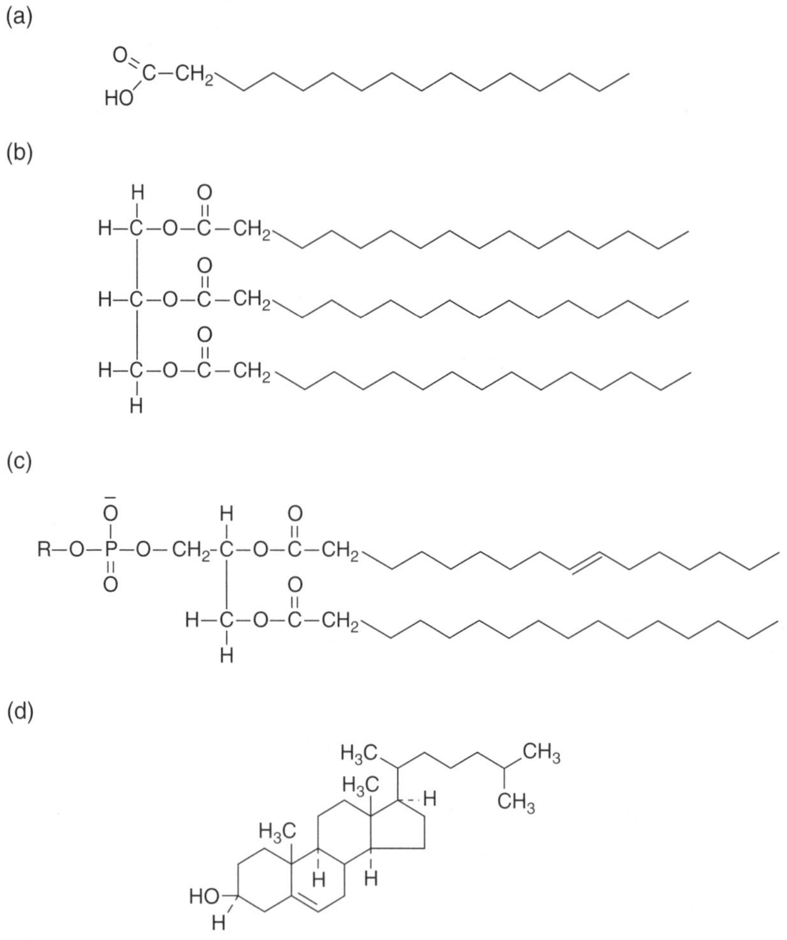

(a)

(b)

(c)

(d)

Figure 12.8 Chemical structures of lipids. (a) Palmitic acid, a long-chain fatty acid; (b) tripalmitin, a triacylglycerol; (c) general structure of a phospholipid; (d) cholesterol, a sterol.

The melting point of a fat depends on the amount of unsaturation in the fatty acids. Fats with a preponderance of unsaturated fatty acids have melting points below room temperature and are referred to as oils (e.g. olive oil). However, the melting point of a natural fat may be increased by hydrogenation. The fat is exposed to hydrogen at a high temperature and in the presence of a catalyst. During this process some double bonds are converted into single bonds and other double bonds are converted from a *cis* to a *trans* configuration (Chapter 2, Section 2.4.3). In both cases, the effect is to straighten out the molecules so they can lie closer together and become solid rather than liquid.

Triacylglycerols undergo hydrolysis when boiled with acids or bases or when acted upon by lipase enzymes, such as present in pancreatic juices. Hydrolysis of triacylglycerols with an alkali, a process called *saponification*, yields a mixture of fatty acid soaps and glycerol. *Waxes* are related to the triacylglycerols. These compounds are naturally occurring esters of long-chain carboxylic acids (C_{16} or greater) with long chain alcohols (C_{16} or greater).

Three major classes of lipids are found in biological membranes: *phospholipids*, *glycolipids*, and *cholesterol*. Phospholipids (Figure 12.8c) can be derived from glycerol and include phosphorus in the form of phosphoric acid. The major phospholipids found in cells contain two fatty acid molecules, joined to the first and second hydroxyl groups of the glycerol. These molecules primarily act as structural elements. All phospholipids undergo a change of phase from a solid to a liquid crystalline state. This transition is associated with increased disorder and mobility in the fatty alkyl chains. As is the case for simple long-chain fatty acids, the temperature of the transformation depends on the chain length, the number of double bonds, and the nature of the head group. Phosphatidylcholine, with two palmitic acid (C_{16}) chains, has a transition temperature of 42 °C. However, if the chain lengths are both reduced by two carbons (to myristic acid chains), then the transition temperature is reduced to 23 °C. The presence of one *cis* double bond also causes a marked reduction in the transition temperature. Since most of the naturally-occurring phospholipids in mammalian cells have at least one such double bond, they will be above the phase transition at body temperature. Glycolipids are sugar-containing lipids, examples including cerebroside, ganglioside. They contain a polar, but uncharged head group.

Certain lipids are nonsaponifiable, i.e. heating with an alkali cannot hydrolyze them. There are two major groups of these lipids: *steroids* and *terpenes*. While most of these occur in only trace amounts in cells, one type of steroid, the sterols, is extremely abundant. Cholesterol, shown in Figure 12.8d, is the major sterol in animal tissues.

Polar lipids readily disperse to form micelles (Chapter 8, Section 8.2.2). These are spherical structures in which the hydrocarbon tails of the lipids are on the inside, hidden from the aqueous environment, and the charged hydrophilic heads are on the outside. If the head group of the lipid is not strong compared with the hydrophobic part, the molecules can form spherical *vesicles* in which the double layers form a shell, with water on both the inside and outside. Phospholipids also form bilayer structures, particularly at the interface between two aqueous surfaces. Such bilayers provide the fundamental framework for natural cell membranes (Section 12.6). Schematic diagrams of micelles, vesicles, and bilayers are shown in Figure 12.9.

12.3 Nucleotides

Nucleotides are the recurring structural units of the nucleic acids. These molecules contain three characteristic components: a nitrogenous base, a five-carbon sugar and phosphoric acid. The nucleotides are the compounds responsible for storing and transferring genetic information and are enormous molecules made up of long strands of subunits, called bases, arranged

(a)

(b)

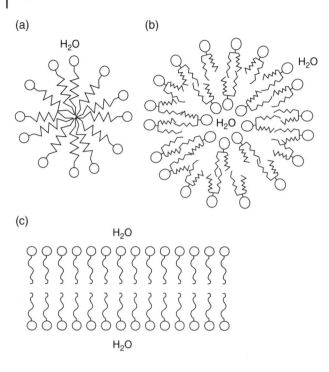

Figure 12.9 Cross-sections of (a) a micelle, (b) a vesicle, and (c) a bilayer formed by amphiphilic molecules.

(c)

in a precise sequence. The bases are 'read' by other components of the cell and used as a guide in making proteins. Two important types of nucleic acid are *ribonucleic acid* (RNA) and *deoxyribonucleic acid* (DNA).

12.3.1 Bases

The bases found in nucleotides are of two types. These are derivatives of two parent heterocyclic compounds pyrimidine and purine, which are themselves not found in nature (Figure 12.10). Three pyrimidine bases are common in nucleic acids: uracil, thymine, and cytosine, universally abbreviated as U, T, and C. Uracil is generally found in RNA and thymine in DNA; cytosine is found in both RNA and DNA. There are two common purine bases, found in both RNA and DNA: adenine (A) and guanine (G). The pyrimidine and purine bases are almost flat molecules that are relatively insoluble in water.

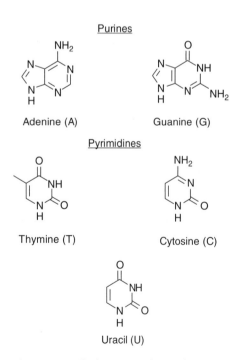

Figure 12.10 The bases in nucleic acids. Thymine occurs in DNA but not RNA and uracil in RNA but not DNA.

12.3.2 DNA

In 1953, James Watson and Francis Crick established the structure of DNA, which in

1962 led to the award of the Nobel Prize for Medicine (together with Maurice Wilkins). The DNA molecule is composed of two long strands in the form of a double helix (Figure 12.11). The strands are made up of alternating phosphate and sugar molecules. The nitrogen bases provide links between these strands, holding them together. Each base is attached to a sugar molecule and is linked by a hydrogen bond to a complementary base on the opposite strand. Adenine always binds to thymine (A to T), and guanine always binds to cytosine (G to C). To make a new, identical copy of the DNA molecule, the two strands need only unwind and separate at the bases (which are weakly bound); with more nucleotides available in the cell, new complementary bases can link with each separated strand, and two double helixes result. If the sequence of bases were AGATC on one existing strand, the new strand would contain the complementary, or 'mirror image', sequence TCTAG. In nature, the DNA backbone is tightly coiled up. This packing is now known to be based on minute particles of protein known as *nucleosomes*. The DNA is wound around each nucleosome in succession to form a beaded structure. The structure is then further folded so

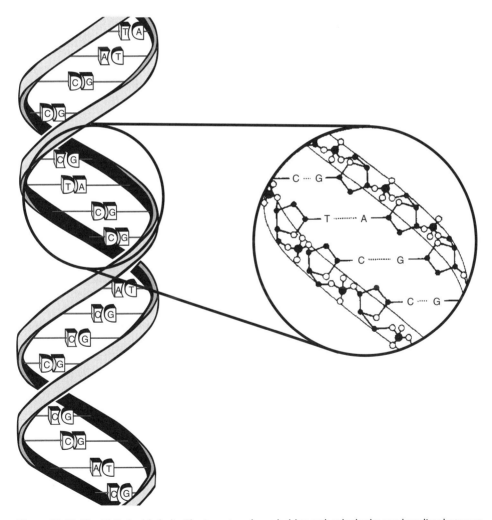

Figure 12.11 The DNA double helix. The two strands are held together by hydrogen bonding between complementary base pairs.

that the beads associate in regular coils. Thus, the DNA has a coiled-coil configuration, like the filament of some electric light bulbs.

12.3.3 RNA

Translation of the genetic code from base sequences to amino acid sequences does not occur in a single step. This is the role of the other nucleic acid, RNA. RNA is a molecule that links the two worlds of DNA and proteins. DNA's letter Ts are made from RNA's letter Us. RNA makes up about 5–10% of the total weight of the cell. There are three major types of ribonucleic acids: *messenger RNA* (mRNA), *ribosomal RNA* (rRNA), and *transfer RNA* (tRNA). Messenger RNA contains only four bases A, G, C, and U. It is enzymatically synthesized in the cell nucleus in such a way that the base sequence of the mRNA molecule is complementary to the base sequence of one of the strands of the DNA molecule. Each mRNA molecule carries the code for one or more protein molecules. Transfer RNAs are relatively small molecules that act as carriers of specific amino acids during protein synthesis. Each of the 20 amino acids found in proteins has one or more corresponding tRNAs. The tRNA molecule may exist in its free form or attached to its specific amino acid. Finally, ribosomal RNA is the most abundant type of RNA. It plays an important role in the structure and biological function of *ribosomes* and constitutes up to 65% of their weight. Ribosomes are complexes of RNA and proteins and undertake the job of translating DNA 'recipes' into proteins.

Adenosine triphosphate, ATP

Adenosine diphosphate, ADP

Figure 12.12 Adenosine triphosphate (ATP) and adenosine diphosphate (ADP).

12.3.4 ATP, ADP

Adenosine triphosphate (ATP) is a nucleotide that performs many essential roles in the cell. This molecule may be considered as the energy 'currency' of life. The ATP molecule is composed of three components, as shown in Figure 12.12. At the centre is a sugar molecule, ribose (the same sugar that forms the basis of DNA). Attached to one side of this is the base adenine. The other side of the sugar is attached to a string of phosphate groups. ATP is remarkable for its ability to enter into many coupled reactions and both to extract and provide energy. In animal systems, the ATP is synthesized in the tiny energy factories called *mitochondria*.

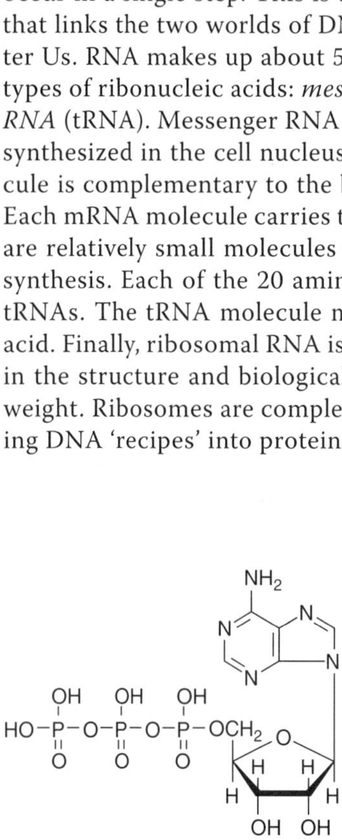

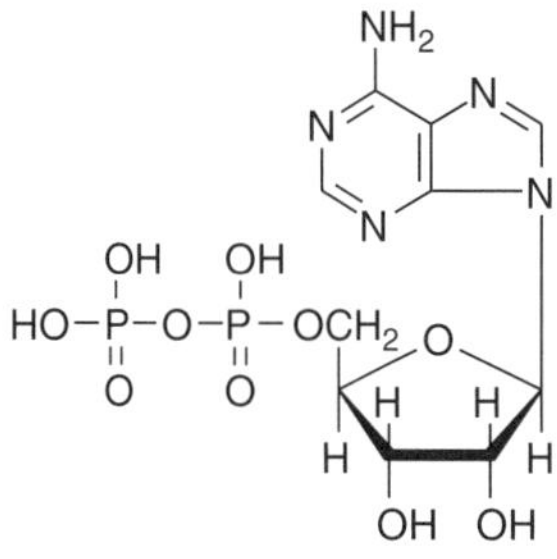

When the third phosphate group of ATP is removed by hydrolysis, a substantial amount of free energy is released; the exact amount depends on the conditions:

$$ATP + H_2O \rightarrow ADP + phosphate \qquad (12.1)$$

where ADP is *adenosine diphosphate* (Figure 12.12).

This conversion from ATP to ADP is a key reaction for the energy supply for life processes. Living things can extract energy from ATP like a battery. The ATP can power reactions by losing one of its phosphorous groups to form ADP, but then food energy in the mitochondria can be used to convert the ADP back to ATP, 'recharging' the battery. In plants, sunlight can be used to convert the less active compound back to the highly energetic form. For animals, the energy from high-energy storage molecules is used to maintain life; these are then recharged to their high-energy state. The oxidation of glucose operates in a cycle called the *Krebs cycle* in animal cells to provide energy for the conversion of ADP to ATP.

12.4   Cells

Some organisms, such as bacteria, are unicellular, consisting of a single cell. Other organisms, such as humans, consist of many cells (multicellular). Humans have an estimated 10^{14} cells; a typical cell has a size of $10\,\mu m$ and a mass of $1\,ng$. Each cell is self-contained and self-maintaining: it can take in nutrients, convert these into energy, carry out specialized functions, and reproduce as necessary. Each cell stores its own set of instructions for carrying out each of these activities. The contents of the cell are confined within a membrane that contains proteins and a lipid bilayer (Section 12.6). This regulates what moves in and out, and maintains the electric potential of the cell.

There are two types of cells, *eukaryotic* and *prokaryotic*. These differ greatly in their size, internal structure, and their genetic and metabolic organization. Prokaryotic cells, which include all the different species of bacteria, are comparatively primitive; these are very small and simple cells. On the other hand, eukaryotic cells, which include those of higher animals and plants, have volumes from 1000–10 000 times greater than those of prokaryotic cells.

Cells that have been attracting much attention are the *stem cells*. These have two important characteristics that distinguish them from other types of cells. First, they are unspecialized cells that renew themselves for long periods through cell division. The second is that under certain physiological or experimental conditions, they can be induced to become cells with special functions such as the beating cells of the heart muscle or the insulin-producing cells of the pancreas. Stem cells are primal cells common to all multi-cellular organisms that retain the ability to renew themselves through cell division and can differentiate into a wide range of specialized cell types.

Two broad categories of mammalian stem cells exist: *embryonic stem cells*, derived from *blastocysts* (a blastocyst is an early-stage embryo – approximately four to five days old in humans and consisting of 50–150 cells) and *adult stem cells*, which are found in adult tissues. In a developing embryo, stem cells are able to differentiate into all of the specialized embryonic tissues. In adult organisms, stem cells and progenitor cells act as a repair system for the body, replenishing specialized cells. As stem cells can be readily grown and transformed into specialized tissues, such as muscles or nerves, through cell culture, their use in medical therapies has been proposed.

The framework of the cell is called the *cytoskeleton*. This helps to maintain the cell's shape during the uptake of external materials and the separation of daughter cells following cell division, and moves parts of the cell in processes of growth and mobility. The eukaryotic cytoskeleton is composed of microfilaments, intermediate filaments, and microtubules. There are a great number of proteins associated with them, each controlling a cell's structure by directing, bundling, and aligning the filaments.

Inside the membrane, a salty *cytoplasm* takes up most of the cell volume. All cells possess DNA and RNA, containing the information necessary to build various proteins. The total length of the entire DNA in a single human cell is around $2\,m$, equivalent to 5.5×10^9 base pairs. A human cell has genetic material in the *nucleus* (the nuclear genome). In humans, the nuclear genome is divided into 46 linear DNA molecules (23 pairs) called *chromosomes*. A *gene* is a segment of a chromosome that codes for a single polypeptide chain of a protein (Section 12.5). A gene may have anywhere from 300–6000 or more nucleotide pairs, giving molecular weights from about 10^5–2×10^6.

Within the cell, *organelles* (little organs) are discrete structures with specialized functions. There are many types of organelles, particularly in the eukaryotic cells of higher organisms. An organelle is to the cell what an organ is to the body. Organelles include mitochondria (Section 12.3.4) and *chloroplasts*, which are involved in photosynthesis. DNA is also present in certain cell organelles, particularly the mitochondria and chloroplasts

12.5   Genetic Coding

The cell's full complement of DNA, i.e. the organism's genetic blueprint, is called its *genome*. As noted above, this does not reside on one large single strand of DNA but is divided up into several sections, each of which is contained in a chromosome. In the case of the human genome, the author Matt Ridley makes the analogy with a book [1]. If one supposes that the human genome is a book, then there are twenty three chapters, called chromosomes. Each chapter contains several thousand stories, the genes. Each story is made up of paragraphs, called *exons*, which are interrupted by advertisements known as *introns*. Each paragraph is made up of words called *codons* and, finally, each word is written in letters called bases. The function of genes is to carry the information required to manufacture the enzyme proteins that orchestrate the body's chemistry.

The information within individual genes resides in a coded form. The chemical structure of a protein can be written down in terms of its constituent amino acids. DNA can be represented as a sequence of base pairs linked by the sugar and phosphate components of nucleotides, which are simply part of the scaffolding. The sequence of base pairs within a gene on a DNA molecule can then represent a protein molecule in coded form.

There are 20 different amino acids that occur in proteins, but only four DNA bases. So a complete DNA-protein code can be established by taking groups of bases to represent each amino acid. A code based on 'one base to one amino acid' clearly will not work. A DNA code in which the characters are groups of just two bases will only provide $4 \times 4 = 16$ different elements – insufficient to encode all of the amino acids. But with groups of three bases, there are 64 possible characters, which is more than enough. The DNA-protein code must therefore use at least three bases to represent each amino acid.

All living organisms employ the same code. Because there is some redundancy in the code – there are 64 different base triplets available to represent 20 amino acids – some amino acids are encoded by more than one triplet. Furthermore, some triplets do not represent amino acids at all, but are control codes, which signify the end of the protein-coding sequence in a gene. Table 12.4 shows this coding structure. Thus alanine (Ala) can be represented by the codons GCU, GCC, GCA, or GCG, and lysine (Lys) can be represented by AAA or AAG.

12.5.1   Replication, Transcription, and Translation

Under the right conditions, the genome can both photocopy and read itself. The photocopying is known as *replication* and the reading as *translation*. Between the replication and translation processes is a further step, *transcription*. Here, the message in the DNA is transcribed into the form of RNA, to be carried by the ribosomes. Replication works because of the fundamental property of the four bases: A likes to pair with T, and G with C. So, a single strand of DNA can copy itself by assembling a complementary strand with Ts opposite all the As, As opposite all

Table 12.4 Sequences of DNA bases encode the information required for the synthesis of proteins from amino acids. The amino acids are those listed in Table 12.1.

		Second position				
		U	C	A	G	
First position	U	Phe	Ser	Tyr	Cys	U
		Phe	Ser	Tyr	Cys	C
		Leu	Ser	Stop	Stop	A
		Leu	Ser	Stop	Trp	G
	C	Leu	Pro	His	Arg	U
		Leu	Pro	His	Arg	C
		Leu	Pro	Gin	Arg	A
		Leu	Pro	Gin	Arg	G
	A	Ile	Thr	Asn	Ser	U
		Ile	Thr	Asn	Ser	C
		Ile	Thr	Lys	Arg	A
		Met	Thr	Lys	Arg	G
	G	Val	Ala	Asp	Gly	U
		Val	Ala	Asp	Gly	C
		Val	Ala	Glu	Gly	A
		Val	Ala	Glu	Gly	G

(Third position: column on the right showing U, C, A, G)

'Stop' is a codon that does not encode an amino acid but represents an instruction for protein synthesis to stop.

the Ts, Cs opposite all the Gs, and Gs opposite all the Cs. The usual state of DNA is, of course, the double helix consisting of the original strand and its complementary pair intertwined.

The first step in the progression from gene to protein is to produce an RNA version of the gene encoded on the DNA molecule, where the sequence of A, T, C, and G nucleotides in the gene is reproduced as a complementary sequence of U, A, G, and C nucleotides in the RNA. The relevant gene-bearing portion of the DNA double helix is unravelled and one of the single strands acts as a template for the construction of the mRNA molecule, as shown in Figure 12.13. The messenger RNA is then detached from the DNA strand, with the protein plan in the form of the sequence of bases.

Proteins are assembled on the mRNA template one amino acid at a time. Each three-base group on the mRNA, which corresponds to an amino acid, represents the codon. The tRNA molecules have one end that becomes anchored to a specific codon; this anchoring region on tRNA consists of a triplet of the base pairs that are complementary to those on the mRNA codon, and it is therefore called an anticodon. For example, the base sequence CGA on the mRNA codon will become anchored to the tRNA anticodon GCU, via the complementary pairs C----G, G----C, and A----U. The other end of the tRNA molecule binds specifically to the amino acid corresponding, in the genetic code, to the mRNA's codon. In the example above, the sequence CGA on mRNA produced from the sequence GCT in a gene on DNA, corresponds to the amino acid arginine (Table 12.4). Thus the tRNA molecule responsible for putting arginine into its place in the protein chain has the anticodon GCU at one end, and binds arginine at the other (Figure 12.14) [2].

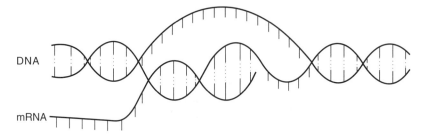

Figure 12.13 The genetic code is translated from DNA sequences to amino acid sequences via the mediation of RNA. RNA molecules containing the information in a single gene, called messenger RNA, or mRNA, are constructed by unwinding sections of the DNA double helix and using one of its unwound strands as a template. *Source:* From Ball [2]. Copyright (1994). Reprinted by permission of Princeton University Press.

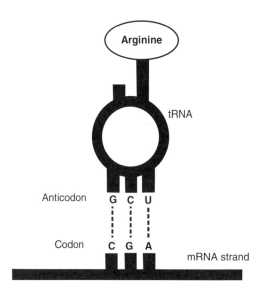

Figure 12.14 Function of transfer RNA or tRNA. One end of the tRNA molecule binds the appropriate amino acid through a complex molecular recognition process. The other end contains a sequence of three bases called an anticodon, which docks into the complementary codon sequence on the mRNA molecule. Here, the codon is that for arginine (CGA); see Table 12.4. *Source:* From Ball [2]. Copyright (1994). Reprinted by permission of Princeton University Press.

The linking together of the tRNA-attached amino acids at the mRNA template requires the assistance of the rRNA, in conjunction with other enzymes and proteins. Several rRNA molecules are bound together with many more proteins in the ribosome, the function of which it is to control this linking process. The ribosome binds to the mRNA and first facilitates the docking of the tRNA anticodon onto the mRNA codon. As successive tRNAs bring their respective amino acids to the mRNA for incorporation into a protein molecule, the ribosome moves along the mRNA chain, one codon at a time, so that it is always ready in the right place to receive the next tRNA in the sequence. The ribosome holds in place two successive tRNAs at a time. One of these will be attached to the growing protein chain, while the other carries the next amino acid to be inserted, via formation of a peptide bond, into the chain, as depicted in Figure 12.15 [2]. The ribosome holds together the end of the polypeptide chain and the next amino acid in just the right position for a peptide bond to form. The formation of this bond transfers the chain to the new tRNA; the old one is then released by the ribosome, which then shunts along to the next codon and is ready to receive the next tRNA. There are base sequences at either end of the mRNA that do not correspond to codons, but instead act as signals to tell the ribosome where to begin and where to end protein synthesis. Once the protein chain is completed, it is detached from the ribosome/mRNA complex, and enzymes destroy the mRNA, having accomplished its task.

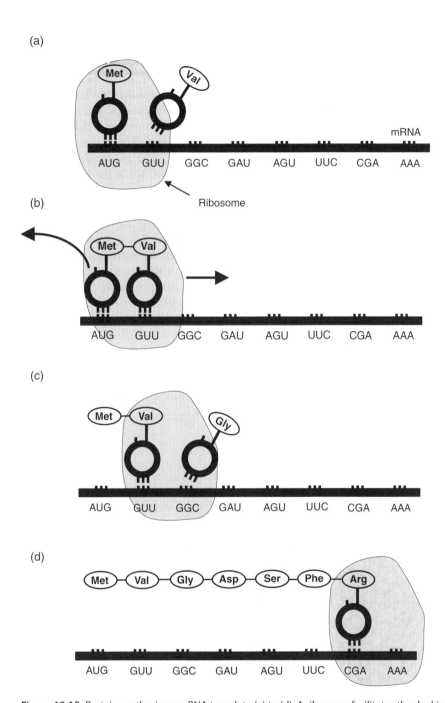

Figure 12.15 Protein synthesis on mRNA template (a) to (d). A ribosome facilitates the docking of amino acid-charged tRNA onto the mRNA codon. The amino acid on the tRNA is then linked to the growing protein chain via a peptide bond. *Source:* From Ball [2]. Copyright (1994). Reprinted by permission of Princeton University Press.

12.6 The Biological Membrane

Membranes are among the most important biological structures. Many of the key functions of living systems, e.g. the ability to maintain steady state conditions, are directly linked to the existence of a membrane around the cell. In 1972, Singer and Nicolson proposed the now widely accepted *fluid mosaic model* of the structure of cell membranes, illustrated in Figure 12.16 [3]. The hydrophilic polar groups associated with a phospholipid bilayer are on the outside, in contact with the aqueous media, while the hydrocarbon chains are in the interior. The model also proposes that integral membrane proteins are embedded in the bilayer. Some of these proteins extend all the way through the bilayer, whilst some only partially across it.

The biological bilayer can be symmetric in terms of the polar head groups but, more commonly, it exhibits asymmetry. Thus, for the red blood cell, sphingomyelin, and phosphatidylcholine are disproportionately located in the outward-facing monolayer, while phosphatidylserine and phosphatidylethanolamine are mainly in the monolayer facing the inside of the cell. Most membranes are electrically polarized with the negative inside the cell (typically −60 mV). The asymmetry will give rise to piezoelectric and pyroelectric behaviour, as noted in Chapter 11, Section 11.3.1 for LB films. The pyroelectric coefficients found in alternating LB films of phospholipids are modest [4], and it is unclear if these are exploited in physiological processes.

The exact relationship between lipid composition, organization, and function in biological systems has not yet been firmly established. Very few membrane proteins, for example, have an absolute requirement for a specific lipid, but their activity can vary considerably in different lipid mixtures. One important feature seems to be that the membranes should have a degree of fluidity. Biological membranes in their native state appear to permit rapid translational and

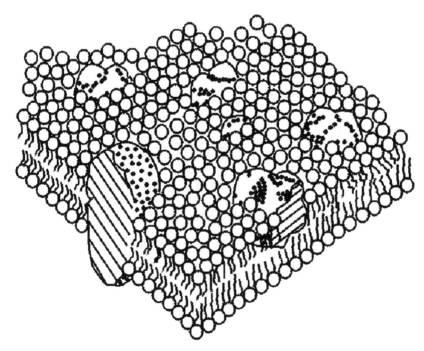

Figure 12.16 Model of a biological membrane. Protein molecules are shown embedded in and traversing the bilayer lipid structure.

rotational motion within the plane of the bilayer. In contrast, movement from one monolayer to the other is much more restricted. Lipids in cell membranes are therefore more likely to be in a liquid-crystalline rather than a condensed solid state. The most influential factor concerning fluidity is the nature of the esterified fatty acid. Shorter chain lengths and higher unsaturation leads to lower transition temperatures (solid to liquid), whilst longer and more saturated fatty acids have higher transition points (Section 12.2.5). The neutral lipid cholesterol (Figure 12.8d) reduces the lipid order in solid crystalline systems, while decreasing disorder in the more fluid situations. This feature is thought to be significant in maintaining the integrity of the biological membrane whilst maintaining its dynamic behaviour.

12.6.1 Transport Across the Membrane

Molecular and Ionic Transport

Membranes are highly selective permeability barriers. The movement of molecules across the lipid bilayer consists of transfer from one aqueous environment to another and is restricted to solute molecules and water (Figure 12.17). Generally, the smaller and less polar the molecule, the more easily it passes through the bilayer. Water is an exception; the small size of H_2O molecules (which offsets their large polarity) leads to a very rapid exchange of water across the bilayer structure. Gases, such as oxygen and carbon dioxide, important in cell metabolism, pass in or out of the cell in a dissolved state and the rates of transfer are determined by the extent to which the gases are soluble in the aqueous environment. Carbon dioxide is very soluble in water and therefore passes freely through membranes. In contrast, oxygen is less soluble and this becomes a limiting factor in cellular metabolism.

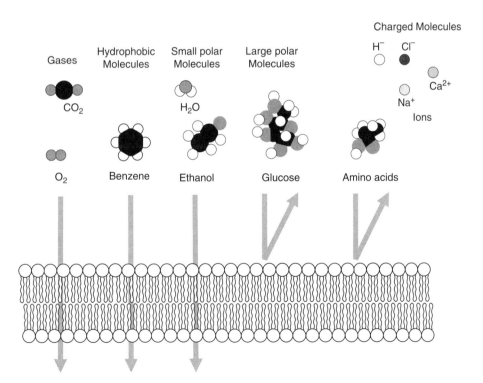

Figure 12.17 Schematic diagram showing the permeability of a lipid bilayer to different molecules. The smaller and less polar the molecule, the more easily it passes through the bilayer. Water is an exception.

The existence of a concentration gradient of solute molecules across a membrane tends to cause a net movement of solute molecules in the direction of this concentration gradient. Of course, transport occurs in both directions and the net flux is the sum of these two movements. In the simplest case, the rate of flow, the flux J (mol m^{-2} s^{-1}) of uncharged molecules in the direction of the gradient can be described by Fick's Law of diffusion (Chapter 2, Section 2.8).

Movements of ionized solutes are also influenced by electrical gradients. The flow of solute may still be described in simple terms by the *Nernst-Planck equation*:

$$J = -\mu c \left(\frac{k_B T}{ce} \frac{dc}{dx} + z \frac{dV}{dx} \right) \tag{12.2}$$

where μ is the mobility of the ion (m^2 V^{-1} s^{-1}), z is number of (electron) charges on the permeating molecule (its valence), and dV/dx is the electrical potential gradient across the molecule. The first term on the right-hand side of Eq. (12.2) is due to diffusion, while the second term originates from drift (electric field effect). The mobility of the ion can also be expressed in terms of the diffusion coefficient, D, using the Einstein relation (Eq. (3.69) in Chapter 3).

Osmosis is a special example of diffusion. This phenomenon is restricted to liquid media and is the diffusion of a solvent (normally water) through a *semipermeable membrane* from a more dilute solution to a more concentrated solution. A semipermeable membrane is a barrier that permits the passage of some substances (by size) but not others. A cell membrane can be considered as being *selectively permeable*, i.e. the membrane 'chooses' what passes through. If a semipermeable membrane separates two solutions of different concentrations, then the solvent will tend to diffuse across the membrane from the less concentrated to the more concentrated solution. Osmosis is an essential process in the natural world; a good example is the absorption of water by plant roots. Diffusion and osmosis are examples of *passive* transport, whereby ions or molecules driven by thermal motion move down concentration gradients set up between solutions separated by biological membranes in living systems.

In some instances, very rapid diffusion of material can take place across a membrane. Here, other constituents of the membrane, usually the proteins, play an important role. In a simple case, the 'transport' protein possesses a specific binding site that recognizes the substance to be translocated. Non-covalent association of the substance triggers structural changes in the protein that effectively allow it to move to the other side of the membrane, illustrated in Figure 12.18. This process is *facilitated diffusion*. The degree of movement of the transported entity on the protein surface may be quite small – a few tenths of a nanometre – and it is probably not correct to envisage a permanent pore or hole through the protein. Where the flow is down a concentration gradient, no energy input is needed, but when the flow is up the gradient, some form of energy is needed to produce the desired conformational state in the protein.

A good example occurs with bacteriorhodopsin, a proton-translocator from the membrane of *Halobacterium halobium* (Section 12.10.1). Like visual rhodopsin, the protein absorbs light energy via bound *cis*-retinal, but unlike the visual pigment, the energy is used to promote proton transfer across the lipid bilayer, through conformational changes in the protein molecule. A mechanism has been proposed whereby the proton is transferred from amino acid side chain to amino acid side chain through the protein. The three-dimensional nature of the protein is constructed such that this transfer occurs readily. A similar process may occur for other proton-transporting systems, but for larger substances such as glucose, this shuttle-type procedure is less likely.

Ion permeabilities of lipid bilayers and of natural membranes can be greatly increased by the incorporation of a range of small molecules called *ionophores* (simple crown ether ionophores have been described in Chapter 10, Section 10.4.5). Such materials have been used as *antibiotics*

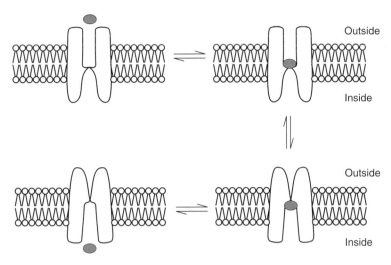

Figure 12.18 Mechanism for material transport across the bilayer. The non-covalent association of a substrate molecule with the transporter protein triggers structural changes that result in transport to the other surface of the membrane. Only a small structural change may be required for this.

(a drug that kills or slows the growth of bacteria). Many ionophores form stable complexes with cations. The nonpolar groups of the ionophore molecule are directed outward so that the ion becomes enclosed in a purse-like structure with a polar lining and a nonpolar exterior. One example is valinomycin, which is capable of selectively complexing with and transporting potassium ions across both biological and synthetic membranes. Complexation is associated with a change in conformation of the ionophore, facilitating this transport. A diagram of the valinomycin molecule and in its complexed state (with potassium), is shown in Figure 12.19. The molecule is a 12-membered macrocyclic ring of alternating D- and L-amino acids and α-hydroxyacids. There are three repeat units, each consisting of L-valine, L-lactic acid, D-valine and D-hydroxyisovaleric acid alternately joined by amide C=O and ester C=O linkages. On complexation with a potassium ion, the valinomycin molecule very subtly changes its conformation, and becomes more hydrophobic. During this process, the water molecules associated with the ions are replaced one by one as the ions become coordinated with the ionophore:

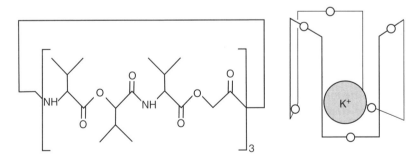

Figure 12.19 Schematic representation of the ionophore valinomycin (left) and its complex with potassium (right). The valinomycin molecule is composed to three identical segments linked to form a 36-membered ring. There are six amide groups and six ester groups arranged alternately around the ring. Nine isopropyl and three methyl side groups are also attached to the ring. The valinomycin-K^+ complex exhibits a so-called 'tennis-ball-seam' conformation.

(a)

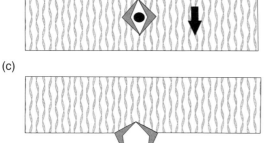

Figure 12.20 Transport of potassium ions (shown as filled circles) across a bilayer membrane. On complexation with a potassium ion (a), the valinomycin molecule subtly changes its conformation (b), and becomes more hydrophobic and moves across the membrane (c).

(b)

(c)

this stepwise displacement of the water of hydration reduces the potential energy barrier to penetration into the membrane. The process of ionic transport across the membrane is illustrated in Figure 12.20.

In other transport systems, ions moving by facilitated diffusion can traverse the cell membrane through channels created by proteins. These embedded transmembrane proteins allow the formation of a concentration gradient between the extracellular and intracellular contents. Ion channels are highly specific filters, even between ions of a similar character, e.g. Na over K, allowing only desired ions through the cell membrane. The specificity of an ion channel is a well-researched topic, although in some cases surprisingly little is known about the precise mechanism of ion channel filters. These ion channels are said to be 'gated' if they can be opened or closed. There are three types of gated ion channels: ligand gated; mechanically gated; and voltage gated. Ligand gated channels open or close in response to the binding of a small signalling molecule or ligand. Some ion channels are gated by extra cellular ligands; some by intracellular ligands. In both cases, the ligand is not the substance that is transported when the channel opens. The binding of neurotransmitter acetylcholine opens sodium channels in certain synapses. Voltage gated channels are found in neurons (Section 12.7) and muscle cells; these open or close in response to voltage changes that occur across the membrane. For example, as an impulse passes down a neuron, the reduction in the voltage opens sodium channels in the adjacent portion of the membrane. This allows the influx of Na^+ into the neuron and thus the continuation of the nerve impulse.

Electron Transport Systems

Many biological processes require the efficient transport of electrons. Very often this occurs in conjunction with, but separate from, the transport of protons. Most of the electron transport systems in biology are involved in some way with bioenergetics, the best characterized structurally being the photosynthetic system (Section 12.10.2).

Electron-transferring reactions, such as those taking place in the respiratory chain, are more generally redox reactions (Chapter 3, Section 3.4.2). These proceed with a transfer of electrons from an electron donor (reducing agent) to an electron acceptor (oxidizing agent). In some oxidation–reduction reactions, the transfer of electrons is made via transfer of hydrogen atoms, each of which carries an electron. In this case, dehydrogenation is equivalent to oxidation and hydrogenation is equivalent to reduction. In other oxidation–reduction reactions, both an electron and a hydrogen atom may be transferred.

Important biological molecules involved in redox processes are the coenzymes nicotinamide adenine dinucleatide (NAD) and its phosphorylated derivative nocotinamide adenine dinucleotide phosphate (NADP). The reduced forms are usually abbreviated NADH and NADHP. These compounds catalyse the following general reactions.

$$\text{reduced substrate} + \text{NAD}^+ \Leftrightarrow \text{oxidized substrate} + \text{NADH} + \text{H}^+ \tag{12.3}$$
$$\text{reduced substrate} + \text{NADP}^+ \Leftrightarrow \text{oxidized substrate} + \text{NADPH} + \text{H}^+$$

The *cytochromes* are a group of iron-containing, electron-transferring proteins that act sequentially in the transport of electrons. These contain iron porphyrin groups and resemble haemoglobin and mysoglobin; all are members of the class of heme proteins. Porphyrins (very similar to the phthalocyanine molecules described in Chapter 5, Section 5.4) are named and classified on the basis of their side-chain substituents. Protoporphyrin IX (Figure 12.21a) is the most abundant and contains four methyl groups, two vinyl groups, and two propionic acid groups. This is the porphyrin that is present in haemoglobin, myoglobin, and most of the cytochromes. Protoporphyrin forms very stable complexes with di- and trivalent metal ions. Such a complex of photoporphyrin with Fe(II) is called hemin, or hematin. The cytochromes undergo Fe(II) → Fe(III) valence changes during their function as electron carriers. In cytochrome *c*, the single iron-protoporphyrin group is covalently linked to the protein, as illustrated in Figure 12.21b.

12.7 Neurons

Neurons (also spelled neurones or called nerve cells) are the electrically-excitable primary cells of the nervous system. A human brain contains 10^{11} neurons. These register information from the

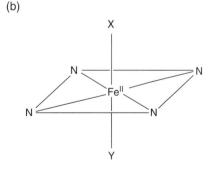

Figure 12.21 (a) Chemical structure of protoporphyrin IX. (b) Binding of an iron atom in cytochrome *c*. The four nitrogen atoms of the porphyrin ring bind to the iron in a planar arrangement. X and Y represent binding groups contributed by the protein.

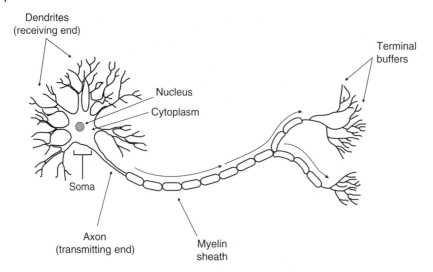

Figure 12.22 Schematic diagram showing the typical structure of a neuron.

environment, integrate and evaluate these data, and then decide whether electrical signals are further transmitted. Many specialized types of neurons exist, and these differ widely in appearance. Neurons are highly asymmetric in shape, and a typical architecture is shown in Figure 12.22. The *soma*, or cell-body, is the relatively large central part of the cell between the *dendrites* and the *axon*; this is the metabolic centre of the cell and the site of protein synthesis and production of energy (ATP). The cell body gives rise to two kinds of cellular extensions: several short dendrites and a single long axon. The dendrites branch out in a tree-like fashion and receive incoming signals from other neurons. Certain neurons in mammals have over 1000 dendrites each, enabling connections with tens of thousands of other cells. The axon is a much finer, wire-like projection, which may extend tens, hundreds, or even tens of thousands of times the diameter of the soma in length. This is the structure which carries nerve signals away from the neuron. Each neuron has only one axon, but this axon may undergo extensive branching and thereby enable communication with many target cells. Axon and dendrites are typically only about a 1 μm thick, while the soma is usually about 25 μm in diameter and not much larger than the cell nucleus it contains. The axon of a human motoneuron can be over 1 m long, reaching from the base of the spine to the toes.

Surrounding the axon is an electrically insulating layer called the *myelin sheath*. This is made up of protein and lipid. The purpose of the myelin sheath is to allow rapid and efficient transmission of impulses along the nerve cells. If the myelin is damaged, the impulses are disrupted; this can cause diseases like multiple sclerosis.

As is the case with every cell in the body, neurons are surrounded by a thin membrane, formed by a phospholipid bilayer. At rest, neurons maintain a difference in the electrical potential on either side of the membrane. The electrical signals are called *action potentials*, which constitute the signals by which the brain receives analyses and conveys information. Action potentials are brief (~1 ms duration) and relatively large amplitude (~100 mV) electrical pulses. The action potential is either present or not. While stimuli that do not reach a certain threshold value of the membrane potential produce no action potential, all stimuli above the threshold invariably generate the same signal.

The action potential is propagated without decay along the axon at a speed of up to $150\,\mathrm{m\,s^{-1}}$. Near to its end, the axon divides into fine branches that make contact to neighbouring neurons.

Figure 12.23 Schematic diagram of a synapse.

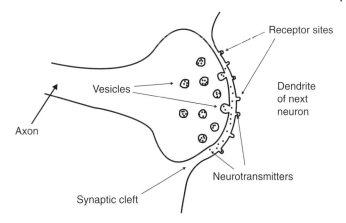

The point of contact between two communicating cells is called the *synapse*, shown schematically in Figure 12.23. The nerve cell transmitting the signal is called the *presynaptic cell*, while that receiving the signal is the *postsynaptic cell*. The arrival of an action potential at the tip of an axon triggers the release of neurotransmitter at a synaptic gap. Neurotransmitters can either stimulate or suppress (inhibit) the electrical excitability of a target cell. An action potential will only be triggered in the target cell if neurotransmitter molecules acting on their post-synaptic receptors cause the cell to reach its threshold potential.

The transmembrane voltage changes that take place during an action potential result from changes in the permeability of the membrane to specific ions, the internal and external concentrations of which cells maintain in an imbalance. In the axon fibres of nerves, depolarization results from the inward rush of sodium ions, while repolarization and hyperpolarization arise from an outward rush of potassium ions. Calcium ions make up most or all of the depolarizing currents at an axon's presynaptic terminus in muscle cells (including those in the heart) and in some dendrites. When a neuron has just generated an action potential, the cell is unable to generate another within a certain time span. This phase lasts a few milliseconds and is caused by the time it takes for the ion channels to recover.

12.8 Biosensors

Biosensors use biological or living materials to provide their sensing functions. In many ways they can be defined as a special type of chemical sensor (Chapter 10). Evolution of species by means of natural selection has led to extremely sensitive organs that can respond to the presence of just a few molecules. Artificial sensors exploit biologically active materials in combination with different physical sensing elements. The bio-recognition element works like a bio-reactor on the top of the conventional sensor. The response of the biosensor will be determined by the diffusion of the analyte, by the reaction products, by the co-reactants or interfering species, and/or by the kinetics of the recognition process. Organisms, tissues, cells, membranes, enzymes, antibodies, and nucleic acid can all be detected by means of a bio-sensor. Biosensors find applications in medicine, food and process control, environmental monitoring, and defence and security. However, the market is driven mainly by medical diagnosis and, in particular, glucose sensors for people with diabetes. The most significant trend likely to impact on development of biosensors is the emergence of personalized medicine.

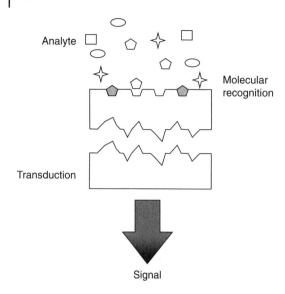

Figure 12.24 Molecular recognition and signal transduction parts of a biosensor.

One of the key issues for biological sensing systems is the immobilization of the active element on the physical transducer. The biologically active material must be confined to the sensing element and kept from 'leaking', while allowing contact with the analyte solution. Furthermore, the reaction products must readily diffuse out of the sensing layer so as not to denature its biologically-active characteristics. Many sensing materials are proteins or contain proteins in their chemical structure. Two techniques employed to immobilize the proteins are binding (adsorption or covalent binding) and retention. Physical retention involves separating the biologically active material from the analyte solution with a layer on the surface of the sensor, which is permeable to the analyte and to any products of the recognition reaction, but not to the biologically active materials. A biosensor usually has two functional parts, for molecular recognition and signal transduction, illustrated in Figure 12.24. To achieve a high selectivity, either a biocatalyst or bioaffinity (immunological) material can be exploited as the molecular recognition element [5, 6].

12.8.1 Biocatalytic Sensors

A biocatalyst recognizes the corresponding substrate (analyte) and immediately generates products by a specific reaction. The complex of the catalyst and the analyte remain stable in a transition state. A change in either the analyte or product is detected in the signal transducing device of the biosensor. Redox enzymes are recognized as the major material in constructing both biocatalytic and bioaffinity sensors. Biocatalytic sensors for glucose, lactate, and alcohol utilize glucose oxidase, lactate dehydrogenase (lactate oxidase), and alcohol dehydrogenase (alcohol oxidase) as the molecular recognizable material. Since these redox enzymes are mostly associated with the generation of electrochemically active substances, many electrochemical enzyme sensors have been developed by linking redox enzymes for molecular recognition with electrochemical devices for signal transduction. Figure 12.25 shows a schematic representation of this process.

An amperometric sensor described in Chapter 10, Section 10.4.2 is an example of an electrochemical sensor. The glucose sensor is probably the most common biosensor of this type. The basic structure consists of either an oxygen or hydrogen peroxide electrode, covered with a glucose oxidase membrane. Glucose oxidase catalyses the following reaction:

$$\beta - \text{D} - \text{glucose} + \text{O}_2 + \text{H}_2\text{O} \rightarrow \beta - \text{D} - \text{gluconate} + \text{H}_2\text{O}_2 \tag{12.4}$$

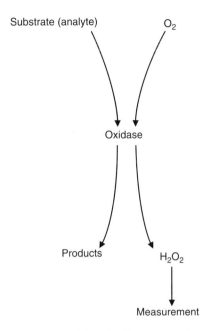

Substrate (analyte) O_2

Oxidase

Products H_2O_2

Measurement

Figure 12.25 Principle of amperometric enzyme sensors.

Glucose in a sample solution is oxidized with a resulting consumption of oxygen when contacted with the membrane-bound glucose oxidase. The decrease of dissolved oxygen is sensitively detected with the oxygen electrode. The output change of the sensor reflects the concentration of glucose in the solution. Hydrogen peroxide is generated in the glucose oxidase-catalysed reaction and the detection of this can be performed using a platinum anode (polarized at ~0.7 V vs. Ag/AgCl). The sensor responds linearly to the hydrogen peroxide, the output current being correlated with the analyte concentration.

Nanotechnology is now making an impact on biosensor development. For example, carbon nanotubes can be blended into a number of sensing layer formulations to improve current densities and performance of enzyme electrodes. However, the most widely used nanomaterial for commercial biosensors is the silver nanoparticle. These have been exploited as a simple electrochemical label in a highly sensitive and inexpensive amperometric immunoassay intended for distributed diagnostics [6].

12.8.2 Bioaffinity Sensors

A bioaffinity sensor involves an antibody, binding protein, or receptor protein, which forms a stable complex with a corresponding ligand. The bioaffinity protein–ligand complex formation is sufficiently stable to result in signal transduction. Immunosensors take advantage of the high selectivity provided by the molecular recognition of antibodies. Because of significant differences in affinity constants, antibodies may confer an extremely high sensitivity to immunosensors in comparison to enzyme sensors. Furthermore, antibodies may be obtained (in principle) for an unlimited number of determinants. Immunosensors are therefore characterized by high selectivity, sensitivity, and versatility.

Immunosensors can be divided in principle into two categories: nonlabelled and labelled immunosensors. Nonlabelled immunosensors are designed so that the immunocomplex, i.e. the antigen–antibody complex, is directly determined by measuring the physical changes induced by the formation of the complex. An example is the use of the surface plasmon resonance technique, described in Chapter 4, Section 4.7.2, to monitor the change in thickness or optical constants on formation of the antigen–antibody complex.

Nonlabelled immunosensors are based on several principles, as illustrated in Figure 12.26. Either the antibody or the antigen can be immobilized on the solid matrix to form a sensing device. The solid matrix should be sensitive enough, in its surface characteristics, to detect the immunocomplex formation. Electrodes, membranes, piezoelectric materials, or optically active surfaces may be used to construct non-labelled immunosensors. The antigen or the antibody to be determined is dissolved in a solution and reacts with the complementary matrix-bound antibody or antigen to form an immunocomplex. This formation changes the physical properties of the surface, such as the electrode potential, the transmembrane potential, the piezoelectric resonant frequency (Chapter 10, Section 10.4.5), or the refractive index. A sufficiently high selectivity may be obtained with non-labelled

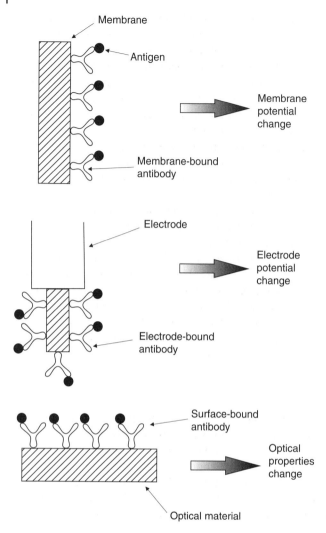

Figure 12.26 Principles of nonlabelled immunosensors.

immunosensors, although such problems as non-specific adsorption onto the matrix-bound antibody surface remain unresolved.

An obvious target for electrochemical affinity sensors is DNA. The advent of the 'DNA chip' has focused attention on alternatives to fluorescence detection. These exploit the fact that short strands of DNA will bind to other segments of DNA that have complementary sequences, and can therefore be used to probe whether certain genetic codes are present in a given specimen of DNA. Different types of micro/nano-fluidic technologies have facilitated DNA purification, amplification, and detection to be integrated into one chip, which combine the advantages of automation, small sizes, much shorter reaction times and reduced cost [7].

In a labelled immunosensor, a sensitively detectable label is incorporated; examples include enzymes, catalysts, fluorophores, electrochemically active molecules, and liposomes.

Optical sensors make use of the effect of chemical reactions on the optical properties of a material (Chapter 10, Section 10.4.5). If the solid surface is sufficiently sensitive to allow changes in its optical properties with immunocomplex formation, optical immunosensors

Figure 12.27 Schematic diagram of an evanescent fibre optic sensor used to excite and collect the fluorescence of surface bound avidin to a biotinylated phospholipid. *Source:* Reprinted with permission from Zhao and Reichert [8]. Copyright (1992) American Chemical Society.

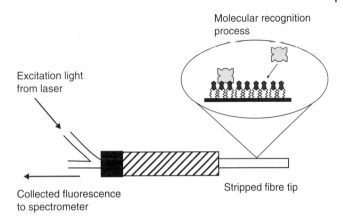

without a label may be constructed. Surface plasmon resonance is so sensitive that an immunocomplex may be detected on the surface of a solid coated with a suitable metal. This principle is exploited in several commercial systems (Chapter 4, Section 4.7.2). Figure 12.27 shows an example of a biosensor based on an optical fibre [8]. The antibody is immobilized on the surface of the fibre core, while a fluorescence compound is used as the label. Both labelled antigen and free antigen react competitively with the bound antibody to form an immunocomplex on the core surface. The surface-bound label can be excited by an evanescent wave that passes through the optical fibre core. On the other hand, fluorescence labels in the bulk solution cannot be excited, even if the excitation beam comes through the optical fibre core. Labels attached to the surface-bound immunocomplex are thus discriminated from labels in solution.

The feasibility of single molecule detection is currently under intense scrutiny. The most widely discussed example is DNA nanopore technology for DNA sequencing [9]. The concept is to observe the ion current across a nanopore, which is extremely sensitive to changes in the shape and size caused by single nucleotides passing through it. More recent work in this area has explored the possibility of sequencing DNA by passing the molecule through nanopores in a sheet of graphene [10].

12.9 DNA Electronics

The study of the electronic behaviour of organic compounds has led some scientists to work on the electrical properties of biological materials. DNA, described above in Section 12.3.2, is arguably the most significant molecule in nature. It can be considered as a potential candidate for nanowires and may also be an important material for molecular electronics applications. Reports into the electronic properties of DNA have already generated controversy in the literature. According to some, DNA is a molecular wire of very small resistance. Others, however, find that DNA behaves as an insulator. These seemingly contradictory findings can probably be explained by the different experimental conditions used to monitor the conductivity [11]. The DC resistivity of the DNA double helix over long length scales ($< 10\,\mu m$) is very high ($\rho > 10^6\,\Omega\,cm$). However, an appreciable AC conductivity can arise from the polarization of the surrounding water molecules [12]. Computer simulations indicate that charge transport along the molecule's long axis is strongly dependent on DNA's instantaneous conformation, varying over many orders of magnitude. It is also suggested that the charge transport can be active over longer length scales than the commonly accepted two to three base pairs [13].

Computations by chemical or biological reactions overcome the problem of parallelism and interconnections in a classical system. If a string of DNA can be put together in the right sequence, it can be used to solve combinational problems. The 'calculations' are performed in test tubes filled with strands of DNA. Gene sequencing is used to obtain the result. For example, Adleman have solved the 'travelling salesman' and other problems to demonstrate the capabilities of DNA computing [14]. Some excellent background reading on this can be found in the book by Amos [15]. Other examples include the application of a DNA-mediated multitasking processor to route planning [16] and an analogue processor that can add, subtract, and multiply [17]. DNA computing on parallel problems potentially provides 10^{14} MIPS (millions of instructions per second) and uses less energy and space than conventional supercomputers. While complementary metal-oxide-semiconductor (CMOS) supercomputers operate 10^9 operations per Joule, a DNA computer could perform about 10^{19} operations per Joule. The von Neumann limit (Chapter 1, Section 1.3.3, Eq. (1.4)) predicts a limit of about 3×10^{20} operations per Joule. Data could potentially be stored on DNA in a density of approximately 1 bit per nm^3, while existing storage media such as Dynamic Random Access Memory (DRAM) requires $10^{12} nm^3$ to store 1 bit.

12.10 Photobiology

12.10.1 Bacteriorhodopsin

Bacteriorhodopsin is one of the most studied proteins for use in bioelectronics and biometric devices. It is a compact molecular machine that pumps protons across a membrane powered by green sunlight. The molecule is the light harvesting protein in the purple membrane of a micro-organism *Halobacterium salinarum* (formerly *Halobacterium halobium*), which lives in extreme conditions, such as salt marches [18, 19]. The bacteriorhodopsin found in halobacteria is in the form of a two-dimensional crystal integrated into their cell membranes. The crystal-line structure causes a substantial increase in bacteriorhodopsin's chemical and thermal stability (up to 140 °C for dry layers).

Sunlight interacts with this protein to pump protons outwards across the cell membranes, making the inside more alkaline than the outside. By this process, light energy is converted to chemical energy and a proton gradient across the membrane is established, which may be used for energy storage. A membrane bound ATPase regenerates ATP from ADP, which then powers the cell.

Bacteriorhodopsin consists of 248 amino acids, arranged in 7 α-helical bundles inside the lipid membrane, which form a cage. At the heart of the bacteriorhodopsin is a molecule of all-*trans* retinal (Vitamin A aldehyde) that is bound deep inside the protein and connected through a lysine amino acid. Retinal contains a string of carbon atoms, and these strongly absorb light. When a photon is absorbed, it causes a change in the conformation of the molecule to its *cis* form (photoisomerization), as depicted in Figure 12.28. This change from a straight form to a bent form drives the pumping of protons.

After the photon absorption and proton transfer, the bacteriorhodopsin recovers to its initial conformation. The entire sequence of transitions is called the photocycle. The precise details of this are a matter of some debate; a simplified version is depicted in Figure 12.29 [18, 19]. Each photocycle lasts 10–15 ms and results in the net translocation of a single proton across the membrane. There are a number of spectrally distinct intermediates, labelled *bR*, *K*, *L*, M_1, M_2, *N*, and *O*; Figure 12.29 indicates the absorption maximum for each state. The initial resting state of the molecule is known as *bR*. Green light transforms this initial state to the intermediate

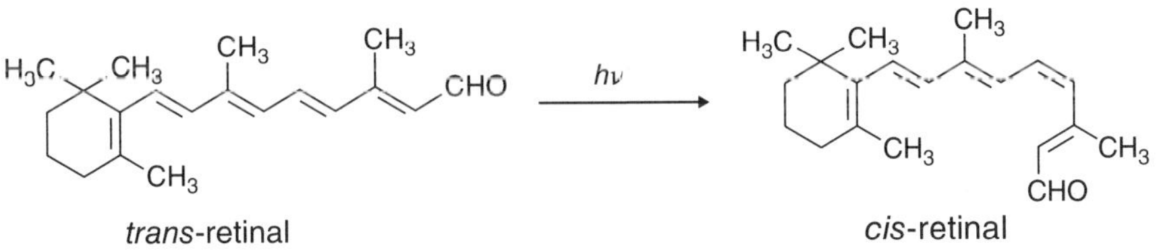

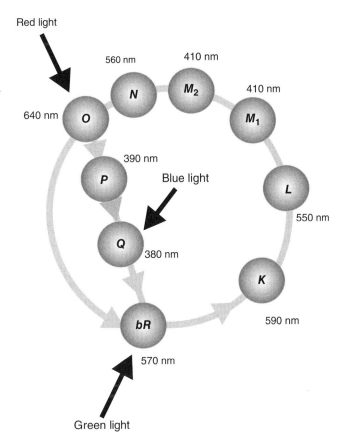

Figure 12.28 Light-induced transformation of retinal from a *trans* to *cis* configuration.

Figure 12.29 Photocycle of bacteriorhodopsin. Green light transforms the resisting *bR* state to the intermediate *K*. Next *K* relaxes, forming *L*, M_1, M_2, *N*, and then *O*. If the *O* intermediate is exposed to red light, a so-called branching reaction occurs. Structure *O* converts to the *P* state, which quickly relaxes to the *Q* state – a form that remains stable almost indefinitely. Blue light, however, will convert *Q* back to *B* [18, 19].

state *K*. Next *K* relaxes, forming other intermediate states *L*, M_1, M_2, *N*, and *O* and finally back to *bR*. If the intermediate state *O* is irradiated with red light, it converts to a further state *P* (a branching reaction), which then relaxes to a very stable state *Q*. Blue light converts *Q* back to the initial *bR* state.

The basic molecular functions of bacteriorhodopsin and their corresponding physical effects are depicted in Figure 12.30 [20]. The proton transport is initialized by proton absorption and a charge separation step on the picosecond timescale. After about 50 µs, the main colour change occurs and after 10–15 ms the proton transport is completed.

The various photochemical changes are under study for exploitation in many optical devices based on bacteriorhodopsin. The cyclicity of the protein (i.e. the number of times that the protein can be photochemically cycled before denaturing) exceeds 10^6, a result of its evolution

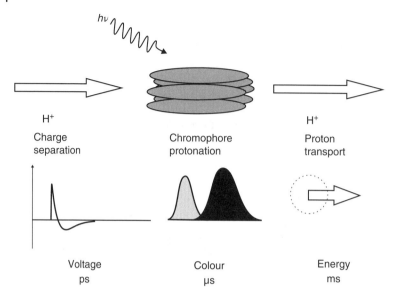

Figure 12.30 Basic molecular functions of bacteriorhodopsin. The proton transport is initialized by photon absorption and a charge separation step on the picosecond time scale. After about 50 μs, the deprotonation of the Schiff base leads to the main photochromic shift during the photocycle. After about 10 ms, the proton transport is completed. *Source:* Reprinted from Hampp [20]. Copyright (2000) American Chemical Society.

in a harsh environment with high salinity. For optical storage applications, the principle is to assign any two long-lasting states of the protein to the binary values of '0' and '1', in order to store the required information. The prospects for three-dimensional data storage seem more promising than for two-dimensional storage (Chapter 11, section 11.6). There are three different types of volume storage under investigation. The first is page-oriented holographic storage; the second is based on the branched photocycle scheme; and the third uses two-photon excitation of the individual data points in the volume of the material. The storage capacity of such memories is very high (10 GByte). The limitation of capacity is mainly connected with problems of lens system and quality of protein. However, it remains unclear whether this type of memory will complete with hard disks or semiconductor memory, as a number of problems in producing commercial optical memories remain.

12.10.2 Photosynthesis

Food and fossil fuel are the products of photosynthesis, the process that converts the energy in sunlight to chemical forms of energy, which can then be used by biological systems. Photosynthesis takes place in many different organisms, from plants to bacteria. The best known example of photosynthesis is that carried out by higher plants and algae, which are responsible for a major part of photosynthesis in oceans. All these organisms convert carbon dioxide to organic material by reducing the gas to carbohydrates in a rather complex set of reactions.

Sunlight is absorbed and converted initially to electronic excitation energy. This starts a chain of electron-transfer events leading to charge separation across a photosynthetic membrane. The resulting potential energy is then used to pump protons across the membrane, generating an osmotic and charge imbalance, which in turn powers the synthesis of ATP. For the sugar glucose (one of the most abundant products of photosynthesis), the relevant equation is

$$6CO_2 + 12H_2O \rightarrow C_6H_{12}O_6 + 6H_2O + 6O_2 \tag{12.5}$$

Light provides the energy to transfer electrons from water to $NADP^+$ forming NADPH; and to generate ATP. ATP and NADPH provide the energy and electrons to reduce CO_2 to organic molecules. Electrons for this reduction reaction ultimately come from water, which is then converted to oxygen and protons.

The first step in photosynthesis is the absorption of light by pigments such as primarily chlorophylls and carotenoids; the chemical structures of chlorophyll *a* and β-carotene are shown in Figure 12.31. Chlorophylls absorb blue and red light and carotenoids absorb blue-green light, typically as shown in Figure 12.32. However, green and yellow light are not effectively

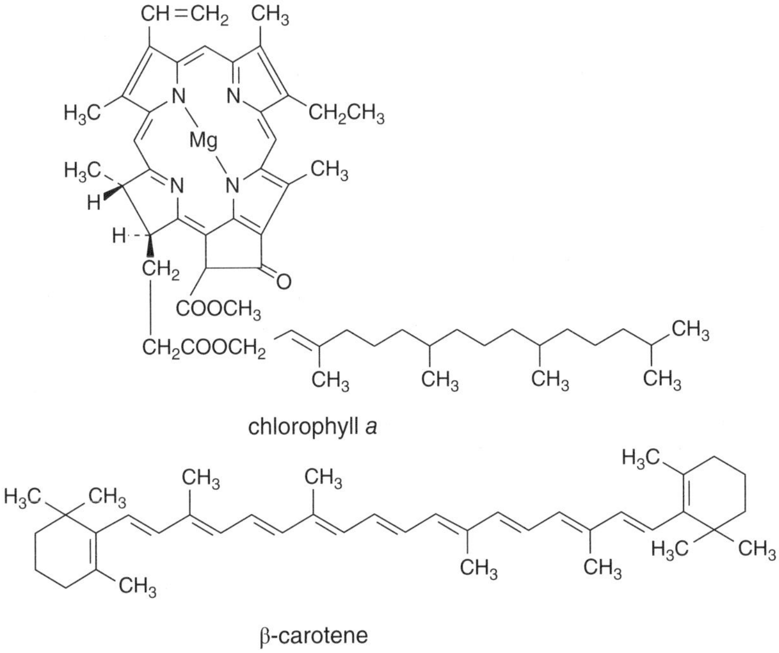

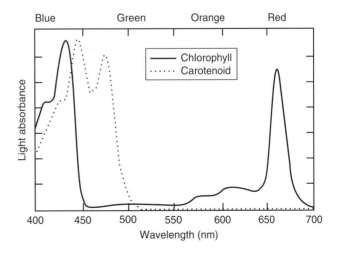

Figure 12.31 Chemical structures of chlorophyll *a*- and β-carotene.

Figure 12.32 Optical absorption spectra of chlorophyll and carotenoids.

absorbed by photosynthetic pigments in plants; therefore, light of these colours is either reflected by leaves or passes through the leaves (why plants are green and carrots are orange).

Other photosynthetic organisms have additional pigments that absorb the colours of visible light, which are not absorbed by chlorophyll and carotenoids. For example, some organisms contain bacteriochlorophyll, absorbing in the infrared in addition to the blue part of the spectrum. These bacteria do not evolve oxygen, but perform photosynthesis under *anaerobic* (oxygen-less) conditions using the infrared light.

Reaction Centres and Antennae

Photosynthetic pigments are normally bound to proteins, which provide the pigment molecules with the appropriate orientation and positioning with respect to each other. Light energy is absorbed by individual pigments, but is not used immediately for energy conversion. Instead, the light energy is transferred to chlorophylls that are in a special protein environment where the actual energy conversion event occurs. The pigments and proteins involved with this primary electron transfer event are together called the *reaction centre*. A large number of pigment molecules (100–5000), collectively referred to as antenna, 'harvest' light and transfer the light energy to the same reaction centre. The purpose is to maintain a high rate of electron transfer in the reaction centre, even at lower light intensities. Antennas permit an organism to increase significantly the absorption cross-section for light without having to build an entire reaction centre and associated electron transfer system for each pigment, which would be very costly in terms of cellular resources. The photosynthetic antenna system is organized to collect and deliver excited state energy by means of excitation transfer to the reaction centre complexes where photochemistry takes place. The process is illustrated in Figure 12.33.

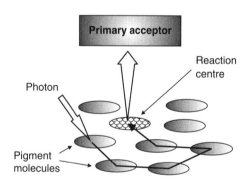

Figure 12.33 Schematic diagram of a photosynthetic reaction centre. A large number of pigment molecules 'harvest' the light and transfer the energy to the same reaction centre.

Many antenna pigments transfer their light energy to a single reaction centre by transmitting the energy to another antenna pigment, and then to another, until the energy is finally trapped in the reaction centre. Each step of this energy transfer must be very efficient to avoid a large loss in the overall transfer process. The association of the various pigments with proteins ensures that transfer efficiencies are high by having the pigments close to each other, and by providing an appropriate molecular geometry of the pigments with respect to each other.

The overall process of photosynthesis is highly complex. Figure 12.34 shows the important reaction steps, which are described in the following sections.

Photosynthetic Electron Transfer

All photosynthetic organisms that produce oxygen have two types of reaction centres, named photosystem II and photosystem I (PS II and PS I), both of which are pigment/protein complexes located in specialized membranes called *thylakoids*. In eukaryotes (plants and algae, Section 12.4), these thylakoids are located in chloroplasts (Section 12.4) and often are found in membrane stacks. Photosystem I, which is activated by light at 680 nm, is associated with chlorophyll *a* and is not involved in oxygen evolution. Photosystem II, which is activated by shorter

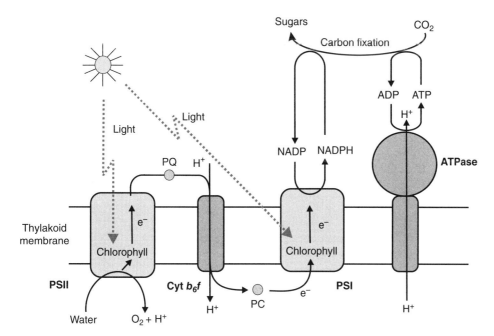

Figure 12.34 Process of photosynthesis.

wavelengths of light, between 500 and 600 nm, appears to be involved in oxygen evolution; it uses a second type of chlorophyll as well as accessory pigments.

Upon oxidation of the reaction centre chlorophyll in PS II, an electron is transferred from a nearby amino acid (tyrosine) that is part of the surrounding protein, which in turn gains an electron from the water-splitting complex. From the PS II reaction centre, electrons flow to free electron-carrying molecules in the thylakoid membrane (plastoquinione–PQ in Figure 12.34) and from there to the cytochrome $b_6 f$ complex. Finally, the electrons are transported to the PS I centre by a small protein (plastocyanin–PC in Figure 12.35).

The other photosystem, PS I, also catalyses light-induced charge separation in a fashion basically similar to PS II: an antenna harvests light, and light energy is transferred to a reaction centre chlorophyll, where light-induced charge separation is initiated. However, in PS I, electrons are transferred eventually to NADP, the reduced form of which can be used for carbon fixation. The oxidized reaction centre chlorophyll eventually receives the electron from the cytochrome complex. Therefore, electron transfer through PS II and PS I results in water oxidation (producing oxygen) and NADP reduction, with the energy for this process provided by light (two quanta for each electron transported through the whole chain).

To summarize the rather complex process of photosynthesis:

a) Light is absorbed by pigments, e.g. chlorophyll.
b) Electron transport is accomplished by membrane-bound protein complexes.
c) Photosystem II absorbs light, oxidizes water, and reduces plastoquinone.
d) Cytochrome $b_6 f$ complex pumps H^+ across the thylakoid during electron transport.
e) Photosystem I absorbs light, reduces $NADP^+$ (to NADPH), and oxidizes plastocyanin.
f) The process of light-driven electron transport and water oxidation generates a H^+ gradient across the thylakoid membrane.
g) The flow of H^+ through an ATP synthase protein drives ATP synthesis.

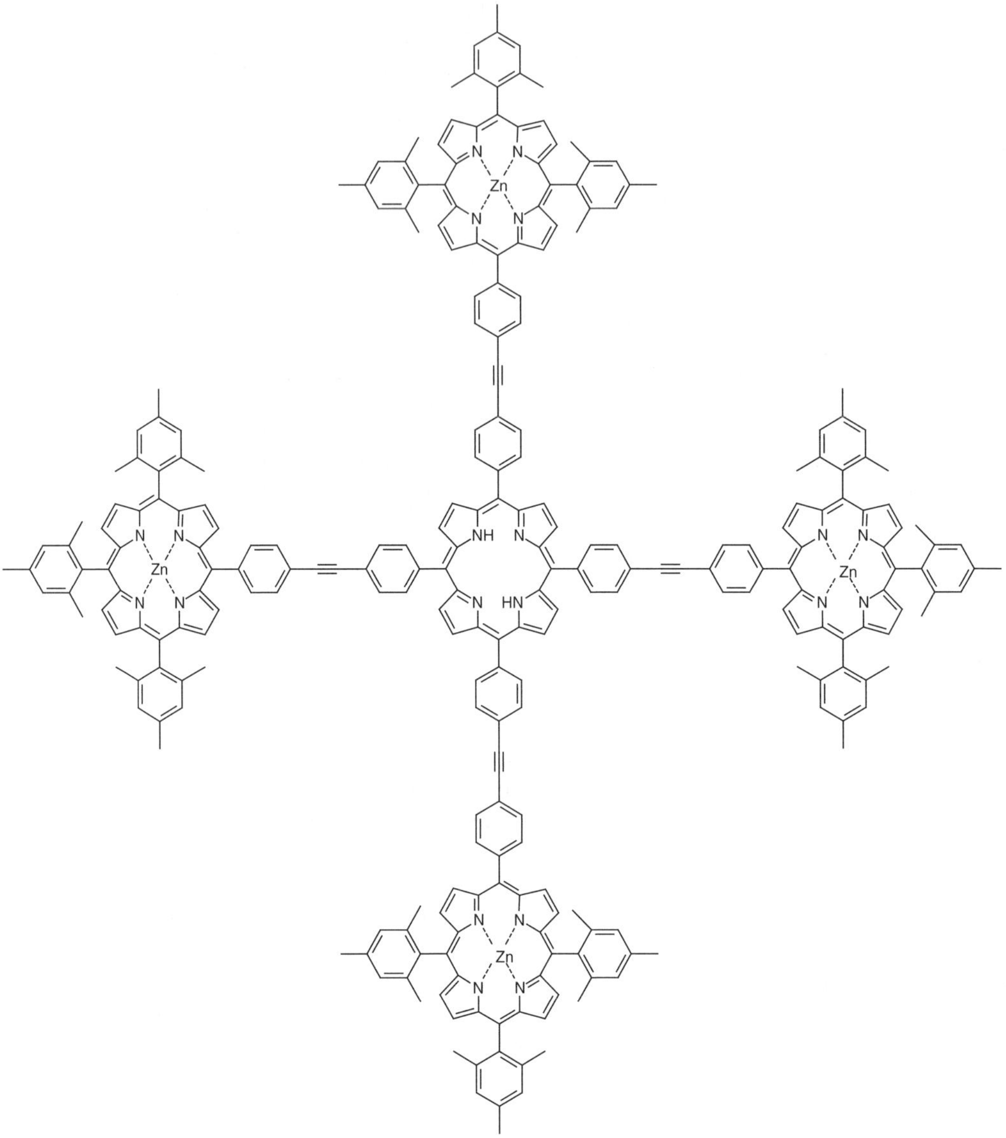

Figure 12.35 A star-like pentameric array of porphyrins for light harvesting [23].

The conversion of carbon dioxide into organic compounds during photosynthesis is called *carbon fixation*. Electron flow from water to NADP requires light and is coupled to generation of a proton gradient across the thylakoid membrane. This proton gradient is used for the synthesis of ATP. ATP and reduced NADP that result from the light reactions are used for CO_2 fixation in a process that is independent of light.

Artificial Photosynthetic Systems

In nature, plants use sunlight to make carbohydrates from carbon dioxide and water. Artificial photosynthesis seeks to use the same inputs – solar energy, water, and carbon dioxide – to produce energy-dense liquid fuels. The estimated maximum efficiency for photosynthesis is 4.5% [21]. Although it may be possible to engineer plants and other types of photosynthetic organisms (algae) as energy-converting machines, the overall efficiency of solar energy conversion is likely to be much less than this figure. However, the efficiencies of the early photochemical and chemical stages of photosynthesis, which are not directly involved in biomass production, are significantly higher and are the subject of much of the research on artificial photosynthesis.

It may be possible to develop a highly efficient, artificial, molecular-based, solar-energy-converting technology that exploits the principles of the 'front-end' of natural photosynthesis. For example, reaction centres are simply assemblies of cofactors (Section 12.2.3) held in the appropriate position and orientation by the protein environment. This natural system has been used as a model to design organic molecules, where the equivalents of the different cofactors are linked together by covalent bonds of various lengths. The result is the creation of a number of sophisticated molecules that serve as artificial reaction centres. The more advanced molecules consist of two chlorophyll-type molecules linked together (one serves as the electron donor, the other as the acceptor) with the electron-accepting molecule linked to two quinones, which serve as electron acceptors in the natural system. The electron donating chlorophyll analogue is linked covalently to a carotenoid, and this can donate an electron to the oxidized chlorophyll. Upon excitation of the chlorophyll, a charge separation occurs resulting in an oxidized carotenoid and a reduced quinone. This charge-separated state is formed with high efficiency. In other artificial reaction centres, porphyrin moieties have been used in place of chlorophyll and fullerenes have been exploited used as the electron acceptor [22].

Porphyrins (Section 12.6.1, Figure 12.21a), the main chromophores of natural photosynthesis, are obvious candidates for the design of artificial antenna systems. Arrays containing porphyrin molecules are certainly the largest class of artificial antennae. In the pentameric array shown in Figure 12.35, efficient energy transfer from the peripheral Zn-containing units to the free-base core has been observed [23].

Much work has been devoted to emulate the water-splitting reaction of photosystem II. The challenge is to devise a water-splitting catalyst that is robust and composed of abundant non-toxic materials that work along similar principles to PS II. Compounds based on ruthenium have shown the greatest promise by exhibiting high catalytic efficiencies and good compatibility with photosensitizer oxidants [24]. Given that ruthenium is not an abundant metal, attention has also been focused on the design and synthesis of water splitting catalysts composed of readily available elements such as Mn, Co, and Fe [21].

Artificial photosynthetic solar-to-fuels cycles typically terminate at hydrogen, with no process installed to complete the cycle via carbon fixation. A development uses a pair of catalysts to split water into oxygen and hydrogen, and feeds the hydrogen to bacteria along with carbon dioxide. The bacteria, a micro-organism that has been bioengineered to specific characteristics, converts the carbon dioxide and hydrogen into liquid fuels [25]. It is suggested that coupling this hybrid device to existing photovoltaic systems would yield a CO_2 reduction efficiency of about 10%, exceeding that of natural photosynthetic systems.

Current inorganic and organic photovoltaic devices for the harvesting solar energy require an energy-intensive production process (Chapter 9, Section 9.7), and even though they have improved significantly over the years in terms of their efficiency, the development of photosynthesis-based technologies for energy collection may prove advantageous. However, photosynthesis and related processes can be applied to areas other than solar energy conversion. For example, there are many possible applications for artificial reaction centres and related molecules in nanotechnology. Synthetic pigments also have found biomedical uses in tumour detection, as they tend to accumulate preferentially in tumours and are highly fluorescent (and thus easily detectable in a patient whom is being operated on to surgically remove a tumour).

12.11 Molecular Motors

12.11.1 Nature's Motors

The development of molecular machines (e.g. based on rotaxanes) has been discussed in Chapter 11, Section 11.11.2). There are a number of examples of both linear and rotating motors to be found in the natural world. Much progress has been made on the understanding of the mechanisms of motion of myosin and kinesin. Myosins are a large superfamily of motor proteins that are involved in muscle contraction. These molecules move along actin (a group of globular multi-functional protein) filaments, while hydrolyzing ATP [26, 27]. Kinesin transports cellular components, such as organelles and signalling molecules, along microtubules.

Structurally, myosin and kinesin are dimeric, having two motor heads, two legs, and a common stalk. The head regions ($7 \times 4 \times 4$ nm in size in the case of kinesin) bind to actin or microtubule filaments. Two different models have been proposed for the movement of the heads along the track. These are depicted in Figure 12.36 [28]. In the hand-over-hand model (Figure 12.36a), ATP binding and hydrolysis causes a conformational change in the forward head (h1) that pulls the rear head (h2) forward, while head 1 stays attached to the track. In the next step, head 2 stays fixed and pulls head 1 forward. In the inchworm model (Figure 12.36b), only the forward head catalyses ATP and leads while the other head follows.

Perhaps the most useful molecule for nanoscale applications is DNA. DNA replication is a required step before cells divide. For example, the processes involved as a ribosome shunts along an mRNA chain (Section 12.5.1 and Figure 12.15) represent a highly coordinated and efficient linear actuation mechanism.

Two examples of biomolecular rotary motors are F_1-ATP synthase (F_1-ATPase) and ATP Synthase (F_0F_1) [28]. These are similar to motors that rotate *flagella* (whip-like organelles that many unicellular organisms, and some multicellular ones, use to move about) in bacteria and appear to have originated in cells about a billion years ago. They have evolved and serve many

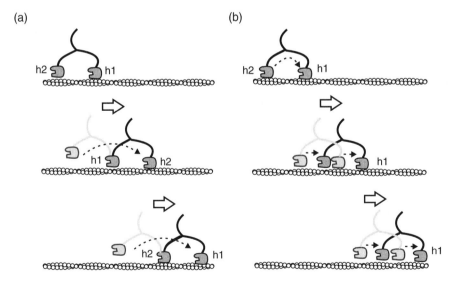

Figure 12.36 Schematic representation of walking mechanisms for natural linear motors. (a) hand-over-hand model. (b) inchworm model. h1 and h2 represent the two motor heads (see text for details). *Source:* Reprinted from Balzani et al. [28]. Copyright (2008), with permission from Wiley-VCH.

different purposes. These motors occur mounted in the wall of a cell and transmit rotational torque and power across the cell wall.

The ATPase molecular motors are found in the membranes of mitochondria, the microscopic bodies in the cells of nearly all living organisms, as well as in chloroplasts of plant cells, where the enzyme is responsible for converting food to usable energy (Section 12.4). The moving part of an ATPase is a central protein shaft (or rotor, in electric-motor terms) that rotates in response to electrochemical reactions with each of the molecule's three proton channels (comparable to the electromagnets in the stator coil of an electric motor), as depicted in Figure 12.37. During ATP hydrolysis, the tail rotates in an anticlockwise direction; it rotates clockwise during ATP synthesis from ADP. Both the F_1 and F_0 portions

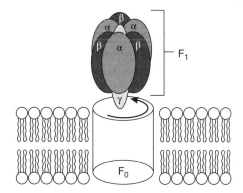

Figure 12.37 Structure of F_0F_1-ATP synthase. This enzyme consists of two principal parts. The asymmetric membrane-spanning F_0 part contains a proton channel, while the soluble F_1 part contains three catalytic sites which cooperate in the synthetic reactions.

are rotary motors. The F_1 unit has been structurally characterized and is made up of α and β subunits arranged around central γ unit. This contacts the F_0 part, which is membrane bound. The rotary activity of F_1-ATPase has been harnessed to develop motor prototypes in a number of laboratories.

Finally, it should be noted that biomotors are not restricted to animal cells. *Forisomes* are plant motors that enable long-distance transport in a natural microfluidics system. These elongated protein bodies of up to 30 μm length are found in the highly specialized cells in the *phloem* (the living tissue that transports the soluble organic compounds made during photosynthesis) of vascular plants, which form a continuous microtube system enabling pressure-driven long-distance transport of photo-assimilates. The forisomes act as reversible stopcocks by undergoing rapid conformational changes, which involve more than three-fold increases in volume. The conformational switch is controlled by Ca^{2+} ions with a threshold concentration in the nM range. These proteins have potential as reversible switches in integrated microfluidic systems.

12.11.2 Artificial Motors

Although the mechanism of some natural motors are well understood, the biggest challenge lies in adapting these systems into usable devices. The functions carried out within a cell by these motors are not very different from what we would expect man-made motors to perform, such as load carrying or rotational movement. Thus, mimicking these motors or adapting them in novel applications may have significant benefits. Eric Drexler has outlined the broad principles of molecular machines in his book *Engines of Creation* [29]. Drexler introduced a nanorobot, which he called an *assembler*. Such a device could, in principle, build almost anything – including copies of itself – atom by atom.

Using techniques from biotechnology, it is possible to make DNA molecules with a sequence of base pairs chosen at will. For example, DNA has been used to build a two-legged walking motor or biped [30]. The motion is depicted in Figure 12.38. Each of the legs in the walker is 36 bases long and is made from two strands of DNA that pair up to form a double helix. At the top, a springy portion of each DNA strand runs across from the left leg to the right, linking them together. At the bottom, one of the two strands extends from the helix to serve as a sticky foot. The track that the walker travels along is also made of DNA, and is designed so that unpaired

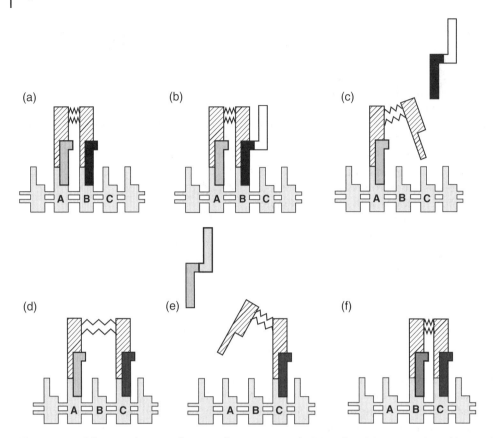

Figure 12.38 Schematic diagrams depicting the movement of a DNA robot. (a) Two DNA-based legs are held fast to DNA footholds A and B by anchors. (b) A free DNA strand attaches to the right anchor. (c) This strand strips the right anchor off, freeing the right foot. (d) An anchor strand for foothold C secures the floating foot to it. (e) Another free DNA strand strips the anchor from the left foot, detaching it. (f) An anchor strand secures the left foot to foothold B. *Source:* Reprinted from Sherman and Seeman. [30]. Copyright (2004) American Chemical Society.

sections of DNA strands stick up likes spikes along its length. These act as footholds for the walker. The feet attach to the footholds via 'anchor' strands of DNA that match up with the foot sequence at one end and with the foothold at the other. Because the left and right foot/foothold sequences are unique, each requires a different anchor. To make the walker take a step, a free piece of DNA is introduced to peel away one of the anchors, thereby releasing the foot. The movement is rather like an inchworm, with one foot edging forward and the other then being dragged up to the same position (Figure 12.26b). A number of other molecular machines based on DNA have also been designed [31, 32].

The twisting motion produced by flagella seems to provide an efficient method for bacteria to move through highly viscous biological fluids. A biomolecular motor, powered by ATP synthase (see above) with tiny metal propellers, has been described [33]. This motor can spin the propeller at eight revolutions per second. The ability of a molecular motor to draw its energy directly from the body has been considered an important step in the development of a new generation of ultra-small medical devices. The molecular motors have propellers about 750 (nm) long and 150 nm in diameter (a human hair is ~1000 nm in diameter). One turn of the motor produces about $120\,pW\,nm^{-1}$. The energy released from the three ATP molecules needed to rotate the motor once is $240\,pN\,nm^{-1}$, giving the motor an efficiency of about 50%.

Problems

12.1 Make an estimate of the information content of the following:
  i) Human brain
  ii) DNA
  iii) Human genome
  iv) Human body

In each case, explain your reasoning. You will need to undertake some research to justify your calculations.

12.2 i) What is the difference between diffusion and osmosis?
     An artificial cell containing an aqueous solution of sugar molecules (0.02 M sucrose, 0.01 M glucose, 0.06 M fructose) is enclosed in a semipermeable membrane and is immersed in a beaker containing a different aqueous sugar solution (0.05 M sucrose, 0.03 M glucose, 0.01 M fructose). The volume of solution in the beaker is initially equal to that inside the artificial cell.
     The membrane is permeable to water and to the simple sugars glucose and fructose, but is completely impermeable to the disaccharide sucrose.

  ii) Which solute(s) will exhibit a net diffusion into the cell?
  iii) Which solute(s) will exhibit a net diffusion out of the cell?
  iv) In which direction will there be a net osmotic movement of water?

12.3 ATP synthase generates sufficient torque to produce three ATP molecules per revolution and can produce 100 molecules per second. How many revolutions per minute (rpm) does this equate to? Rotation at this rate requires about 20 $k_B T$ of work per ATP molecule per second. Estimate the power needed to operate one ATP synthase molecule (body temperature = 37 °C).

12.4 In photosynthesis, the energy of at least eight 'red' photons is required per O_2 molecule released or CO_2 molecule fixed. A typical product of carbon fixation is glucose ($C_6H_{12}O_6$), whose energy content is 2805 kJ per mole (the enthalpy of combustion or energy generated when glucose is burnt). Estimate the efficiency of the photosynthetic process. Why is the practical efficiency likely to be less than this figure?

12.5 A coding system is devised where groups of two bases (A, G, T, C) are used to represent the 16 hexadecimal (hex) numbers 0 to F. In this scheme, the hex number 25 codes as TA, CC and the hex number 9A codes as AC, AA. What is the *decimal* equivalent of the base sequence: TC, CC, GG, AC, CA, CC, TT?

References

1 Ridley, M. (1999). *Genome*. London: Harper Perennial.
2 Ball, P. (1994). *Designing the Molecular World*. Princeton: Princeton University Press.
3 Findlay, J.B.C. (1995). The biological membrane. In: *An Introduction to Molecular Electronics* (ed. M.C. Petty, M.R. Bryce and D. Bloor), 279–294. London: Edward Arnold.
4 Petty, M., Tsibouklis, J., Petty, M.C., and Feast, W.J. (1992). Pyroelectric behaviour of synthetic biomembrane structures. *Thin Solid Films* 210-211: 320–323.

5 Aizawa, M. (1995). Biosensors. In: *An Introduction to Molecular Electronics* (ed. M.C. Petty, M.R. Bryce and D. Bloor), 295–314. London: Edward Arnold.

6 Turner, A.P.F. (2013). Biosensors: sense and sensibility. *Chem. Soc. Rev.* 42: 3184–3196.

7 Wu, K.R., Cao, W., and Wen, W. (2014). Extraction, amplification and detection of DNA in microfluidic chip-based assays. *Microchim. Acta* 181: 1611–1631.

8 Zhao, S. and Reichert, W.M. (1992). Influence of biotin lipid surface density and accessibility on avidin binding to the tip of an optical fibre sensor. *Langmuir* 8: 2785–2791.

9 Howorka, S., Cheley, S., and Bayley, H. (2001). Sequence-soecific detection of individual DNA strands using engineered nanopores. *Nat. Biotechnol.* 19: 636–639.

10 Merchant, C.A., Healy, K., Wanunu, M. et al. (2010). DNA translocation through graphene nanopores. *Nano Lett.* 10: 2915–2921.

11 Dekker, C. and Ratner, M.A. (2001). Electronic properties of DNA. *Phys. World* 14: 29–33.

12 Briman, M., Armitage, N.P., Helgren, E., and Grüner, G. (2004). Dipole relaxation losses in DNA. *Nano Lett.* 4: 733–736.

13 Tan, B., Hodak, M., Lu, W., and Bernholc, J. (2015). Charge transport in DNA nanowires connected to carbon nanotubes. *Phys. Rev. B* 92: 075429.

14 Adleman, L.M. (1994). Molecular computation of solutions to combinatorial problems. *Science* 266: 1021–1024.

15 Amos, M. (2006). *Genesis Machines*. London: Atlantic books.

16 Shu, J.-J., Wang, Q.-W., Yong, K.-Y. et al. (2015). Programmable DNA-mediated multitasking processor. *J. Phys. Chem. B* 119: 5639–5644.

17 Song, T., Garg, S., Mokhtar, R. et al. (2016). Analog computation by DNA strand displacement circuits. *ACS Synth. Biol.* 5: 898–912.

18 Wagner, N.L., Greco, J.A., Ranaghan, M.J., and Birge, R.R. (2013). Directed evolution of bacteriorhodopsin for applications in bioelectronics. *J. R. Soc. Interface* 10: 20130197.

19 Wickstrand, C., Dods, R., Royant, A., and Neutze, R. (2015). Bacteriorhodopsin: would the real structural intermediates please stand up? *Biochim. Biophys. Acta* 1850 (3): 536–553.

20 Hampp, N. (2000). Bacteriorhodopsin as a photochromic retinal protein for optical memories. *Chem. Rev.* 100: 1755–1776.

21 Barber, J. and Tran, P.D. (2013). From natural to artificial photosynthesis. *J. R. Soc. Interface* 10: 20120948.

22 Gust, D., Moore, T.A., and Moore, A.L. (2001). Mimicking photosynthetic solar energy transduction. *Acc. Chem. Res.* 34: 40–48.

23 Prathapan, S., Johnson, T.E., and Lindsey, J.S. (1993). Building-block synthesis of porphyrin light-harvesting arrays. *J. Amer. Chem. Soc.* 115: 7519–7520.

24 Kärkäs, M.D., Verho, O., Johnston, E.V., and Åkermark, B. (2014). Artificial photosynthesis: molecular systems for catalytic water oxidation. *Chem. Rev.* 114: 11863–12001.

25 Liu, C., Colón, B.C., Ziesack, M. et al. (2016). Water splitting-biosynthetis system with CO_2 reduction efficiencies exceeding photosynthesis. *Science* 352: 1210–1213.

26 Kneussel, M. and Wagner, W. (2013). Myosin motors at neural synapses: drivers of membrane transport and actin dynamics. *Nat. Rev. Neurosci.* 14: 233–247.

27 Rief, M., Rock, R.S., Mehta, A.D. et al. (2000). Myosin-V stepping kinetics: a molecular model for processivity. *Proc. Natl. Acad. Sci. USA* 97: 9482–9486.

28 Balzani, V., Credi, A., and Venturi, M. (2008). *Molecular Devices and Machines: Concepts and Perspectives for the Nanoworld*, 2e. Weinheim: Wiley-VCH.

29 Drexler, K.E. (1990). *Engines of Creation: The Coming Era of Nanotechnology*. New York: Anchor Books.

30 Sherman, W.B. and Seeman, N.C. (2004). A precisely controlled DNA walking device. *Nano Lett.* 4: 1203–1207.

31 Wang, Z.-G., Elbaz, J., and Willner, I. (2011). DNA machines: bipedal walker and stepper. *Nano Lett.* 11: 304–309.

32 Jung, C., Allen, P.B., and Ellington, A.D. (2016). A stochastic DNA walker that traverses a microparticle surface. *Nat. Nanotechnol.* 11: 157–164.

33 Soong, R.K., Bachand, G.D., Neves, H.P. et al. (2000). Powering an inorganic nandevice with a biomolecular motor. *Science* 290: 1555–1558.

Further Reading

Alberts, B., Johnson, A., Lewis, J. et al. (2015). *Molecular Biology of the Cell*, 6e. New York: Garland Science.

Bar-Cohen, Y. (ed.) (2011). *Biomimetics: Nature-Based Innovation*. Boca Raton: Taylor & Francis.

Birge, R.R. (1994). *Molecular and Biomolecular Electronics*. Washington, DC: American Chemical Society.

Blankenship, R.E. (2014). *Molecular Mechanisms of Photosynthesis*, 2e. Oxford: Wiley-Blackwell.

Jones, R.A.L. (2004). *Soft Machines: Nanotechnology and Life*. Oxford: Oxford University Press.

Karunakaran, C., Bhargava, K., and Benjamin, R. (2015). *Biosensors and Bioelectronics*. Amsterdam: Elsevier.

Kumar, C.S.S.R. (ed.) (2007). *Nanomaterials for Biosensors*. Weinheim: Wiley-VCH.

Nicolini, C. (ed.) (1996). *Molecular Manufacturing*. New York: Plenum.

Nicolini, C. (2016). *Molecular Bioelectronics: The 19 Years of Progress*, 2e. New Jersey: World Scientific.

Appendix

Constants

velocity of light in free space	c	$2.998 \times 10^8 \, \mathrm{m \, s^{-1}}$
permittivity of free space	ε_0	$8.854 \times 10^{-12} \, \mathrm{F \, m^{-1}}$
electronic charge	e	$1.602 \times 10^{-19} \, \mathrm{C}$
Planck's constant	h	$6.626 \times 10^{-34} \, \mathrm{J \, s}$
Boltzmann's constant	k_B	$1.381 \times 10^{-23} \, \mathrm{J \, K^{-1}}$
Avogadro's number	N_A	$6.022 \times 10^{26} \, \mathrm{kilomole^{-1}}$
		$(= 6.022 \times 10^{23} \, \mathrm{mol^{-1}})$
universal gas constant	R	$8.314 \times 10^3 \, \mathrm{J \, kilomole^{-1} \, K^{-1}}$
Faradays constant	F	$9.649 \times 10^7 \, \mathrm{C \, kilomole^{-1}}$
acceleration due to gravity	g	$9.807 \, \mathrm{m \, s^{-2}}$

Useful Relationships

1 electronvolt (eV) = $1.602 \times 10^{-19} \, \mathrm{J}$

For vacuum, energy in eV = 1.240/(wavelength in μm)

$1 \, \mathrm{eV} = 8066 \, \mathrm{cm^{-1}}$

1 eV per particle = $23\,060 \, \mathrm{kcal \, kilomole^{-1}}$ ($= 23.06 \, \mathrm{kcal \, mole^{-1}}$)

1 calorie = $4.186 \, \mathrm{J}$

At 300 K, kT $\approx 1/40 \, \mathrm{eV}$

1 atmosphere = $1.013 \times 10^5 \, \mathrm{N \, m^{-2}}$

Properties of Selected Elements

Element	Symbol	Atomic number	Atomic weight (amu)	Outer electron configuration	Density of solid, ~20°C (kg m^{-3}) × 10^3	Crystal structure	Melting point (°C)
Hydrogen	H	1	1.008	1s	–	–	−259
Helium	He	2	4.003	1s^2	–	–	−272 (at 26 atm)
Lithium	Li	3	6.94	2s	0.534	bc cubic	181
Beryllium	Be	4	9.012	2s^2	1.85	hexagonal	1287
Boron	B	5	10.81	2s^{2}2p	2.34	rhombohedral	2076
Carbon	C	6	12.011	2s^{2}2p^2	2.27	hexagonal	sublimes at 3727
Nitrogen	N	7	14.007	2s^{2}2p^3	–	–	−210
Oxygen	O	8	16.00	2s^{2}2p^4	–	–	−219
Fluorine	F	9	19.00	2s^{2}2p^5	–	–	−220
Neon	Ne	10	20.18	2s^{2}2p^6	–	–	−249
Sodium	Na	11	22.99	3s	0.968	bc cubic	98
Magnesium	Mg	12	24.31	3s^2	1.74	hexagonal	650
Aluminium	Al	13	26.98	3s^{2}3p	2.70	fc cubic	660
Silicon	Si	14	28.09	3s^{2}3p^2	2.33	fc cubic	1414
Phosphorus	P	15	30.97	3s^{2}3p^3	1.82	triclinic	44
Sulphur	S	16	32.06	3s^{2}3p^4	2.07	orthorhombic	115
Chlorine	Cl	17	35.45	3s^{2}3p^5	–	–	−102
Argon	Ar	18	39.95	3s^{2}3p^6	–	–	−189
Potassium	K	19	39.10	4s	0.89	bc cubic	63
Calcium	Ca	20	40.08	4s^2	1.55	fc cubic	842
Titanium	Ti	22	47.87	3d^{2}4s^2	4.51	hexagonal	1668
Vanadium	V	23	50.94	3d^{3}4s^2	6.0	bc cubic	1910
Chromium	Cr	24	52.00	3d^{5}4s	7.15	bc cubic	1907
Manganese	Mn	25	54.94	3d^{5}4s^2	7.21	bc cubic	1246
Iron	Fe	26	55.85	3d^{6}4s^2	7.86	bc cubic	1538
Cobalt	Co	27	58.93	3d^{7}4s^2	8.9	hexagonal	1495
Nickel	Ni	28	58.69	3d^{8}4s^2	8.91	fc cubic	1455
Copper	Cu	29	63.55	3d^{10}4s	8.96	fc cubic	1085
Zinc	Zn	30	65.41	3d^{10}4s^2	7.14	hexagonal	420
Gallium	Ga	31	69.72	4s^{2}4p	5.91	orthorhombic	30
Germanium	Ge	32	72.64	4s^{2}4p^2	5.32	fc cubic	938
Arsenic	As	33	74.92	4s^{2}4p^3	5.73	rhombohedral	817
Selenium	Se	34	78.96	4s^{2}4p^4	4.81	hexagonal	221
Bromine	Br	35	79.90	4s^{2}4p^5	–	–	−7
Rubidium	Rb	37	85.47	5s	1.53	bc cubic	39
Niobium	Nb	41	92.91	4d^{4}5s	8.57	bc cubic	2477

(Continued)

Element	Symbol	Atomic number	Atomic weight (amu)	Outer electron configuration	Density of solid, ~20 °C $(kg\,m^{-3}) \times 10^3$	Crystal structure	Melting point (°C)
Molybdenum	Mo	42	95.94	$4d^5 5s$	10.2	bc cubic	2623
Palladium	Pd	46	106.42	$4d^{10}$	12.0	fc cubic	1555
Silver	Ag	47	107.87	$4d^{10} 5s$	10.5	fc cubic	962
Cadmium	Cd	48	112.41	$4d^{10} 5s^2$	8.65	hexagonal	321
Indium	In	49	114.82	$5s^2 5p$	7.31	tetragonal	157
Tin	Sn	50	118.71	$5s^2 5p^2$	7.27	tetragonal	232
Antimony	Sb	51	121.76	$5s^2 5p^3$	6.70	rhombohedral	631
Tellurium	Te	52	127.6	$5s^2 5p^4$	6.24	hexagonal	450
Iodine	I	53	126.90	$5s^2 5p^5$	4.93	orthorhombic	114
Caesium	Cs	55	132.91	$6s$	1.93	bc cubic	28
Barium	Ba	56	137.33	$6s^2$	3.51	bc cubic	727
Tantalum	Ta	73	180.95	$5d^3 6s^2$	16.7	bc cubic	3017
Tungsten	W	74	183.84	$5d^4 6s^2$	19.3	bc cubic	3422
Platinum	Pt	78	195.08	$5d^9 6s$	21.5	fc cubic	1768
Gold	Au	79	196.97	$5d^{10} 6s$	19.3	fc cubic	1064
Mercury	Hg	80	200.59	$5d^{10} 6s^2$	–	–	−39
Lead	Pb	82	207.2	$6s^2 6p^2$	11.3	fc cubic	327
Bismuth	Bi	83	208.98	$6s^2 6p^3$	9.78	rhombohedral	272

Index

Organic and Molecular Electronics: From Principles to Practice, Second Edition. Michael C. Petty.
© 2019 John Wiley & Sons Ltd. Published 2019 by John Wiley & Sons Ltd.
Companion website: www.wiley.com/go/petty/molecular-electronics2